Restoration of Puget Sound Rivers

Edited by
David R. Montgomery
Susan Bolton
Derek B. Booth
Leslie Wall

Center for Water and Watershed Studies

in association with

University of Washington Press
Seattle & London

We would like to thank the College of Forest Resources, the National Marine Fisheries Service, and the Society for Ecological Restoration Northwest Chapter for partial funding of this book.

Printed in the United States of America

11 10 09 08 07 06 05 6 5 4 3 2

University of Washington Press
PO Box 50096, Seattle, WA 98145-5096, U.S.A.
www.washington.edu/uwpress

Library of Congress Cataloging-in-Publication Data

Restoration of Puget Sound rivers / edited by David Montgomery . . . [et al.].
p. cm.
This volume is the result of a spring 2000 meeting of the Society for Ecological Restoration's Northwest Chapter.
ISBN 0-295-98295-0 (pbk.: alk. paper)
(Published in association with the Center for Water and Watershed Studies, Seattle, Washington)
1. Stream restoration—Washington (State)—Puget Sound Region—Congresses. 2. Pacific salmon—Habitat—Washington (State)—Puget Sound Region—Congresses. I. Montgomery, David R., 1961–
QH76.5.W2 R47 2003
333.95′28153′09795—dc21 2002034767

The paper used in this publication is acid-free and 90 percent recycled from at least 50 percent post-consumer waste. It meets the minimum requirements of American National Standard for Information Sciences—Permanence of Paper for Printed Library Materials, ANSI Z39.48-1984. ♾♻

Contents

1. Puget Sound Rivers and Salmon Recovery

David R. Montgomery, Derek B. Booth, and Susan Bolton

A symposium on Restoration of Puget Sound Rivers at the spring 2000 meeting of the Society for Ecological Restoration's Northwest chapter presented an opportunity to synthesize regional expertise on river and stream restoration into a single volume. Largely drawn from presentations at the conference, the chapters of this book span a wide range of backgrounds and interests, including public policy, riparian forestry, stream ecology, hydrology, geomorphology, geology, and civil engineering. Chapters of the book proceed from geological and geomorphological controls on river and stream characteristics and dynamics, to the biological aspects of river systems in the region, to chapters that address social constraints and the application of fluvial geomorphology, civil engineering, riparian ecology, and aquatic ecology to regional river restoration projects and programs. While we recognize that the material presented herein could not be comprehensive given the broad scope of the subject, these chapters have been selected to provide a compilation of state-of-the-art considerations and approaches for developing river restoration programs.

The recent listing of various runs and stocks of Pacific salmon under the Endangered Species Act (ESA) has focused national attention on the condition of rivers and streams of the Pacific Northwest (PNW). In the Puget Sound region, recent ESA listings triggered statewide efforts to improve channel habitat involving ongoing expenditure of many millions of dollars annually and resulted in preparation of a "Statewide Strategy to Recover Salmon" by the Governor's Salmon Recovery Office (GSRO 1999). Public and governmental response to the listing of Puget Sound salmon stocks is a key national test of the ESA, as it represents a listing of an economically, ecologically, and culturally important species in a major metropolitan area.

A multitude of natural processes and human actions influence salmon abundance, and reversing the ongoing decline of salmon in the Puget Sound region is complicated by the problem that the collective effect of human activities alters the character and dynamics of rivers and streams and therefore aquatic ecosystems. Factors influencing salmon abundance are often generalized into the "Big Four" influences of harvest, hydropower (dams), habitat, and hatcheries (Figure 1). In contrast to the Columbia River system where the era of dam construction (1938-1975) coincides with a precipitous decline in the salmon fishery (Figure 2), dams are not the primary cause of declines in salmon stocks in Puget Sound. Moreover, historically the most productive Puget Sound salmon streams and rivers were low-gradient channels in major river valleys downstream of major dams. The declines in Puget Sound salmon populations are due to the combined effects of habitat degradation and loss, together with over-harvest exacerbated by hatchery practices that adversely impacted wild salmon. Isolating the relative contribution of these impacts on salmon abundance is further complicated by the effect of climate variability on marine productivity and the survival of adult salmon (Pearcy 1997).

A credible salmon recovery strategy emphasizing river restoration as a key program component should clearly articulate the role for river restoration within the context of an overall recovery strategy that addresses each of the primary impact factors (i.e., all four Hs). Although often overlooked in scoping river recovery efforts, the role of a fifth "H," history, is particularly important because river restoration programs need to assess whether the target rivers in fact can be restored. In addition, an understanding of the extent of recent modifications to river systems is central to identifying what it would take to restore rivers so that society can evaluate whether it is willing to accept that cost. Although answers to these questions remain unclear for Puget Sound rivers, a regional program of river manipulation is now proceeding due to the

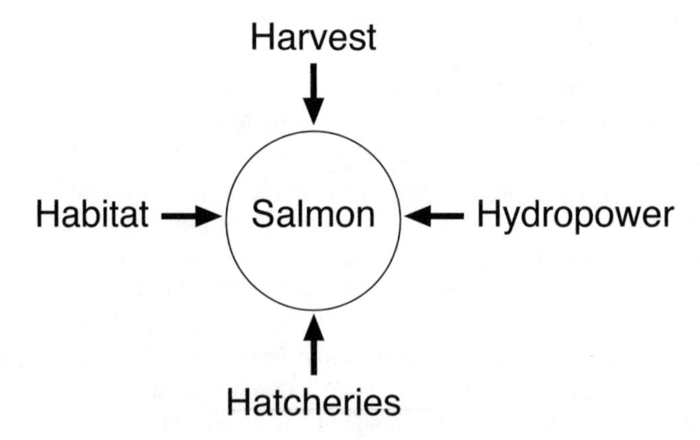

Figure 1. The "Big Four" influences on salmon abundance in the Pacific Northwest.

political need and public desire to act and due to the widespread belief among policy-makers that causes of salmon declines are sufficiently understood so as to be addressed readily.

Yet how well do we know how to restore Puget Sound rivers, let alone the salmon that live in them? What were Puget Sound rivers like prior to European influences? What constraints now exist due to changes in the hydrologic regime from extensive urbanization? How should recovery actions be prioritized or sequenced? This book introduces and discusses various options, opportunities, and constraints on restoration of Puget Sound rivers and streams as part of regional salmon recovery efforts.

RESTORATION

Restoration means returning something to a prior state. Rivers are dynamic systems in which specific attributes are continually created, altered, and destroyed. Consequently, river restoration means not only reestablishing certain prior conditions but also reestablishing the processes that create those conditions. Although the importance of defining what constitutes the desired state for a restoration effort is obvious, how to define that state is not.

The inevitable lack of data on pre-historic river conditions means that an understanding of the nature, scope, and extent of historical changes is needed to define a reference against which to set restoration objectives. Historical

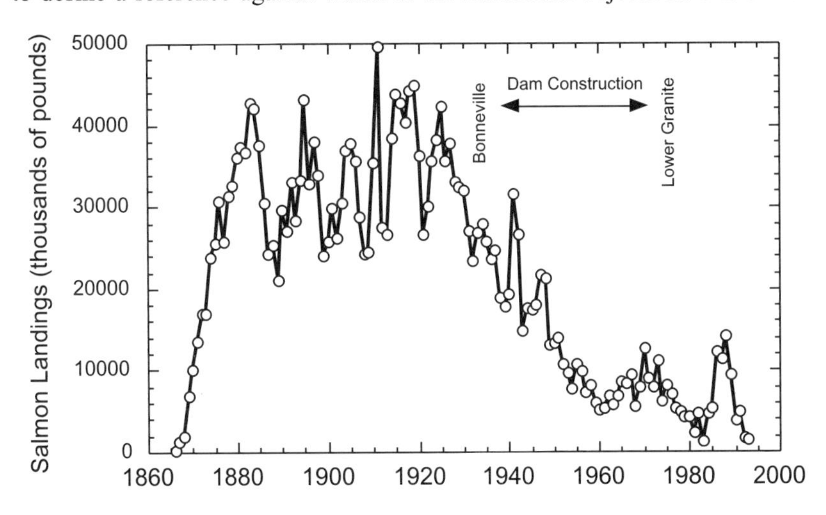

Figure 2. Relation of Columbia River commercial salmon landings (WDFW 1994) to the era of dam construction.

changes to river systems can be interpreted through analysis of historical documents such as maps, surveyors notes, photographs, journals, and other published and unpublished records. Most Puget Sound rivers experienced sweeping disturbances and sustained profound (and in some cases irreversible) changes as Europeans settled the region. Channel-spanning log jams, some of which had been stable enough to support old-growth forests, were cleared in the 1860s through 1880s by early settlers to facilitate up-river navigation and to alleviate local flooding. The Army Corps of Engineers pulled tens of thousands of snags from Puget Sound rivers in a river clearing program that was most active from 1870 to 1920 (Chapter 4). Levee building and diking of wetlands profoundly altered valley bottoms on major rivers from the 1880s through at least the 1940s. Finally, rapid urbanization from the 1950s through 1990s affected hydrologic regimes and channel characteristics (Booth 1991; Chapter 11).

Today's rivers and streams bear little resemblance to those that drained the original forested environment of the Puget Lowlands (Figures 3-6). Although channel conditions have experienced sweeping changes since the last glaciation (Chapter 2), conditions before the extensive anthropogenic changes initiated in the late 19th century define the obvious target for regional river restoration efforts today. Yet each river in the Puget Sound region had different characteristics and dynamics, which have been masked to varying degrees by the extent of subsequent changes. Consequently, watershed-specific reconstruction of historic river conditions provides a useful foundation for river restoration efforts.

Recent work documenting the importance of woody debris recruited from large trees as a source of key members for stable log jams (Abbe and Montgomery 1996) shows the critical importance of riparian forest stand conditions to channel processes and ecosystem functions in Puget Sound rivers. In particular, research on rivers that provide reasonable analogs for river conditions prior to European contact shows that wood-rich river systems had an anastomosing network of floodplain channels controlled by stable log jams, between which a very dynamic main channel periodically avulsed in response to log jam formation (Chapter 4). Because many of the natural habitat-forming processes were mediated or catalyzed by large log jams, restoring Puget Sound rivers to their original dynamic state would require reestablishing recruitment of trees and logs large enough to function as key members and catalyze formation of stable log jams. The fundamental implication therefore is that restoring Puget Sound rivers requires restoring floodplain forests (Chapter 10).

Are we are willing to do what it takes to restore Puget Sound rivers? Some of the earliest historic changes to Puget Sound rivers would be extraordinarily

Figure 3. Riparian buffer in industrial forest land, Tolt River.

Figure 4. Duwamish River in urbanizing King County.

Figure 5. Channel serving as a drainage ditch on the floodplain of the Skagit River.

Figure 6. Historical photograph believed to be Ravenna Creek in Seattle.

difficult to undue. No politician would suggest restoring Olympia Marsh, which once covered much of the floodplain of the Skagit River, even though it was a major salmon nursery before the rivers were cleared of channel-damming log jams. Nor would they offer serious proposals to restore the original drainage outlets of the Cedar and Black Rivers, which were drastically rearranged in the early twentieth century by the lowering of Lake Washington (Chrzastowski 1981). Restoration of floodplain forests is certainly possible in some areas, but it is just as clearly not a viable option for other areas, such as the lower Duwamish River surrounded by paved facilities associated with the Port of Seattle. While restoration is a laudable goal, a more realistic appraisal of constraints and potential opportunities will in many instances lead to adoption of more modest goals—those of river rehabilitation.

Rehabilitation

River rehabilitation aims to improve river conditions but does not necessarily seek re-establishment of natural conditions and dynamics. Given the extensive historic changes to rivers, and the resulting constraints, most projects billed as "river restoration" actually achieve only a form of partial river rehabilitation. Many different types of projects can be considered river rehabilitation efforts. Examples from the Puget Sound region include introduction of spawning gravel to the Cedar River below the Landsburg dam, construction of engineered log jams to retard bank erosion (Chapter 17), and various in-channel structures to promote habitat diversity and channel stability (Chapters 15 and 16). An archetypal example of an urban stream rehabilitation is the proposal to daylight Ravena Creek through the University Village shopping mall in Seattle. While the habitat value of the unearthed channel would be significantly enhanced relative to its present culverted state, the stream would still be flowing through a parking lot. The aim of rehabilitation projects is to improve river conditions, but rehabilitation does not carry the same burden of re-establishing self-sustaining natural conditions associated with the goal of restoration.

Regulation

Regulation means controlling the access to, or the use of, something in order either to prevent undesirable consequences or to promote desirable outcomes. Regulation of the salmon rivers and fisheries of England and Scotland can be traced back before the Norman invasion (Netboy 1974). The earliest known legislation restricting salmon fishing was probably the edict issued by King Malcolm II of Scotland in 1030 that established a closed season for taking "old salmon." Regulating the effect of dams on salmon fisheries has a long history; an act passed in 1318 during the reign of King Robert the First forbade the erection of fixtures that would prevent the progress of salmon up and down a river. Concern over habitat quality also is long-standing. A statute dating from the reign of Richard the Lion-Hearted declared that rivers be kept free of obstructions so as to permit a well-fed three-year-old pig, standing sideways in the stream, not to touch either side. In other words, the effects of over-fishing, dams, and habitat degradation on salmon have been regulated for centuries. Unfortunately, attempts to recover and restore salmon stocks, dating at least as far back as an act passed in 1712 during the reign of Queen Anne, did not prove particularly successful at preserving English salmon runs.

In the Pacific Northwest today, the political cornerstone of salmon recovery efforts is an emphasis on local control using voluntary measures. In particular, the mantra of "stakeholder involvement" is deeply embedded in our contemporary political landscape, but it is not clear whether salmon are also considered stakeholders with standing equal to economic interests. One way to evaluate the likely effectiveness of salmon recovery efforts founded on such an approach is to examine the past effectiveness of other attempts at local control for salmon management. In New England, for example, acts passed as early as 1741 (by the Colonial legislature) provided for inspection of dams to ensure the adequacy of fish passage. Between 1820 and 1880 over one hundred and fifty fishery laws relating to anadromous species were passed by the state of Maine alone. In these efforts, enforcement of fish passage was provided for at the local level, yet "little or nothing was ever done to implement this legislation—mill dams and weirs multipled [sic] and the fish were locked out in one river after another" (Netboy 1974, p. 180). If there is a lesson here for the Pacific Northwest in regard to salmon recovery efforts, it is that reliance on local control and voluntary measures needs to be guided by an overriding strategy and backed up by regulatory authority, able and willing to evaluate progress and to enforce salmon-conservation measures needed to ensure success of the overall plan.

A final key issue that needs to be incorporated into salmon recovery planning is the demographic trend of ever more people coming into the Puget Sound region (Figure 7). Associations between salmon abundance and land use for Puget Sound rivers indicate that urbanized areas and agricultural lands host low densities of salmon compared to rural and forested areas (Chapter 5). Curiously, most regulatory attention so far has focused on the upland forest environment, even though such areas were probably marginal salmon habitat historically for coho and chinook salmon, the species of greatest contemporary concern. Salmon recovery strategies have not addressed how to manage lowland habitats for long-term salmon recovery in the face of increasing development. Unless such pressures are addressed, river restoration and rehabilitation programs in place today will simply delay rather than prevent further declines in salmon abundance.

Regulation is the only viable way to address these larger trends. The cumulative effect of individual decisions may not reflect even a strong societal consensus to save salmon, because the emergent outcome of many decisions based on individual or site-specific criteria are likely to produce outcomes at odds with broader societal objectives and desires. Regulatory intervention could be in the form of incentives or outright controls and restrictions, but only the government has the authority to develop and the ability to implement such policies. Although centralized decision making is both vulnerable to slack enforcement and unpopular in today's political climate, decentralized decision-making allows the gradual degradation of river conditions through cumulative compromises. Hence, the region faces a dilemma in attempting to craft a viable, credible long-term strategy for salmon recovery.

Recovery of Puget Sound Rivers and Salmon

As concern over recovery of salmon stocks is a driving motivation behind efforts to restore or rehabilitate Puget Sound rivers, it is appropriate to examine the relation of such efforts to the regional salmon recovery strategy. Unfor-

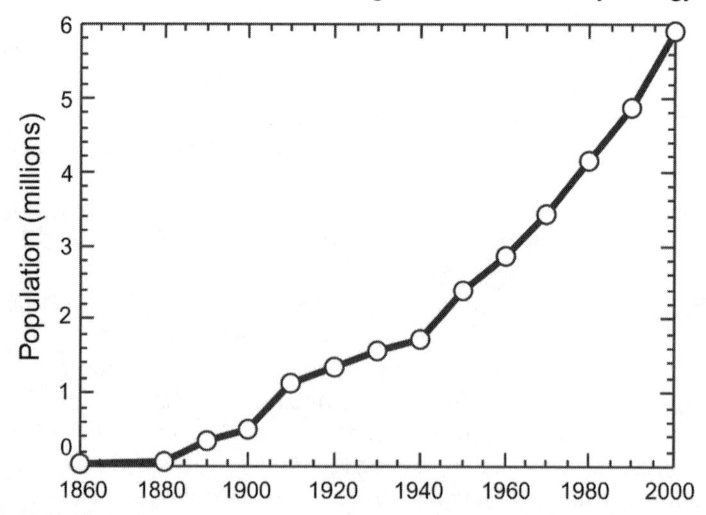

Figure 7. Human population growth in Washington State since 1860.

tunately, this examination is rather difficult because the current "strategy" is only a loose collection of tactics rather than an integrated plan (Currens et al. 2000). Although there are many valuable and ongoing salmon-conservation efforts relevant to Puget Sound rivers, no adopted salmon recovery strategy defines and defends the intended relative emphasis among reduced fishing pressure, river restoration, river rehabilitation, modified hatchery practices, and regulation of watershed development. Yet, a credible salmon recovery strategy needs to prioritize and integrate changes to all of the primary factors that impact salmon abundance, in order to ensure that efforts to address one factor are neither futile nor compromised by inattention to other factors.

An effective regional salmon recovery program needs to include habitat protection, restoration guided by an understanding of historical states, rehabilitation efforts based on assessments of current opportunities, and regulation both to prevent further degradation and to guide or drive implementation of the overall strategy. If salmon *recovery* (rather than a slower rate of decline) is the goal, then any strategy must allow no further increase in the impact resulting from any of the four Hs, unless such impacts would be demonstrably mitigated by reduced impacts from one of the other factors. A recovery plan should, at worst, allow no further harm, because reversing a declining trend obviously first requires changing the direction of the rate of change.

What can be done to reverse current trends and actually increase salmon abundance in the Puget Sound region, in spite of the projected doubling of the human population in the next half century? Two potentially complementary options for Puget Sound rivers are available: (1) reducing fishing pressure; and (2) protecting, restoring, and rehabilitating enough habitat to sustain viable salmon populations. A ban on commercial salmon fishing might produce a rapid response if harvest is a primary contributing factor to suppression of salmon populations—but this option is not even discussed in salmon recovery plans. Of the second set of options, habitat protection can be viewed as an insurance policy for longer-term habitat restoration efforts, as sustained commitment and a substantial investment of time and resources will be needed to trigger significant change and required to evaluate the success (or failure) of habitat-based salmon recovery programs.

A strategic plan consistent with longer-term salmon recovery would be to create a series of greenways along major river valleys. In the Puget Lowland, wide valley bottoms coincide with floodplains on which societal subsidies, in the form of flood control measures and levees, are usually required to sustain economic activity. Re-establishment of valley-bottom riparian forests within a corridor extending beyond the outer envelope defined by channel meanders would allow restoration of natural channel dynamics. Establishment of a system of open-space preserves along river corridors could provide long-term refugia to anchor regional salmon recovery. This strategy could be implemented gradually through a floodplain buyout program, restrictions on development within historically active river corridors, and levee removal. Such efforts would not substitute for efforts to minimize the impacts of upland land use, but they could help ensure that salmon recovery efforts would not ultimately prove futile.

Degradation of Puget Sound rivers and salmon populations occurred progressively over 150 years, as the result of both deliberate and inadvertent effects of evolving decisions and policies. Restoration of Puget Sound rivers could take even longer, but rehabilitation could improve important aspects of river conditions much more rapidly. The degree to which society is willing to give back space in the landscape to rivers will define the degree to which rivers can eventually recover their natural ecological processes, functions, and dynamics. If we are neither willing nor able to provide the space needed to restore dynamic rivers, then rehabilitation efforts will take on greater importance in long-term salmon conservation strategies. Yet if we don't rehabilitate or restore salmon habitat, salmon will continue their slide toward regional extirpation even if we have not chosen deliberately to sacrifice them for short-term economic gain.

Restoration of Puget Sound rivers is not a fanciful daydream. After an absence of well over a century, salmon are returning to the Thames River, once one of the worst sewers of Europe and far more degraded than any of this region's watercourses. The Puget Sound region still has options for allowing salmon to co-exist with a large and growing human population. But these options are increasingly constrained with each passing year. Nonetheless, this region can still choose a vision with abundant salmon in its future and could design policies to achieve that vision. Whatever the plan, recovery of Puget Sound salmon requires clear vision, forceful leadership, and strategic thinking not yet in evidence. The chapters that follow provide perspective and insight from people working to restore Puget Sound rivers.

References

Abbe, T.B. and D.R. Montgomery. 1996. Interaction of large woody debris, channel hydraulics and habitat formation in large rivers. *Regulated Rivers: Research & Management* 12:201-221.

Booth, D.B. 1991. Urbanization and the natural drainge system—Impacts, solution, and prognosis. *The Northwest Environmental Journal* 7:93-118.

Chrzastowski, M. 1981. Historical changes to Lake Washington and route of the Lake Washington ship canal, King County, Washington. U.S. Geological Survey Open-File Report 81-1182.

Currens, K.P., H.W. Li, J.D. McIntyre, D.R. Montgomery, and D.W. Reiser. 2000. Review of "Statewide Strategy to Recover Salmon: Extinction is Not an Option." Report 2000-1, Independent Science Panel, Olympia.

(GSRO) Governor's Salmon Recovery Office. 1999. *Statewide Strategy to Recovery Salmon: Extinction is Not an Option*. State of Washington, Governor's Salmon Recovery Office.

Netboy, A. 1974. *The Salmon: Their Fight for Survival*. Houghton Mifflin Co., Boston, Massachusetts.

Pearcy, W.G. 1997. Salmon production in changing ocean domains. In D.J. Stouder, P.A. Bisson, and R.J. Naiman (eds.) *Pacific Salmon and Their Ecosystems*. Chapman & Hall, New York. pp. 331-352.

WDFW (Washington Department of Fish and Wildlife). 1994. Status Report. Columbia River Fish Runs and Fisheries 1938-93. Washington Department of Fish and Wildlife, Olympia, Washington.

2. The Geology of Puget Lowland Rivers

Derek B. Booth, Ralph A. Haugerud, and Kathy Goetz Troost

ABSTRACT

Regional and local geologic conditions encompass many of the influences that watersheds impose on their channels: watershed size, water and sediment delivery to channels, the geometry of channels, and the responsiveness of a stream or river to change. Restoration or rehabilitation projects that do not consider the limitations and opportunities posed by the geologic setting are likely to fail, because the newly created features will not be maintained by the geologic conditions and geomorphic processes active in particular locales. The framework of the Puget Lowland has been established by a long history of tectonic and depositional processes, but the major features that control stream and watershed processes are primarily the result of the last ice-sheet advance, culminating about 16,000 years ago, coupled with more localized postglacial modification of the landscape. The effects of geologic history and geologic deposits are expressed across a range of spatial scales, which in turn result in varying biological conditions in different parts of the river network and varying responses to watershed disturbance. As a result of these geologic differences, not all streams have the same level of biological activity, and so not all stream restoration efforts should strive for the same goals and objectives. Even the best of human intentions are likely to fall short of their goals if the tools of geology are not used to recognize those conditions that can limit the success of rehabilitation.

Introduction

Stream and river channels are commonly described as "products of their watersheds," but this platitude does not convey the tremendous range of spatial and temporal scales over which a watershed influences channel form and channel processes. Regional and local geologic conditions, although rarely described with their fluvial consequences in mind, encompass many of those influences that watersheds impose on their channels: runoff patterns, sediment sources, channel gradient, hydraulic roughness, valley form, watershed size, and the responsiveness of a stream or river to change. The geologic setting of watersheds, and of individual stream reaches, will determine what types of channel morphology and habitat features occurred under natural conditions, and thus what restoration or rehabilitation objectives are appropriate and achievable. Conversely, rehabilitation projects that do not consider the limitations and opportunities posed by the geologic setting are doomed to failure—instream structures will be undermined by channel migration, imported gravel will be buried by the watershed contribution of mobile sediment, and the form of reconstructed channels will not be maintained by the geomorphic processes active in particular locales.

Yet the influence of geology on rivers and streams is not always straightforward. As a fluvial system evolves it becomes an agent of geologic change itself, modifying the same landscape that once determined its behavior. Escarpments are incised by gullies and streams; lowlands are filled with alluvium; and the topographic form of the landscape imposed by tectonic, volcanic, and glacial activity becomes modified by patterns of fluvial erosion and deposition along the drainage network.

The Puget Lowland (Figure 1) shows the interplay of recent fluvial activity superimposed on a much longer history of tectonic, volcanic, and glacial action. It now contains one of North America's premier sheltered waterways, Puget Sound, most of the major population centers of the Pacific Northwest, and once-abundant runs of Pacific salmon. Its counterparts to the north and south, the Georgia Depression and the Willamette Lowland, share many of the same elements of geologic history and fluvial activity. This discussion, although focused on the Puget Lowland, is therefore relevant throughout the humid western region of the Pacific Northwest.

Overview of Puget Lowland geology

Tectonic Setting and Bedrock Framework

The rocks and unconsolidated deposits of Washington (Figure 2a) record more than 100 million years of earth history. Knowledge of this history has been gained from over a century of careful study, and the story is still unfolding as a result of continued geologic research. The foundation of this landscape is incompletely displayed by rocks now exposed in the North Cascades along the eastern boundary of the Puget Lowland. They record a history of oceans, volcanic island arcs, and subduction zones, mostly of Mesozoic age (the geologic period 220 to 65 million years ago—Ma) but including some late Paleozoic components (in western Washington, as old as 275 Ma) (Frizzell et al. 1987; Tabor and Haugerud 1999; Tabor et al. 2001) (Figure 2b).

The original deposits of these now-metamorphosed Paleozoic and Mesozoic rocks of the Cascade Range are quite varied, suggesting that they formed in widely separated environments and probably far from ancestral North America. (The closest bona-fide "old" North American rocks crop out near Spokane.) These old Cascade deposits were brought together by late Mesozoic plate convergence at the western edge of North America (via subduction) and subsequent translation along the continental margin (via strike-slip fault motion), reflecting the persistence of tectonic activity in the Pacific Northwest at the leading edge of the North American continent.

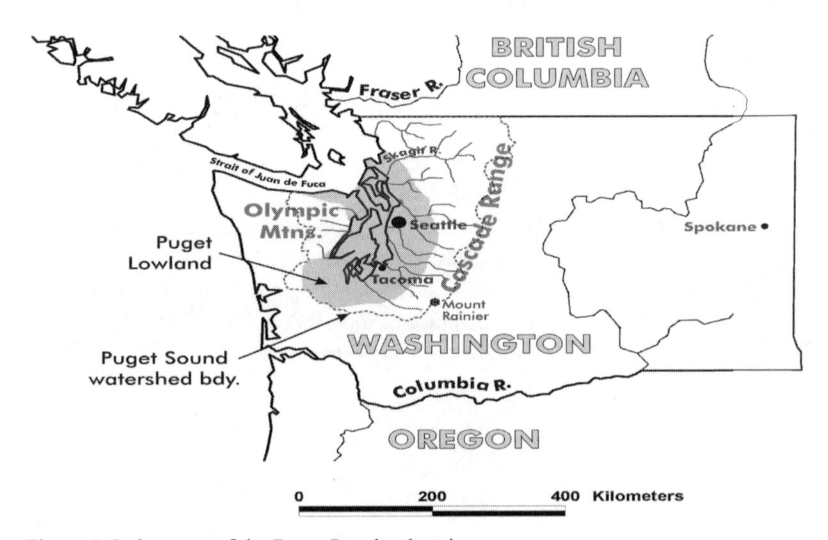

Figure 1. Index map of the Puget Lowland region.

A

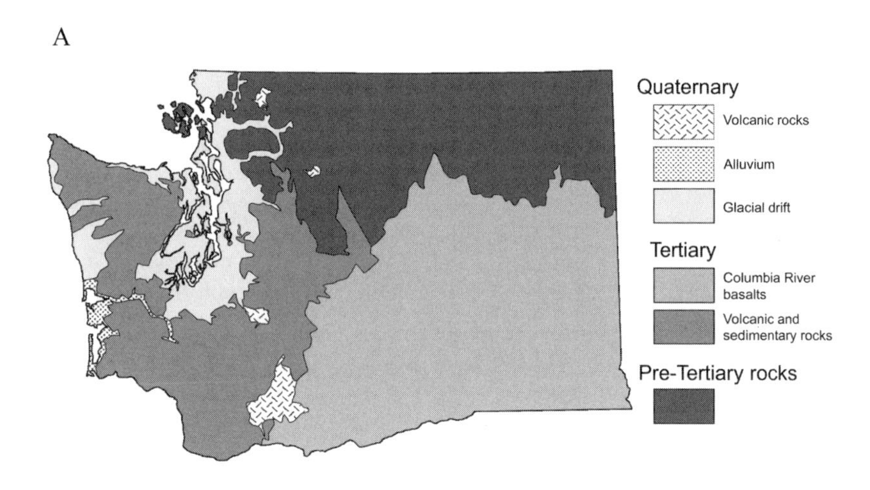

B

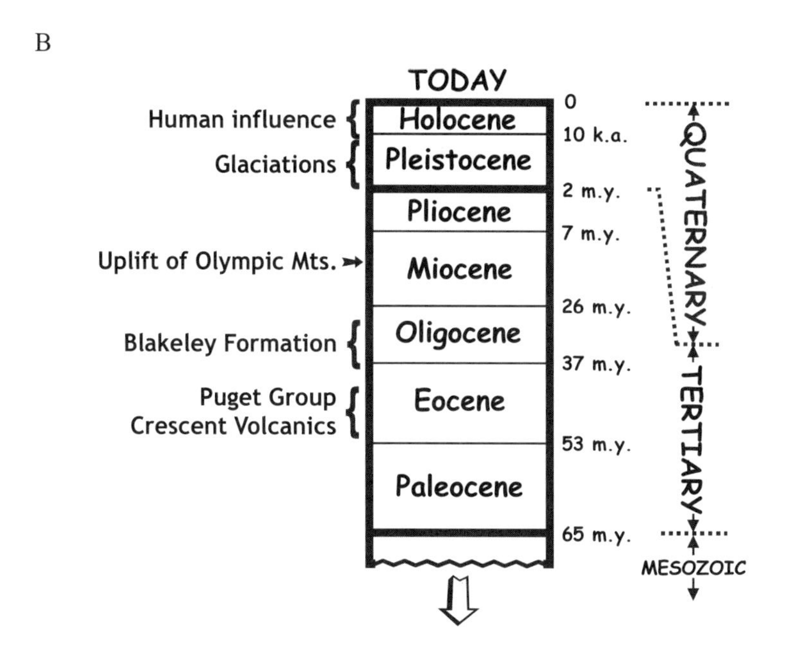

Figure 2. (A) Simplified geologic map of Washington State. (B) Time scale of geologic deposits and events for the Puget Lowland.

Middle and late Eocene (ca. 50 to 42 Ma) sandstone and volcanic rocks overlie these older rocks. During this time, large rivers flowed across an extensive (and subsiding) coastal plain that lay west of the modern Cascade Range and east of the modern Olympic Mountains (and probably east of Puget Sound as well) (Figure 3). This ancient river system produced the rocks of the Puget Group, whose relatively good resistance to erosion is responsible for the prominence of the Newcastle Hills in the central Puget Lowland, and whose abundant plant debris resulted in coal deposits that helped shape the nineteenth century economy and history of the region. Under the rigors of fluvial transport, however, these rocks tend to break down rather quickly, which is why the gravels of modern Puget Lowland rivers and streams overwhelmingly consist of glacially transported sediment derived from more resistant rock bodies farther north.

Subsequent reorganization of tectonic plates in the northeastern Pacific Ocean resulted in renewed plate convergence, subduction, and volcanism

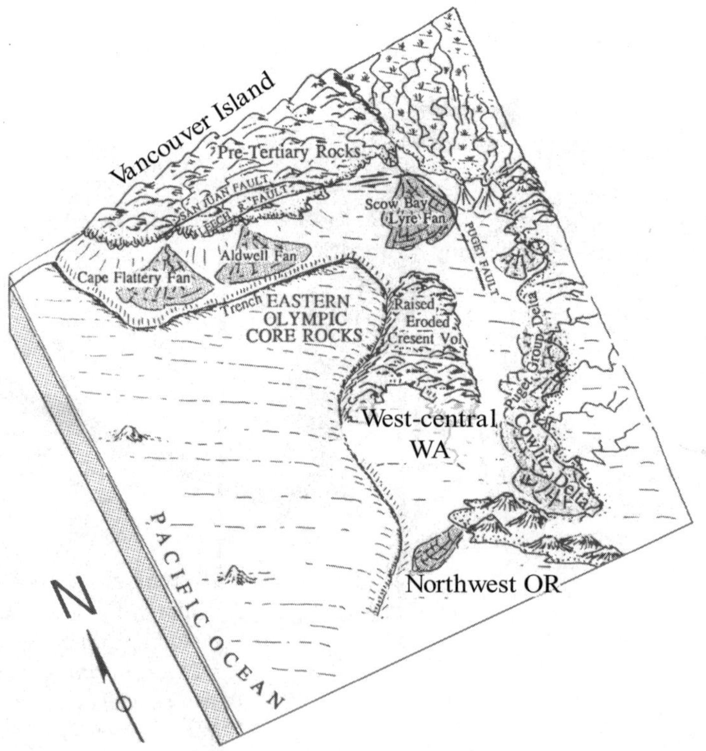

Figure 3. Eocene physiography of the Pacific Northwest (modified from Christiansen and Yeats, 1992).

along the Cascade arc in earliest Oligocene time (about 35 Ma), followed by sedimentary deposition of the Blakeley Formation sandstone across the central Puget Lowland on what is now Seattle and Bainbridge Island. The modern-day form of the Cascade Range is not a direct descendant of this interval of tectonic and volcanic activity—the modern mountain range was uplifted less than 6 Ma (Cheney 1997), but it expresses a similar style of tectonic activity that has been episodically active over the last several tens of million years. Especially in southernmost Washington and in Oregon, the Cascade arc continues to be active, and volcanoes of the arc dominate the landscape (e.g., Mount Baker, Mount Rainier, and Mount Hood). They produce dramatic topographic relief but contribute only relatively weak and easily degraded gravel clasts to downstream rivers.

The Olympic Mountains, most completely described by Tabor and Cady (1978), form the western boundary of the Puget Lowland. They are part of a topographic high built by compression and shortening above the Cascadia subduction zone, driven by the convergence of the North American and Pacific plates. A horseshoe of high peaks along the southern, eastern, and northern part of the range (including Mt. Washington, The Brothers, Mt. Constance, and Mt. Angeles) is underlain by mostly submarine basalt of the Crescent Volcanics, erupted 55 to 50 Ma during rifting of the seafloor at the western edge of the continental margin. These rocks form the abrupt western boundary of the Puget Lowland as seen from the central part of the lowland.

In the core of the Olympic Mountains, enclosed by this horseshoe, lightly metamorphosed deep-water sandstone and shale underlie Mt. Anderson, Mt. Olympus, and the low country farther west. These sediments were largely derived from North America and are younger than (and were transported across) the Crescent Volcanics on their way to deeper water. They were finally deposited on the continental slope above the slowly deforming seafloor and then subducted east and beneath the same terrain across which they had just been transported. Radiometric dating and stratigraphic relationships suggest that uplift of the Olympic Mountains, which eventually raised these ocean-bottom rocks to their modern elevation by some 2 or 3 km, was underway by about 14 Ma (see Brandon et al. 1998, and references therein).

Why is the Olympic Mountains segment of the Pacific-North American subduction zone so much higher than segments to the north and south? The distribution of deep earthquakes indicates that beneath and east of the Olympics, where the trend of the continental margin changes from north to northwest, there is an arch in the subducted oceanic plate. Some geologists have suggested that this arch causes the greater uplift of the Olympics. Or perhaps the Olympics are higher because subduction forces more sediment beneath this part of the continental margin, a plausible consequence of the nearby outfalls of the Fraser and Columbia rivers. If this alternative proves to be correct, fluvial and landscape-forming processes are in fact interacting at a scale beyond any suggested in this chapter.

Glaciations

Quaternary History

Although the evolution of the tectonic and bedrock framework of the Puget Lowland continues to the present day, most of these influences had been established by the beginning of the Quaternary period, that segment of geologic time defined as the last 2 million years of Earth history. During this period, oscillations in the earth's orbit around the sun have caused alternate warming and cooling episodes that have given rise to an extensive record of continental-scale glaciations. At least six invasions of glacial ice into the Puget Lowland have left a discontinuous record of Pleistocene glacial and interglacial intervals (Blunt et al. 1987). Originating in the mountains of British Columbia, this ice was part of the Cordilleran ice sheet of northwestern North America. During each successive glaciation, ice advanced into the lowland as a broad tongue first called the *Puget lobe* by Bretz (1913). All but the most recent of these advances are older than can be accurately dated with the ^{14}C technique, and so their detailed history is somewhat indeterminate.

This most recent ice-sheet advance into western Washington (Figure 4) was named the Vashon stade of the Fraser glaciation by Armstrong et al. (1965) (a *glaciation* is a climatic period of extensive glacial advance and retreat; a *stade* is a period of secondary glacial advance and retreat within a glaciation). Ice occupied the Puget Lowland about 18–15,000 calendar years ago; at its maximum, Seattle was buried by at least 900 m (3000 ft) of ice (Booth 1987; Porter and Swanson 1998). Most Puget Lowland topography is a direct product of, or at least shows a strong imprint from, this period of ice-sheet occupation. By comparison, recent geologic processes such as stream and wave erosion, landsliding, tectonic deformation, and volcanic eruptions have subsequently modified this topography only slightly.

Glacial Deposits

The spatial extent (and the rates) of sedimentation during glacial times was widespread and voluminous. Nonglacial periods (which include the present day), in contrast, have experienced active deposition primarily in river valleys and at the base of steep slopes. Thus the modern landscape is underlain primarily by glacial sediments. The thickness of these deposits varies greatly, but across much of the Puget Lowland it exceeds several hundred meters, and locally the deposits extend over a kilometer below modern sea level where they rest upon an irregular bedrock surface. They are thus of critical importance in understanding the interaction of modern rivers and streams with their watersheds.

Most, although not all, of the glacial deposits exposed at the ground surface in the Puget Lowland are products of the advance and retreat of the Puget lobe during the Vashon stade. These sediments (Figure 5), collectively named Vashon Drift (*drift* refers to any deposit of glacial origin, whether deposited by ice or water), are divided into several units: *advance deposits*—lacustrine silt and clay deposited into proglacial lakes, followed by well-sorted sand and gravel ("outwash") carried by streams flowing from the ice sheet as it spread south; *till*—unsorted sand, gravel, silt and clay deposited beneath the ice sheet; *ice-contact* and *ice-marginal deposits*—comprising sorted and unsorted debris deposited adjacent to, or in some cases on top of, the ice sheet; and *recessional deposits*—well-sorted sand and gravel deposited by streams draining from the ice as the ice-front receded, as well as silt and clay deposited

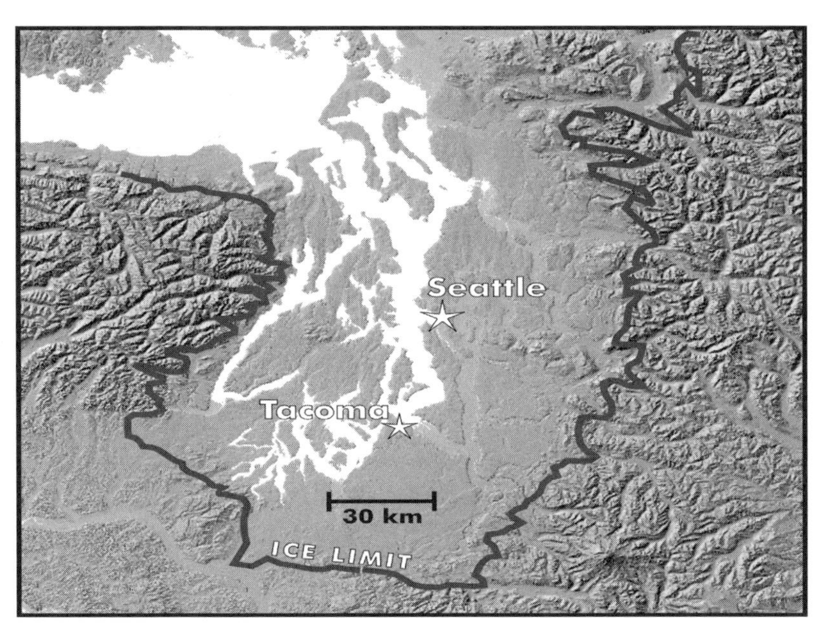

Figure 4. Ice limit of the Puget lobe during the most recent ice-sheet advance, the Vashon stade (redrawn from Thorson 1980 and Booth 1987).

in lakes dammed by the receding ice. In the central and northern Puget Sound, extensive deposits of *glaciomarine drift*—nonsorted sediment deposited off the front or the base of the ice sheet into marine waters—are also found.

These glacial deposits have a wide range of physical properties. From the perspective of hydrologic processes and stream-channel response, two of these properties—permeability and consolidation—are particularly important. Outwash deposits (both advance and recessional) compose the vast majority of permeable sediments found across the Lowland. They permit rapid infiltration and groundwater movement, they lack cohesion and thus are susceptible to fluvial erosion, and they supply abundant bedload-sized sediment to rivers and streams. Lacustrine deposits and till are the most widespread low-permeability materials, providing the dominant controls on groundwater flow and determining whether, and where, infiltrating groundwater will reach deeper aquifers. On sloping terrain, the superposition of highly permeable outwash over low-permeability lacustrine sediments determines the location of hillside seeps and springs; it gives rise to the geologic setting most closely associated with landsliding and other forms of mass-wasting (Tubbs 1974), which in turn provides the dominant source of sediment delivered to lowland rivers and streams (Reid and Dunne 1996). At shallower depths, a similar geologic setting is commonly found where permeable Holocene soils overlay the uppermost unweathered Pleistocene deposit, commonly till (and, locally, bedrock).

In contrast to permeability, consolidation is associated not with depositional environment but with stratigraphic position. Those sediments overrun

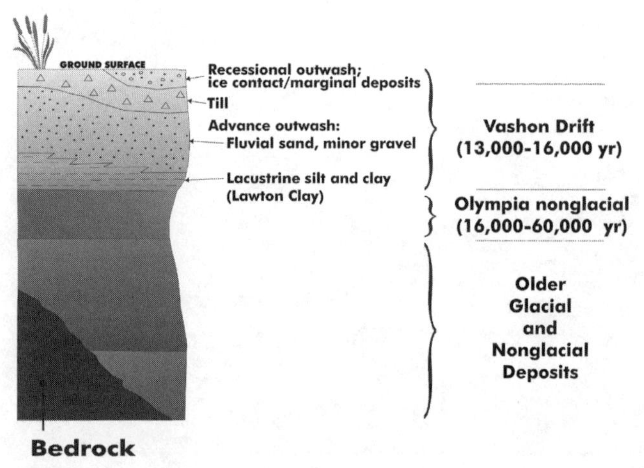

Figure 5. Stratigraphy of Vashon and pre-Vashon glacial and nonglacial deposits of the Puget Lowland.

by one (or more) ice-sheet advance display measurable overconsolidation and relatively high density (Olmstead 1969; Laprade 1982), important physical properties for determining erosion susceptibility and stream-channel morphology. Even noncohesive outwash, if dense and overconsolidated beneath Vashon till, will stand in dry near-vertical faces for many years or decades; cohesive overconsolidated deposits, such as till, may resist even active stream-channel erosion.

Pre-Vashon strata, both glacial and nonglacial, are consistently overconsolidated, but because they are exposed only sporadically across the Puget Lowland, they exert relatively little direct influence on rivers and streams. Where present at depth, they do have a variety of effects on groundwater flow and channel morphology because of their widely varying physical and hydrologic properties, a consequence of their equally varied origins in the multiple glacial and nonglacial periods preceding the Vashon stade (Troost 1999). They also display great variety in their geometry: buried river channels can provide groundwater conduits deep below the surface; lowland lake deposits may provide an effective barrier to groundwater flow but extend laterally only as far as the pre-Vashon lake in which they formed. Where modern stream channels flow over slopes that expose these strata, knickpoints can develop on the more resistant, typically fine-grained layers, imposing steps in the longitudinal profile of the channel that may persist for decades or longer.

In the mountains adjacent to the Puget Lowland, bedrock is the predominant geologic influence on channel morphology, but the valleys are commonly floored with a mosaic of stream deposits, talus, landslide debris, colluvium, and till. These deposits in large part date from an early, pre-Vashon stade of the Fraser glaciation named the "Evans Creek," but some deposits may have been formed in part during minor alpine glacial re-advances during the late Holocene (i.e., in post-glacial time). In the lower reaches of many of these same alpine valleys, valley-bottom deposits are overlain by fine sand, silt, and clay deposited by *upvalley*-flowing streams and in lakes dammed by the ice sheet that occupied the Lowland, beyond the valley mouths, during the Vashon stade.

Postglacial Processes and Deposits

Rivers, landslides, waves, and volcanic mudflows have continued to modify the Puget Lowland landscape and to create deposits younger than those of the last glacial advance. Although not influential at the same scale as continental glaciations, these processes and their resulting products are locally significant. The sediments are uniformly unconsolidated but permeabilities vary widely, reflecting the variety of depositional process and parent materials.

The most dramatic of these processes has been the Osceola mudflow near the town of Enumclaw, deposited after the catastrophic collapse of the northeast side of Mount Rainier about 5,700 years ago. Subsequent erosion and redeposition of the abundant post-Osceola supply of volcanic debris eventually filled the previously marine Green/Duwamish River valley (Dragovich et al. 1994) from about the present city of Auburn north to Seattle's Elliott Bay.

Humans have also modified the Earth's surface in many parts of the Puget Lowland in postglacial time. Some of these modifications have had profound consequences for the region's rivers and streams, either by direct modification of the channels themselves or by alteration of the hydrologic and sedimentological regimes of their watersheds. As separate topics in their own right, these issues are left for discussion in other chapters.

Distribution of Deposits in the Puget Lowland

Although a great variety of Quaternary deposits is found throughout the Puget Lowland, the characteristics of glacial and postglacial erosion and deposition have given rise to broadly predictable patterns of geologic materials across the modern landscape. In plan view, the landscape is mostly mantled by *till*, the compact heterogeneous deposit laid down most recently at the base of the Vashon-age ice sheet. Locally the till is thin or absent, revealing the underlying deposits, most commonly the sandy *Vashon advance outwash* (but locally some older and generally less permeable sediments as well). Extensive advance outwash deposits are also exposed in the walls of channels cut through the till, most commonly during ice-sheet retreat when voluminous meltwater discharges off of the ice sheet combined with alpine rivers draining the Cascade Range and Olympic Mountains.

Glaciomarine drift is also broadly exposed but its geographic extent is more limited, a consequence of the late-glacial history of the Lowland. This deposit, a product of subaqueous melting of the terminus of an ice sheet or the bottom of a floating ice shelf, is common but only in areas that were below sea level during ice retreat. Although global sea level (termed *eustatic* sea level) continued to rise following deglaciation of the Lowland, the earth's crust in the Puget Lowland also rose after deglaciation (Thorson 1989) in response to the weight lost by melting hundreds to thousands of meters of ice (termed *isostatic rebound*), just as it had depressed several thousand years earlier when the ice first advanced. The initial depression (and so also the subsequent uplift) was greater to the north than to the south because the ice was thicker to the north. At about the latitude of Seattle, isostatic rebound equaled eustatic sea-level rise. Thus marine deposits dating from the first opening of Puget Sound to marine waters after deglaciation are common only in the northern Puget Lowland and are not recognized south of Seattle (Pessl et al. 1989; Yount et al. 1993).

In contrast to the distribution of these deposits across broad areas of the Lowland, the distribution of the recessional outwash deposits is more narrowly focused. As the ice sheet retreated from its maximum position, meltwater drained into the axis of the Lowland but could not follow what would become its modern drainage path north and west out the Strait of Juan de Fuca, because the strait was still filled with many hundreds or even thousands of meters of ice. Instead, meltwater was diverted south along the margins of the retreating ice sheet, coalescing into ever-broader rivers. Channels and locally broad plains of *Vashon recessional outwash* now form much of the landscape in these ice-marginal locales and recessional river valleys. These landforms can be traced downstream to their glacial-age spillway out of the Puget Lowland, south through the valley of Black Lake near Olympia and then along the valley of the modern Chehalis River west to the Pacific Ocean. The distribution of extensive postglacial deposits is even more restricted, with river valleys the most common depositional sites and those draining the active Cascade volcanoes displaying a particularly voluminous legacy of Holocene mudflows.

The vertical distribution of sediments is also broadly predictable. Vashon till is laterally extensive but commonly just 1 or 2 meters thick. Below it, Vashon advance outwash provides the bulk of the modern landscape lying above sea level, with thickness of many tens of meters common. The top of the advance outwash is relatively flat, notwithstanding abundant superimposed channels and elongated hills; it records the level of the vast braided-river outwash plain formed during the advance of the Cordilleran ice sheet into the Puget Lowland (Booth 1994). Beneath the advance outwash, deposits are less commonly exposed and more variable in origin and in character. In the Seattle area, early Vashon-age lake deposits (the Lawton Clay) are common and can be found up to elevations of about 60 m; farther south, these sediments are almost entirely absent. Even older glacial and nonglacial deposits are present at low elevations throughout the region, particularly on Whidbey Island (Pessl et al. 1989) and in the Seattle-Tacoma metropolitan area (Troost 1999).

Holocene Changes in Base Level

Rapid changes in both sea level and the ground surface following the retreat of the last ice sheet affected not only the present-day exposure of glaciomarine deposits but also the base level of the major rivers and streams of the Puget Lowland that flow into the Sound. Marine waters reentered the newly scoured and deglaciated Sound as soon as ice-sheet retreat reopened the Strait of Juan de Fuca and Admiralty Inlet, but this was followed almost immediately by isostatic rebound, which ranged from negligible near Olympia to about 200 m in the northern Puget Lowland (Thorson 1989). After this relatively abrupt change in local base levels, global sea-level rise from the melting of ice sheets world-wide continued for about 7,000 years, accounting for approximately 90 m of sea level rise that slowed dramatically only about 5,000 years ago. At this time, global sea level was within 5 m of its present altitude (Matthews 1990).

In the last 5,000 years, global sea level has not been entirely static but its continued rise was nearly complete by 2,000 years ago (Clague and Bobrowsky 1990). If this pattern of global late-Holocene rates are broadly applicable to the Puget Lowland, they should be reflected in a gradual aggradation in the lower reaches of rivers and progressive upstream deposition of deltaic and estuarian sediments more rapid than explainable by the sediment load of the rivers alone. Although tectonic-driven changes in land-surface elevation undoubtedly complicated this simple picture (Eronen et al. 1987), well-dated aggrading shoreline features on West Point, a constructional beach spit extending into Puget Sound from Seattle, confirm the existence of the analogous process in the marine environment (Troost and Stein 1995).

Late-Holocene sea-level rise has been sufficiently slow, however, that other processes are likely to have overwhelmed its manifestation in many other geologic settings. Slow upstream aggradation in estuaries and river mouths was surely obliterated in those valleys where volcanic mudflows have occurred during the Holocene. Of the Cascade Mountain drainages entering the Puget Lowland, only the Snohomish River has been unaffected by such events.

Tectonic movement can also impose abrupt vertical changes in relative base level, of which the Seattle fault is the most prominent source (Figure 6). During an earthquake 1100 years ago (Bucknam et al. 1992), the south side of the fault rose 7 m and the north side dropped at least 1.5 m in the center of the Lowland. The magnitude of the vertical offset is unknown to the east and west of Puget Sound, as are the exact number of events and total vertical offset that may have accumulated from earlier earthquakes during the Holocene. The fault's importance to the base level of several major river systems is undisputed, however, because it passes just upstream of the mouth of the Duwamish River, between the inlet and outlet of Lake Sammamish, and below the bedrock-controlled falls of the Snoqualmie River.

Influence of Geology on Channel Conditions and Responses—the Effects of Scale

Geologic conditions affect many aspects of watershed and fluvial processes; deposits compose the landscape itself, and so their effects are pervasive. One way to organize these effects is to group them by spatial scale, which in western Washington also corresponds to the geologic process(es) that have

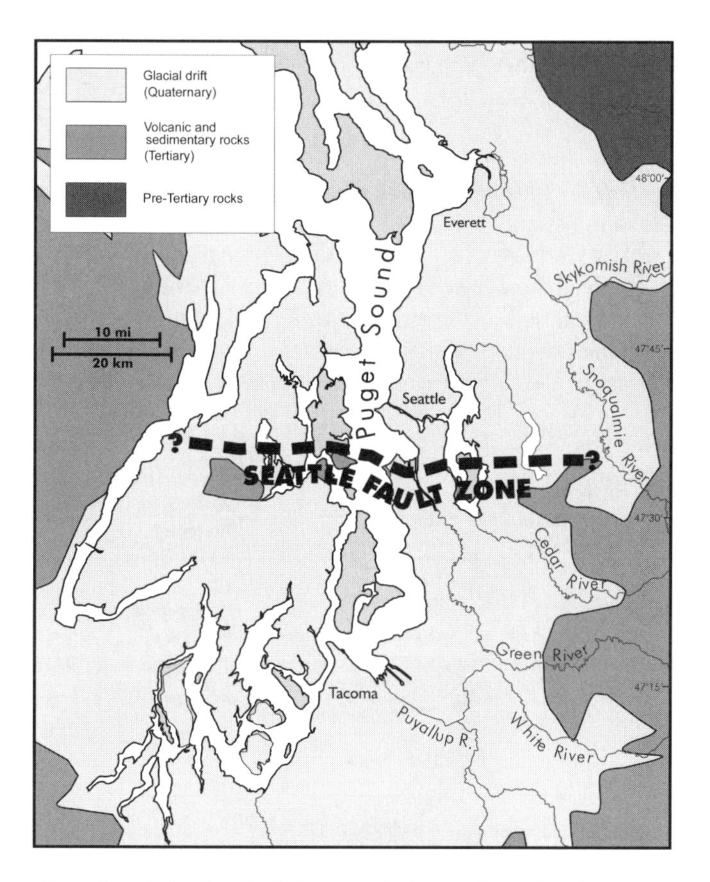

Figure 6. Location of the Seattle fault zone relative to the major rivers of the Puget Lowland. During the last great earthquake 1100 years ago on this fault, the south side rose as much as 7 m in the center of the Lowland; the north side subsided about 1 m.

given rise to them (Figure 7). From a fluvial and hydrologic perspective, these scales correspond to conditions relevant to the broad distribution of rivers and watersheds across the Puget Lowland, to the specific location and character of those watercourses as they traverse the lowland, and to the processes active on a single hillside that deliver water and sediment to the channel below. Geologically, these scales correspond to the tectonic framework of the Lowland, the glacial deposition and scour of the last continental ice sheet that occupied this region, and the resulting deposits (and their history of Holocene weathering) that form the modern ground surface. These divisions are in part arbitrary, but they help our identification of the conditions of greatest relevance to particular settings.

Large Scale (10s–100s of km)—The Imposition of Watershed Topography

The Puget Lowland owes its existence to many tens of millions of years of tectonic activity at the western edge of the North American continent, which in turn has driven both the creation of rocks and their subsequent uplift to form the lateral boundaries of this distinctive topographic region (element "1" of Figure 7). These bordering mountains have imposed a typical maxi-

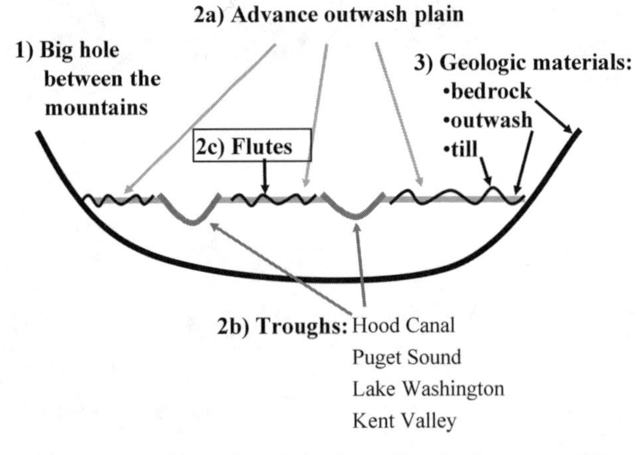

Figure 7. Major topographic scales of the Puget Lowland, expressed in a simplified west-to-east cross section. The "big hole," advance outwash plain, and troughs are at the scale of topography imposed on Lowland watersheds; the flutes provide a finer scale of topographic control on the location of rivers and streams; and the geologic materials of hillslope deposits determine the fate of runoff and groundwater flow.

mum size of approximately 4000 km^2 on the main rivers of the Cascade Range (such as the White, Green, Cedar, Snohomish, and Stillaguamish) and their east-flowing counterparts from the Olympic Mountains as they reach Puget Sound. They also form a largely (but not entirely) effective barrier to other rivers that might otherwise have entered this region from a much larger drainage area. Only the Skagit River, with headwaters east of the axis of the Cascade Range, and the Columbia River, with sufficient power from a much greater watershed to erode a path through the actively uparching Cascade Range in late Tertiary time, have breached these mountainous barriers (the Skagit River watershed is about twice the size of its next-largest Cascade drainage; the Columbia River watershed is over 100 times larger). This geologic setting also ensures that the major rivers will follow a relatively uniform downstream progression: rocky alpine headwaters, precipitous decline into confined mountain valleys, and emergence into broad low-gradient lowland valleys where the channels themselves are walled by unconsolidated fluvial sediments.

Medium Scale (1s–10s of km)—The Influence of Ice-Sheet Glaciation

This spatial scale is represented by the overall topographic pattern within the Puget Lowland itself—namely, a moderately dissected but still quite recognizable plateau, extending almost 100 km east-west from the Cascade Range to the Olympic Mountains and even farther north and south from beyond the Canadian border to the city of Olympia and beyond (element "2a" of Figure 7). The discontinuities in that plateau are of two kinds: bedrock prominences, notably the San Juan Islands and the hills that stretch west-to-east across the Puget Lowland immediately south of the Seattle fault (Green and Gold mountains, and the Newcastle Hills); and the marine (and once-marine) channels of Puget Sound (element "2b" of Figure 7). These channels are of nearly the same dimension as the upland plateau itself: narrow, but extending longitudinally as much as 150 km within Washington and several times that distance up past the eastern shore of Vancouver Island, scoured by overriding ice and subglacial meltwater. The plateau is also superimposed with smaller streamlined hills and valleys (element "2c" of Figure 7). Many of these glacial-age valleys continue to guide the modern flow of not only small streams but also the major lowland rivers, particularly as those rivers emerge from the Cascade rangefront. In combination, these influences establish the downstream progression of channel gradient, valley morphology, and sediment-delivery processes that primarily determine channel form and function (see Chapter 3).

Puget Lowland River Valleys

The major landscape-altering geologic processes of the Puget Lowland, glaciations and (to a lesser but still noteworthy extent) volcanic activity, have both contributed to the character of the modern drainage network. Invasions of the continental ice sheet created the predominant north-south grain of the topography which now guides the flow of nearly every major river in the southern lowland, imposed on their eastward (Olympic) or westward (Cascade) flow out of the mountains.

Through-going north-south valleys are common near the lateral boundaries of the lowland and particularly along the Cascade rangefront northeast and southeast of Seattle. Numerous subglacial and proglacial valleys provided passageways for voluminous meltwater discharges between the Skykomish, Snoqualmie, Cedar, and Green river valleys and their associated tributaries. Although now perched far above the modern drainage network, these relict river valleys are still evident along much of the Cascade front (Booth and Hallet 1993).

The interplay of glacial and volcanic activity is also evident in river-basin morphology, well-exemplified in nearly every channel with a Holocene volcano in its watershed. One of the best examples is found between the Stillaguamish and Skagit watersheds, where the Sauk River traverses an anomalous south-to-north valley of substantially greater width than that immediately upstream of Darrington. Its neighbor to the immediate south and west, the North Fork Stillaguamish River, also occupies a downstream valley dramatically (and abruptly) broader than its immediate upstream counterpart (Figure 8). These relationships owe their existence to the glacial and postglacial history of the region, initiated by glacial drainage of the upper Skagit River south along what is today the valley of the lower Sauk River by virtue of subglacial drainage pathways (Beechie et al. 2001; Tabor et al. 2001). All of the upper Skagit River, Suiattle River, and Sauk River drained out the valley of the North Fork Stillaguamish River during late-glacial time, diverted first by the ice sheet itself and later by a plug of Vashon-age sediment that temporarily blocked the Skagit River valley downstream of its modern confluence with the Sauk. The valley of the North Fork Stillaguamish River shows the effects of an extensive history of such drainage, with a broad valley carved through bedrock walls now several kilometers apart. The modern North Fork Stillaguamish River itself, however, is clearly an "underfit" channel, occupying a valley previously excavated by a larger river.

The agent of postglacial diversion of the Sauk and Suiattle rivers was volcanic eruptions from Glacier Peak (Vance 1957; Beget 1982). Lahars and lahar-derived fluvial sediment choke the mouth of the Suiattle River and

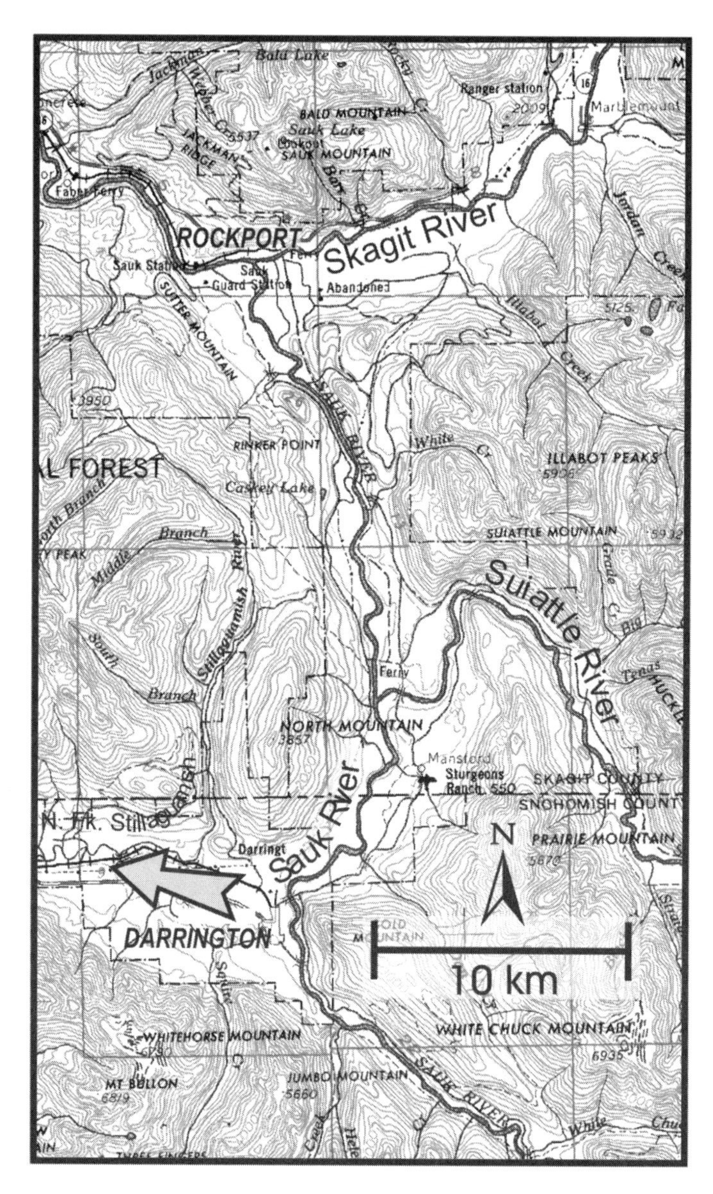

Figure 8. Topography of the lower Sauk River area, where volcanic and glacial modifications to the landscape have diverted major river channels in the Holocene and late Pleistocene Valley (base map from USGS Darrington and Rockport 7.5' quadrangles).

form the drainage divide at Darrington, and they fill much of the lower North Fork Stillaguamish valley. Drainage of the Sauk River is thus a function of local deposition of sediment where the valley widens at Darrington, with the local alluvial-fan topography determining whether the Sauk River will flow west towards the Stillaguamish or north towards the Skagit. The North Fork Stillaguamish River and the lower Sauk River have nearly identical gradients, suggesting that the Sauk established the gradient in both valleys by successive occupations during the Holocene.

The Suiattle River is similarly influenced by laharic deposition at its mouth, but its flow direction is constrained by the Sauk River. In late glacial time, the Sauk would have flowed south, but as the ice receded it would have reversed to flow north into the Skagit. Because of this reversal of drainage, the Suiattle River experienced a base-level change from perhaps 150 m altitude at Darrington (to the south) to 60 m altitude at Rockport (to the north). In response to this lower base level, and thus significantly steeper gradient, the Suiattle has incised as much as 50 m in its lower valley. The highest laharic surface recognized at the mouth of the Suiattle River is at 160 m elevation, barely sufficient to exceed the present valley divide at Darrington, over 10 km distant. As a result, the river is unlikely to drain down the North Fork Stillaguamish River again, even if the Sauk River was so diverted and a new lahar filled the Suiattle River. Westward diversion of the Sauk River thus is quite plausible in a future major eruption of Glacier Peak; southward diversion of the Suiattle River, however, is likely to require another glaciation.

Channeled and Elongated Topography of the Puget Lowland

Many of the watersheds that originate wholly within the Puget Lowland comprise three distinctive geologic and topographic zones—the upland plateau; a steep sideslope; and a base level defined by a lake, major river, or Puget Sound. In contrast to the monotonic decline not only in elevation but also in topographic gradient typical of the major riverine watersheds that begin at the Cascade or Olympic crest, the three watershed "zones" of the Lowland do not display any simple relationship between drainage area, gradient, and substrate. The upper headwaters are *not* always steepest, and the steepest areas are *not* always underlain by the most resistant sediment (or bedrock).

The headwater zones of these lowland watersheds lie on the upland plateau and are underlain by glacial sediment, primarily Vashon till with a thin (or absent) mantle of recessional outwash. The topography in these areas is one of elongated hills and troughs in the down-glacier direction, with gentle slopes and an overall relief of less than 30 m in most areas. This is the surface of the glacially overridden Vashon-age advance-outwash plain, described above. Low-gradient areas or closed depressions that developed beneath the ice sheet have now been partly or completely filled in by lakes, wetlands, or isolated patches of glacial outwash, deposited as the ice was retreating from the region during deglaciation (Figure 9).

As these headwater streams drain off the upland plateau, they flow either *parallel* to the ice-flow direction (generally to the north or south) down a gently to moderately sloping surface, or *across* the ice-flow direction (generally to the east or west) and over a more steeply descending slope. These steeper east-west slopes constitute the second topographic "zone," a result of subglacial scour followed by postglacial erosion. They flank the major subglacial or recessional drainage channels of the Vashon ice sheet, which have captured and are now occupied by the major rivers (*e.g.*, the Snoqualmie and the Cedar), by major lakes (Sammamish and Washington), or by Puget Sound itself.

In this zone, channel erosion has commonly exposed outwash sand and minor gravel derived from the advancing ice sheet overlying silt and clay of the early-glacial lake beds, defining a modern geologic environment characterized by both severe erosion and severe landsliding. In par-

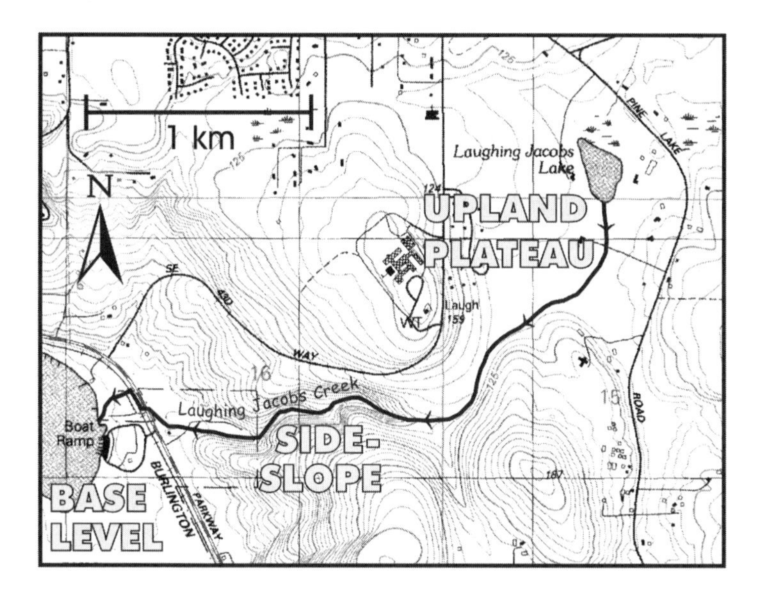

Figure 9. Example of the topography of typical lowland streams, where low-gradient headwaters drop steeply over the sideslope of the upland plateau to a base level set by a major river, Puget Sound, or (in this case) a lowland lake (base map from USGS Issaquah 7.5' quadrangle).

ticular, the advance outwash deposit contains mainly sand and relatively little gravel; thus stream-channel incision, leading to the catastrophic loss of channel grade, can proceed quite rapidly, particularly where new upland development has disrupted the hydrologic regime and culvert outfalls have created new discharge points (Booth 1990). Where gravel is more abundant, the rate of channel downcutting is somewhat less, but the net results appear to be qualitatively equivalent. Landslide susceptibility is enhanced by the perching of groundwater at the top of the older, less permeable deposits marked by the surface appearance of seeps and springs. Increased pore pressures above this boundary reduce the strength of the overlying material and increase the likelihood of mass movement.

The third topographic zone of typical Puget Lowland watersheds lies at the base of the upland plateaus, where channels approach their modern base level. A final type of modern stream channel is common here, flowing in a broad, flat-bottomed valley whose size and gradient has been inherited from glacial activity. The present stream is thus underfit, its valley carved by much more voluminous meltwater either beneath the ice sheet or by subaerial rivers during ice-sheet retreat. As a result of this legacy, the modern channel is incapable of moving all of the sediment delivered to it by the tributary streams, and the valley as a whole is a site for very long-term storage of eroded upland sediment. Olalla Valley on the Kitsap Peninsula (Figure 10) is an excellent example.

Where the stream first transitions from the steeper sideslopes to the valley bottom, much of the sediment load of the channel is deposited as an *alluvial fan*. These features, ubiquitous lobate landforms wherever channels abruptly lose gradient and thus sediment-transport capability, are formed during the lateral migration of sediment-laden discharges. They are poor choices for land development, because they can be made safe for human habitation only by anchoring the location of the channel, exacerbating instream sediment accumulation, and the channels themselves will rapidly bury instream structures and most habitat modifications that obstruct sediment passage in any way. Alluvial fans built by small streams may cover only a few hectares; larger ones may occupy one or more square kilometers and are associated with zones of rapid channel migration and avulsion. Although these landforms are most clearly expressed by relatively small channels undergoing abrupt changes in gradient, many of the same geomorphic processes occur at a larger scale in the mainstem rivers of the Olympic Mountains and the Cascade Range as they emerge from the mountain front into the Puget Lowland. Here, the imposed topography also establishes a change in gradient that results in zones of enhanced sediment deposition and rapid channel migration. Although a "hazard" from the perspective of human occupation of the nearby floodplain,

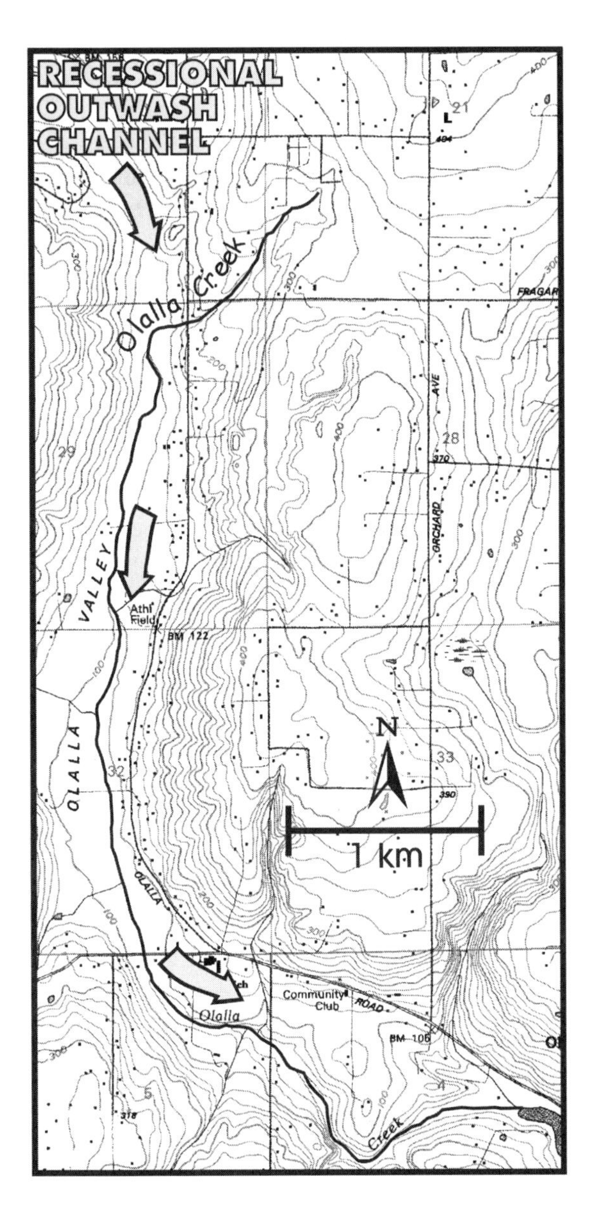

Figure 10. Example of an under-fit modern stream, occupying the late-glacial recessional outwash channel of Olalla Valley (base map from USGS Olalla 7.5' quadrangle).

these processes are also responsible for creating some of the most productive side-channel fish habitat in these river systems.

Channel Types and Channel Classification

Geomorphologists and biologists have been organizing and categorizing the many types of stream channels for about a century. In the mountain-to-lowland forested watersheds of the Pacific Northwest, the classification of Montgomery and Buffington (1997) explicitly characterizes the dominant geomorphic processes found in a typical downstream progression: steep headwater catchments, underlain by bedrock, to larger and more gently sloping watershed areas in broad alluvial valleys. This sequence is a product of watershed geology, as is its counterpart in the central Puget Lowland where headwater reaches can be quite flat and steeper reaches located only farther downstream. Sediment-delivery processes, channel-roughness elements, and incision susceptibility are thus very different in lowland channels than in nearby mountainous channels, and all depend on the geologic controls over watershed processes (see Chapter 3).

Small Scale (Hillslopes and Channel Reaches)—
The Influence of Geologic Deposits

This spatial scale reflects the conditions specific to a hillside or a valley segment, where the hydrologic response of a watershed and the local behavior of a channel are first determined. The range of possible conditions is as broad as the variety of deposits across the Puget Lowland, and the possible extent of channel responses is equally varied.

Hillslope Hydrology—Runoff and Infiltration

When rain and meltwater reach the surface of the ground, they encounter a filter that is of great importance in determining the path by which hillslope runoff will reach a stream channel. The paths taken by water, be they over or under the ground surface, determine many of the physical characteristics of a landscape, the generation of runoff, and the response of stream channels to the climatic regime and to watershed land uses. These factors, in turn, will determine the frequency and magnitude of high flows and channel erosion and deposition, the level of summer base flows, the hillslope processes that deliver sediment to stream channels, and ultimately the suitability of the channel for biota. Stream restoration cannot proceed without knowing the hydrologic regime of a channel, both now and in the future; in a region where forestry and urbanization continue to alter watershed land cover, the hydrologic regime of a channel is a direct product of the hydrologic processes on the hillslopes.

The most common, and most important, analytical task in characterizing hillslope hydrology is normally to "partition" the precipitation amongst runoff, evapotranspiration, infiltration, and groundwater recharge. The overriding determinants of this process are the vegetation and the character of the ground surface and soil layer (Chapter 11), which in turn depends in large measure on the properties of the underlying geologic deposit. The relative magnitudes of runoff and infiltration will determine which process(es) of storm runoff will occur (Dunne and Leopold 1978):

- *Horton overland flow* (surface runoff over unsaturated ground where the rainfall intensity exceeds infiltration capacity),
- *subsurface stormflow* (shallow groundwater flow in direct response to rainfall),
- *return flow* (shallow groundwater that reemerges to flow over the ground surface because of saturated subsurface conditions), and
- *direct precipitation onto saturated areas.*

The last two processes are also known together as *saturation overland flow* (Figure 11).

In *undisturbed* watersheds of the Puget Lowland and across the Pacific Northwest, the underlying geology probably has only a secondary control on these storm runoff processes. In the 15,000 years since deglaciation, weather-

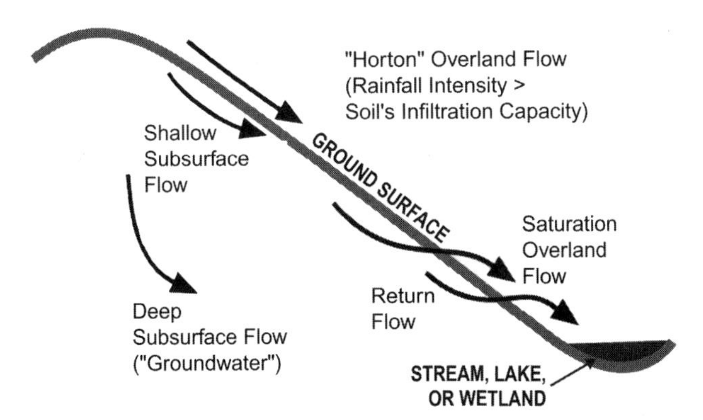

Figure 11. Schematic diagram of flow paths on a hillslope.

ing and biological activity have developed soils that are typically 1–2 m thick over even the most impervious of geologic deposits (or bedrock). These soils have infiltration capacities that are several orders of magnitude greater than their parent material (Snyder et al. 1973) and that greatly exceed typical rainfall intensities. Thus subsurface flow typically will be the dominant runoff process, almost independent of geologic substrate, during all but the largest storms.

Subsurface geology, however, becomes far more critical where natural erosion or human disturbance has thinned, compacted, or stripped the surficial soil. The permeability of the underlying geologic materials across the Puget Lowland varies by more than four orders of magnitude (Olmstead 1969). Typical rainfall intensities will readily exceed the infiltration capacity of the most widespread of these geologic materials, glacial till and bedrock, wherever they are exposed and unweathered. In such areas, delivery of water and sediment can be altered from that of a subsurface flow regime to a Horton overland flow regime, with attendant changes to peak discharges, sediment delivery, and water chemistry.

In contrast, where deep permeable deposits such as glacial outwash are present, erosion of the overlying soil is unlikely to impose substantial hydrologic changes. Yet if urban development covers these areas of once-permeable substrate with pavement, tremendous relative increases in discharges can result (see Chapter 11). Ironically, this geologic setting should also be the easiest to use the intrinsic permeability of the underlying deposit to minimize the increases in runoff delivered to downstream channels, through the infiltration of stormwater. Current stormwater management practices have been slow to recognize these geologic differences at the scale of an individual hillside and thus to take advantage of their opportunities for reducing downstream impacts.

Sources of Bed Material and Roughness

Most of the geologic deposits of the Puget Lowland have been transported to their present-day position by either glacial ice or by voluminous meltwater-fed rivers. They comprise a wide range of grain sizes, from clay- and silt-sized particles to boulders, and so their erosion can deliver a wide range of particle sizes to stream and river channels. At a watershed scale, this allows fluvial adjustment of bed-material sizes in response to changing discharge or sediment loads (Chapter 3). At a site scale, it will determine the susceptibility for channel erosion and incision, the character of the channel morphology, and the suitability of a reach for benthic organisms and fish.

Two exceptions to the general condition of wide-ranging particle sizes merit note, however. The most important arises downstream of extensive hillslope deposits of Vashon advance outwash, exposed beneath the Vashon till as the layers of a cake are exposed beneath its frosting. Although the advance outwash is commonly described as being "increasingly gravelly near the top of the deposit" (e.g., Mullineaux et al. 1965), the volumetric contribution of gravel to this deposit is quite small—it is, overwhelmingly, medium-grained sand throughout the Lowland, fluvially deposited many kilometers in front of the advancing ice sheet (Booth 1994). Where channels are floored by this deposit, channel roughness must be derived from gravel delivered (and subsequently winnowed) from the overlying till, or from large woody debris (Booth 1990). The former is not abundant, and the latter, although historically abundant and quite sufficient to maintain channel stability, is prone to human-induced loss (Montgomery et al. 1995; Booth et al. 1997).

The other common geologic deposits that lack coarse sediment are the lacustrine silts and clays that were also deposited in front of the advancing ice sheet (but at greater distances than the advance outwash). Sporadic "dropstones," gravel clasts rafted out into the proglacial lake within icebergs and subsequently released by melting into the silt and clay of the lake-bottom sediments, compose much less than one percent of this deposit. The remaining sediment is cohesive, sometimes weathering or eroding into gravel-sized fragments, which persist only briefly under the hardships of fluvial transport. Where unfractured, however, the resistance of this deposit to fluvial erosion is many times greater than that of the sandy advance outwash.

Susceptibility to Channel Expansion and Incision

Stream-channel expansion is a natural part of drainage-basin evolution, but its modern importance in the Puget Lowland is a direct consequence of human activity, particularly watershed urbanization. Channel incision is a particularly damaging response to increased discharge because it eliminates instream habitat, isolates the channel from its floodplain and so further increases in-channel flood flows, and releases tremendous quantities of sediment into the downstream channel network. Nearly every watershed subject to urban development responds to the increases in runoff by some degree of channel expansion, but not all suffer from dramatic incision.

Many of the simple physical parameters of a stream or of the watershed, such as slope or imperviousness, show no obvious value in predicting the magnitude of channel change in response to watershed development (Booth and Henshaw 2001). The role of geologic materials, however, shows a consistent relationship with incision susceptibility and magnitude. Incision generally requires relatively steep topographic gradients where human structures (such as culverts or weirs) do not provide local grade control that would limit vertical adjustments of the channel bed. In these settings, a clear difference is observed between cohesive silt-clay substrates, which generally permit only low rates of channel adjustment, and granular hillslope deposits (normally medium sand, most commonly of the Vashon advance outwash), which offer very little resistance to downcutting once the channel's pre-disturbance source of channel roughness (normally large woody debris) is physically removed or undermined by increased discharges.

Consequences of the Geologic Influences on Watersheds and Channels

Spatial variations in geologic conditions result in profound differences not only in the physical but also in the biological conditions of stream channels. Although aquatic biota depend on many factors (Karr and Chu 1999) of which only a subset are influenced by watershed geology, that influence is pivotal through the imposition of channel gradient and in-channel conditions. Channel gradient is largely determined by position on the landscape, which in the relatively young landscape of the Puget Lowland is determined almost entirely by its geologic history. Salmonids and other fish species exhibit preferred gradients for different life stages; the regional geology thus establishes broad but identifiable zones where the various life histories will be found (e.g., Montgomery et al. 1999). In-channel conditions, also critical to the life histories of many species, are a direct product of geologic conditions and fluvial processes. Watershed sediment sources will partly determine the bed texture of the stream and determine if certain preferred sediment sizes, notably gravel, are available in sufficient quantities. Certain deposits, notably Vashon till and Vashon recessional outwash, have these sediment sizes in abundance; other deposits are almost entirely lacking in gravel. The presence of a hyporheic zone around the channel and the upwelling of groundwater into the channel are both products of valley morphology, floodplain sediments, and subsurface stratigraphy. If our knowledge of these processes is ever to improve, it will require integration of the geologic framework of a channel network with the hydrologic conditions of the surrounding landscape.

The varying temporal history of geologic processes, notably the history of drainage-network development and disturbance, is also important to any understanding of biological conditions. Ice-sheet glaciation of the Puget Lowland has provided ice-free conditions for only the last 15,000 years, a legacy that is abundantly displayed in the genetic stocks of native fish (see Chapter 5).

Because of these diverse physical and biological influences on watershed processes and conditions, aspects of the regional and local geology must be understood for stream restoration or rehabilitation to be successful. Some of these geologic aspects are obvious and readily determined—measuring channel gradient, for example, needs no specialized geologic information—but others, such as the delivery of sediment, the likely long-term disturbance history, or the flux of groundwater into and out of the channel, can be evaluated only with knowledge of geologic materials, history, and processes. Predicting the response of a stream to human manipulation demands conceptual models of water and sediment delivery to the fluvial system, and of the physical framework through which those fluxes occur. The geologic setting of a river or stream also imposes intrinsic natural constraints on what can be accomplished through human intervention.

Whether evaluated or not, these geologic conditions have *always* imposed constraints on the nature of channel networks, and on the biological productivity of their rivers and streams. Not all streams have the same level of biological activity, and so not all stream restoration efforts should strive for the same goals and objectives. If the tools of geology are not used to recognize those conditions that can limit the success of rehabilitation, even the best of human intentions are likely to fall short of goals and to waste resources in the process. Furthermore, other opportunities may be passed unrecognized, even where stream and watershed conditions hold the promise of a favorable response. We cannot afford such ignorance.

Acknowledgments

Our thanks to fellow geologists Susan Perkins and John Bethel, who provided very helpful perspectives from extensive experience on the application of geologic information to stream and watershed processes. Partial funding for this investigation was provided by the Center for Urban Water Resources Management at the University of Washington and the U. S. Geological Survey.

References

Armstrong, J.E., D.R. Crandell, D.J. Easterbrook, and J.B. Noble. 1965. Late Pleistocene stratigraphy and chronology in southwestern British Columbia and northwestern Washington. *Geological Society of America Bulletin* 76:321-330.

Beechie, T.J., B.D. Collins, and G.R. Pess. 2001. Holocene and recent geomorphic processes, land use, and salmonid habitat in two north Puget Sound river basins. In J. M. Dorava, D. R. Montgomery, B. Palcsak, and F. Fitzpatrick (eds.) *Geomorphic Processes and Riverine Habitat*. American Geophysical Union, Washington, DC. pp 37-54.

Beget, J.E. 1982. Postglacial volcanic deposits at Glacier Peak, Washington, and potential hazards from future eruptions. U. S. Geological Survey Open-File Report 82-830.

Blunt, D.J, D.J. Easterbrook, and N.W. Rutter. 1987. Chronology of Pleistocene sediments in the Puget Lowland, Washington. *Washington Division of Geology and Earth Resources Bulletin* 77:321-353.

Booth, D.B. 1987. Timing and processes of deglaciation along the southern margin of the Cordilleran ice sheet. In W. F. Ruddiman and H. E. Wright, Jr. (eds.) *North America and Adjacent Oceans During the Last Deglaciation*. The Geology of North America v. K-3. Geological Society of America, Boulder, Colorado. pp 71-90.

Booth, D.B. 1990. Stream-channel incision following drainage-basin urbanization. *Water Resources Bulletin* 26:407-417.

Booth, D.B. 1994. Glaciofluvial infilling and scour of the Puget Lowland, Washington, during ice-sheet glaciation. *Geology* 22:695-698.

Booth, D.B. and B. Hallet. 1993. Channel networks carved by subglacial water: Observations and reconstruction in the eastern Puget Lowland of Washington. *Geological Society of America Bulletin* 105:671-683.

Booth, D.B. and P.C. Henshaw. 2001. Rates of channel erosion in small urban streams. In M. Wigmosta and S. Burges (eds.) *Land Use and Watersheds: Human Influence on Hydrogeology and Geomorphology in Urban and Forest Areas*. AGU Monograph Series, Water Science and Application Vol. 2, pp. 17-38.

Booth, D.B., D.R. Montgomery, and J. Bethel. 1997. Large woody debris in urban streams of the Pacific Northwest. In L.A. Roesner (ed.) *Effects of watershed development and management on aquatic ecosystems*. Engineering Foundation Conference Proceedings, Snowbird, Utah, August 4-9, 1996. pp. 178-197.

Brandon, M.T., M.K. Roden-Tice, and J.I. Garver. 1998 Late Cenozoic exhumation of the Cascadia accretionary wedge in the Olympic Mountains, northwest Washington State. *Geological Society of America Bulletin* 110: 985-1009.

Bretz, J.H. 1913. Glaciation of the Puget Sound region. *Washington Geological Survey Bulletin* No. 8.

Bucknam, R.C., E. Hemphill-Haley, and E.B. Leopold. 1992. Abrupt uplift within the past 1700 years at southern Puget Sound. *Science* 258:1611-1614.

Cheney, E.S. 1997. What is the age and extent of the Cascade magmatic arc? *Washington Geology* 25:28-32.

Christiansen, R.L., R.S. Yeats, S.A. Graham, W.A. Niem, A.R. Niem, and P.D. Snavely, Jr. 1992. Post-Laramide geology of the U.S. Cordilleran region. In Burchfiel B.C., P.W. Lipman, and M.L. Zoback. (eds.) *The Cordilleran Orogen: Conterminous U.S.* Geology of North America v. G-3. Geological Society of America, Boulder, Colorado. pp. 261-406.

Clague, J.J. and P.T. Bobrowsky. 1990. Holocene sea level change and crustal deformation, southwestern British Columbia. In Geological Survey of Canada, Current research, Part E—Cordillera and Pacific margin: Geological Survey of Canada Paper 90-1E, pp. 245-250.

Dragovich, J.D., P.T. Pringle, and T J. Walsh. 1994. Extent and geometry of the mid-Holocene Osceola Mudflow in the Puget Lowland: Implications for Holocene sedimentation and paleogeography. *Washington Geology* 22:3-26.

Dunne, T. and L.B. Leopold. 1978. *Water in Environmental Planning* W. H. Freeman and Company, New York.

Eronen, M., T. Kankainen, and M. Tsukada. 1987. Late-Holocene sea-level record in a core from the Puget Lowland, Washington. *Quaternary Research* 27:147-159.

Frizzell, V.A., Jr., R.W. Tabor, R.E. Zartman, and C.D. Blome. 1987. Late Mesozoic or early Tertiary melanges in the western Cascades of Washington. In J. E. Schuster (ed.) *Selected Papers on the Geology of Washington*. Washington Division of Geology and Earth Resources Bulletin 77:129-148.

Karr, J.R. and E.W. Chu. 1999. *Restoring Life in Running Waters*. Island Press, Washington, DC.

Laprade, W.T. 1982. Geologic implications of pre-consolidated pressure values, Lawton Clay, Seattle: Washington. *Proceedings, 19th Annual Engineering Geology and Soils Engineering Symposium*, Idaho State University, Pocatello, Idaho. pp. 303-321.

Matthews, R.K. 1990. Quaternary sea-level change. *Sea-Level Change*. National Academy Press, Washington, DC. pp. 88-103.

Montgomery, D.R. and J.M. Buffington. 1997. Channel reach morphology in mountain drainage basins. *Geological Society of America Bulletin* 109:596-611.

Montgomery, D.R., J.M. Buffington, R.D. Smith, K.M. Schmidt, and G. Pess. 1995. Pool spacing in forest channels. *Water Resources Research* 31:1097-1105.

Montgomery, D.R., G. Pess, E.M. Beamer, and T.P. Quinn. 1999 Channel type and salmonid spawning distributions and abundance. *Canadian Journal of Fisheries and Aquatic Sciences* 56:377-387.

Mullineaux, D.R., H.H. Waldron, and M. Rubin. 1965. Stratigraphy and chronology of late interglacial and early Vashon time in the Seattle area, Washington. *U. S. Geological Survey Bulletin* 1194:1-10.

Olmstead, T.L. 1969. Geotechnical aspects and engineering properties of glacial till in the Puget Lowland, Washington. *Proceedings, 7th Annual Engineering Geology and Soils Engineering Symposium*, Moscow, Idaho. pp. 223-233.

Pessl, F., Jr., D.P. Dethier, D.B. Booth, and J.P. Minard. 1989. Surficial geology of the Port Townsend 1:100,000 quadrangle, Washington. U.S. Geological Survey Miscellaneous Investigations Map I-1198F.

Porter, S.C. and T.W. Swanson. 1998. Radiocarbon age constraints on rates of advance and retreat of the Puget lobe of the Cordilleran ice sheet during the last glaciation. *Quaternary Research* 50:205-213.

Reid, L.M. and T. Dunne. 1996. *Rapid Evaluation of Sediment Budgets*. Catena, Verlag, Germany.

Snyder, D.E., P.S. Gale, and R.F. Pringle. 1973. Soil survey of King County area, Washington. U. S. Department of Agriculture, Soil Conservation Service.

Tabor, R.W., D.B. Booth, J.A. Vance, and A.B. Ford. 2001. Geologic map of the Sauk River 1:100,000 quadrangle, Washington. U.S. Geological Survey Miscellaneous Investigations Map I-2592.

Tabor, R.W. and W.M. Cady. 1978. Geologic map of the Olympic Peninsula, Washington. U.S. Geological Survey Miscellaneous Investigations Series Map I-994, 2 sheets, scale 1:125,000.

Tabor, R.W. and R.A. Haugerud. 1999. *Geology of the North Cascades—A mountain mosaic.* The Mountaineers, Seattle, Washington.

Thorson, R.M. 1989. Glacio-isostatic response of the Puget Sound area, Washington. *Geological Society of America Bulletin* 101:1163-1174.

Thorson, R.M. 1980. Ice sheet glaciation of the Puget Lowland, Washington, during the Vashon stade (late Pleistocene). *Quarternary Research* 13:303-321.

Troost, K.G. 1999. The Olympia nonglacial interval in the southcentral Puget Lowland, Washington. Master's thesis. University of Washington. Seattle, Washington.

Troost, K.G. and J.K. Stein. 1995. Geology and geoarchaeology of West Point. In L.L. Larson, and D.E. Lewarch. (eds.) *The Archaeology of West Point, Seattle, Washington*. Report to King County Department of Metropolitan Services, v. 1, pt. 1, pp. 2-1 to 2-78.

Tubbs, D.W. 1974. Landslides in Seattle. Information circular 52. State of Washington, Dept. of Natural Resources, Olympia, Washington.
Vance, J.A. 1957. The geology of the Sauk River area in the northern Cascades of Washington. Ph.D. dissertation. University of Washington. Seattle, Washington.
Yount, J.C., J.P. Minard, and G.R. Dembroff. 1993. Geologic map of surficial deposits in the Seattle 30' x 60' quadrangle, Washington. U.S. Geological Survey Open-File Report 93-233, 2 sheets, scale 1:100,000.

3. Fluvial Processes in Puget Sound Rivers and the Pacific Northwest

John M. Buffington, Richard D. Woodsmith, Derek B. Booth, and David R. Montgomery

Abstract

The variability of topography, geology, climate, vegetation, and land use in the Pacific Northwest creates considerable spatial and temporal variability of fluvial processes and reach-scale channel type. Here we identify process domains of typical Pacific Northwest watersheds and examine local physiographic and geologic controls on channel processes and response potential in the Puget Sound region. We also review the influence of different channel types on opportunities and limitations for channel restoration. Finally, we develop regime diagrams that identify typical combinations of channel characteristics associated with different alluvial channel types. These diagrams can be used to set target values for creating or maintaining desired channel types and associated habitats or to assess the stable channel morphology for imposed watershed conditions. Regime diagrams that are based on explicit physical models also can be used to predict likely trends and magnitudes of channel response to natural or anthropogenic disturbances (such as restoration activities). Moreover, spatial linkages of processes and the potential for distal disturbances to propagate through channel networks means that local restoration efforts that do not address larger scale watershed processes and disturbances may be ineffective or costly to maintain.

Introduction

Millions of dollars are being spent in the United States on river and stream restoration projects. In the Columbia River basin alone, the Bonneville Power Administration spent an average of $44 million a year on habitat restoration projects and related research during the 5-year period from 1996-2000 (BPA 2001). Now, a comparably ambitious program for river restoration is developing in the Puget Sound region driven by concerns over salmon recovery under the Endangered Species Act. Despite the enormous capital investment in such efforts—and the legal and social mandates that underlie them—channel restoration efforts throughout the Pacific Northwest remain largely uncontrolled experiments with little pre-restoration analysis and even less post-restoration monitoring and assessment. Consequently, it is difficult to assess the success of these projects and to advance restoration practice in a systematic fashion.

Restoration projects focused on fish habitat can be traced back at least as far as the 1930s in the United States (Reeves et al. 1991). Critical reviews of habitat restoration projects report a mixed record of successes and failures (Frissell and Nawa 1992; Beschta et al. 1994), due in part to (1) incomplete understanding of fluvial processes, (2) project designs inappropriate for local channel processes, and (3) a focus on local conditions without consideration of the larger watershed context. The latter consideration is particularly important for restoration planning because river channels integrate watershed processes and translate natural and anthropogenic disturbances downslope through the landscape. For example, it would make little sense to "restore" an equilibrium channel form to a channel poised to receive an increased sediment load from an upslope legacy of past disturbance. Consequently, a holistic understanding of channel and watershed processes is needed for effective management and restoration of riverine ecosystems.

Preconceived notions of natural channel conditions underpin many river restoration efforts. The public frequently has an idyllic image of natural channels as tree-lined, meandering gravel-bed rivers with high quality, abundant aquatic and riparian habitat. While environmentally and esthetically desirable, that sort of channel can be supported only under very specific physical conditions, which may be of limited extent in a given watershed. This chapter examines controls on channel morphology and fluvial processes typical of Pacific Northwest watersheds and reviews the influences of different channel types on opportunities and limitations for channel restoration.

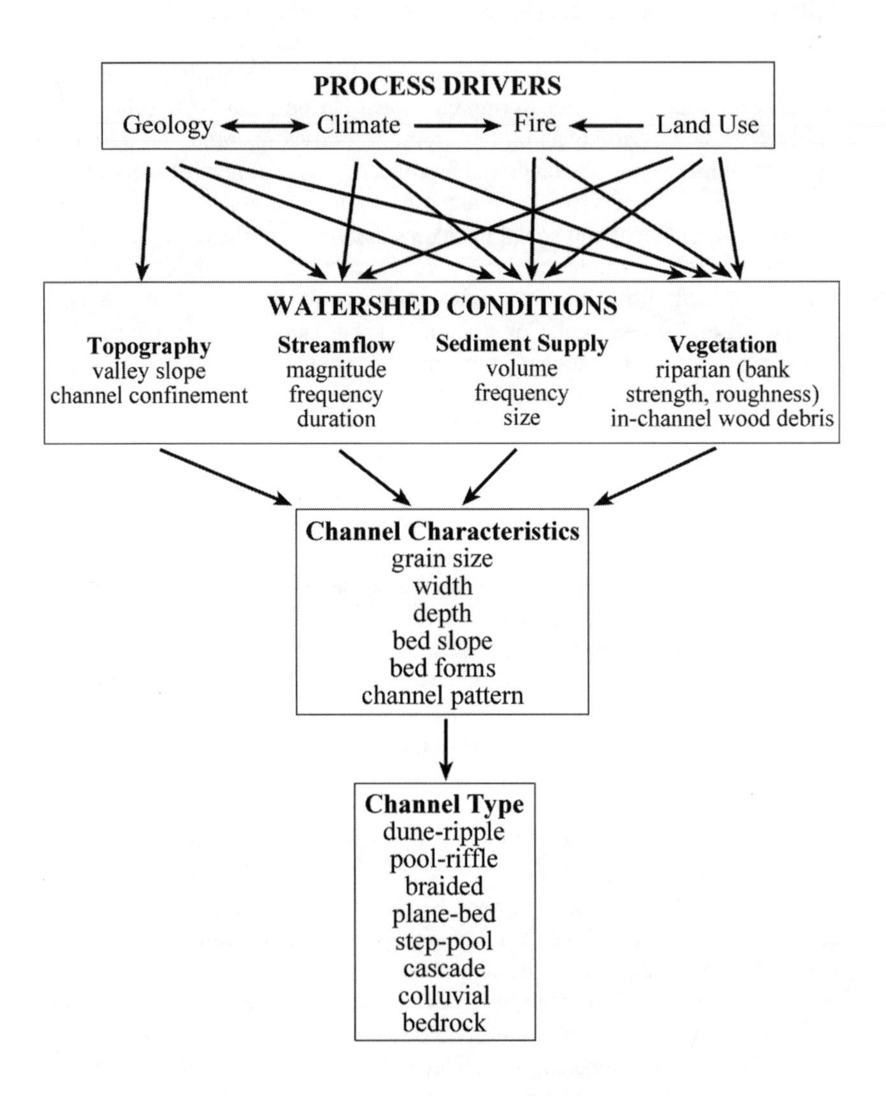

Figure 1. Controls on fluvial processes and channel morphology in Pacific Northwest watersheds. Arrows indicate interaction amongst different factors.

Controls on Channel Morphology

Physical processes within Pacific Northwest watersheds are driven by several primary factors: geology, climate, fire, and land use (Figure 1). These process drivers, in turn, impose a suite of watershed conditions on the fluvial system: topography, streamflow, sediment supply, and vegetation. The imposed watershed conditions influence channel characteristics: grain size, width, depth, bed slope, bed forms, and channel pattern. Mutual adjustment of channel characteristics for different combinations of imposed watershed conditions gives rise to different reach-scale channel types (or morphologies) that differ in habitat properties and resilience to disturbance. The sequential relationship between watershed conditions, channel characteristics, and channel type (Figure 1) has been recognized by many investigators (Mollard 1973; Schumm 1985; Kellerhals and Church 1989), but the role of differences in large-scale controls on channel conditions and response potential are not as widely recognized. In particular, the physiography and geologic history of a region may exert a dominant influence on channel processes and response potential (Chapter 2).

Physiography and Watershed Conditions

The Pacific Northwest contains several physiographic provinces (Figure 2) characterized by differences in process drivers and physical conditions. To-

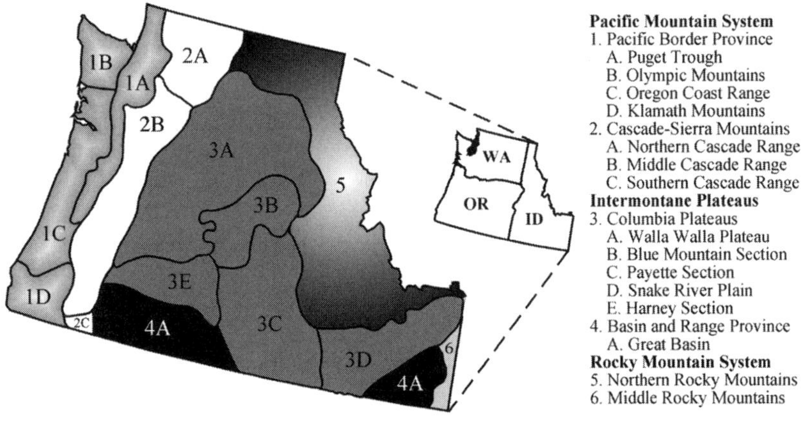

Figure 2. Physiography of the conterminous U.S. Pacific Northwest (after Powell [1896]). Fenneman's (1931) classification system is used to divide physiographic regions (bold items) into finer scale provinces (numbered items) and sections (lettered items).

pography in the Pacific Northwest varies from low-gradient glacial outwash plains of the Puget Lowland (north end of the Puget Trough) and basalt plateaus of the Columbia and Snake rivers to steep mountainous terrain of the Olympic Mountains, Cascade Range, and Northern Rocky Mountains. Rock types are equally diverse, including soft marine sediments, resistant basaltic lavas and metamorphic rocks, and granitic plutons that upon weathering produce large quantities of sand. Climate also varies from lush, coniferous rain forests on the western flanks of the Cascade Range to semi-arid, high elevation deserts east of the Cascades. Streamflow regimes range from the storm-driven winter flood regime west of the Cascade crest to the snowmelt-driven spring floods more typical east of the Cascade crest.

Even within the more restricted area of the Puget Sound, the diverse topography, geology, and glacial history of the region impart considerable spatial and temporal variability to fluvial processes through their influence on vegetation, sediment supply, and stream discharge (Chapter 2). The Puget Sound region can be subdivided into four physiographic sections: the northern and middle Cascade Ranges, the Olympic Mountains, and the Puget Lowland portion of the Puget Trough. The Cascade Ranges and Olympic Mountains, which together are defined here as the Puget Upland, are characterized by steep, mountainous terrain and rapid changes in slope over short length scales, giving rise to substantial spatial variability of channel morphology and fluvial processes within individual watersheds. In contrast, the Puget Lowland has relatively subdued topography in which local geology more strongly influences channel processes and response than do differences in slope. In mountain drainage basins of the Puget Sound region, and the Pacific Northwest in general, channel morphology typically ranges from steep, confined channels that are sediment-limited with boulder and bedrock beds (Figure 3a), to low-gradient alluvial channels that are typically unconfined and sediment-loaded, with sand and gravel beds (Figure 3b). The same suite of channel types occurs across both the Puget Lowland and Upland, but differences in watershed conditions and processes lead to very different conditions, dynamics, and responses between otherwise comparable channels. In particular, specific characteristics of Puget Lowland and Upland channels differ due to the huge supply of glacial sediment in the Puget Lowland and the inherited glacial lowland topography.

The range of channel morphologies found in Pacific Northwest landscapes can be related qualitatively to watershed conditions of streamflow, sediment supply, valley gradient, and channel confinement (Figure 4). Within this framework, physical domains can be identified for channels formed by fluvial versus mass wasting processes (domains 2 and 1 on Figure 4), bedrock versus alluvial channels (2b and 2a, respectively), and various alluvial chan-

A

B

Figure 3. Photographs of (A) a cascade channel (North Fork Payette River, Idaho) and (B) the confluence of a dune-ripple and a pool-riffle channel (South and Middle Forks of the Payette River, Idaho; photograph courtesy of Carter Borden).

nel types (within domain 2a). Characteristics of these alluvial channel types are summarized in Table 1 and further described elsewhere (Montgomery and Buffington 1997, 1998). Figure 4 provides a conceptual framework for the interactions between watershed conditions, channel characteristics, and channel type discussed previously (Figure 1). For example, greater valley slope and channel confinement create channels with steeper bed slopes, larger particle sizes, and lower width-to-depth ratios, giving rise to systematic changes in channel morphology (e.g., changes in alluvial channel type from dune-ripple through cascade morphologies).

Although not shown in Figure 4, riparian vegetation and in-channel woody debris can significantly influence channel characteristics and morphology. Woody debris and bank vegetation can alter channel hydraulics, rates of sediment transport, storage, and supply, grain size, bed and bank topography, bed slope, and channel width, depth and pattern (e.g., Hogan 1986; Bisson et al. 1987; Smith et al. 1993a,b; Keller et al. 1995; Buffington and Montgomery 1999a; Abbe 2000). By modifying channel characteristics and local watershed conditions (streamflow, sediment supply, and topography), riparian vegetation and woody debris can force different alluvial channel types in

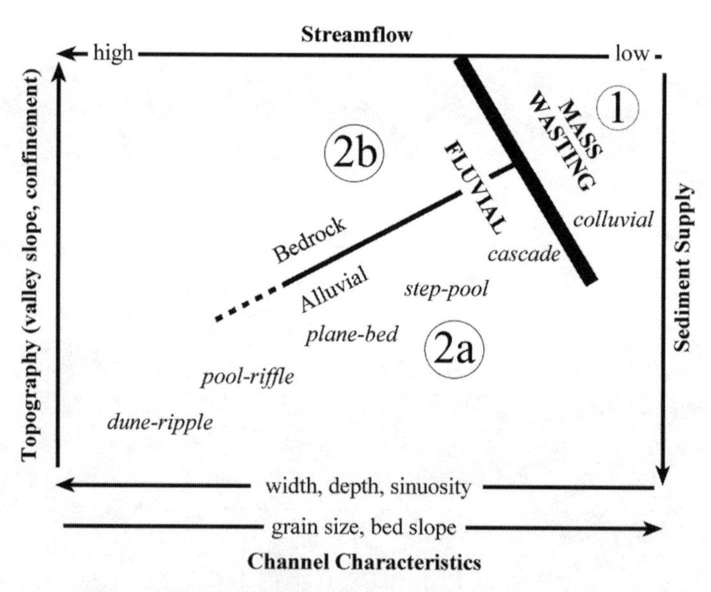

Figure 4. Influence of watershed conditions (streamflow, sediment supply, and topography) and channel characteristics on channel reach morphology (after Mollard [1973]). Numbered items indicate process domains: mass wasting (1), fluvial (2), alluvial channels (2a), and bedrock channels (2b). See text for further discussion.

Table 1. Alluvial channel types

Type, Typical Slope (m/m), and Bed Material	*Description*
 dune-ripple, <0.001, sand	Low-gradient, unconfined, sand-bed rivers occupying large alluviated valleys and typically decoupled from hillslopes. Variety of mobile bed forms (ripples, dunes, sand waves, plane-bed, and antidunes) that depend on Froude number and transport intensity. Well-defined floodplain morphology with a bankfull recurrence interval of roughly 1-2 years. Bankfull discharge also is the effective discharge (that which transports the most sediment over the long term). Transport-limited, with a low threshold for bedload transport (ratio of bankfull shear stress to critical stress for incipient motion [τ/τ_c] is on the order of 10-100).
 pool-riffle, 0.001-0.02, gravel and cobble	Alternating pool and bar topography caused by oscillating lateral flow that forces local flow convergence (pool scour) and divergence (bar deposition). Moderate- to low- gradient, unconfined channels, with coarse bed material and typically extensive floodplains. Two-phase bedload transport, characterized by supply-limited transport of fine grains over an immobile armor during low flows, and transport-limited motion of most available particle sizes during high flows that mobilize the armor. Bankfull discharge is the effective discharge and has an approximately 1-2 year recurrence interval. Potential for extensive salmonid spawning and rearing habitat.

Table 1. Alluvial channel types (continued).

Type, Typical Slope (m/m), and Bed Material	*Description*
 plane-bed, 0.01-0.4, gravel, cobble, and some boulder	Long reaches of glide, run, or riffle morphology lacking significant pool or bar topography. Low width-to-depth ratios and moderate values of relative submergence (ratio of bankfull flow depth to median particle size) damp lateral flow oscillations that would otherwise create an alternate bar morphology. Susceptible to obstruction-forced pool formation. Moderate-gradient channels with coarse bed materials and variable floodplain extent. Bed surface is typically armored, with a near-bankfull threshold for significant bedload transport (τ/τ_c is on the order of 1). Two-phase transport. Bankfull discharge is likely the effective discharge, with an approximately 1-2 year recurrence interval. Potential for extensive salmonid spawning habitat.
 braided, <0.03, sand to boulder	Multi-thread rivers with large width-to-depth ratios and moderate slopes. Individual threads may have a pool-riffle morphology or a bar-riffle morphology lacking pools. Pool scour commonly occurs where braid threads converge. Braiding results from high sediment loads or channel widening caused by bank destabilization. Braided channels in the Pacific Northwest commonly occur 1) as glacial outwash channels, 2) in locations overwhelmed by a locally high sediment supply, 3) in alluvial valleys where banks have been destabilized by riparian cutting or livestock trampling, or 4) in semi-arid regions with insufficient riparian vegetation to stabilize banks composed of cohesionless sediments.

Table 1. Alluvial channel types (continued).

Type, Typical Slope (m/m), and Bed Material	*Description*
 step-pool, 0.02-0.08, cobble and boulder	Repeating sequences of steps and plunge pools formed by wood debris, resistant bedrock, or by boulders that accumulate either as kinematic waves or as macroscale antidunes. Steep-gradient, confined channels, with little floodplain development, and directly coupled to hillslopes. High transport capacities that efficiently transport cobble- to sand-sized material on an annual basis. Amplitude and wavelength of steps and pools may be adjusted to maximize hydraulic resistance and stabilize channel form. Alternatively, hydraulic roughness provided by bed topography may be adjusted so as to equilibrate rates of sediment supply and bed load transport capacity, thereby providing a mechanism for channel stability. Quality of pool habitat depends on plunge pool geometry and associated hydraulics. Limited salmonid spawning extent (backwater environments and pool tails).
 cascade 0.04-0.25 boulder	Chaotic arrangement of boulder-sized bed material and continuous macroscale turbulence. Typically confined by valley walls and directly coupled to hillslopes. Steep gradients and relatively deep, concentrated flow allow efficient transport of cobble- to sand-sized sediment during annual floods, but movement of the channel-forming boulders requires infrequent large floods. Little sediment storage due to shallow depth to bedrock and lack of floodplain development. Limited salmonid spawning sites (channel margins and backwater environments). Infrequent, turbulent pools of small volume.

portions of the landscape where they would not otherwise occur (Montgomery et al. 1995; 1996a; Montgomery and Buffington 1998).

Relative differences in the role of fluvial versus hillslope processes and the extent of bedrock versus alluvial control on channel characteristics naturally divide Pacific Northwest channel systems into three generalized process domains (headwater channels, confined channels in steep valleys, and unconfined alluvial channels in low-gradient valleys) (Montgomery 1999). However, differences in physiography and geologic history cause channel processes and response potential to differ between similar process domains in Lowland and Upland channels of the Puget Sound.

Headwater Channels in the Cascade Range and Olympic Mountains

Channels and valleys in headwater regions of the Puget Upland are characterized by steep slopes and are strongly influenced by hillslope processes, particularly mass wasting. Sediment shed from hillslopes gradually accumulates to form colluvial wedges in topographic hollows and valley fills along ephemeral streams in headwater valleys (Dietrich et al. 1982, 1986). Fluvial sediment transport in these colluvial channels is weak and ineffective (Montgomery and Buffington 1997), and consequently these channels are shallow surficial features that do not significantly influence valley form and landscape evolution. Instead, headwater valleys are maintained by catastrophic failure of accumulated colluvial soils during storm events. The resulting debris flows typically scour downslope colluvial channels to bedrock before depositing a slug of sediment once they reach slopes of 3° to 6°, encounter objects they cannot entrain, or lose momentum rounding tight corners through the channel network (Benda and Cundy 1990). Soil-mantled hillslopes are particularly susceptible to failure following loss of vegetative root strength due to fire or timber harvest. Root strength offered by vegetation is a primary factor holding soils on steep slopes in many Pacific Northwest landscapes (Schmidt et al. 2001). Depending on the recency of debris flow activity, headwater valleys may host either colluvial or bedrock channels.

Headwater Channels in the Puget Lowland

Unlike their counterparts in the Upland, headwater channels of the Puget Lowland are commonly very low gradient, originating on glacial till-mantled plateaus that perch shallow groundwater and locally support lakes or wetlands as the upstream-most expression of surface water. Sediment delivery to these headwater reaches is slow, and even in free-flowing reaches the stream power is very low, limiting both sediment transport and development of alluvial channel morphology. In the 13,000 years since deglaciation, these channels have accomplished little modification of this topography, but their morphology can be quite responsive to changes in either discharge or sediment loading as a consequence of watershed disturbance. Increased discharge, a typical result of urban development, commonly results in channel expansion and offers the potential of far more catastrophic incision farther downstream where gradients steepen (Booth 1990). Increased sediment loading, from either urban or agricultural activities, can result in rapid aggradation of headwater channels. The hillslope processes common in headwater mountain drainage basins of the Pacific Northwest, notably debris flows, rapid stormflow response over thin soils, and delivery of large bedrock clasts to the channel, are almost entirely absent in headwater channels of the Puget Lowland. Hence, headwater channels in the Puget Lowland are more sensitive to changes in basin hydrology and surface erosion processes than are headwater channels in the Puget Upland.

Confined Channels in the Cascade Range & Olympic Mountains

Farther downslope, drainage area increases and fluvial processes increasingly dominate channel morphology in Puget Upland channels. Confined fluvial channels in the Cascade Range and Olympic Mountains can exhibit either bedrock or alluvial channel types (cascade and step-pool morphologies, Table 1). Bedrock channels formed by fluvial processes are located in steep- to moderate-gradient portions of the network that occupy bedrock-walled valleys. Bedrock channels formed by fluvial processes occur where transport capacity is greater than bed load sediment supply, whereas alluvial channels occur where sediment supply exceeds transport capacity (Gilbert 1914). Montgomery et al. (1996a) demonstrated that a slope and drainage-area framework could delineate the occurrence of bedrock versus alluvial channel types in channels that are not strongly influenced by woody debris. By equating relations for transport capacity and sediment supply, they solved for an inverse relationship between drainage area and the critical slope needed to maintain a bedrock channel free of alluvial cover. The resulting critical-slope function is region-specific and depends on local conditions of geology, climate, and sediment supply (both volume and size) (Massong and Montgomery 2000).

The presence of woody debris along a bedrock channel may in some instances trap enough sediment to convert the channel to a forced alluvial channel floored by gravel (Montgomery et al. 1996a). The nature of the sediment supplied to the channel (and in particular its durability) strongly influences the extent of bedrock and alluvial channel types in steep, confined channels lacking woody debris, and thereby the sensitivity of these channels to loss of wood. The sedimentary rocks of the Olympic Peninsula rapidly disintegrate into fine sediment once introduced into the channel. Consequently steep, confined channels supplied with such sediment will tend to form bedrock channels unless sufficient woody debris is available to maintain an alluvial bed. In contrast, bedrock reaches are less common in the Washington Cascade Range where durable volcanic rocks form the primary sediment supply for confined channels in steep valleys.

Confined Channels in the Puget Lowland

In many parts of the Puget Lowland, stream channels drop steeply over an eroded edge of the headwater plateaus and enter a high-gradient, confined ravine where the intrinsic ability of the flow to transport sediment, and the rate at which sediment is delivered to the channel from adjacent hillslopes, both increase dramatically in comparison to upstream reaches. This zone commonly corresponds to parts of the underlying geologic strata dominated by noncohesive sandy sediment that is very easily eroded by running water and relatively poor in coarse gravel that might otherwise help armor the streambed and reduce the rate of vertical incision (Chapter 2). In undisturbed settings, vertical channel stability has been achieved primarily through abundant wood debris that can dissipate a significant amount of the shear stress applied to the channel bed and banks. Where logs are no longer present, or where their contact with the streambed has been undermined by increased flows, vertical stability can be lost rapidly. The resulting channel incision can proceed for up to many tens of meters until some combination of reduced channel gradient, resistant geologic layer at depth, or increased sediment delivery from oversteepened channel banks reestablishes an equilibrium profile.

Unconfined Alluvial Channels in the Puget Sound Region

Characteristics of low-gradient unconfined alluvial channels (dune-ripple and pool-riffle morphologies, Table 1) vary with differences in watershed conditions (e.g., sediment supply and stream discharge) in the Puget Sound region. Unconfined channels draining the Puget Upland are supplied with a mixture of glacial sediments and igneous and metamorphic rock fragments, whereas those draining the Puget Lowland tend to be supplied with abundant quantities of glacial sediments that span a wide range of grain sizes from silt to cobble. Moreover, differences in physiography and geologic history influence valley formation and degree of channel confinement, which in turn affect channel type and associated habitat characteristics in low-gradient alluvial channels. In some cases, low-gradient rivers in the Puget Sound region occupy broad troughs carved by glacial meltwater that have filled with river deposits since deglaciation (Chapter 2). In other cases, low-gradient rivers have carved valleys into the regional outwash plain. In contrast to steep confined channels where vertical stability of the bed is an important issue, lateral stability is a primary concern in low gradient channels. These channels build their floodplains by both overbank deposition during floods and bedload deposition as the channel moves laterally. Hence, the entire active floodplain may be considered the overflow channel and generally defines the channel-migration zone.

Restoring Riverine Ecosystems

Recognition of the characteristic ranges of channel conditions associated with each channel type is critical for successful restoration. Restoration projects that impose channel morphologies in environments that are outside of their characteristic ranges will not be self-maintaining and may prove unstable. For example, bar deposition and creation of a self-formed pool-riffle morphology typically will not occur on stream gradients greater than about 2% (Kinoshita and Miwa 1974; Church and Jones 1982; Florsheim 1985). Consequently, a pool-riffle morphology placed on a stream gradient >2% is unlikely to be maintained unless pool scour and bar deposition are forced by in-channel flow obstructions (Lisle 1986). Similarly, for a given discharge and grain size, there is a critical slope above which meandering channel patterns cannot be maintained (Leopold and Wolman 1957; Ferguson 1987; Knighton and Nanson 1993).

Field data from North America and Europe demonstrate that alluvial channel morphology varies systematically with channel gradient, width-to-depth ratio, and relative submergence (ratio of bankfull flow depth to median grain size) (Figure 5), supporting the hypothesis that different channel morphologies result from mutual adjustment of channel characteristics to imposed watershed conditions (Figures 1 and 4). Although there is considerable overlap, each alluvial channel type has characteristic distributions and combinations of channel gradient, relative submergence, and width-to-depth ratio that co-vary with one another (Figure 6). Knowledge of these or other

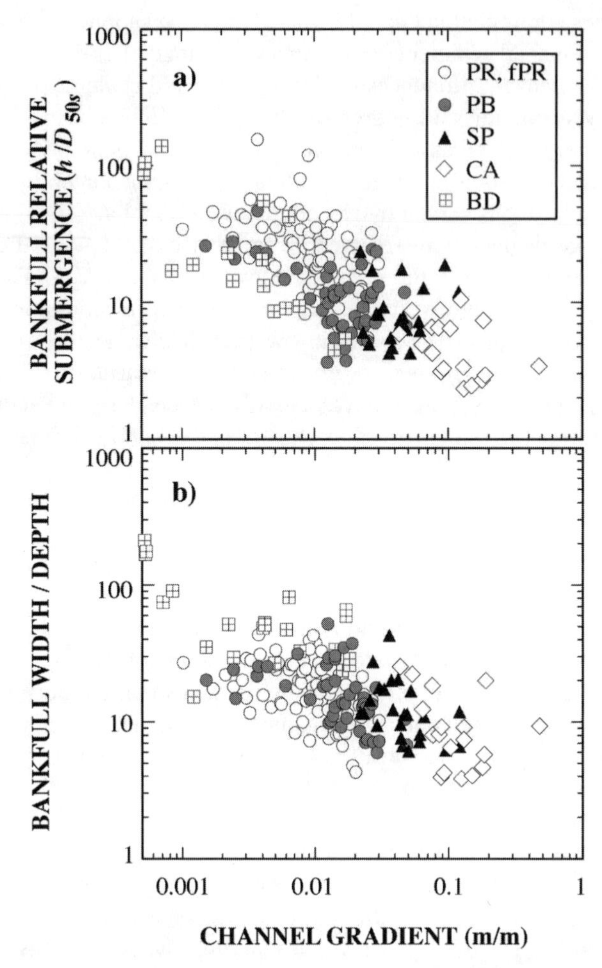

Figure 5. Slope versus a) relative submergence and b) width-to-depth ratio for pool-riffle (both self-formed (PR) and obstruction-forced (fPR); reference numbers 6, 8-13), plane-bed (PB; 6, 8-13), step-pool (SP; 6, 9, 11, 14), cascade (CA; 11), and braided (BD; 1, 3-5, 7) channel morphologies. Numbers following morphologic codes indicate data sources: (1) Leopold and Wolman (1957) (mean annual discharge); (2) Fahnestock (1963); (3) Emmett (1972); (4) Burrows et al. (1981); (5) Prestegaard (1983); (6) Florsheim (1985); (7) Ashworth and Ferguson (1989); (8) Buffington and Montgomery (unpublished data); (9) Montgomery et al. (1995); (10) Montgomery et al. (1996b); (11) Montgomery and Buffington (1997); (12) Buffington and Montgomery (1999a); (13) Montgomery et al. (1999); and (14) Traylor and Wohl (2000). (Modified from Buffington et al. [in press]).

appropriate channel characteristics can help to define restoration objectives and limitations. In particular, one could use observed ranges of channel characteristics for different channel types to develop target conditions for designing a desired channel type, or to assess what channel type is likely to be the stable morphology for given watershed conditions.

Regime Diagrams for Alluvial Channels

Regime diagrams (also referred to as state diagrams) can be used to quantify physical controls on reach-level channel type by dividing different channel morphologies into distinct domains describing the physical regime (state) that gives rise to a given channel type. Regime diagrams have been used for a variety of purposes, such as: (1) to stratify morphologic phases of sand-bed channels (Gilbert 1914; Shields 1936; Simons and Richardson 1966; Ikeda 1989); (2) to examine limits of bar formation in gravel-bed rivers (Florsheim 1985); and (3) to distinguish controls on straight, meandering, and braided channel patterns (Leopold and Wolman 1957; Parker 1976).

Here, we separate alluvial channel types as a function of Ikeda's (1989) channel form index and flow intensity. The form index is defined as the product of channel slope (S) and bankfull width-to-depth ratio (W/h). Flow intensity is defined as the ratio of the bankfull shear velocity to the critical value for initiating bed load transport (u^*/u^*_{c50}, similar to Olsen et al.'s (1997) bed stability index)

$$\frac{u^*}{u^*_{c50s}} = \sqrt{\frac{\tau}{\tau_{c50s}}} = \sqrt{\frac{\rho g h S}{\tau^*_{c50s}(\rho_s - \rho) g D_{50s}}} = \sqrt{\frac{h^* S}{\tau^*_{c50s} R}} \qquad (1)$$

where τ is the total bankfull boundary shear stress determined as a depth-slope product, (ρghS) τ_{c50} is the critical shear stress for motion of the median surface grain size (D_{50s}) and is determined from the Shields (1936) equation, ρ and ρ_s are the fluid and sediment densities, respectively, g is gravitational acceleration, τ^*_{c50} is the dimensionless critical stress for incipient motion of D_{50s} (set equal to 0.03 [Buffington and Montgomery 1997]), h^* is the relative submergence (h/D_{50s}), and R is the submerged specific gravity of sediment [$(\rho_s-\rho)/\rho$, set equal to 1.65].

Ikeda's regime diagram was developed for sand-bed morphologies, but it also does a reasonable job of partitioning the variety of channel types found in the Pacific Northwest (Figure 7). Although there is considerable overlap amongst channel types, distinct fields are identified for self-formed pool-riffle, braided, step-pool, and cascade channel types. Plane-bed channels

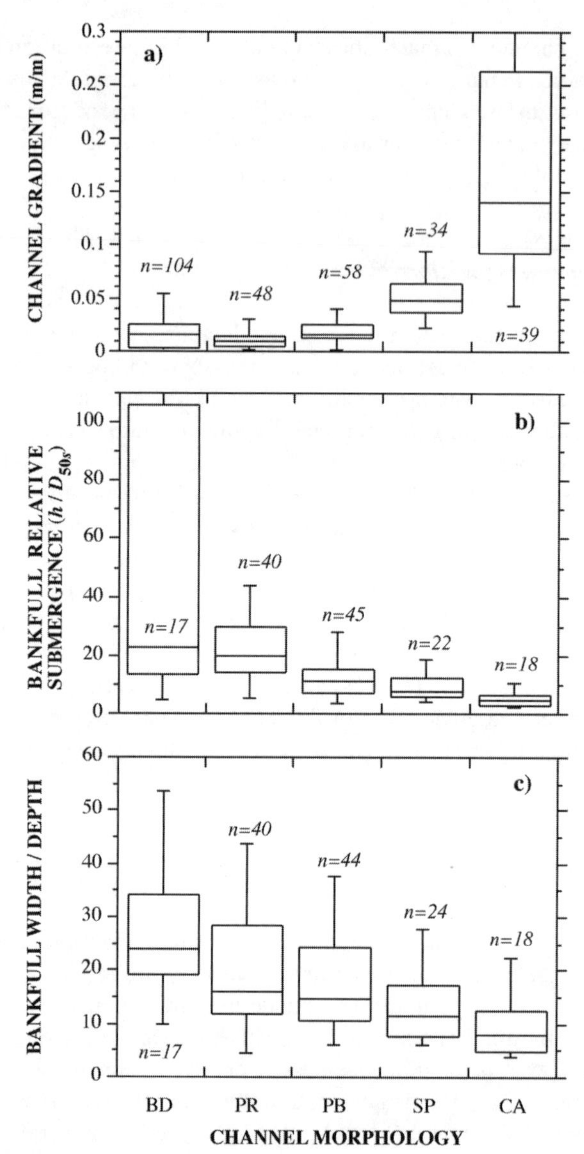

Figure 6. Distributions of channel gradient, relative submergence, and width-to-depth ratio for data of Figure 5. The line within each box indicates the median value, box ends are the inner and outer quartiles and whiskers are the inner and outer tenths of the distribution. Additional slope data from Fahnestock (1963) are added for braided reaches in panel a. (modified from Buffington et al. [in press]). See Figure 5 caption for definition of channel-type abbreviations.

define a subfield within the pool-riffle space and, on average, exhibit a narrower and higher range of form index than pool-riffle channels. Obstruction-forced pool-riffle channels occur across the entire pool-riffle space, including the plane-bed subfield. The occurrence of forced pool-riffle channels in both the plane-bed and self-formed pool-riffle fields is consistent with Montgomery et al.'s (1995) hypothesis that woody debris can create a forced pool-riffle morphology in channels that would otherwise have either a self-formed pool-riffle or plane-bed morphology. The sloping regime boundaries between channel types in Figure 7 are discriminated by both the channel form and flow intensity indices, whereas flow intensity alone appears to separate braided channels from step-pool and cascade morphologies.

Regime diagrams are useful for identifying physical domains of different channel morphologies, but most regime diagrams are not complete physical models. For example, there is no hypothesized relationship between the dimensionless parameters used in Figure 7, nor is there any *a priori* prediction of how different channels might plot within the framework of the figure. In contrast, Parker (1990) developed a more physically complete regime diagram that couples equations for streamflow, bed load transport, and channel characteristics (grain size, flow depth, and slope).

The Parker framework relates dimensionless bed load transport rate (q_b*) to dimensionless streamflow per unit width (q*), reminiscent of Lane's (1955) proportional relationship between discharge and bed load transport rate. The dimensionless bed load transport rate is defined here from the Meyer-Peter and Müller (1948) equation as:

$$q_b{}^* = 8(\tau^* - \tau^*_{c50s})^{1.5} \qquad (2)$$

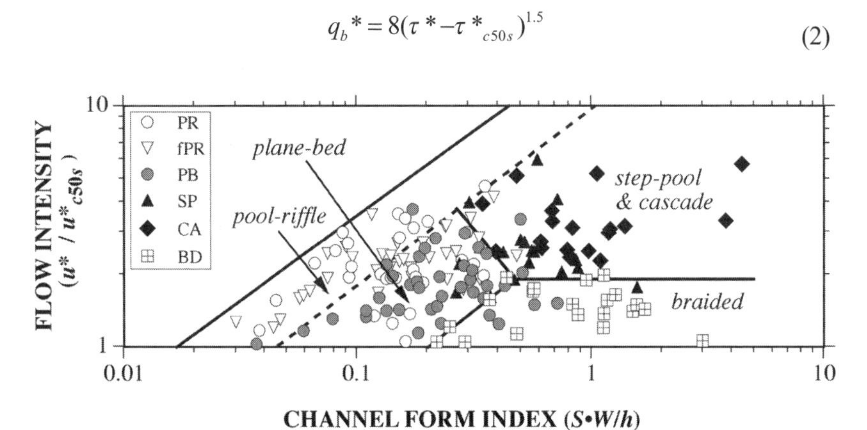

Figure 7. Ikeda-type regime diagram for data of Figure 5. See Figure 5 caption for definition of channel-type abbreviations.

where τ^* is the bankfull Shields stress ($\tau^*=hS/RD_{50s}=h^*S/R$). In this framework, q_b^* is the equilibrium transport rate (input equals output), and thus an indicator of sediment supply as well. The dimensionless specific streamflow is defined as

$$q^* = \frac{<u>h}{\sqrt{RgD_{50\,s}}\,D_{50\,s}} \tag{3}$$

where $<u>$ is the vertically-averaged velocity determined here from the law of the wall (Keulegan 1938):

$$<u> = \frac{u^*}{k}\ln\left(\frac{0.4h}{z_0}\right) \tag{4}$$

In (4), κ is von Karman's constant [0.408, Long et al. (1993)] and z_0 is the height above the bed where the velocity profile goes to zero. The Whiting and Dietrich (1990) approximation is used to define z_0 as $0.1D_{84}$, where D_{84} is the surface grain size for which 84% of the sizes are smaller. For a log-normal grain size distribution, D_{84} can be expressed in terms of the median grain size and the grain-size standard deviation (σ)

$$D_{84} = 2^{-(\phi_{50}-\sigma_\phi)} = D_{50}2^{\sigma_\phi} \tag{5}$$

where ϕ_{50} is the median grain size in $\log_2$ phi units (Krumbein 1936) and σ_ϕ is the grain-size standard deviation in the same units. The σ_ϕ value is set equal to 1.21 ± 0.01, which is an average value for rivers with median grain sizes in the range of 8 to 256 mm (Buffington 1999). Inserting (4) into (3), with the above definitions, yields

$$q^* = \frac{\sqrt{\tau^*}}{k}h^*\ln\left(\frac{4h^*}{2^{-\sigma}}\right) = \frac{\sqrt{h^*S/R}}{k}h^*\ln\left(\frac{4h^*}{2^{-\sigma}}\right) \tag{6}$$

Within this framework, different channel types exhibit subparallel trends of q_b^* as a function of q^* (Figure 8). The data stratify themselves by differences in channel gradient (S), relative submergence (h^*), and excess shear stress (τ^*/τ^*_{c50s}), paralleling systematic differences in those values previously discussed (Figure 5). In particular, as one moves from pool-riffle to cascade channel types, in a direction perpendicular to the trend of each data set, there is a general increase in channel gradient, excess shear stress, and dimensionless bed load transport rate (q_b*), while dimensionless specific discharge (q*) decreases. For a given value of dimensionless discharge, higher bed load transport rates are achieved in step-pool and cascade channels through greater values of both channel gradient and excess shear stress (Figure 8). Conversely, lower-gradient pool-riffle and plane-bed channels can achieve bed load transport rates similar to steeper-gradient cascade and step-pool channels by having larger values of specific discharge due to greater relative submergence and less hydraulic resistance.

Although the above analysis remains a simplified representation of channel processes, it provides some insight regarding the mutual adjustment of channel characteristics, streamflow, and equilibrium transport rate amongst different channel types. In particular, Figure 8 demonstrates a quantitative linkage between these factors that supports the hypothesis that different channel types result from mutual adjustment of channel characteristics (S, h*) for imposed watershed conditions (q*, q_b*). Hence, these different fields formalize the trade-offs in S, h*, and channel type in response to variations in q* and q_b*.

Channel Response Models

River and stream restoration projects are experiments in channel response. Channels are nonlinear systems that dynamically respond to restoration activities, which frequently involve large scale manipulation of channel dimensions, substrate, or planform. Consequently, an understanding of the potential magnitude and style of channel response is needed to address how a channel should be modified to produce the desired objectives and how the project site and neighboring channel reaches will respond to the planned restoration activities.

Many workers have proposed channel response models in which channel characteristics and watershed conditions (streamflow and sediment supply) are related to one another via empirical or theoretical proportionalities (Lane 1955; Schumm 1971; Santos-Cayudo and Simons 1972; Nunnally 1985; Clark and Wilcock 2000). Although qualitative channel response models are useful because they are easy to apply, quantitative models are attractive because they allow numeric prediction of response magnitudes. These predictions are not possible with simple proportionalities, such as the relationship of stream power to sediment size and flux developed by Lane and others.

Parker's (1990) state diagram is an example of an explicit model that can be used to examine mutual interactions between channel characteristics and to identify likely trends and magnitudes of channel response to specific restoration actions. To illustrate this framework, we review and elaborate upon several of Parker's disturbance scenarios. Parker's regime diagram was originally intended for gravel-bed rivers, but the physical equations and response concepts are applicable to alluvial channels in general (Figure 8). The model assumes a straight, wide channel.

In Scenario 1 (shown by Vector 1 in Figure 8), an increased sediment load (larger q_b*) is imposed on the channel, while streamflow (q*) remains constant. According to the model predictions, this is accomplished, in part, by increasing channel slope (S), which is a typically observed response to a large increase in sediment supply. A high sediment supply may overwhelm the channel transport capacity, causing aggradation and a gradual increase in channel slope. Given sufficient time, slope-driven increases in boundary shear stress and channel capacity will match the imposed sediment load, resulting in a new state of channel equilibrium (Gilbert 1917). Increased sediment load also may cause textural fining (decreased grain size) that smoothes the bed and allows greater bed load transport (larger value of excess shear stress, τ^*/τ^*_{c50s}), thereby providing an additional mechanism for equili-

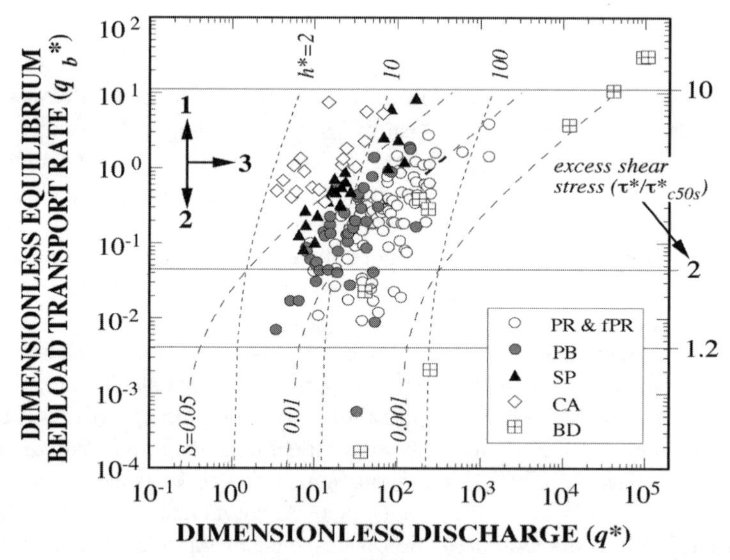

Figure 8. Parker-type regime diagram for data of Figure 5. See Figure 5 caption for definition of channel-type abbreviations. S is channel slope, h* is relative submergence (ratio of flow depth to median surface grain size, h/D_{50s}), and τ^*/τ^*_{c50s} is excess shear stress. Vectors 1-3 indicate disturbance scenarios discussed in the text.

brating rates of sediment supply and bed load transport (Dietrich et al. 1989; Buffington and Montgomery 1999b). In general, the expected response for Scenario 1 is aggradation and textural fining. Conversely, a reduced sediment supply (Scenario 2) is expected to cause channel degradation and textural coarsening if discharge remains constant. This type of response is commonly observed below dams (Gilbert 1917; Komura and Simmons 1967; Williams and Wolman 1984).

In Scenario 3, an increase in streamflow is imposed with no change in the volume or size distribution of the bed load supply (constant q^*_b). The elevated discharge initially causes a transport capacity in excess of sediment supply, resulting in surface coarsening (armoring) and channel degradation (decline in S) similar to Scenario 2. Decreasing slope, in turn, increases flow depth and relative submergence (larger value of h^*), thereby reducing bed roughness and increasing channel conveyance (discharge capacity).

Parker's framework also can be used to examine more complex disturbance scenarios, such as simultaneous perturbations of both discharge and sediment supply (following vectors between those shown in Figure 8), or nonlinear disturbance paths. As presented here, the model cannot be used to examine potential changes in channel width because discharge and sediment supply are nondimensionalized by width. However, the model could be reformulated to explicitly account for channel width. Channel responses predicted from Parker's framework are comparable to those obtained from qualitative response models, but Parker's approach has the advantage of being able to predict specific magnitudes of channel response.

Restoration Limitations Imposed by Channel Type, with Special Reference to Salmonids

Each channel type imposes characteristic physical processes and boundary conditions that must be considered when assessing channel condition and designing restoration projects. For example, the pools and gravel substrates that compose important components of salmonid habitat are limited to specific channel types and may be difficult to create or maintain in certain other channel types or in certain locations within a watershed. The absence of pools or suitable spawning gravels may be a natural condition in some channel types and thus would require heroic efforts to change and maintain.

Table 2 identifies the relative potential for different channel types to produce some of the physical components of salmonid habitat (pools, spawning "gravels," and side-channel refugia). The magnitude and frequency of naturally occurring habitat disturbances are also assessed based on associa-

Table 2. Relative potential to produce specified components of salmonid habitat, and typical frequency and magnitude of habitat disturbances due to natural processes.

Channel Type	*Pools*	*Spawning "gravel"*	*Side Channels*	*Habitat Disturbances Frequency*	*Habitat Disturbances Magnitude*
dune ripple	moderate	none	high	high	low
pool-riffle	high	high	high	moderate	moderate
plane bed	low	high	moderate	moderate	moderate
step-pool	high	low	low	low	high
cascade	low	low	low	low	high
colluvial	low	low	none	low	high
bedrock	low	low	none	low	high
braided	high	high	low	moderate	moderate

tion of channel type with typical process domains. For example, steep-gradient cascade and step-pool channels are prone to infrequent, catastrophic disturbance (e.g., by debris-flow passage) that may extirpate a local salmonid population. In contrast, effects of catastrophic events in any given headwater basin become progressively diffused as they move down the channel network to lower-gradient pool-riffle and dune-ripple channels. Consequently, lower-gradient channels typically experience frequent, but only low to moderate-magnitude disturbances for salmonids.

Suitable habitat for salmonids is predominantly found in the lower gradient alluvial channel types (dune-ripple, pool-riffle, and plane-bed), with habitat becoming progressively more marginal in step-pool and cascade channels (Montgomery et al. 1999). Although salmonid habitat is limited in these steeper channels, those channels nevertheless make up the majority of the stream network in mountain drainage basins. Thus a large percentage of the total available habitat will likely be found in these channel types.

Pool-riffle and plane-bed channels are particularly responsive to obstruction-forced scour, and thus they are likely candidates for restoration projects focused on increasing the number and diversity of pool habitats. In forested environments, wood is an effective flow obstruction that creates a variety of pool types and hydraulic conditions (see review by Buffington et al. [in press]). Moreover, pool spacing is inversely related to wood piece frequency across a broad range of physiographic environments of western North America (Figure 9). Although pool spacing varies tremendously for a given wood-debris frequency, the data tend to stratify by physiographic province and section, likely reflecting region-specific differences in hydrology, sediment supply, land use, and wood characteristics (tree species and size). Urban channels of the Puget Lowland, in particular, have some of the lowest wood frequencies and highest pool spacings in the Pacific coastal region (Horner et al. 1997). Although the relationship between pool spacing and obstruction frequency has received considerable attention in recent land management practice (Chapter 14), pool scour is influenced by a variety of other factors, including obstruction size and type, sediment supply, and channel dimensions (width, depth, slope, and grain size), each of which must be considered when designing restoration and management strategies for pools in alluvial channels (Buffington et al. in press).

The quality of spawning gravels offered by different alluvial channel types depends on local sediment supply (volume and size), absolute shear stress of the flow (and thus channel competence and potential bed-material size), and the spatial variability of shear stress within a reach. Plane-bed reaches may

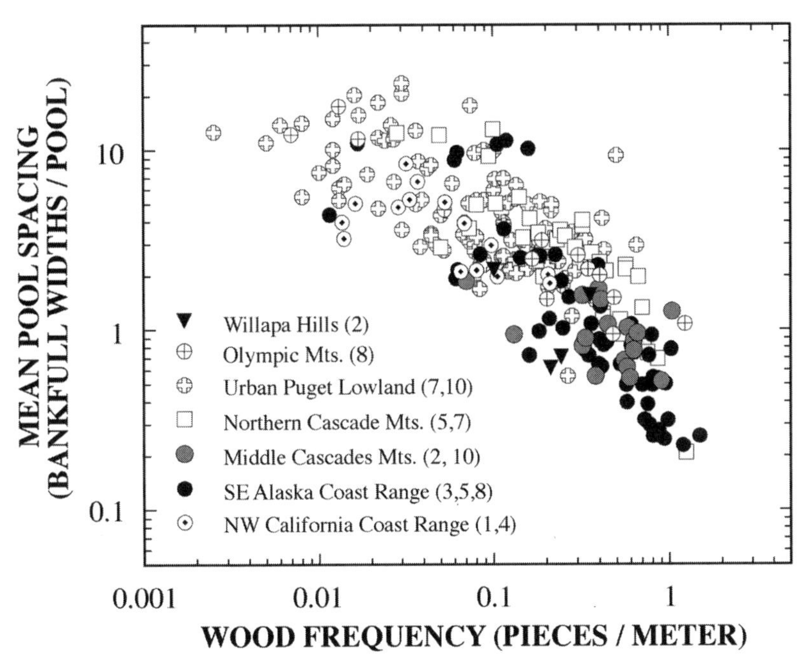

Figure 9. Average pool spacing as a function of wood piece frequency in pool-riffle and plane-bed channels. Numbers in parentheses indicate data sources: (1) Florsheim (1985), (2) Bilby and Ward (1989; 1991) (slopes <2%), (3) Buffington and Montgomery (unpublished data), (4) Keller et al. (1995), (5) Montgomery et al. (1995), (6) May (1996), (7) Beechie and Sibley (1997) (slopes <2%), (8) Buffington and Montgomery (1999a), (9) Larson (1999), and (10) Turaski (2000).

offer abundant gravels and cobbles, but they are typically monotextural (Buffington and Montgomery 1999a), reflecting the relatively uniform shear stress acting along the channel length. In contrast, pool-riffle and wood-forced pool-riffle channels are characterized by bed surfaces composed of textural patches (grain-size facies) that provide a range of particle sizes and associated habitats (Buffington and Montgomery 1999a). Textural patches likely result from spatial divergence of shear stress and sediment supply forced by channel obstructions (such as wood debris and bed and bank topography).

Pools and backwater gravel deposits can be enhanced in step-pool and cascade channels, although the effort required to modify the boulder-sized bed materials make such projects costly. Moreover, the physical processes acting in these channels (high transport capacities, flashy hydrographs, and potential debris-flow passage) make enhancement projects more prone to failure and therefore costly to maintain.

Identification of Restoration Opportunities

Recognition of the physical controls on reach-scale morphology facilitates rapid assessment of where to focus restoration or maintenance efforts. For example, digital elevation models can be used to predict channel type as a function of stream gradient. This information might be used to identify potential locations of existing salmonid habitat, or to develop strategies for optimizing habitat in unrealized areas. For example, studies of forest channels in Washington and Alaska indicate that bar and wood roughness may reduce channel competence and surface grain size to levels usable for salmonid spawning in channels that would otherwise be too coarse (Buffington and Montgomery 1999a). Application of a model for predicting bed surface grain size using digital elevation models indicates that hydraulic roughness due to woody debris has the potential to significantly increase spawning habitat availability in mountain drainages basins of western Washington (Buffington 1998). Digital elevation models can also be used to predict the occurrence of bedrock versus alluvial channel types as a function of slope and drainage area (Montgomery et al. 1996a; Massong and Montgomery 2000). Such predictions could be used either to assess current alluvial habitat, or to examine the potential for "reclaiming" alluvial habitat through introduction of wood in otherwise bedrock reaches. Ranges in channel gradient also could be used to identify reaches potentially susceptible to conversion from a forced pool-riffle to plane-bed morphology upon loss of wood.

Spatial Linkages and Temporal Variations

Drainage basins are composed of landscape elements (such as hillslopes, lakes, alluvial valleys, and tidal zones) that are connected to one another via the channel network. The types and arrangement of landscape units and their characteristic process domains influence biological systems and community structure (Montgomery 1999; Rieman et al. in press). Analysis of how different components of a watershed are connected to and influenced by one another over multiple spatial and temporal scales is necessary for accurate interpretation of watershed conditions and for developing strategies to maintain or restore riverine ecosystems. Restoration projects frequently focus on the immediate, local problem but neglect the larger watershed processes and linkages. This tends to result in a reactive approach that provides short-term solutions for local issues but may not address underlying problems. Because river channels link watershed elements and their processes, the root cause of a problem may be distant both in space and time. Consequently, a holistic approach is needed for understanding how disturbances propagate through a basin and for understanding what factors limit production of desirable morphologies and aquatic habitats.

Identifying the multiple spatial and temporal scales of events and processes influencing a particular location within a watershed is not a trivial problem. Some of the larger influences on watershed processes are readily discernable, such as the control of Pleistocene glaciation on river systems in the Puget Lowland (Chapter 2), or the influence of the Bonneville Flood on the Snake River (O'Connor 1993). Other effects are more subtle and not as easily identifiable. In addition, processes and morphologies may be oscillating over time due to periodic disturbances, as well as evolving toward new states due to longer and larger forcing (e.g., response to geologic and climatic disturbances).

Restoration of Puget Sound Rivers

Many Puget Sound rivers have been so altered by urbanization, agriculture, timber harvest, channelization, and flow regulation that it is difficult to envision their historic appearance, let alone quantitatively reconstruct those conditions. Even if it were possible to understand completely the historic physical and biological processes of Puget Sound rivers, it is unlikely that those systems could be restored (*sensu stricto*) to their naturally functioning state given the enormous social and economic costs that would be required to relocate homes, businesses, and infrastructure that currently blanket the landscape. Consequently, land managers and environmental engineers working in the Puget Sound region are faced with limitations concerning the location and extent of possible restoration activities. These limitations are ancillary to the natural physical controls on channel morphology and watershed processes and may complicate restoration design in the Puget Sound region.

Consequently, planning and design of river restoration programs in the Puget Sound, as elsewhere throughout the Pacific Northwest, rests on three fundamental components. First, an understanding of the physical setting and potential of the channel in question is essential. Second, knowledge of the historical context and changes to both the river and its watershed are required. Third, clear policy objectives are necessary for using these first two components to develop programs likely to achieve desired objectives. Restoration programs that neglect any of these three elements are less likely to succeed.

References

Abbe, T.B. 2000. Patterns, mechanics and geomorphic effects of wood debris accumulations in a forest river system. Ph.D. dissertation. University of Washington. Seattle, Washington.

Ashworth, P.J., and R.I. Ferguson. 1989. Size-selective entrainment of bed load in gravel bed streams. *Water Resources Research* 25:627-634.

Beechie, T.J. and T.H. Sibley. 1997. Relationship between channel characteristics, woody debris, and fish habitat in northwestern Washington streams. *Transactions of the American Fisheries Society* 126:217-229.

Benda, L.E. and T.W. Cundy. 1990. Predicting deposition of debris flows in mountain channels. *Canadian Geotechnical Journal* 27:409-417.

Beschta, R.L., W.S. Platts, J.B. Kauffman, and M.T. Hill. 1994. Artificial stream restoration—Money well spent or an expensive failure? In *Environmental Restoration.* Universities Council on Water Resources, Big Sky, Montana. pp. 76-104.

Bilby, R.E. and J.W. Ward. 1989. Changes in characteristics and function of woody debris with increasing size of streams in western Washington. *Transactions of the American Fisheries Society* 118:368-378.

Bilby, R.E. and J.W. Ward. 1991. Characteristics and function of large woody debris in streams draining old-growth, clear-cut, and second-growth forests in southwestern Washington. *Canadian Journal of Fisheries and Aquatic Sciences* 48:2499-2508.

Bisson, P. A., R. E. Biliby, M. D. Bryant, C. A. Dolloff, G. B. Grette, R. A. House, M. L. Murphy, K. V. Koski and J. R. Sedell. 1987. Large woody debris in forested streams in the Pacific Northwest: Past, present, and future. In E.O. Salo and T.W. Cundy (eds.) *Streamside management: Forestry and Fishery Interactions*. University of Washington Institute of Forest Resources, Seattle, Washington. pp. 143-190.

Booth, D.B. 1990. Stream-channel incision following drainage-basin urbanization. *Water Resources Bulletin* 26:407-417.

Bonneville Power Administration (BPA). 2001. Fish and Wildlife Group fiscal aspects. http://www.efw.bpa.gov/EW/FISCAL/fiscal1.html.

Buffington, J.M. 1998. The use of streambed texture to interpret physical and biological conditions at watershed, reach, and subreach scales. Ph.D. dissertation. University of Washington. Seattle, Washington.

Buffington, J.M. 1999. Variability of sediment sorting in coarse-grained rivers: A case of self-similarity. *EOS, American Geophysical Union Transactions* 80:449.

Buffington, J.M., T.E. Lisle, R.D. Woodsmith, and S. Hilton. In press. Controls on the size and occurrence of pools in coarse-grained forest rivers. *River Research and Applications.*

Buffington, J.M. and D.R. Montgomery. 1997. A systematic analysis of eight decades of incipient motion studies, with special reference to gravel-bedded rivers. *Water Resources Research* 33:1993-2029.

Buffington, J.M. and D.R. Montgomery. 1999a. Effects of hydraulic roughness on surface textures of gravel-bed rivers. *Water Resources Research* 35:3507-3522.

Buffington, J.M. and D.R. Montgomery. 1999b. Effects of sediment supply on surface textures of gravel-bed rivers. *Water Resources Research* 35:3523-3530.

Burrows, R.L., W.W. Emmett, and B. Parks. 1981. Sediment transport in the Tanana River near Fairbanks, Alaska, 1977-79. U.S. Geological Survey Water-Resources Investigations 81-20.

Church, M. and D. Jones. 1982. Channel bars in gravel-bed rivers. In R.D. Hey, J.C. Bathurst, and C.R. Thorne (eds.) *Gravel-bed Rivers*. John Wiley & Sons, Chichester. pp. 291-338.

Clark, J.J. and P.R. Wilcock. 2000. Effects of land-use change on channel morphology in northeastern Puerto Rico. *Geological Society of America Bulletin* 112:1763-1777.

Dietrich, W.E., T. Dunne, N.F. Humphrey, and L.M. Reid. 1982. Construction of sediment budgets for drainage basins. In F.J. Swanson, R.J. Janda, T. Dunne, and D.N. Swanston (eds.) *Sediment Budgets and Routing in Forested Drainage Basins*. USDA Forest Service General Technical Report, PNW-GTR-141, Pacific Northwest Forest and Range Experiment Station,

Portland, Oregon. pp. 5-23.

Dietrich, W.E., J.W. Kirchner, H. Ikeda, and F. Iseya. 1989. Sediment supply and the development of the coarse surface layer in gravel-bedded rivers. *Nature* 340:215-217.

Dietrich, W.E., C.J. Wilson, and S.L. Reneau. 1986. Hollows, colluvium, and landslides in soil-mantled landscapes. In A.D. Abrahams (ed.) *Hillslope Processes*. Allen and Unwin, Boston, Massachusetts. pp. 361-388.

Emmett, W.W. 1972. The hydraulic geometry of some Alaskan streams south of the Yukon River. U.S. Geological Survey Open-File Report.

Fahnestock, R.K. 1963. Morphology and hydrology of a glacial stream—White River, Mount Rainier, Washington. U.S. Geological Survey Professional Paper 422A.

Fenneman, N.M. 1931. *Physiography of Western United States*. McGraw-Hill, New York.

Ferguson, R.I. 1987. Hydraulic and sedimentary controls of channel pattern. In K.S. Richards (ed.) *River Channels: Environment and Process*. Blackwell, Oxford. pp. 129-158.

Florsheim, J.L. 1985. Fluvial requirements for gravel bar formation in northwestern California. Master's thesis. Humboldt State University. Arcata, California.

Frissell, C.A. and R.K. Nawa. 1992. Incidence and causes of physical failure of artificial habitat structures in streams of western Oregon and Washington. *North American Journal of Fisheries Management* 12:182-197.

Gilbert, G. K. 1914. The transportation of débris by running water. U.S. Geological Survey Professional Paper 86.

Gilbert, G. K. 1917. Hydraulic-mining debris in the Sierra Nevada. U.S. Geological Survey Professional Paper 105.

Hogan, D.L. 1986. Channel morphology of unlogged, logged, and debris torrented streams in the Queen Charlotte Islands. BC Ministry of Forests and Lands, Land Management Report no. 49.

Horner, R.R., D.B. Booth, and A.A. Azous. 1997. Watershed determinants of ecosystem functioning. In L.A. Roesner (ed.) *Effects of Watershed Development and Management on Aquatic Ecosystems: Proceedings of an Engineering Foundation Conference*. American Society of Civil Engineers, New York. pp. 251-274.

Ikeda, H. 1989. Sedimentary controls on channel migration and origin of point bars in sand-bedded meandering channels. In S. Ikeda, and G. Parker (eds.) *River Meandering*. American Geophysical Union, Washington, D.C. pp. 51-68.

Keller, E.A., A. MacDonald, T. Tally, and N.J. Merritt. 1995. Effects of large organic debris on channel morphology and sediment storage in selected tributaries of Redwood creek. U.S. Geological Survey Professional Paper 1454-P.

Kellerhals, R. and M. Church. 1989. The morphology of large rivers: Characterization and management. In D.P. Dodge (ed.) *Proceedings of the International Large River Symposium.* Canadian Special Publication of Fisheries and Aquatic Sciences. pp. 31-48.

Keulegan, G.H. 1938. Laws of turbulent flow in open channels. *Journal of Research of the National Bureau of Standards* 21:707-741.

Kinoshita, R. and H. Miwa. 1974. River channel formation which prevent downstream translation of transverse bars. *Shinsabo* 94:12-17.

Knighton, A.D. and G.C. Nanson. 1993. Anastomosis and the continuum of channel pattern. *Earth Surface Processes and Landforms* 18:613-625.

Komura, S. and D.B. Simmons. 1967. River-bed degradation below dams. *Journal of the Hydraulics Division, American Society of Civil Engineers* 93:1-14.

Krumbein, W.C. 1936. Application of logarithmic moments to size frequency distributions of sediments. *Journal of Sedimentary Petrology* 6:35-47.

Lane, E.W. 1955. The importance of fluvial morphology in hydraulic engineering. *Proceedings of the American Society of Civil Engineers* 81:745:1-17.

Larson, M.G. 1999. Effectiveness of large woody debris in stream rehabilitation projects in urban basins. Master's thesis. University of Washington. Seattle, Washington.

Leopold, L.B., and M.G. Wolman. 1957. River channel patterns: Braided, meandering, and straight. U.S. Geological Survey Professional Paper 282-B.

Lisle, T.E. 1986. Stabilization of a gravel channel by large streamside obstructions and bedrock bends, Jacoby Creek, northwestern California. *Geological Society of America Bulletin* 97:999-1011.

Long, C.E., P.L. Wiberg, and A.R.M. Nowell. 1993. Evaluation of von Karman's constant from integral flow parameters. *Journal of Hydraulic Engineering* 119:1182-1190.

Massong, T.M. and D.R. Montgomery. 2000. Influence of lithology, sediment supply, and wood debris on the distribution of bedrock and alluvial channels. *Geological Society of America Bulletin* 112:591-599.

May, C.W. 1996. Assessment of cumulative effects of urbanization on small streams in the Puget Sound Lowland ecoregion: Implications for salmonid resource management. Ph.D. dissertation. University of Washington. Seattle, Washington.

Meyer-Peter, E. and R. Müller. 1948. Formulas for bed-load transport. In *Proceedings of the 2nd Meeting of the International Association for Hydraulic Structures Research.* International Association for Hydraulics

Research, Delft. pp. 39-64.

Mollard, J.D. 1973. Air photo interpretation of fluvial features. In *Fluvial Processes and Sedimentation*. National Research Council of Canada, Subcommittee on Hydrology, Inland Waters Directorate, University of Alberta, Edmonton, Alberta. pp. 341-380.

Montgomery, D.R. 1999. Process domains and the river continuum. *Journal of the American Water Resources Association* 35:397-409.

Montgomery, D.R., T. B. Abbe, J.M. Buffington, N.P. Peterson, K.M. Schmidt, and J.D. Stock. 1996a. Distribution of bedrock and alluvial channels in forested mountain drainage basins. *Nature* 381:587-589.

Montgomery, D.R., E.M. Beamer, G.R. Pess, and T.P. Quinn. 1999. Channel type and salmonid spawning distribution and abundance. *Canadian Journal of Fisheries and Aquatic Sciences* 56:377-387.

Montgomery, D.R. and J.M. Buffington. 1997. Channel-reach morphology in mountain drainage basins. *Geological Society of America Bulletin* 109:596-611.

Montgomery, D.R. and J.M. Buffington. 1998. Channel processes, classification, and response. In R. Naiman, and R. Bilby (eds.) *River Ecology and Management*. Springer-Verlag, New York. pp. 13-42.

Montgomery, D.R., J.M. Buffington, N.P. Peterson, D. Schuett-Hames, and T.P. Quinn. 1996b. Streambed scour, egg burial depths and the influence of salmonid spawning on bed surface mobility and embryo survival. *Canadian Journal of Fisheries and Aquatic Sciences* 53:1061-1070.

Montgomery, D.R., J.M. Buffington, R.D. Smith, K.M. Schmidt, and G. Pess. 1995. Pool spacing in forest channels. *Water Resources Research* 31:1097-1105.

Nunnally, N.R. 1985. Application of fluvial relationships to planning and designing of channel modifications. *Environmental Management* 9:417-426.

O'Connor, J.E. 1993. Hydrology, hydraulics and geomorphology of the Bonneville flood. Geological Society of America Special Paper 274 83.

Olsen, D.S., A.C. Whitaker, and D.F. Potts. 1997. Assessing stream channel stability thresholds using flow competence estimates at bankfull stage. *Journal of the American Water Resources Association* 33:1197-1207.

Parker, G. 1976. On the cause and characteristic scales of meandering and braiding in rivers. *Journal of Fluid Mechanics* 76:457-480.

Parker, G. 1990. Surface-based bedload transport relation for gravel rivers. *Journal of Hydraulic Research* 28:417-436.

Powell, J.W. 1896. Physiographic regions of the United States. In J.W. Powell (ed.) *Physiography of the United States*, National Geographic Society Monographs. American Book Co., New York. pp. 65-100.

Prestegaard, K.L. 1983. Bar resistance in gravel bed steams at bankfull stage. *Water Resources Research* 19:473-476.

Reeves, G.H., J.D. Hall, T.D. Roelofs, T.L. Hickman, and C.O. Baker. 1991. Rehabilitating and modifying stream habitats. In W.R. Meehan (ed.) *Influences of Forest and Rangeland Management on Salmonid Fishes and Their Habitats*. American Fisheries Society Special Publication 19, Bethesda, Maryland. pp. 519-557.

Rieman, B.E., J.B. Dunham, and J.L. Clayton. In press. Emerging concepts for management of river ecosystems and challenges to applied integration of physical and biological sciences in the Pacific Northwest, USA. In P. Goodwin (ed.) *New Paradigms in River and Estuarine Management*. North Atlantic Treaty Organization, Brussels.

Santos-Cayudo, J. and D.B. Simons. 1972. River response. In H.W. Shen (ed.) *Environmental Impact on Rivers*. H.W. Shen, Fort Collins, Colorado. pp. 1-1:1-25.

Schmidt, K.M., J.J. Roering, J.D. Stock, W.E. Dietrich, D.R. Montgomery and T. Schaub. 2001. The variability of root cohesion as an influence on shallow landslide susceptibility in the Oregon Coast Range. *Canadian Geotechnical Journal* 38:995-1024.

Schumm, S.A. 1971. Fluvial geomorphology: Channel adjustment and river metamorphosis. In H.W. Shen (ed.) *River Mechanics*. H.W. Shen, Fort Collins, Colorado. pp. 5-1:5-22.

Schumm, S.A. 1985. Patterns of alluvial rivers. *Annual Review of Earth and Planetary Sciences* 13:5-27.

Shields, A. 1936. Anwendung der Aehnlichkeitsmechanik und der Turbulenzforschung auf die Geschiebebewegung. *Mitteilungen der Preussischen Versuchsanstalt fur Wasserbau und Schiffbau* 26:26.

Simons, D.B., and E.V. Richardson. 1966. Resistance to flow in alluvial channels. U.S. Geological Survey Professional Paper 422-J.

Smith, R.D., R.C. Sidle and P.E. Porter. 1993a. Effects on bedload transport of experimental removal of woody debris from a forest gravel-bed stream. *Earth Surface Processes and Landforms* 18:455-468.

Smith, R.D., R.C. Sidle, P.E. Porter and J.R. Noel. 1993b. Effects of experimental removal of woody debris on the channel morphology of a forest gravel-bed stream. *Journal of Hydrology* 152:153-178.

Traylor, C.R., and E.E. Wohl. 2000. Seasonal changes in bed elevation in a step-pool channel, Rocky Mountains, Colorado, USA. *Arctic, Antarctic, and Alpine Research* 32:95-103.

Turaski, M.R. 2000. Temporal and spatial patterns of stream channel response to watershed restoration, Cedar Creek, southwest Oregon. Master's thesis. University of Wisconsin. Madison, Wisconsin.

Whiting, P.J., and W.E. Dietrich. 1990. Boundary shear stress and roughness over mobile alluvial beds. *Journal of Hydraulic Engineering* 116:1495-1511.

Williams, G.P., and M.G. Wolman. 1984. Downstream effects of dams on alluvial rivers. U.S. Geological Survey Professional Paper 1286.

4. Reconstructing the Historical Riverine Landscape of the Puget Lowland

Brian D. Collins, David R. Montgomery, and Amir J. Sheikh

Abstract

Human activities in the last 150 years greatly altered the riverine landscape and salmonid habitats of the Puget Lowland. Archival investigations together with field studies of relatively undisturbed rivers make it possible to describe the landscape prior to settlement by Euro-Americans. Landforms, dynamics, and habitats in lowland river valleys and estuaries varied broadly with differences in regional geologic history. Rivers that incised a Holocene valley through Pleistocene glacial sediments typically had an anastomosing pattern with multiple channels, floodplain sloughs, and frequent channel-switching avulsions, due in large part to wood jams. In contrast, rivers in broader, lower-gradient valleys created by runoff below Pleistocene glaciers generally had a single-channel meandering pattern, with oxbow lakes, infrequent meander-cut-off avulsions, and vast floodplain wetlands. Because wood appears to have strongly influenced riverine dynamics at a wide range of scales, floodplain forests are central to river restoration. Archival sources can characterize species and diameters of trees in historical forests and the geomorphic, hydrologic, and geographic variables influencing them; process studies indicate conditions and wood characteristics necessary for jam formation. Regional differences in channel morphologies, processes, suites of valley-bottom landforms, and forests, combined with different land-use histories, have important implications for the rationale, approach, and land area needed in restoring lowland river and forest ecosystems.

River History and the Puget Lowland

A century and a half of development since European settlement has transformed the appearance and function of Puget Sound's riverine landscape. Human inhabitance has been most extensive and landscape change most noticeable in lowland river valleys, eradicating or degrading much of the region's historically richest and most abundant salmonid habitat (Sedell and Luchessa 1981; Beechie et al. 1994, 2001; Collins and Montgomery 2001). This river-use history is not unique to Puget Sound. The same has occurred worldwide: as riverine landscapes were more intensively inhabited, "civilized" rivers became physically simplified and biologically impoverished (e.g., Vileisis 1997; McNeil 2000). However, the relatively-recently settled Puget Lowland is unusual in having remnant natural areas and a wealth of archival sources describing pre-settlement conditions (we use the term "pre-settlement" as an abbreviated reference to "prior to settlement by Euro-Americans"). These circumstances make it possible to reconstruct on paper the historical river as an aid to undertaking river rehabilitation or restoration.

The problem of reconstructing badly degraded landscapes or landscapes that no longer exist spans the intersection of diverse disciplines including archaeology, ecology, landscape ecology, ethnobotany, palynology, and history, to form the field of historical ecology (e.g., Egan and Howell 2001). Environmental history, which includes a focus on understanding the political, social, and cultural forces behind landscape change and how those changes in turn shape society, overlaps with and complements historical ecology (e.g., White 1992; Whitney 1996). Reconstructing the riverine environment of Puget Sound can draw on the methods of these disciplines but also must be grounded in geology and process geomorphology, because the region's riverine landscape is geologically young and physically dynamic, and its ecosystems are closely linked to physical processes.

The Geologic Setting

Seven major watersheds drain the western Cascade Range to Puget Sound (Figure 1), ranging in size from the 1,770 km^2 Stillaguamish to the 7,800 km^2 Skagit. Steep mountain headwater slopes lessen in mountain valleys and decrease dramatically in the Puget Lowland. In the lowland, deep, generally north-south trending troughs either partially filled with sediments or by Puget Sound or other water bodies, are a dominant topographic feature (Chapter 2). Repeated advances by the Puget Lobe of the Cordilleran ice sheet created these valleys at least in part by subglacial fluvial runoff (Booth 1994). Sev-

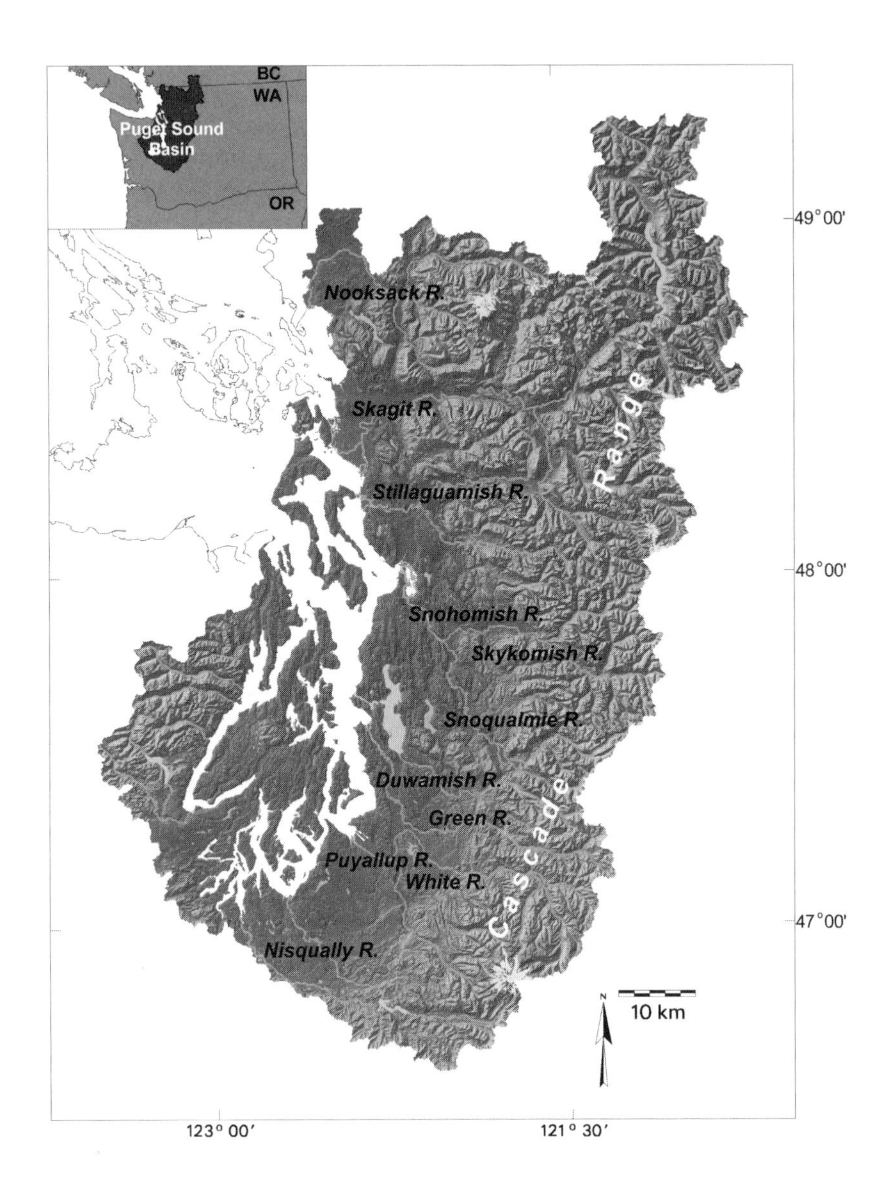

Figure 1. Location of watersheds and rivers in eastern Puget Sound.

eral major rivers have found a post-glacial course through these Pleistocene troughs, including the lower Nooksack, the Snohomish, Snoqualmie, Sammamish, Duwamish, and Puyallup Rivers (Figure 1). These valleys have a low gradient and are typically 3–5 km wide. Other rivers cut across the lowland glacial fabric, and incised steeper and narrower (1–2 km wide) post-glacial (Holocene) valleys. These include the upper Nooksack, the Stillaguamish, and Nisqually Rivers.

Various post-glacial forces modified (and continue to modify) this Pleistocene legacy. Holocene fluctuations of sea level and isostatic rebound changed the extent of subaerial valley bottom (Beechie et al. 2001), especially in north Puget Sound where isostatic effects were greatest (Thorson 1989). Voluminous lahars from eruptions of Glacier Peak volcano inundated the Sauk, Skagit, and Stillaguamish Rivers (Beget 1982); remnants of lahar deposits since incised by fluvial erosion can be found in each of these three valleys. These lahars also extended the Skagit River delta greatly seaward (Dragovich et al. 2000). At least 60 Holocene lahars moved down valleys heading on Mount Rainier (Hoblitt et al. 1998), many of which traveled into the White and Puyallup Rivers (Figure 1). Mount Rainier's National Lahar traveled from the Nisqually River to Puget Sound less than 2,200 ybp (Hoblitt et al. 1998).

These lahars have been most influential in shaping channels and habitats in valleys that are transitional—geographically, as well as in gradient and width—between the mountains and the lowland. In north Puget Sound, these include the Skagit, Sauk, North Fork Stillaguamish, and Skykomish; each valley includes terraces of Pleistocene glacial and Holocene lahar sediments, through which the river has incised (Beechie et al. 2001). In the less-glaciated south, such lowland-to-mountain transitional valleys as the White and upper Puyallup are more heavily influenced by the presence of lahar deposits. For example, the White River is cutting a deep canyon through deposits of the 5,600 years before present (Hoblitt et al. 1998) Osceola Mudflow. In this chapter, we concentrate on lowland rivers, but many of the concepts we develop can be applied to these smaller, transitional, mountain-valley streams as well.

Can We Know the Past with any Certainty?

Our knowledge of historical environments, especially those greatly changed by anthropogenic forces, is inherently uncertain. This uncertainty reflects the incomplete views through the available windows onto past landscapes as well as the spatial and temporal variability of landscape processes. In light of this uncertainty, how reliably should historical reconstructions be viewed?

The answer depends on the methods used. Using independent methods with overlapping temporal and spatial scales, and cross-referencing between archival studies and field investigations can define and reduce uncertainty. By using both cross-referenced and multi-scaled methods, we can hope to see the past clearly enough to confidently develop and evaluate restoration objectives.

Our reconstruction of the landscape is necessarily limited to conditions that existed in the mid 19th century, or around the time when non-native settlers arrived, because for the time prior to the written record we can only make broader, less detailed descriptions using indirect field methods. However, it is possible to supplement this snapshot-in-time with inferences about long-term (Holocene) landscape and ecosystem evolution and more rapid change. For example, forest composition in the region probably attained modern characteristics approximately 6,000 ybp (Barnosky 1981; Leopold et al. 1982; Cwynar 1987; Brubaker 1991). The interplay of isostatic uplift, river incision and sea level change is slow and causes only minor change over the time frame of a few centuries. We have recent analogs to draw on for understanding the effects of intermittent, dramatic disturbances to river valleys such as volcanic lahars (e.g., from the 1980 eruptions of Mt. St. Helens) and earthquake-associated uplift. Archival and field studies can adequately characterize the changes occurring on decadal and more frequent time scales. Moreover, many agents of anthropogenic change over the last ~150 years have been much more rapid than natural processes.

We refer to the "historical" landscape rather than the "natural" environment, because people have inhabited the Puget Lowland at least since the glaciers last retreated. While there are ethnographic studies of fisheries management, few studies exist on native practices that would have modified the ecology or morphology of the riverine landscape, such as by native plant cultivation or gathering (e.g., Gunther 1973; Turner 1995) or burning practices. It should be understood that the landscape we seek to reconstruct was indeed a landscape, resulting from a fusing of cultural and natural influences that included native land-management practices.

Methods for Reconstructing the Historical Landscape

The Synergy of Archival and Field Studies

Archival studies and field investigations both contribute toward understanding the historical landscape of Puget Sound. For example, a reach of the lower Nisqually River that passes through the Fort Lewis Army Base and the Nisqually Indian Reservation has retained natural banks and a mature valley-bottom forest throughout the last 150 years, and it displays many of the morphological and biological characteristics of river valleys in their pre-settlement condition (Collins and Montgomery 2001; Collins et al. 2002). It can thus function as an historical analog. Field studies are useful for providing information at a small scale, giving insight into form and process that cannot typically be discerned from archival sources.

However, without an archival reference standard, it is difficult to be certain whether and in what ways this (or any) field site represents pre-settlement conditions. First, an isolated landscape fragment may not necessarily include processes and features that formerly operated at larger scales, for which archival sources may help to generate hypotheses. Second, a single 10 km-long reach cannot represent the variation in river dynamics, geologic setting, and forest conditions throughout the entire region, making archival investigations necessary for describing the historical geographic variability.

The two approaches are thus complementary (Figure 2): Archival sources help in generating hypotheses about processes that formerly operated throughout the historical landscape, for which field studies can then generate particular process models. Such process models provide the basis for designing effective river management and habitat restoration and conservation schemes, and thus the basis for applying models to a given location. Without the perspective of archival sources, there is the risk of focusing only on insights from contemporary process studies and overlooking landscape-scale features and processes that no longer exist. For example, a focus on a 15 m-wide streamside buffer on a leveed, lowland river would neglect the fact that most riverine habitat may historically have been in sloughs, ponds, and wetlands hundreds

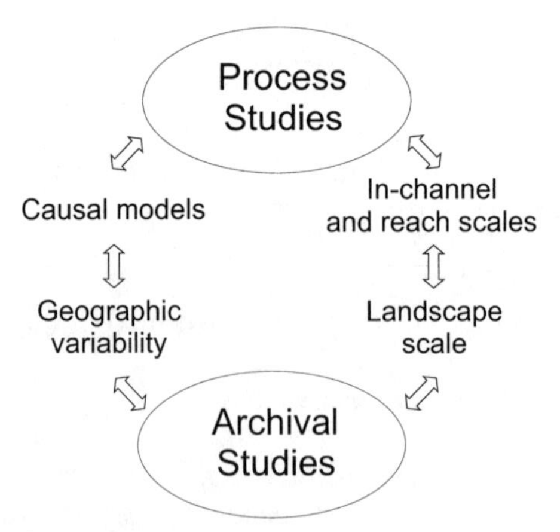

Figure 2. Synergy of archival and field studies in characterizing historical riverine processes and habitats.

or thousands of meters from the river. Without an understanding of the larger scale, management efforts risk managing the microcosm, instead of addressing structures or processes that exist(ed) or operate(d) at a landscape scale. Because of this complementarity of field and archival studies—their different emphases of scale, and process versus regional variability—the two work in an iterative and synergistic way toward characterizing historical riverine processes and environments.

Mapping a Forgotten Landscape

We are developing a methodology for mapping historical river landscapes and their aquatic habitat that brings archival materials into a geographic information system (GIS), supplementing that data with modern digital elevation models (DEMs), aerial photography, and process understanding gleaned from field studies. The approach synthesizes historical materials and the modern tools of GIS and remotely-sensed imagery (Figure 3).

Maps and field notes of the General Land Office (GLO), which conducted a cadastral survey of the Puget Lowland between about 1850 and 1890, are a fundamental resource. Carried out in nearly all river valleys (and uplands) prior to and in preparation for the arrival of settlers, this survey preceded widespread building of sea or river dikes and stream clearing and floodplain logging. It is a unique resource for characterizing riverine conditions prior to Euro-American settlement.

The GLO field notes include information on natural vegetation, which botanists have used since at least the 1920s (Sears 1925) to reconstruct presettlement vegetation cover (for reviews see Whitney 1996; Whitney and DeCant 2001; for recent examples see Galatowitsch 1990; Nelson et al. 1998). The same information is also useful for characterizing riparian and valley bottom forests, including the size and species of recruitable wood and for mapping and characterizing riverine wetlands (North and Tevarsham 1984; Collins and Montgomery 2001), prairie or savannah areas (Radeloff et al. 1999), and changes to channel widths (Knox 1977).

These data include "bearing" or "witness" tree records from reference points at the corners of mile-square sections and half way between corners ("quarter corner" points), where surveyors measured the distance and compass direction to several nearby trees. Surveyors were instructed to identify four witness trees at section corners and two at quarter-corner boundaries. If there were no trees nearby, surveyors built a mound of earth. In their field notes, surveyors recorded the diameter and common name of each witness tree and the distance and bearing to it. In addition to these regularly-spaced points, survey-

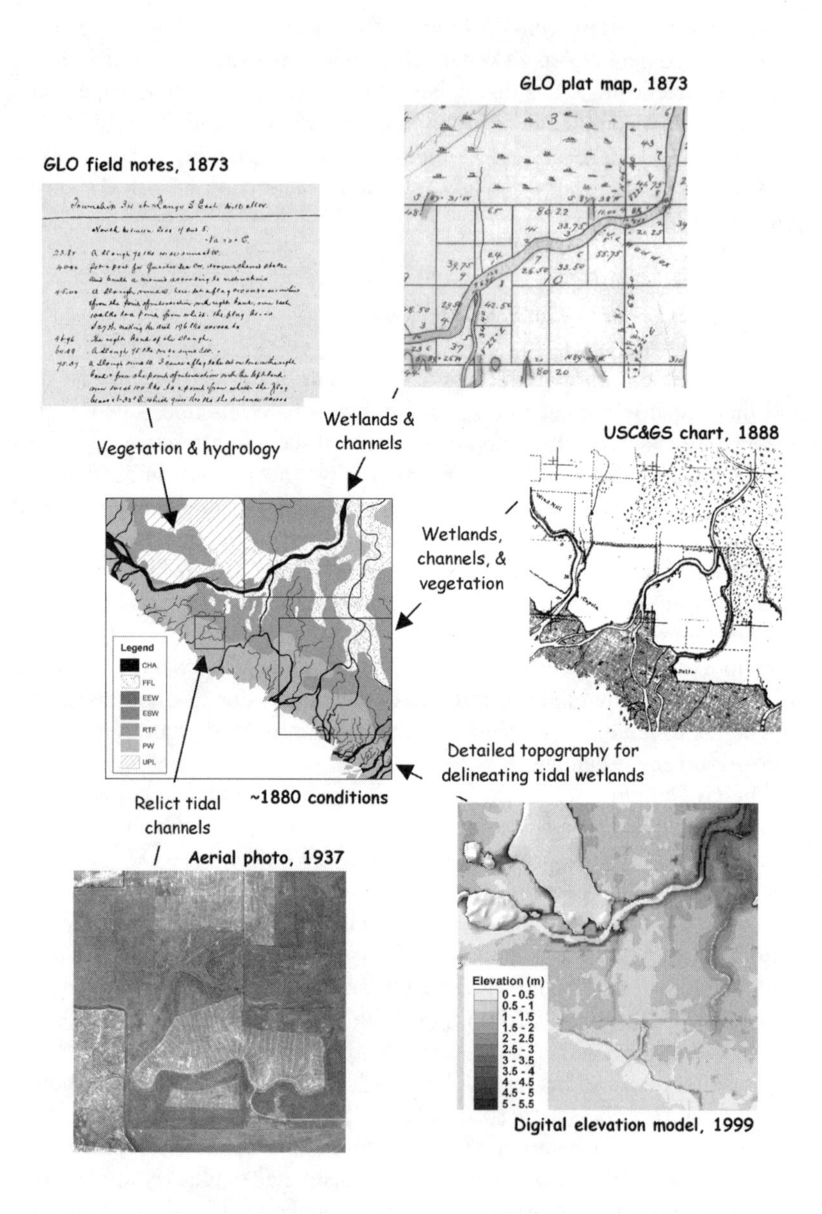

Figure 3. Example of the use of archival materials, field studies, aerial photographs, and digital elevation models (DEMs) in a GIS to map the historical riverine environment.

ors also established "meander corner" points where section lines intersected the banks of navigable rivers and sloughs and documented two bearing trees. These meander points allow us to characterize separately streamside trees from other valley-bottom trees.

Because instructions issued by the GLO evolved over time (see White [1991] for a compilation of instructions to surveyors), field notes must be interpreted in light of the instructions current for that time and region. For example, various criteria for selecting bearing trees were published, which in turn might have differed from actual field practice (see Collins and Montgomery [2002] for details on interpreting field notes in the Puget Lowland). One important bias in characterizing species frequency and size results from bearing trees being greater than 7.5 cm in diameter. This means that bearing tree records under-represent smaller-diameter species (e.g., vine maple [*Acer circinatum*] and willow [*Salix spp.*]). On the other hand, we found that bearing tree records accurately characterize species frequency based on basal area (the percent of the sum of cross-sectional area of all trees accounted for by the cross-sectional area of any one species). We determined this by relocating 1873 survey points in the Nisqually valley bottom and establishing bearing trees following our interpretation of the instructions to surveyors in effect for the 1873 survey, and comparing results to plots in which we recorded species and diameter of all trees larger than 0.01 m in diameter (Collins and Montgomery 2002).

In addition to recording witness trees, surveyors were instructed to note land and water features they encountered, including major changes to the plant community, streams and marshes, and the width of all "water objects." Springs, lakes and ponds and their depths, the timber and undergrowth, bottomlands, visual signs of seasonal water inundation, and improvements were also to be noted along section lines. The completeness of this information varies from surveyor to surveyor, but it nonetheless provides important secondary data for interpreting the landscape. For example, the date at which observations of water depth were made by surveyors, and their notes on indicators of seasonal water depths can be used to characterize summer and winter water depths in wetlands.

The GLO maps and field notes reflect field observation only along section boundaries and navigable channels. Within sections, the maps include many wetlands with indeterminate boundaries, and wetlands or smaller (non-navigable) channels that are drawn speculatively, sometimes in locations that are improbable or impossible when compared to modern topographic mapping. Other archival sources and modern information can be used to map wetlands and small channels within section interiors and also to confirm or add data along section lines. For example, early U. S. Coast and Geodetic Survey

(USC&GS) charts of coastlines and coastal rivers provide more spatially continuous data to the up-stream limit of navigation (generally a few tens of kilometers inland) than do the GLO maps. The charts also delineate forest, salt marsh, freshwater marsh, and cultivated fields. In eastern Puget Sound, the USC&GS made detailed and accurate charts in the late 1870s to late 1880s at a scale of 1:10,000 or 1:20,000. Although most of the charts post-date some amount of tidewater diking, they are the basis for estimates of estuarine wetland loss in Puget Sound (Bortleson et al. 1980); these earlier estimates thus cannot take into account wetland areas diked prior to the USC&GS mapping.

Beginning in 1876 the U.S. Army Engineers filed annual reports on field investigations of western Washington rivers (Annual Reports of the Chief of Engineers, U.S. War Department; hereafter abbreviated U.S. War Department). Their river descriptions highlighted wood because it created hazards for, or often completely blocked rivers to, steamboat navigation. After 1880, army engineers began clearing this wood and by the end of that decade developed a regular program of "snagging" that continues to this day. Other useful sources of historical information include U.S. Department of Agriculture and Bureau of Soils reports and maps (e.g., Nesbit et al. 1885; Mangum et al. 1909); settlers accounts; contemporary histories (e.g., Interstate Publishing Company 1906); photographs (the earliest useful photographs we have located are from the 1880s); and U.S. Geological Survey (USGS) topographic maps, starting in the 1890s.

More recent imagery and mapping add spatial resolution and accuracy. Modern vertical stereo aerial photography in western Washington began in the 1930s. These early aerial photographs show, for example, relict swales and vegetation patterns indicative of channels filled in during the previous half century, or relict patches of wetland or forest that—when georeferenced and brought into a GIS—can be interpreted, in conjunction with archival map sources. Recent soils mapping (e.g., Debose and Klungland 1983) can also offer clues to historical vegetation and wetlands. Digital elevation models made from aerial photogrammetric data or LIDAR (Light Distance and Ranging) in the last decade, by providing detailed, spatially continuous topography, help to delineate depressional wetlands, or estuarine or riverine-tidal wetlands with elevation-related boundaries (Figure 3). (In describing wetlands, we follow the system of Cowardin et al. [1985], although we use "riverine-tidal" to refer to wetlands created by tidal backwater effects.)

The resulting map interpretations of the historical landscape allow us to strip away the last 150 years of diking, draining, ditching, and channel clearing to gain a new view of river and floodplain morphology, including how valley morphology and river pattern varied throughout the region in response to the Pleistocene glacial legacy.

Influence of Pleistocene Glaciation on River Pattern

Removing the modern cultural overprint from the riverine landscape reveals how the effects of Pleistocene glaciation fundamentally influence the nature and distribution of present-day (and historical) fluvial landforms and dynamics. The pattern of rivers in Pleistocene valleys created by subglacial runoff strongly contrasts with those in Holocene valleys that have been fluvially eroded. The Snoqualmie River exemplifies the former. This meandering river has a distinct meander belt several meters *higher* in elevation than the surrounding floodplain (Figure 4A). The elevated meander belt results from Holocene fluvial deposition as the river has built its gradient in the broad, low-gradient glacial valley. Topographic maps and DEMs show the same morphology in the Snohomish valley, which was also formed by Pleistocene subglacial runoff (Booth 1994).

Topography of the lower Nisqually River, by contrast, typifies steeper valleys created by post-glacial (Holocene) fluvial incision into glacial deposits (Figure 4B). The lower Nisqually has an anastomosing (or branching, multiple-channel) pattern, with local relief of 2-4 m created by multiple channels and forested islands. Historical maps and photographs and relict topography indicate that other rivers in Holocene valleys, such as the Stillaguamish River, formerly had a similar pattern (see later in this discussion; Figure 6).

The contrast between the two valley types, and the overriding importance of the erosional and depositional effects of Pleistocene glaciation on valley topography and river morphology, is clear in the Nooksack River (Figure 4C). A lobe of the Cordilleran ice sheet that extended southward into the Nooksack valley through the Sumas River drainage (Dragovich et al. 1997) sculpted the lower Nooksack River valley. Consequently, in the Nooksack downstream of the Sumas River drainage, the valley is broader and lower in gradient than upstream. Additionally, the floodplain of the lower Nooksack has extensive areas that are lower in elevation than the river channel, similar to the Snoqualmie River in Figure 4A, whereas upstream of the Sumas, the elevation varies across the valley bottom in association with multiple channels and islands, as in the Nisqually in Figure 4B.

Both types of river—aggrading, meandering rivers in Pleistocene valleys and anastomosing rivers in Holocene valleys—are responding to Pleistocene glaciation, but their responses are opposite. The first type is depositing sediment and building its grade within the gently sloping Pleistocene valleys, while the second is incising into the general Pleistocene drift surface.

These two different river patterns are also associated with very different river dynamics and associated floodplain landforms. For example, in the meandering Snoqualmie River, there are many oxbow ponds and wetlands, but

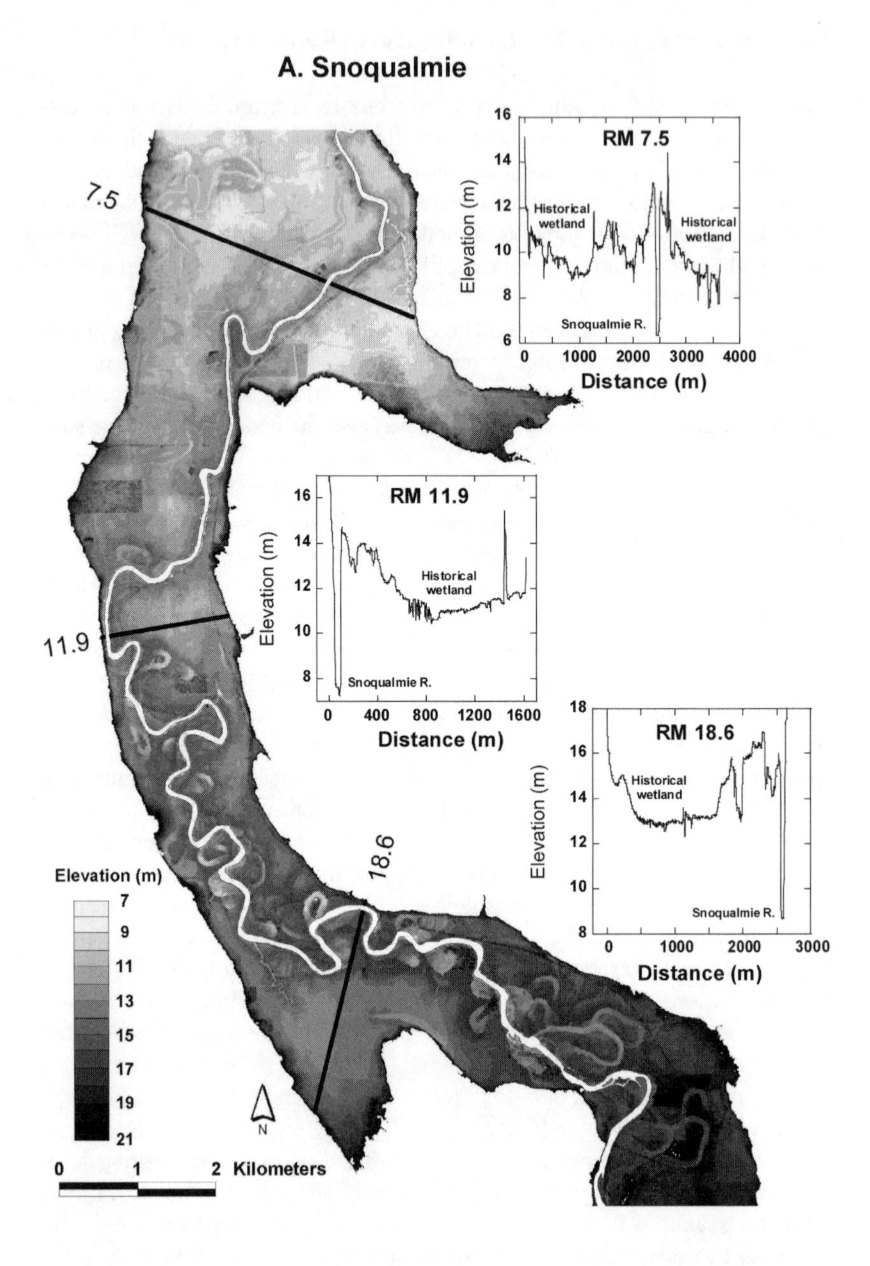

Figure 4. Topography and representative valley cross-sections in: (A) the Snoqualmie River (DEM created from LIDAR imagery).

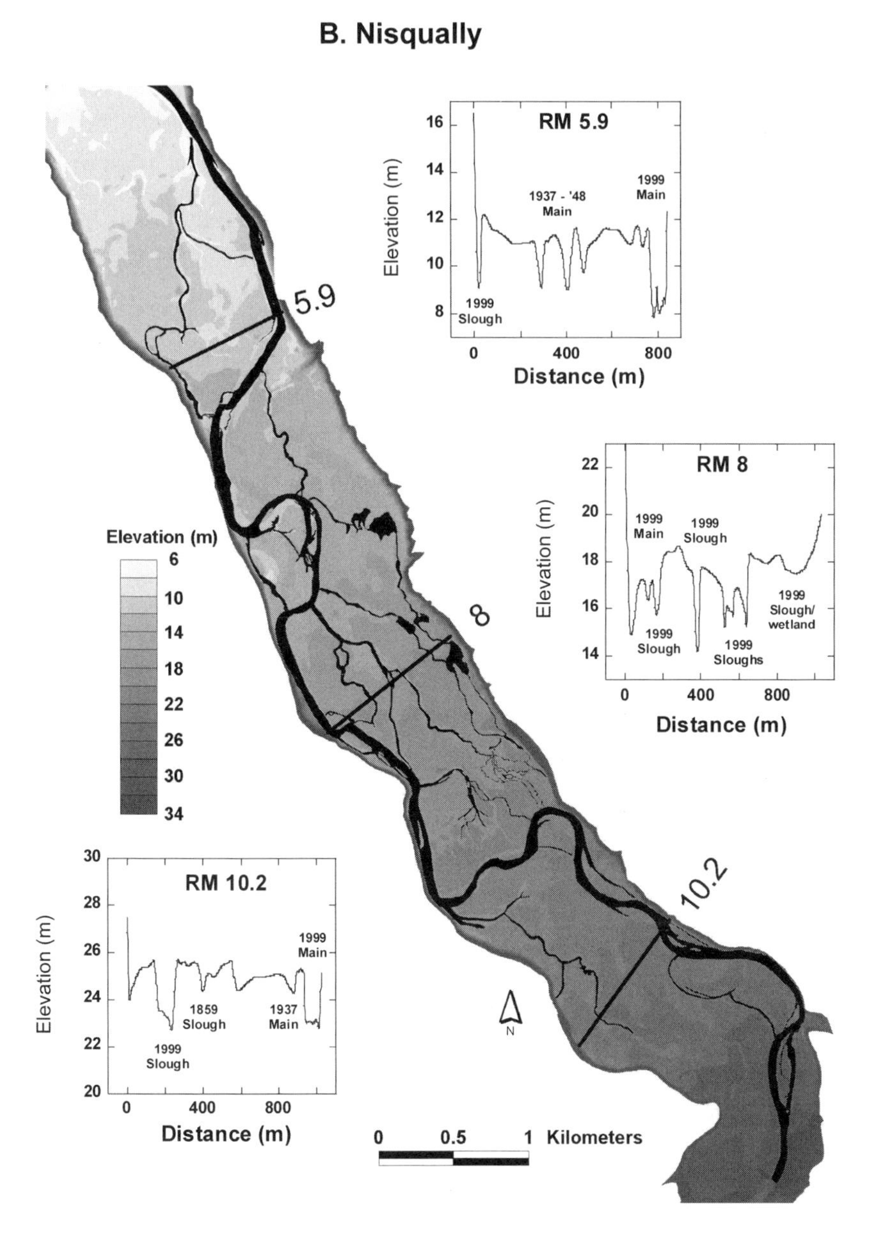

Figure 4 (continued). (B) the Nisqually River (DEM created from topographic mapping from aerial photos).

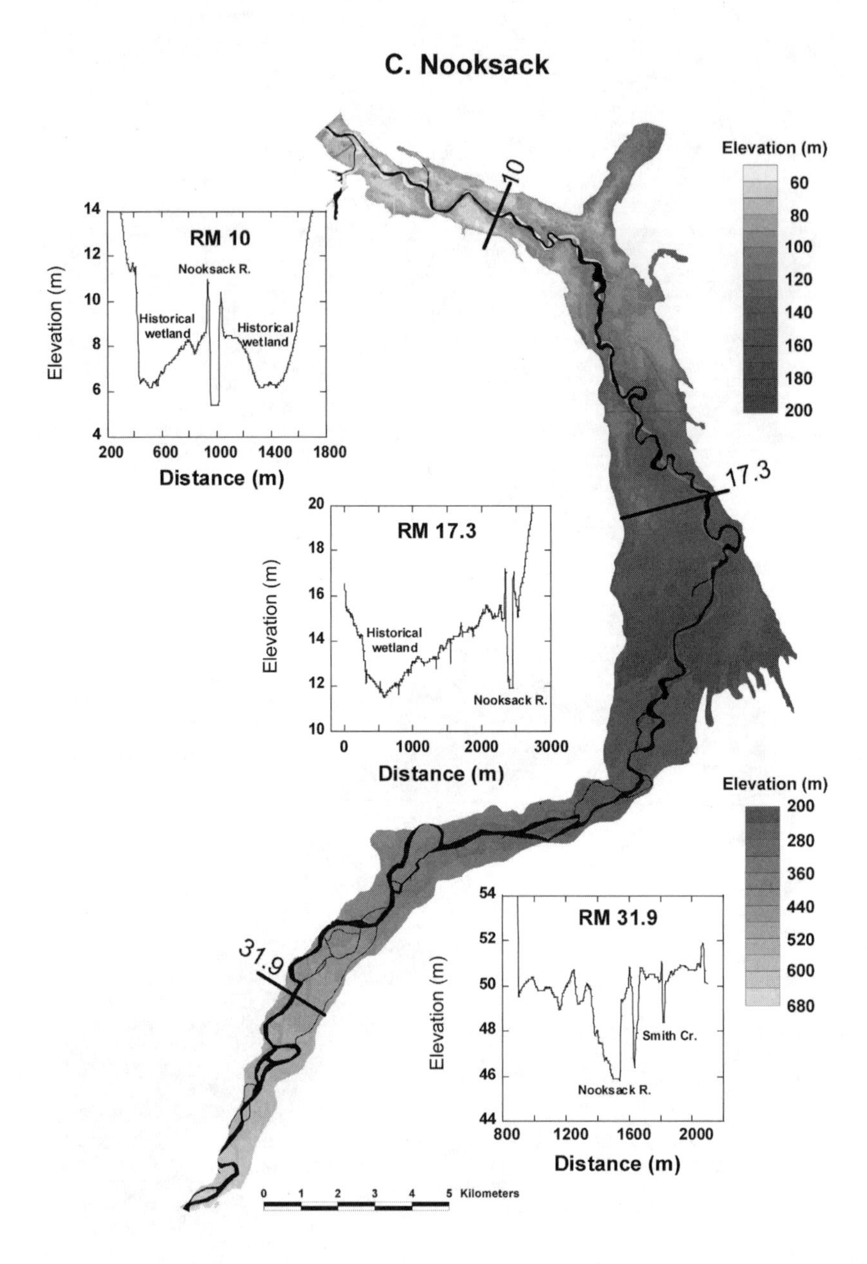

Figure 4 (continued). (C) Nooksack River valley (DEM created from photogrammetric data).

there has been little change in the river or oxbows in the 130 years since the earliest mapping (Figure 5A). Most oxbow lakes now present on the valley were present in 1870 (marked by "1" in Figure 5A). The river appears to migrate slowly and meanders to avulse—the process responsible for creating oxbows—infrequently. The greatest change to the Snoqualmie River valley over the period of historical record is not to channels or to ponds and wetlands formed in oxbows, but is instead the diminution of formerly extensive valley wetlands in low-elevation areas and the clearing of the valley bottom forest (see later discussion of Figure 13).

In contrast, channel positions mapped for the anastomosing Nisqually River over nearly the same period illustrate that river's more frequent course changes (Figure 5B). In some areas the river migrates and river bends are cutoff, as in the Snoqualmie, and these cutoffs become sloughs. However, a second and more common type of avulsion is the river's switching back and forth between multiple channels or from main channel to floodplain slough (we use floodplain slough to refer to a smaller, perennial stream that departs from and rejoins the main river, and which is generally formed in a relict main channel). Wood jams are integral to maintaining the Nisqually's multiple-channel pattern and in causing and mediating avulsions. Preliminary analysis of aerial photographs from 1937 through 1999 shows that flow splits can form at a migrating river bend, when the river intersects an abandoned main channel, diverting flow into it. Jams commonly form at that split, thereby stabilizing it. In addition, the growth of jams at such splits can gradually reduce flow to one branch, eliminating it or reducing it to a perennial slough. Jams also cause avulsions by accumulating in and plugging channels, diverting flow into a relict channel, which then becomes the main channel. Jams at the mouth of the now-abandoned channel then regulate flow into it, causing it to flow perennially as a floodplain slough.

This "metering" of flow into floodplain sloughs, which also mediates the frequency of avulsions, is common. In our study reach in 1998, we field-identified 18 channels that received water from the main river during low-flow discharge. Each of these floodplain channels had a jam associated with its inlet (Figure 6A). In each case, the jam regulated flow into the slough, preventing or delaying the river from avulsing into it. Most of these sloughs were located in what could be identified as a relict main channel on earlier aerial photographs.

The prevalence of this channel-switching dynamic over more gradual channel migration is due in part to the presence on the floodplain of patches of mature forest. These patches remained uneroded by the Nisqually River during the 130-year period of map and photo record, the river instead avulsed around them. On the Queets River, Abbe (2000) found similar "hard points,"

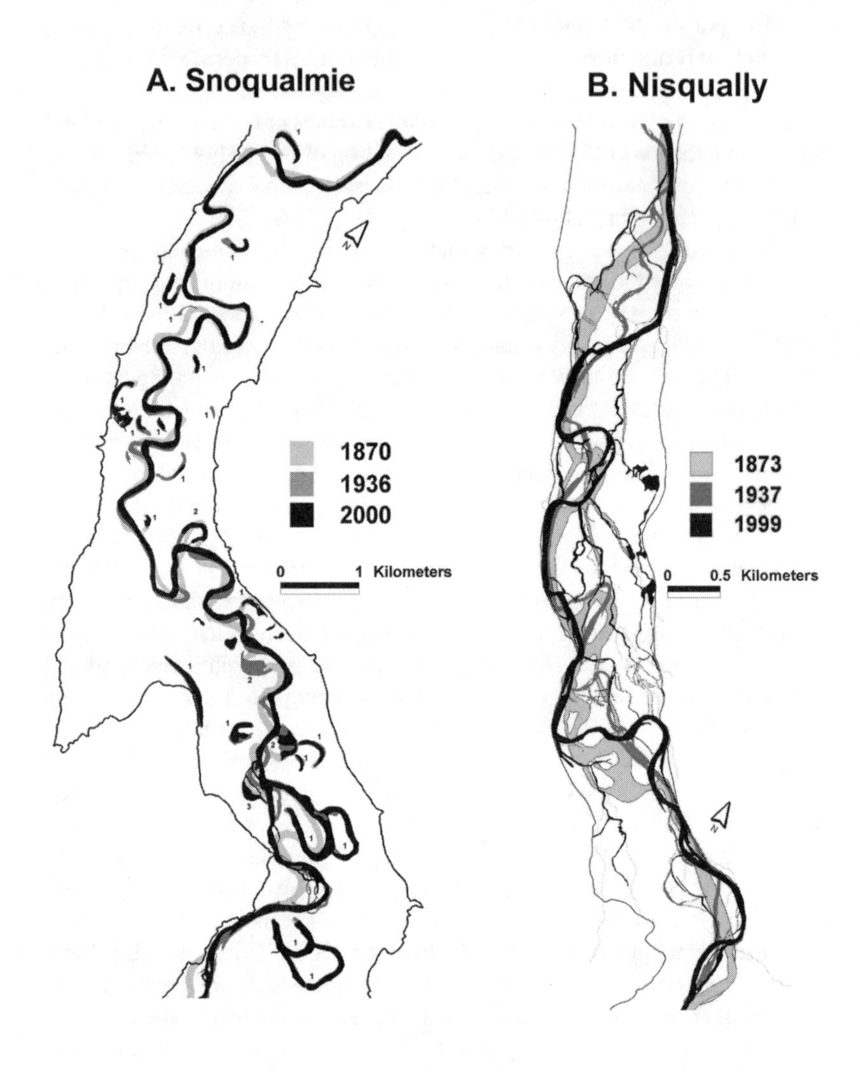

Figure 5. (A) Channel and oxbows in the Snoqualmie River in 1870, 1936 and 2000, and (B) locations of the Nisqually River, 1873, 1937 and 1999. Channel locations in 1870 and 1873 from General Land Office maps; other years are from aerial photos. Numbers in (A) represent year oxbows were first apparent: 1=1870; 2=1936; and 3=2000.

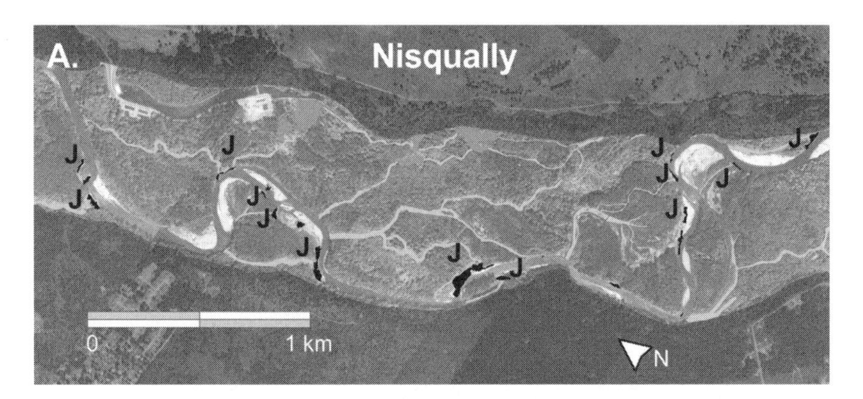

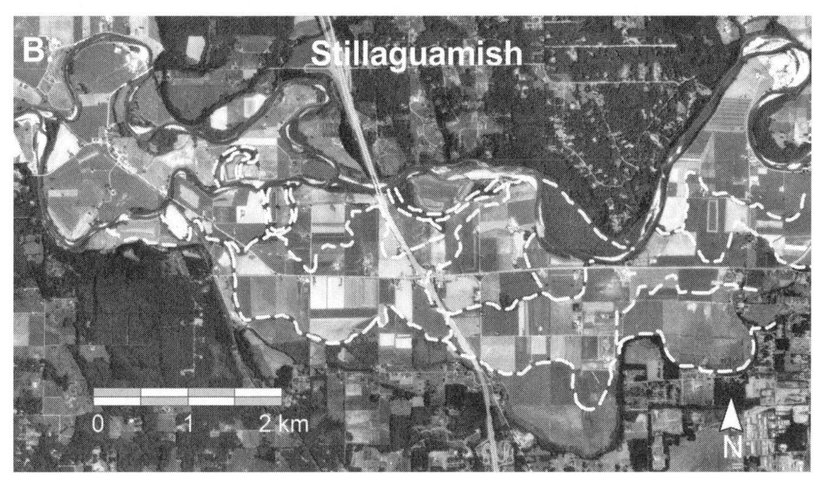

Figure 6. (A) The Nisqually River from 1999 aerial photographs. Those floodplain channels having flowing water in summer 2000, and which are obscured by tree cover on the aerial photographs, are shown with gray, and were mapped from field work in 1998 and 2000 onto 1999 1:12,000-scale ortho-photographs. Large log jams shown with black. "J" indicates a jam that is associated with the inlet to a floodplain slough. (B) The Stillaguamish River from 1990 aerial photographs. Dashed lines indicate relict floodplain sloughs that are no longer present but were shown on 1930 and 1941 maps. Flow in both panels is from right to left. Modified from Collins and Montgomery (2001).

or patches that maintained their stability for up to centuries, created by stable, persistent wood jams at their upstream ends.

Reconstructing channel locations from older maps and mapping relict topography on more recent photographs suggests that the pre-settlement Stillaguamish River had a similar anastomosing pattern as the Nisqually (Figure 6B). Archival sources indicate that wood jams were associated with many of these channel splits in the Stillaguamish. Sedell and Frogatt (1984) showed a correspondence in time between wood removal and simplifications in the Willamette River's pattern. Wood jams, the result of many trees contributed by fluvial erosion of the surrounding forest, appear to have been critical to the dynamics of such anastomosing rivers.

Forests, Rivers, and Wood

The Historical Forest

Puget Sound's dense river-bottom forest, among the most productive on Earth, has been almost entirely cleared. However, field notes from a century and a half ago include information sufficient to recreate that forest in the abstract: tree diameters, species frequency and distribution, preferred growth environments, and geographic ranges. Besides providing a unique glimpse of the region's pre-logging riverine forest, this information can help guide forest restoration planning.

The mid-19th century, mixed hardwood-conifer, riverine forest was heavily weighted toward hardwoods. Of the approximately 7,000 GLO bearing trees we have georeferenced, 71% are hardwoods (Figure 7). This was especially the case for streamside forests, which were composed of 84% hardwoods (Figure 7B). (These percentages underestimate the relative abundance of hardwoods; as described previously, bearing trees under-represent small-diameter species, which are more commonly hardwoods such as vine maple [*Acer circinatum*], willow [*Salix spp.*], and red alder [*Alnus rubra*].) While less abundant, conifers accounted for the majority of biomass as indicated by basal area (Figure 7D-F). Several coniferous species grew quite large. For example, documented cedar (western redcedar, *Thuja plicata*; on first usage we refer to the common names recorded by land surveyors, and provide the probable species) had a mean diameter of 76 cm (median = 61 cm) and included individuals as large as 381 cm in diameter (Figure 8A). Spruce (Sitka spruce, *Picea sitchensis*) in field notes was as large as 282 cm in diameter (mean = 62 cm, median = 50 cm). Several hardwood species also attained a large diameter; maples (bigleaf maple, *Acer macrophyllum*) were as large as 183 cm

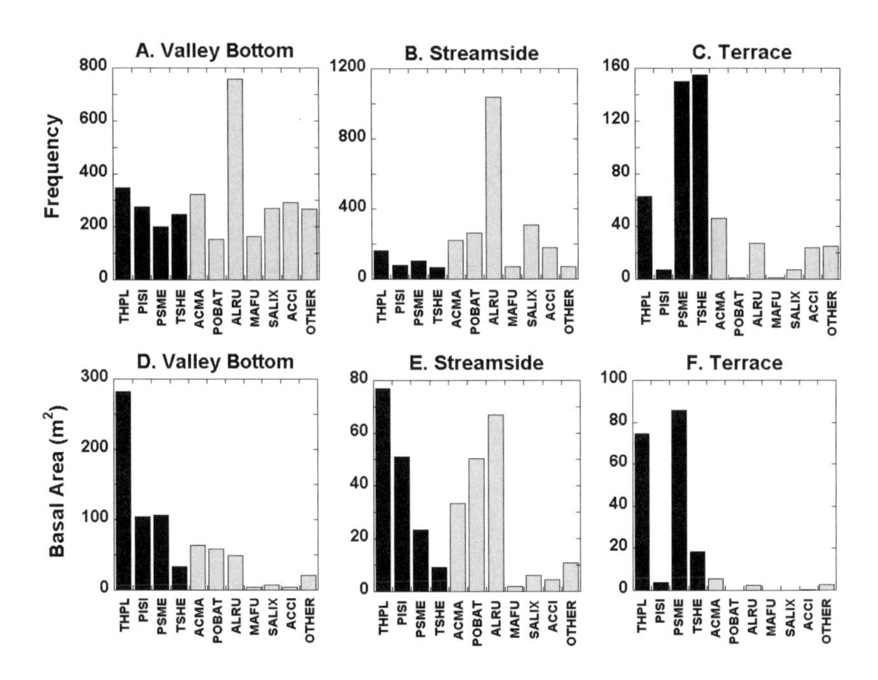

Figure 7. Data on bearing trees from GLO field notes, in eastern Puget Sound river valleys from the Nooksack south to the Nisqually. Frequency of trees in (A) valley bottom forest, (B) stream-adjacent forest, and (C) river terraces. Cumulative basal area in (D) valley bottom forest, (E) stream-adjacent forest, and (F) river terraces. Coniferous species have dark-shaded bar. N=7,348. THPL: western redcedar (*Thuja plicata)*; PISI: Sitka spruce (*Picea sitchensis)*; PSME: Douglas fir (*Pseudotsuga menziesii)*; TSHE: western hemlock (*Tsuga heterophylla)*; ACMA: bigleaf maple (*Acer macrophyllum)*; POBAT: black cottonwood (*Populus trichocarpa)*; ALRU: Red alder (*Alnus rubra)*; MAFU: Pacific crabapple (*Malus fusca)*; SALIX: Willow (*Salix spp.);* ACCI: vine maple (*Acer circinatum*). "Other" species include: white fir (grand fir, *Abies grandis*), ash (Oregon ash, *Fraxinus latifolia),* dogwood *(*western flowering dogwood*, Cornus nuttallii)*, birch (paper birch, *Betula papyrifera*); hazel (beaked hazelnut*, Corylus cornuta var. californica*); bearberry or barberry (uncertain, possibly Oregon grape, *Mahonia aquifolium*); chittemwood (cascara, *Rhamnus purshiana*), cherry (bitter cherry, *Prunus emarginata);* elder (red elderberry*, Sambucus racemosa*); aspen (quaking aspen, *Populus tremuloides*).

(mean = 35 cm, median = 25 cm), and cottonwoods (black cottonwood, *Populus trichocarpa*) as large as 203 cm (mean = 47 cm, median = 30 cm) (Figure 8A).

Various riverine trees in eastern Puget Sound occurred within distinct elevation and latitude ranges and landforms. Sitka spruce, for example, was the lowest-elevation conifer (Figure 8B), a common (and typically the only) large conifer in tidewater areas; it was less common in the southern Sound (Figure 8C). In contrast, western hemlock, which is the potential climax species throughout the Puget Sound region (Franklin and Dyrness 1988), occurred mostly at higher elevations (Figure 8B), was uncommon in the southern study area (Figure 8C), and was only abundant on river terraces (Figure 7C). Douglas fir and western redcedar occurred throughout the area (Figures 8B and 8C). Neither tree was common in streamside areas (Figure 7B), although the great size of cedar caused it to account for the greatest proportion of streamside arboreal biomass (Figure 7E). Both were somewhat more common in valley-bottom forests outside the immediate streamside area (Figure 7A), but both achieved a dominant frequency only on river terraces (Figure 7C). Trees also had identifiable ranges in elevation relative to the streambank. For example, among bearing trees in the Snoqualmie River valley, spruce was the conifer most tolerant of seasonal flooding, growing 1-2 m below the river bank; alder and willow grew as much as nearly 4 m below the riverbank (Figure 8D). At the other extreme, western hemlock generally occurred several meters above the banks, above the threat of flooding.

Rivers of Wood

Rivers transported not only water but also vast amounts of wood. While settlers' accounts are often more colorful than accurate, they suggest the staggering amounts of wood that choked rivers in flood. For example,

> "The amount of drift which floats down one of these rivers in a freshet is astonishing. It is not unusual, when a river is bank full and the current running 6 miles an hour, to see the channel covered with drift, and the flow kept up twenty-four hours with scarcely a break. Such a flow of drift may be repeated several times in a year on a stream like the Skagit or Snohomish." (Morse, in Nesbit et al. 1885, p. 76)

Not only did wood challenge the conveyance of rivers (and the prose of observers), dead trees were so abundant and well-lodged in riverbeds that logging and upstream settlement was stymied until settlers and the Army Engineers could pull, blast, and cut wood from rivers in the 1870s–1890s

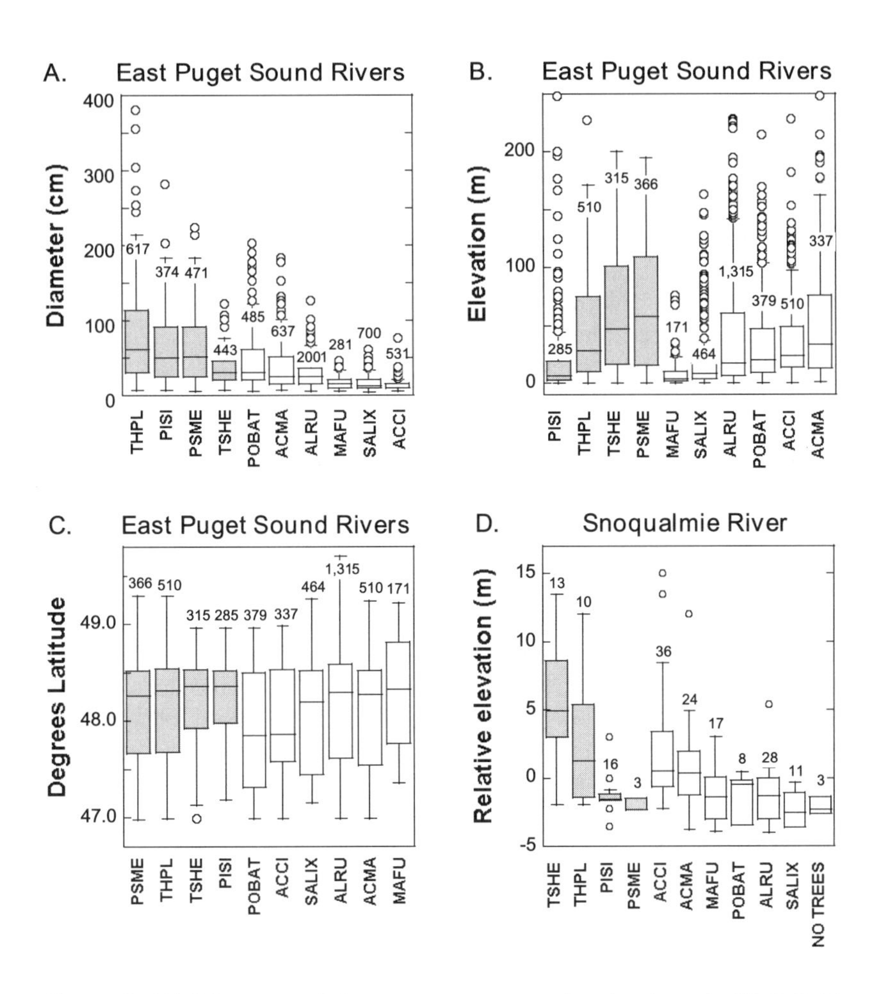

Figure 8. (A) Diameters of the ten most common bearing trees in GLO field notes from eastern Puget Sound river valleys, and their range in (B) elevation and (C) latitude. (D) Elevation of GLO bearing tree species relative to the riverbank elevation in the Snoqualmie River valley. Conifers have shaded bars. Numbers are sample size for each species except "no trees" in panel D, where it refers to number of sites. Species abbreviations are as in Figure 7. Each box encloses 50% of the data. Horizontal line within box represents median. The lines extending from the top and bottom of boxes indicate minimum and maximum values, excepting outlier values (circles) greater than the inner quartile plus 1.5 times the inner two quartiles.

(Sedell and Luchessa 1981; Collins et al. 2002). Logging spread up-valley, and logs could be driven down-river in rafts, once rivers were cleared of blocking wood jams and snags that menaced navigation. Logging and settlement progressed up-valley so rapidly that lowland river valleys had been cleared of nearly all forest by 1900 (Plummer et al. 1902). Thus the evidence of wood in rivers is even more obscured by time and human activities than is that of the historical valley-bottom forests.

We investigated the potential effects of the late nineteenth century removal of riverine wood by collecting field data in 1998 from the Nisqually River and similar data from the Snohomish and Stillaguamish Rivers; from the latter two rivers, wood has been systematically removed, the floodplain forest cut down and converted to agriculture and other uses, and the river banks leveed and hardened. We also used archival sources to determine whether field data from the Nisqually is a reasonable surrogate for historical conditions, to provide information on wood accumulations for which there are no existing analogs, and to describe the geographic variation in wood characteristics (Collins et al. 2002).

In 1998, the Nisqually River had far more wood per channel width than the other two rivers—approximately 8 and 21 times more than the Snohomish and Stillaguamish, respectively. Most of this difference is accounted for by the abundance of wood in jams in the Nisqually River (Figure 9). Excluding jams from the Nisqually, wood abundance was comparable to the other two rivers. We suspect that few jams occur in the Snohomish and Stillaguamish rivers for two reasons. One is the absence of long, large-diameter pieces with rootballs, which in the Nisqually River act as key pieces that initiate and stabilize jams. Large wood pieces with rootballs are no longer present in the Snohomish and Stillaguamish rivers because they have lacked mature riparian forests for more than a century. The other reason is that the two rivers recruit far less wood than the Nisqually because the leveed rivers cannot erode the floodplain, which also generally lacks a riparian forest. The presence of two upstream dams on the Nisqually River makes that river's accumulation of wood all the more striking and points to the importance of local wood recruitment. In contrast, neither the Stillaguamish nor Snohomish have dams, and thus have no limit on wood transport from upstream.

Very little recently recruited wood is found in the Stillaguamish and Snohomish rivers compared to the Nisqually. Most of the older wood in the Stillaguamish and Snohomish is decay-resistant cedar, presumably relict from before forests were cleared a century ago. This reflects, in part, a decrease in wood recruitment from historical conditions. Also, without jams, the rivers retain far less wood. The lack of retention is reinforced by recently recruited wood being small in diameter and readily transported.

The Army's snagging records supplement these field data and provide a quantitative indicator of historical wood abundance in regional rivers and its change through time (Figure 10). Nearly all snagging occurred after huge raft jams had been cleared, and much of the riparian forest had already been cut down, so the number of snags removed would have been less than the amount present under pre-settlement conditions. Yet snags remained a problem. In a 1907 report on the White River (Figure 1), the Army Corps' Major Hiram Chittenden wrote:

> "...channels are strewn with immense trunks, often two hundred feet long, with roots, tops, and all ...[forming] jams, which frequently block the channels altogether. This drift constitutes the gravest feature of the flood problem, for the supply is practically unlimited, and the quantity carried by a great flood is such that very little can be done with it at the time by human agency. Levees or other protection works are of little avail in the presence of these drift jams, and it seems like an almost useless expense to built such works so long as they are menaced by so great a certainty of being destroyed or otherwise rendered useless." (Chittenden 1907)

Between 1880 and 1980, 150,000 snags were removed from five rivers, including the Stillaguamish and Snohomish, with more than one-half of these from the Skagit. A total of 30,000 snags were removed from the lower Skagit

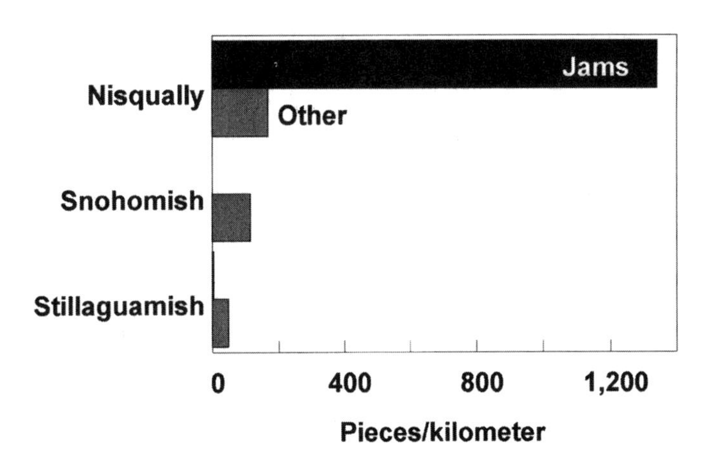

Figure 9. Wood abundance in the Nisqually, Snohomish, and Stillaguamish Rivers, field measured in 1998. Modified from Collins and Montgomery (2001).

River between 1898 and 1908. A diminishing rate of snag removal after 1900 (Figure 10) reflects the decline in recruitment of wood large enough to lodge in the riverbed and remain stable. This in turn presumably reflects the effects of riparian logging (and particularly the removal of very large trees from the valley-bottom forest), leveeing, and bank protection. Snag-boat captains' records indicate that very large pieces were represented in the wood load—the annual maximum snag diameter between 1889 and 1909 ranged from 3.6 to 5.3 m (U. S. War Department 1889–1909), diameters which are confirmed by engineers' observations (e.g., U. S. War Department 1895).

These accumulations were major influences on river channels. Raft jams, the largest accumulations and first to be removed, could be kilometers long, channel spanning, and persist for hundreds of years. For example, a Skagit River jam at the present-day site of Mount Vernon existed for at least a century. A pioneer had learned from the native people that its surface supported live trees two to three feet in diameter (Interstate Publishing Company 1906, p. 206). The jam was packed solidly enough that it could be crossed "at almost any point." The jam was described as 9 m deep, consisting of "from five to eight tiers of logs, which generally ranged from three to eight feet in diameter" (Interstate Publishing Company 1906, p. 114). Beneath the previously described Mt. Vernon raft jam were in some places "furious cataracts,"

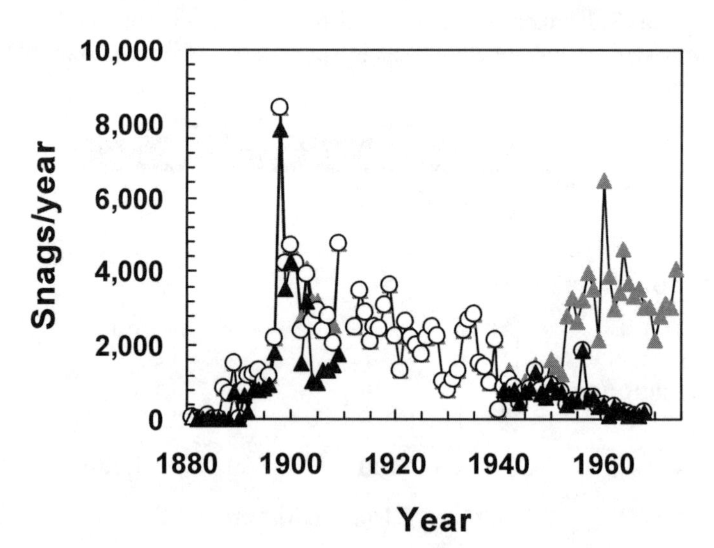

Figure 10. Snags removed from Puget Sound rivers, 1881–1970. Hollow circles: all Puget Sound rivers; solid triangles: Skagit River only; gray triangles: all Puget Sound rivers and harbors. Modified from Collins et al. (2002).

and in others "deep black pools filled with fish" (Interstate Publishing Company 1906, p. 106). The river was as deep as 7 m below the jam at the lowest water stage.

The Geomorphic and Ecological Importance of Wood

Wood accumulations were important to river dynamics at a range of spatial and temporal scales (Figure 11). At the largest scale, raft jams routed water and sediment onto floodplains and deltas. For example, contemporary accounts and map evidence suggest the Mt. Vernon raft jam had a dominant influence on landscape-scale flooding patterns on the lower Skagit River (see Collins et al. 2002). Resulting wetlands would have provided extensive habitat, including habitat ideally suited for salmonid rearing. In the American Midwest, such raft jams retained vast amounts of sediment and dammed tributaries, transforming the valley-bottom environment (Triska 1984).

At the reach scale, as previously described, wood jams in some rivers maintained multiple channels and islands, and created and maintained floodplain sloughs. The historical reduction in the total amount of channel edge in the Stillaguamish River (Beechie et al. 2001) and to a lesser extent in the Snohomish River (Haas and Collins, unpublished data) reflects the simplification of channel pattern. Field studies in the Skagit River indicate that wood, particularly wood jams, along banks significantly increases the fish habitat value of riverbanks (Beamer and Henderson, Skagit System Cooperative, LaConner, WA, unpublished data).

Pools exemplify the role of wood at a smaller scale. In autumn 1998, we measured 85 pools in the Nisqually River study reach and found a pool spacing of 1.4 channel widths (CW) per pool (Figure 12). Wood was the dominant factor forming 61% of pools, including 26% associated with mapped, stable jams. This finding is similar to that of Abbe and Montgomery (1996, Figure 3), who found wood formed 70% of observed pools in a 25 km-long reach of the Queets River in Olympic National Park. In the Nisqually, pools associated with jams were considerably deeper than other pools, the mean depth being three times greater than free-formed pools. Jam-associated pools were twice as deep as pools formed by individual pieces having attached rootballs, augmented by wood or formed by banks.

In contrast to the Nisqually, pool spacing measured on the Stillaguamish ranged between 3 and 5 CW/pool in the three reaches, or two to three times less frequent than in the Nisqually. Only one-ninth (11%) of pools in the Stillaguamish were formed by wood. More than one-half formed along riprap-armored banks. Although deep (Figure 12), these pools lacked cover and would provide considerably less habitat value than pools associated with wood. Similarly, in an 8.5 km-long reach of the Snohomish River, beginning at the confluence of the Snoqualmie and Skykomish rivers, wood created only one relatively shallow pool, and the pool spacing of 3 CW/pool was twice that of the Nisqually River, indicating one-half as many pools. Comparing pool data from the Snohomish and Stillaguamish Rivers to the Nisqually suggests that in the Puget Lowland historically freely migrating rivers with mature floodplain forests had two to three times more pools than contemporary, leveed rivers with little riparian recruitment. While artificially hardened banks appear to create deep pools, pool depth alone is not sufficient to create high-quality habitat because wood also provides cover, complexity, and nutrient-rich substrate to pools, increasing their habitat value (e.g., Bjornn and Reiser 1991).

The historical change in size and quantity of recruitable wood may account for the idea that "wood is more easily transported in large channels [of the Pacific Northwest], leading to a reduction in the amount and aggregation of the remaining pieces" (Bilby and Bisson 1998). In the Nisqually River, by contrast, wood abundance is much greater than predicted by data from smaller streams (see Figure 13.2 in Bilby and Bisson 1998). Thus to some degree the generalization that wood plays a diminishing role in channel structure as streams increase in size is simply a reflection of the cumulative historical effect of human actions.

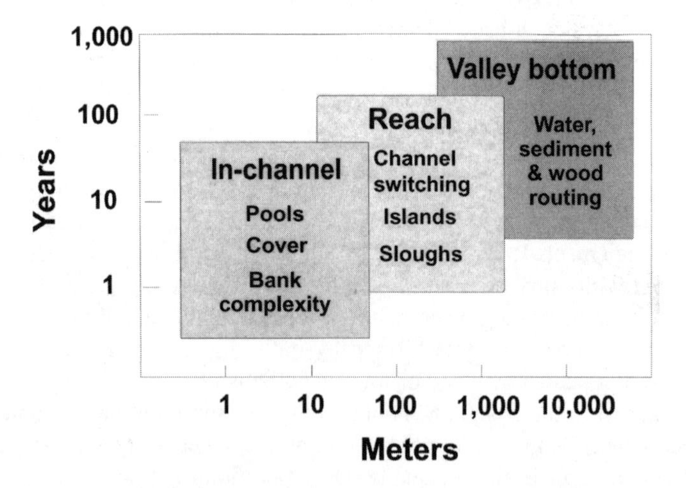

Figure 11. Temporal and spatial scales at which wood accumulations influenced lowland Puget Sound rivers. Modified from Collins et al. (2002).

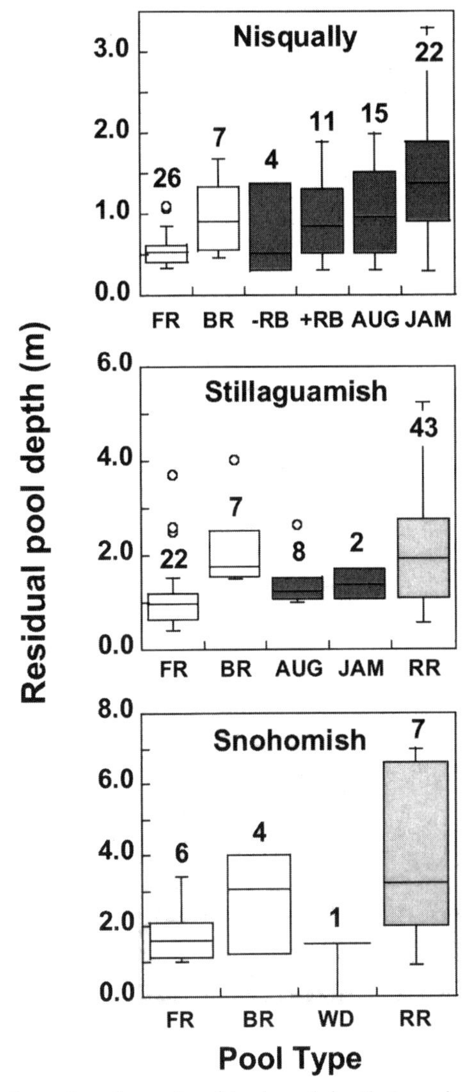

Figure 12. Number of pools, and residual pool depth, by primary pool-forming factor, in the Nisqually River, Stillaguamish River, and Snohomish River. Pool-forming factors: FR: free-formed alluvial; BR: bedrock forced; -RB: wood piece without rootball; +RB: wood piece with rootball; WA: wood augmented; JAM: wood jam; (all wood-related pools have dark-shaded bars); RR: riprap armored bank (light-shaded bars). Numbers on plots are the sample size. Modified from Collins et al. (2002).

Where the Habitat Was (and Where it Went)

In the Puget Sound basin, the majority of channel and wetland area accessible to salmonids was in the larger rivers and floodplains of the Puget Lowland (Sedell and Luchessa 1981; Beechie et al. 1994; Collins and Montgomery 2001). The historical abundance of habitat in lowland floodplains and deltas can only be determined from archival materials, because so little remains of these habitats. We are currently creating GIS maps of historical riverine environments of Puget Sound, beginning with the north Sound, and in the following discussion draw on this work in progress to illustrate the nature and distribution of habitats.

In the Snohomish River valley, prior to widespread landscape modifications by settlers, the majority of land area was either channel or wetland (Figure 13A). Vast floodplain wetlands and extensive estuarine marshes accounted for nearly two-thirds (62%) of the valley bottom. By the 19th century's end, much of this wetland had been diked, ditched, and drained; by the end of the 20th century, only small patches remained (Figure 13B). Wetlands were extensive in the Snohomish basin because the Pleistocene glacial-meltwater-shaped valley has a low gradient, a great width, and low elevation relative to the river banks, all favoring extensive wetlands. Especially notable in the Snohomish basin were the vast riverine-tidal wetlands, or freshwater wetlands influenced by the tides. The extensive freshwater "Marshland" wetland (Figure 13A) formed in a portion of the floodplain lower than the river, topographically similar to the Snoqualmie River (Figure 4A), which also had extensive low-elevation wetlands (Figure 13A).

The distribution of habitats in the lower Nooksack mainstem (Figure 14A) was similar to those in the Snohomish basin. The Nooksack's low-gradient delta-estuary had extensive riverine-tidal freshwater wetlands, and upstream of the estuary, the channel meandered between expansive freshwater wetlands on the lower-elevation floodplain. As in the Snohomish basin, most of these wetlands have long been drained and ditched (Figure 14B).

In the upper Nooksack mainstem, upstream of the influence of continental glaciation (Figure 4C), the valley was narrower, steeper, and the channel anastomosed and lacked the wetlands of the lower river (Figure 14B). Similarly in the Snohomish basin, the Skykomish River (Figure 13A) contrasts with the Snohomish and Snoqualmie, in having a multiple-channel pattern. Both the Snohomish and Nooksack basins illustrate the contrasting channel patterns—meandering single thread compared to anastomosed—and the different valley landforms—oxbows and extensive low-elevation wetlands compared to floodplain sloughs—that historically existed in Pleistocene compared to Holocene river valleys.

Both the upper Nooksack mainstem and the Skykomish River now have a simpler channel pattern than they did historically (Figures 13B and 14B). Landscape reconstructions of the upper Nooksack River mainstem in the intervening period (~1910 and 1938) shows that early in the 20th century the river took on a braided pattern, presumably in part because streamside logging weakened banks and contributed coarse sediment to the river. Over the rest of the century, levees confined the channel, creating the present-day relatively straight, confined channel. Meanwhile in the lower Nooksack mainstem, the cutting off of meanders and construction of levees also created a relatively straight confined channel. Land use changes have thus caused the historically very different channel patterns of the upper and lower Nooksack mainstem to converge into relatively similar patterns today.

The Skagit-Samish delta (see later, Figure 15G) is unique in the region because of its large size and unique origin from mid-Holocene (~5,000 ybp) Glacier Peak lahars (Dragovitch et al. 2000). The delta is also unique in its historical quantity and variety of wetland and channel habitats. Estuarine wetlands were extensive in the low-gradient, spreading delta, totaling more than twice as much as those on the other three north-Sound deltas (Nooksack, Stillaguamish, and Snohomish) combined; riverine-tidal wetlands were second in extent only to the Snohomish estuary; and the extent of palustrine wetlands dwarfed those in other estuaries. Numerous distributary sloughs bisected the delta. Most of the wetland habitats were diked and drained by the end of the 19th century excepting a portion of primarily estuarine emergent wetland (Figure 15H)—which is the largest remaining estuarine wetland in Puget Sound—and most of the distributary sloughs closed off to water influx by dikes.

Differences between these North Sound rivers demonstrate the important role of archival sources in characterizing the abundance and variation of aquatic habitats. They also demonstrate that the region's geologic history created distinct types of valleys and estuaries with broadly similar habitats.

Using Historical Information in Restoration, Rehabilitation, and Conservation Planning

Toward What Restoration Needs does Historical Analysis Point?

We use "restoration" to mean re-establishing a self-sustaining, dynamic riverine landscape closely resembling the pre-settlement condition. We use "rehabilitation" to refer to re-establishing certain historical processes or features, or certain habitats, which probably involves on-going intervention, engi-

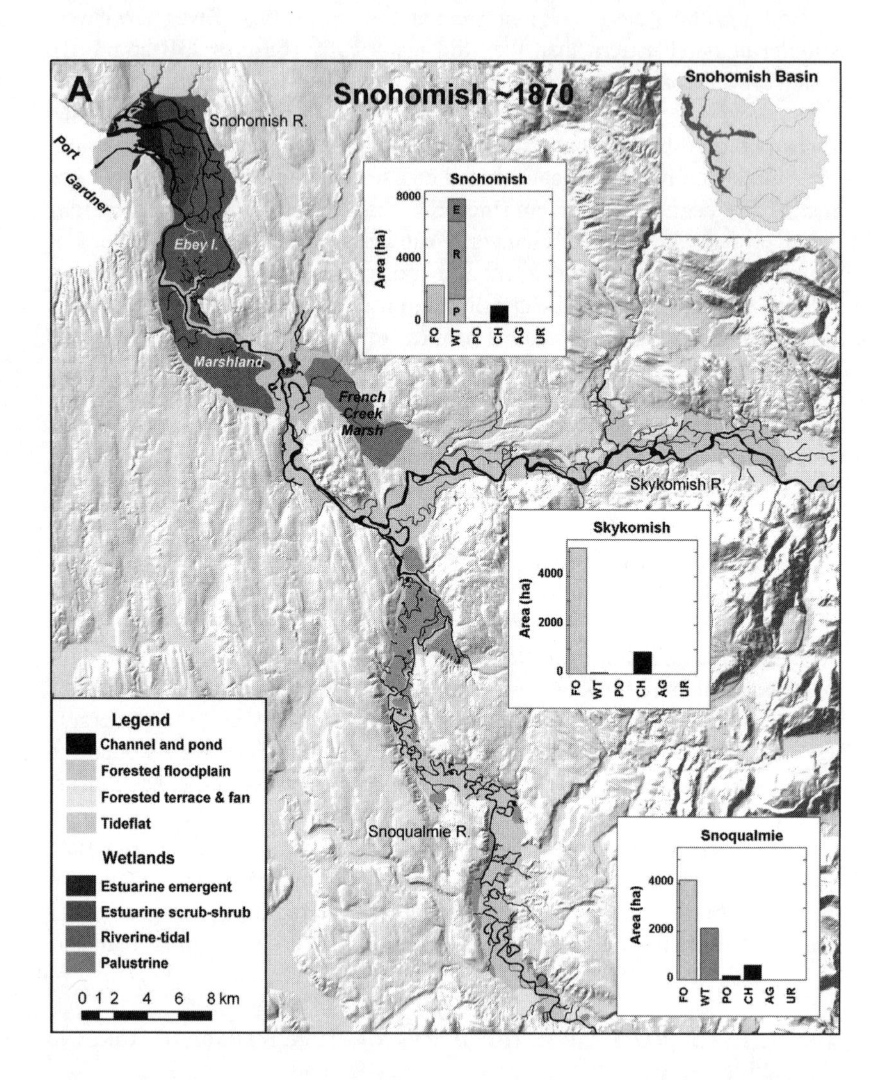

Figure 13. (A) Channels and wetlands in the valleys of the Snohomish, Snoqualmie, and Skykomish Rivers in ~1870, or prior to widespread landscape modifications by settlers, as interpreted from archival sources, primarily GLO field survey records and USC&GS charts. Bar graphs show floodplain area (terraces and fans are excluded) in following categories: FO = forested floodplain; WT = wetland (E = estuarine; R = riverine-tidal; P = palustrine); PO = pond; CH = channel; AG = agriculture/cleared land; urban = urban.

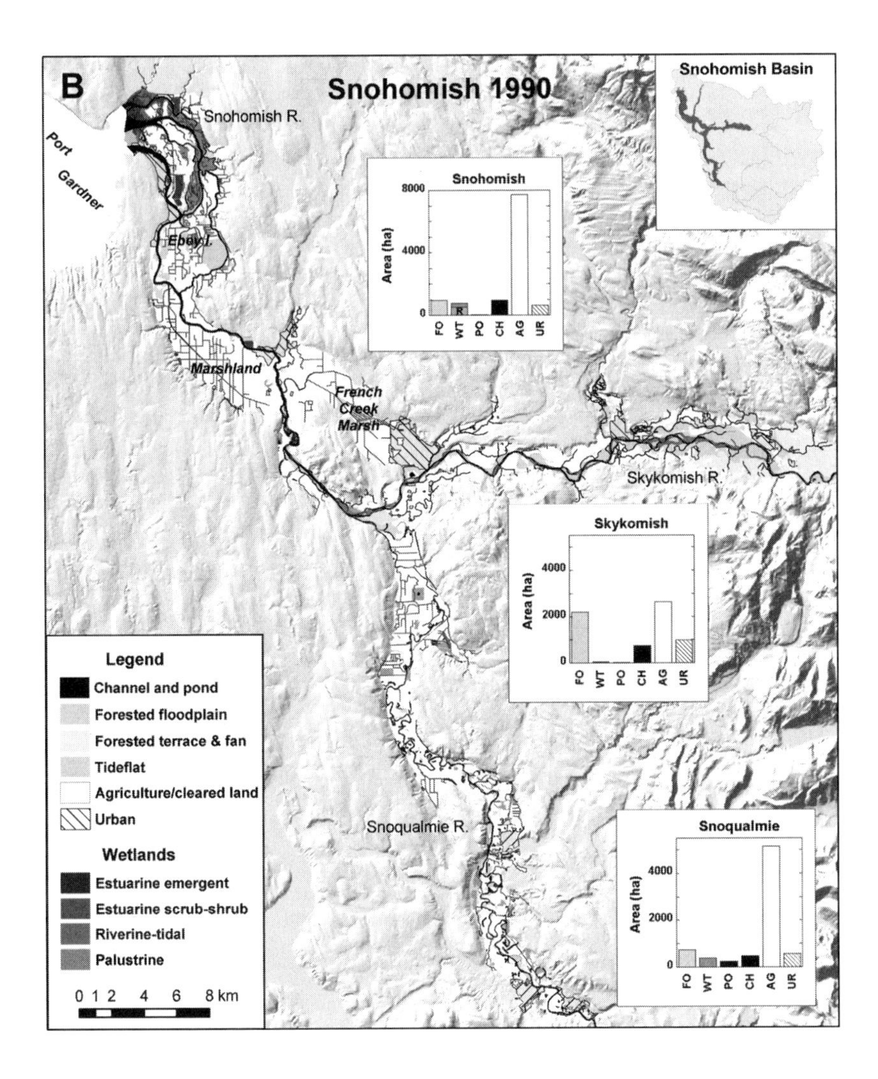

Figure 13 (continued). (B) Conditions in 1990, primarily from aerial photographs, supplemented with hydrography and wetlands from Washington Department of Natural Resources and National Wetland Inventory, and USGS land use and land cover mapping. Abbreviations for bar graphs are the same as in Figure 13A; channel category includes gravel bar (unshaded) and low-flow channel (black).

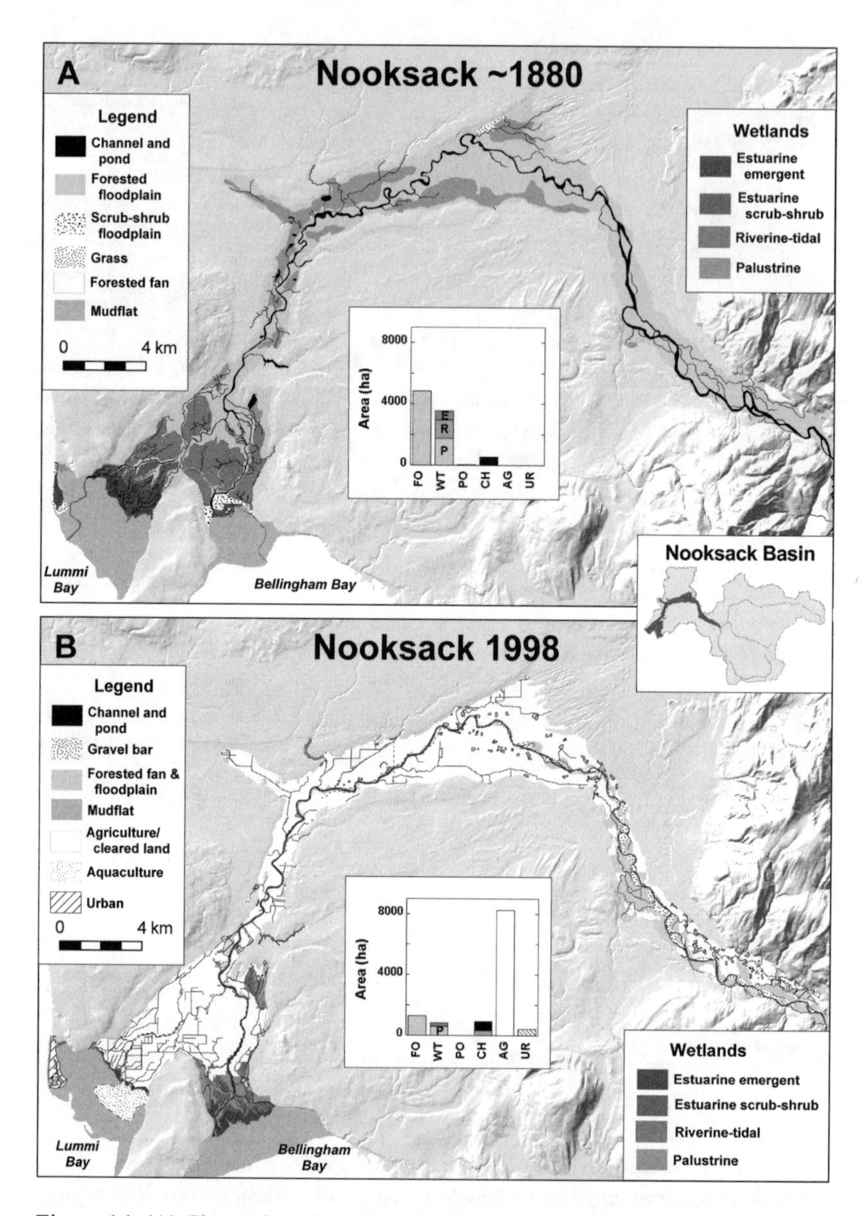

Figure 14. (A) Channels and wetlands in the mainstem Nooksack River valley in ~1880, as interpreted from archival sources, primarily GLO field survey records and USC&GS charts. (B) Conditions in 2000, mapped as in Figure 13B. Abbreviations in bar graphs are the same as in Figure 13.

neering, or maintenance. Historical studies can reveal opportunities and constraints, which in turn may dictate the choice of restoration versus rehabilitation. Because the processes that create riverine landforms, dynamics, and habitats vary between and along rivers, initial planning for restoring a river includes identifying the historically dominant processes. Doing so points toward the appropriate conceptual models of riverine function and restoration. Describing the historical locations and types of riverine habitats associated with these processes then makes it possible to set restoration or rehabilitation targets.

Identifying Primary Elements, Issues, and Opportunities

Restoration opportunities and constraints differ not only with geologic setting and physical dynamics but also with land use history (Table 1). In Pleistocene valleys such as the Snoqualmie, critical opportunities include: connecting oxbow ponds and wetlands to the river; re-establishing riparian forests along oxbows and channels; and re-establishing historically extensive valley wetlands (Table 2). In the Snoqualmie, the riverine system is only moderately degraded because the channel has generally not been hardened and oxbows have not been filled (Figure 15 A-B). On the other hand, channelization has more significantly altered the nearby Sammamish River (Figure 15C-D), which has a similar geologic setting to the Snoqualmie (Table 1). Restoring the Sammamish River would begin with the more intensive task of re-establishing the historical meandering pattern (Table 2). In addition, urban development in part of the Sammamish valley would likely pose more constraints than might the Snoqualmie River valley's agricultural uses. Interbasin water transfers can represent more far-reaching (and more challenging) alterations. For example, in 1916 most all of the Black River's (Figure 15E-F) inflow was eliminated when opening of the Lake Washington Ship Canal lowered the lake's water level (Chrzastowski 1981). Somewhat less radically, the present-day 1250-km^2 area of the Duwamish River is much smaller than its historical 4250-km^2 watershed, when the basin included the watersheds of the Cedar, Sammamish, and White rivers (Blomberg et al. 1988).

In the Nisqually- and Stillaguamish-type of river (i.e., formerly anastomosing rivers in Holocene river valleys), opportunities include re-establishing floodplain sloughs (Table 2). Such sloughs, and beaver ponds commonly associated with them, were critical rearing habitats for coho salmon (Beechie et al. 2001). However, in contrast to the Snoqualmie, Sammamish, or Black, rivers such as the Stillaguamish were more dynamic, having more rapid river migration and a dynamic shifting of flow from main channel to slough and

Table 1. Categories of river restoration situations, organized by the geomorphic setting and degree of anthropogenic change, and representative reaches.

	Extent of Restoration Activity Needed (1) Vegetation; (2) Channel; (3) Channel-floodplain connectivity		
Puget Lowland Landform Type	*A. Less* 1. Mature forest or natural vegetation 2. Natural banks 3. Floodplain channels hydraulically connected to river	*B. More* 1. Widespread vegetation clearing 2. Some hardening and levees 3. Floodplain ditched, drained; secondary channels blocked	*C. Most* 1. Most native vegetation removed 2. Widespread hardening and levees; channel straightened 3. Floodplain ditched, drained; secondary channels filled
I. Delta-Estuary	Nisqually	Skagit Nooksack	Puyallup Duwamish
II. Pleistocene Glacial Troughs	Snoqualmie (Snoqualmie Falls to Skykomish R.)	Sammamish	Puyallup (lower) Duwamish Black
III. Holocene Fluvial Valleys	Nisqually (Fort Lewis and Nisqually Indian Reservation)	Stillaguamish (mainstem)	Cedar (lower)

from slough back to main channel. This dynamic behavior presents greater challenges for reconciling valley-bottom land uses with river restoration than in relatively static, meandering rivers such the Snoqualmie, because a self-sustaining restoration would involve removing or setting back levees and re-establishing extensive forests. Rehabilitation opportunities, on the other hand, which include engineering flow into static floodplain sloughs, may be more easily compatible with existing land uses. In any case, rivers have commonly incised since historical meander cutoffs or straightening, which could complicate reconnecting floodplain sloughs. For example, channel bed surveys in the Stillaguamish River in the vicinity of several floodplain sloughs currently proposed for reconnection shows that the river downcut by 1 to 2 m between 1929 when sloughs were disconnected and a channel survey in 1991. Elsewhere, many former floodplain sloughs and distributaries have been

Table 2. Typical restoration and rehabilitation actions for rivers in different landforms or geomorphic setting, and resources possibly required.

Geomorphic Setting	*Restoration Actions*	*Restoration Resources*		
		Land	*Financial*	*Time*
I. Delta-Estuary	Reconnect distributary sloughs	Small	Moderate	Short
	Breach or remove dikes to re-establish tidal or freshwater flow	Extensive	Large	Short
II. Pleistocene Glacial Troughs	Re-establish meanders (channelized rivers)	Moderate	Large	Short
	Riparian planting	Small	Small	Mod-Long
	Connect oxbows to river	Small	Small	Short
	Passive wetland restoration	Extensive	Small	Mod-Long
III. Holocene Fluvial Valleys	Re-establish meanders (channelized rivers)	Moderate	Moderate	Short
	Levee removal or pullback	Extensive	Large	Short
	Floodplain reforestation	Extensive	Moderate	Mod-Long

ditched or filled, in some cases with toxic materials, complicating efforts to restore flow into them.

Still different opportunities exist in estuaries. While the relative amount differs among different estuary types, there is potential to recoup habitat in all estuaries which had extensive distributary and blind tidal channel habitats (Table 2). In moderately altered systems such as the Nooksack (Figure 14) or Skagit basins (Figure 15G-H), there are opportunities to restore flow, including freshwater to now-diked-off distributary sloughs, or tidal flow to estuarine marsh now blocked by sea dikes. Removing, setting back, or breaching dikes or engineering flow through dikes can initiate restoration or rehabilitation of these environments. Analogous to the river incision that may complicate reconnecting floodplain sloughs, land subsidence in historically diked and drained estuarine marshes can complicate marsh restoration (e.g., Zedler 1996). On the other hand, patterns and rates of estuarine sedimentation can change through time, causing historically diked-off marshes to have greater elevations than when they were initially diked. This is the case in the Nooksack River delta (Figure 14), where the Lummi River was formerly the dominant flow channel until around 1860 when changes to a log jam diverted water into the Nooksack, after which the Lummi River gradually dried up. High sediment loads from the watershed have extended the Nooksack delta more than a mile outward, and most modern wetlands are recently created. In the highly industrial estuaries of the Duwamish and Puyallup (Figure 15I-K) rivers, severe constraints created by infrastructure and reshaping of the hydrology

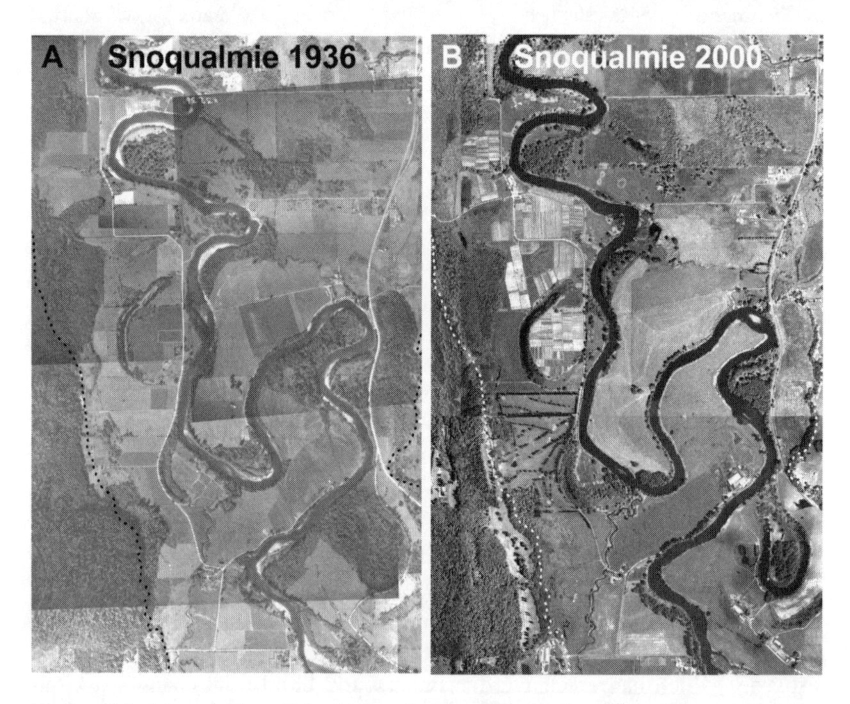

Figure 15. Map and aerial photo views of rivers in Table 3, representing different geologic settings and extent of historical land-use modification: (A) Patches of riparian forest and valley-bottom wetlands remained in 1936 along the Snoqualmie River; (B) In 2000, the river channel and oxbow wetlands were relatively unchanged, but wetlands and forests diminished. Dashed lines show limit of valley bottom.

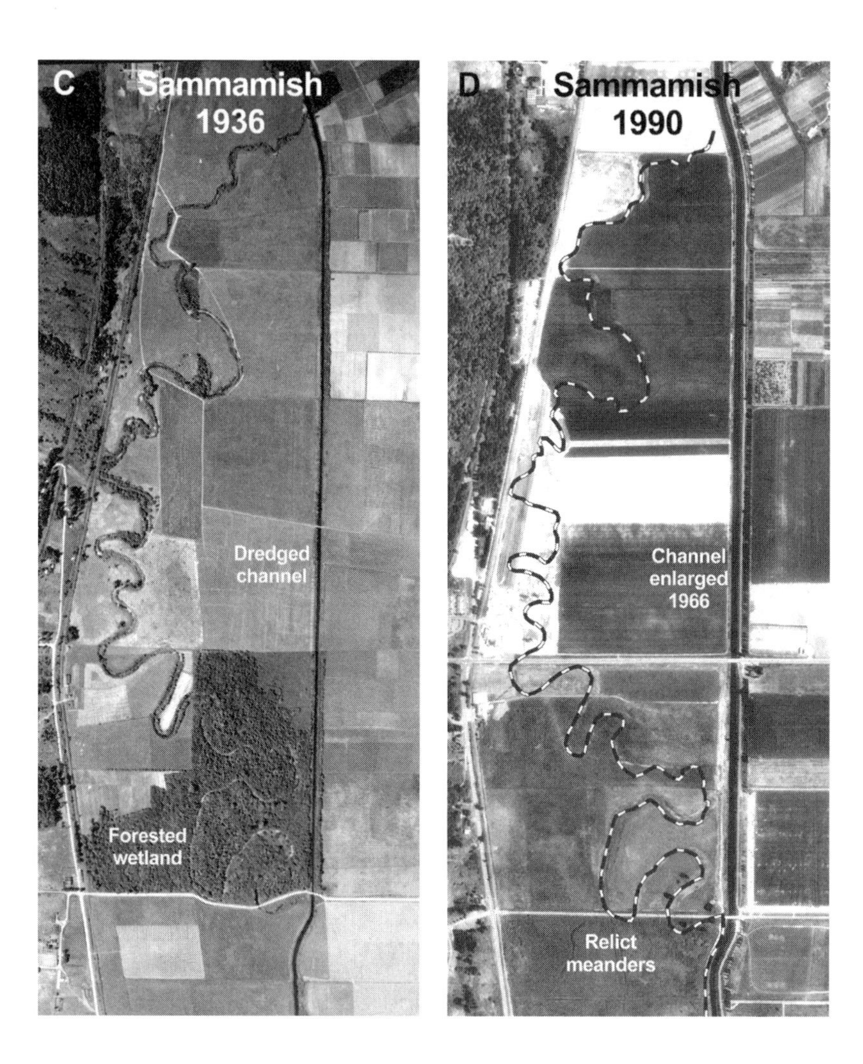

Figure 15 (continued). (C) Small patches of former wetlands that historically nearly filled the Sammamish River valley remained in 1936, along with relict, meandering riverbeds that still functioned during floods. (D) By 1990, the relict meanders were subdued swales in agricultural fields, and an earlier-cut-off ditch had been enlarged to contain the river and most floods.

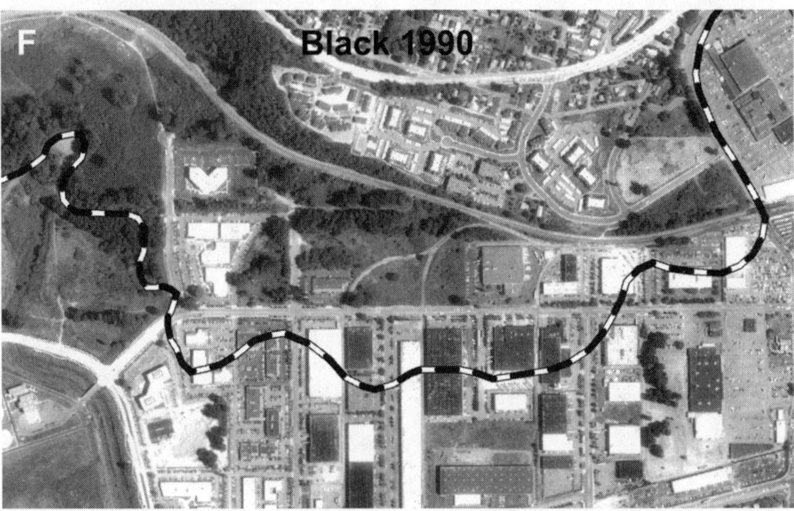

Figure 15 (continued). (E) The Black River retained a natural channel in 1940 with greatly reduced flow. (F) By 1990, the lower river retained a natural environment, but moving successively upstream, the former riverbed is covered by buildings, then becomes coincident with a street, and farthest upstream (out of view) buried by aviation runway.

limit restoration potential to creating or rehabilitating functional habitat elements (Simenstad and Thom 1992).

Regional and within-watershed differences in channel morphologies, processes, and suites of valley-bottom landforms have important implications for whether the central task is to restore valley-bottom forests and river migration or to restore hydrologic connection. For less dynamic rivers in Pleistocene glacial troughs, the forest-river dynamic is less critical. There, the *hydrologic* connection is more important. Emphasis is on restoring the flow of water to valley-marginal wetlands, which can be restored in a "passive" way because of their subdued topographic position relative to the river channel, and on restoring the hydraulic connection between the river and floodplain oxbow lakes. In deltas and estuaries, emphasis is also on hydrologic connection. For restoring moderately degraded anastomosing rivers, such as the Holocene valleys, *riparian forests*, and their connections with the channel are

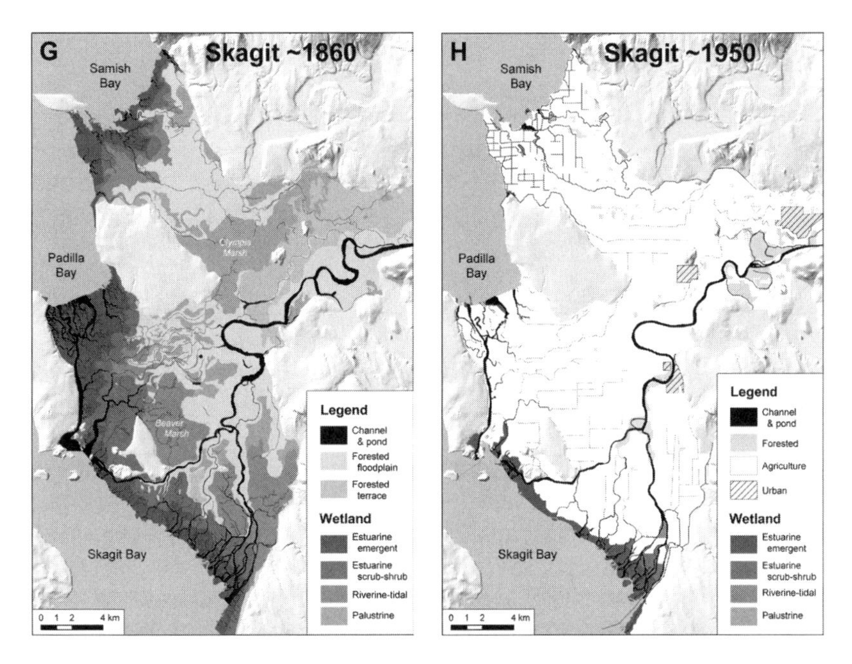

Figure 15 (continued). (G) The Skagit River delta had extensive estuarine, riverine-tidal, and freshwater wetlands prior to ditching and diking that began in the 1860s. Mapping sources are as in Figures 13A and 14A. (H) By the middle of the next century, most wetlands had been diked and drained, and many distributary sloughs closed off. Mapping from USGS topographic maps and aerial photographs.

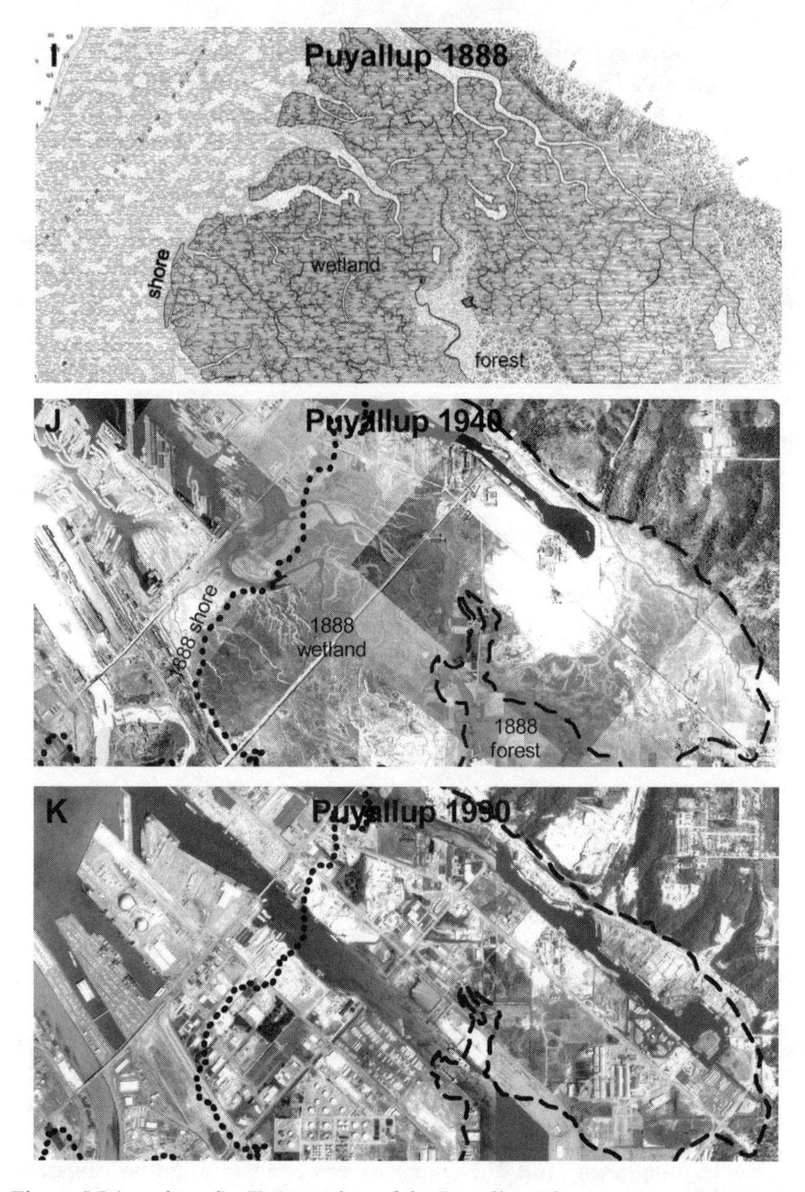

Figure 15 (continued). (I) A portion of the Puyallup River estuary as shown on an 1888 USC&GS chart. (J) Many of the estuarine wetland channels are still present as recently as 1940, when infrastructure had mostly been created upon made-land seaward of the historical delta shoreline. (K) Dense industrial development and dredging transformed the estuary in the late 20th century.

Table 3. Conceptual framework for restoring wood jams and river dynamics to channels in which wood is a dominant element, such as the "Holocene fluvial valleys" in Table 1. From Collins and Montgomery (2002).

	Steps in Restoring Wood Jams and River Dynamics			
Years:	*0–10*	*1–50*	*50–100*	*100+*
Actions:	Riparian reforestation: Includes fast-growing species. Levee set-back or removal.	Instream structures: Includes placing key pieces or building wood jams.	Naturally-recruited logjams: Fast-growing species form key pieces.	Naturally-recruited logjams: Slower-growing species form key pieces.
Results & Functions:	Initiate future supply of wood. Restore lateral erosion and avulsion.	Short-term pool-forming and channel-switching functions. Stable sites for forest regeneration.	Long-term, sustainable supply of wood jams. Long-term, sustainable pool-forming and channel-switching functions.	

important. There, restoration would require removing or pulling back levees and replanting forests (Table 2). However, the sequencing, time required, and overall strategy for doing so can be informed by historical studies.

Planning Riverine Reforestation and Wood Reintroduction

The importance of wood accumulations to fluvial processes (Chapter 3) argues that planning for sustainable large-river restoration in temperate, forested regions such as the Pacific Northwest include the recovery of in-channel wood, and how the composition and extent of the riparian forest translates into the quantity and function of wood, particularly wood jams. Wood large enough to function as key pieces is critical to jam formation and hence to river restoration (see also Chapters 16 and 17). To create jams, rivers must also have access to a large number of trees recruitable by bank erosion and avulsion. Structural approaches (e.g., building wood jams) are not sustainable without continued intervention. A supply of wood large enough to form key pieces implies the presence of large trees in the riparian forest, a dynamic flow regime capable of eroding forested floodplain, and banks that will allow channel migration. Such restoration requires sufficient riparian land area, which may not be available in heavily populated areas, where it may only be possible to rehabilitate selected features or functions (e.g., Kern 1992; Brookes 1996; de Waal et al. 1998).

Based on our analysis of the Nisqually River system (Collins and Montgomery 2002) and experience with other large Pacific Northwest rivers, we propose the following outline for a strategy to reestablish a self-sustaining, dynamic river morphology and habitat in wood-depleted areas. First steps include levee setbacks and riparian planting, including tree species near the river that will rapidly develop a large size (Table 3). The forested corridor width needed to provide a sufficient, long-term source of wood and to allow for channel migration and avulsion depends on the local geomorphic context. In the first few decades of a restoration plan, engineered solutions may provide short-term functions and hasten riparian forest regeneration. Such actions include placing key pieces in systems with adequate recruitment but an absence of pieces large enough to form key pieces, or in systems lacking both, constructing wood jams (Chapter 17) which may provide short-term functions and hasten riparian forest regeneration. Within 50 to 100 years, self-sustaining wood jams should develop if key pieces of sufficient size and racked pieces of sufficient quantity are available. While differences in durability between hardwood species and conifers have recently been shown to be less in submerged conditions (Bilby et al. 1999) than in terrestrial conditions (Harmon et al. 1986), few key pieces we observed were fully submerged, and thus key pieces of deciduous wood would be expected to be considerably less durable than conifers. This durability may be inconsequential for the primary function of key pieces, because a jam is likely to be incorporated into the floodplain within 1–2 decades as the river migrates or avulses away from the jam, and forest trees colonize it. However, the river is also likely to eventually re-entrain wood from most such abandoned jams, thereby allowing key pieces to be "recycled" into the river. Hardwoods may only be durable enough to function once as key pieces. For this reason, in the longer term, slower growing and more durable species are also important sources of key pieces.

This framework calls into question common assumptions about river restoration in the Pacific Northwest. First, most restoration efforts have focused on static habitat creation ("instream structures" in Table 3) rather than reestablishing processes (Reeves et al., 1991; Frissell and Ralph 1998); forest restoration is a critical additional component to sustainable river restoration. Second, because conifers have been logged from essentially all lowland rivers in the region, it is likely that riparian hardwoods are now more common in riparian areas than they were historically, and restoration strategies commonly include converting hardwoods to conifers. However, large trees are necessary to provide key pieces for jams, suggesting that riparian forests be managed at least initially to produce large trees from a mix of species. In fact, historical land survey records show that hardwoods dominated most river valleys historically, and several hardwood species could grow to be quite large (e.g., see Figure 7). Third, riparian restoration plans often assume a timeframe of a century or centuries, which can be the time needed to develop large western redcedar. However, riparian reforestation that includes fast-growing species can produce large trees that are essential for creating key pieces within a shorter time frame. River restoration can be accomplish in stages, from engineered jams (1–10 years), to jams initiated by fast-growing, largely deciduous pieces (50–100 years), followed in the longer term (100+ years) by slower growing but more durable pieces. The strategy outlined above defines a new approach to coupling river and forest restoration that relies on a "restoration succession" that seeks to restore key processes on the way to achieving restoration objectives rather than attempts to create desired conditions through direct intervention.

Conservation Planning

Many processes and environments of lowland river valleys in the Puget Sound basin are inherently buffered from upstream inputs. For example, increased sedimentation or flooding caused by headwater land uses would not markedly affect floodplain habitats because of the immense flood storage capacity and because increased sediment deposition would concentrate in the channel. For this reason, while watershed restoration professionals often emphasize the need to first restore headwater processes, such as reducing erosion associated with logging roads, prior to undertaking restoration projects downstream, lowland restoration does not necessarily depend on prior headwater restoration. Restoring lowland habitats has unique challenges, such as that posed by the inherent invasibility of riparian habitats (Planty-Tabacchi et al. 1996) combined with the abundance of exotic species in lowland environments.

This inherent buffering from upstream impacts, in combination with the historical abundance and variety of lowland habitats, suggests that these mostly vanished or badly degraded habitats, if restored or rehabilitated, could logically serve as refugia for salmonids, supplemental to disturbance-prone headwater areas that are now the focus of conservation planning (e.g., Frissell et al. 1993; Doppelt et al. 1993).

The Role of Historical Reconstruction

Prioritizing and ordering restoration activities within a watershed and understanding the interactive mechanics of geomorphic processes and riverine habitat and how they differ throughout the region are central to effective restoration planning (see Chapter 9). Historical studies in the Puget Lowland suggest answers to these problems that are contrary to commonly held assumptions about the historical distribution of habitats, the processes that generate them, and how, where, and in what order habitats might best be restored. These counter-intuitive insights support the argument for the importance of undertaking historical analysis early in restoration, rehabilitation, and conservation planning.

It is likely that many of the insights from historical studies in the Puget Lowland are relevant for other regions. For example, wood jams were formerly important to large rivers not only in the Pacific Northwest but also throughout forested temperate regions of the world. Forests and in-channel wood were cleared from the eastern United States and Europe (for review, see Montgomery et al. in press), and a general approach relevant to restoring riverine forests and wood in the Pacific Northwest may have potential for broader application; the same may be true of other facets of reconstructing the Puget Sound's riverine landscape.

Historical landscape studies have application to ecosystems studies and management worldwide. The particular collection of historical materials, methods, and field approaches uniquely useful in the Puget Lowland does not necessarily transfer directly to another environment. For example, in Europe the imprint of agricultural and industrial civilization extends millennia into the past; forest disturbance in Europe has been significant for at least 6,000 years (Williams 2000), and river clearing and engineering date to the Roman era (Herget 2000). Central to European river history are the archival methods of the historian and archaeological approaches unique to the regions' challenges (e.g., Haslam 1991). Nonetheless, while the types of source materials and methods may vary among regions and landscapes, a variety of temporal- and spatial-scale appropriate, cross-referenced and multi-scaled approaches are available for revealing the historical riverine landscape in any environment (e.g., Egan and Howell 2000).

Acknowledgments

Work drawn upon for this chapter was funded by: the Quaternary Research Center, Puget Sound Regional Synthesis Model (PRISM), the Center for Streamside Studies and the Royalty Research Fund at the University of Washington; the National Marine Fisheries Service Northwest Fisheries Science Center; the Seattle District of the U.S. Army Corps of Engineers; USDA Forest Service Pacific Northwest Forest and Range Research Station; King County Department of Natural Resources; the Bullitt Foundation; Skagit System Cooperative; and the Nooksack Tribe. LIDAR data used in Figure 4A is from the Puget Sound LIDAR Consortium. The concept for Figure 3 is from Figure 17.2 by Robin Grossinger in Egan and Howell (2001). Thurston County provided digital photogrammetric elevation data used to create the DEM in Figure 4B. Paula Cooper of the Whatcom County Department of Public Works provided the topographic survey data, and Harvey Greenberg, University of Washington Department of Earth and Space Sciences, created the DEMs used in figures 1 and 4C.

References

Abbe, T.B. 2000. Patterns, mechanics, and geomorphic effects of wood debris accumulations in a forest river system. Ph.D. Dissertation. University of Washington. Seattle, Washington.

Abbe, T.B. and D.R. Montgomery. 1996. Large woody debris jams, channel hydraulics and habitat formation in large rivers. *Regulated Rivers: Research & Management* 12:201-221.

Barnosky, C.W. 1981. A record of late Quaternary vegetation from Davis Lake, southern Puget Lowland, Washington. *Quaternary Research* 16:221-239.

Beechie, T.J., B.D. Collins, and G.R. Pess. 2001. Holocene and recent changes to fish habitats in two Puget Sound basins. In J.M. Dorava, D.R. Montgomery, B. Palcsak, and F. Fitzpatrick (eds.) *Geomorphic Processes and Riverine Habitat*. American Geophysical Union, Washington, D.C., pp. 37-54.

Beechie, T., E. Beamer, and L. Wasserman. 1994. Estimating coho salmon rearing habitat and smolt production losses in a large river basin, and implications for restoration. *North American Journal of Fisheries Management* 14:797-811.

Beget, J.E. 1982. Postglacial volcanic deposits at Glacier Peak, Washington, and potential hazards from future eruptions. U.S. Geological Survey Open-file Report 82-830.

Bilby, R.E., J.T. Heffner, B.R. Fransen, J.W. Ward, and P.A. Bisson. 1999. Effects of immersion in water on deterioration of wood from five species of trees used for habitat enhancement projects. *North American Journal of Fisheries Management* 19:687-695.

Bilby, R.E. and P.A. Bisson. 1998. Function and distribution of large woody debris. In R.J. Naiman and R.E. Bilby (eds.) *River Ecology and Management: Lessons from the Pacific Coastal Ecoregion*. Springer, New York, New York, pp. 324-346.

Bjornn, T.C. and D.W. Reiser. 1991. Habitat requirements of salmonids in streams. In W.R. Meehan (ed.) *Influences of forest and rangeland management on salmonid fishes and their habitats*. American Fisheries Society Special Publication 19:83-138.

Blomberg, G., C. Simenstad, and P. Bickey. 1988. Changes in Duwamish River estuary habitat over the past 125 years. In *Proceedings, First Annual Meeting on Puget Sound Research*. Puget Sound Water Quality Authority, Seattle, Washington, pp. 437-454.

Booth, D.B. 1994. Glaciofluvial infilling and scour of the Puget Lowland, Washington, during ice-sheet glaciation. *Geology* 22:695-698.

Bortleson, G.C., M. J. Chrzastowski, and A.K. Helgerson. 1980. Historical changes of shoreline and wetland at eleven major deltas in the Puget Sound region, Washington, U.S. Geological Survey Hydrological Investigations Atlas HA-617.

Brookes, A. 1996. Floodplain restoration and rehabilitation. In M. G. Anderson, D.E. Walling, and P.D. Bates (eds.) *Floodplain Processes.* John Wiley & Sons, Chichester and New York, New York. pp. 553-576.

Brubaker, L.B. 1991. Climate and the origin of old-growth Douglas-fir forests in the Puget Lowland. In L.F. Ruggiero, K.B. Aubry, A.B. Carey, M.H. Huff (eds.) Wildlife and vegetation of unmanaged Douglas-fir forests. USDA Forest Service General Technical Report PNW-GTR-285:17-24.

Chittenden, H.M. 1907. *Report of an Investigation by a Board of Engineers, of the Means of Controlling Floods in the Duwamish-Puyallup Valleys and Their Tributaries in the State of Washington*. Lowman & Hanford S. and P. Company, Seattle, Washington.

Chrzastowski, M. 1981. Historical changes to Lake Washington and route of the Lake Washington Ship Canal, King County, Washington. U. S. Geological Survey Water Resources Investigations Open File Report 81-1182.

Collins, B.D. and D.R. Montgomery. 2001. Importance of archival and process studies to characterizing pre-settlement riverine geomorphic processes and habitat in the Puget Lowland. In J.M. Dorava, D.R. Montgomery, B. Palcsak, and F. Fitzpatrick (eds.) *Geomorphic Processes and Riverine Habitat*. American Geophysical Union, Washington, D.C. pp. 227-243.

Collins, B.D. and D.R. Montgomery. 2002. Forest development, wood jams and restoration of floodplain rivers in the Puget Lowland. *Restoration Ecology.* 10:237-247.

Collins, B.D., D.R. Montgomery, and A.D. Haas. 2002. Historical changes in the distribution and functions of large wood in Puget Lowland rivers. *Canadian Journal of Fisheries and Aquatic Sciences* 59:66-76.

Cowardin, L.M., V. Carter, F.C. Golet, and E.T. LaRoe. 1985. Classification of wetlands and deepwater habitats of the United States. U. S. Fish and Wildlife Service, Report FWS/OBS-79/31.

Cwynar, L.C. 1987. Fire and forest history of the North Cascade range. *Ecology* 68:791-802.

Debose, A. and M.W. Klungland. 1983. Soil Survey of Snohomish County Area, Washington. U.S. Department of Agriculture, Soil Conservation Service. Washington, D.C.

de Waal, L C., A. R. G. Large, and P. M. Wade. 1998. *Rehabilitation of Rivers: Principles and Implementation.* John Wiley & Sons, New York, New York.

Doppelt, B., M. Scurlock, C. Frissell, and J. Karr. 1993. *Entering the Watershed: A New Approach to Save America's River Ecosystems.* Island Press, Covelo, California.

Dragovich, J.D., D.T. McKay, Jr., D.P. Dethier, and J.E. Beget. 2000. Holocene Glacier Peak lahar deposits in the lower Skagit River valley, Washington. *Washington Geology* 28:19-21, 59.

Dragovich, J.D., D.K. Norman, Jr., R.A. Haugerud, and P.T. Pringle. 1997. Geologic map and interpreted geologic history of the Kendall and Deming 7.5-minute quadrangles, western Whatcom County, Washington. Washington Division of Geology and Earth Resources Open File Report 97-2.

Egan, D. and E.A. Howell. 2001. Introduction. In D. Egan and E. A. Howell (eds.) *The Historical Ecology Handbook: A Restorationist's Guide to Reference Ecosystems*. Island Press, Springer, Washington, D.C. pp. 1-23.

Franklin, J.F. and C.T. Dyrness. 1988. *Natural Vegetation of Oregon and Washington*. Oregon State University Press, Corvallis, Oregon.

Frissell, C.A. and S.C. Ralph. 1998. Stream and watershed restoration. In R. J. Naiman and R. E. Bilby (eds.) *River Ecology and Management: Lessons from the Pacific Coastal Ecoregion,* Springer, New York, New York. pp. 599-624.

Frissell, C.A., W.J. Liss, and D. Bayles. 1993. An integrated, biophysical strategy for ecological restoration of large watersheds. In N.E. Spangenbery and D. F. Potts (eds.) *Changing Roles in Water Resources Management and Policy.* American Water Resources Association, Herndon, Virginia. pp. 449-456.

Galatowitsch, S.M. 1990. Using the original land survey notes to re-construct pre-settlement landscapes in the American West. *Great Basin Naturalist* 50:181-191.

Gunther, E. 1973. *Ethnobotany of Western Washington: The Knowledge and Use of Indigenous Plants by Native Americans*. University of Washington Press, Seattle, Washington.

Harmon, M.E., J.F. Franklin, F.J. Swanson, P. Sollins, S.V. Gregory, J.D. Lattin, N. H. Anderson, S. P. Cline, N. G. Aumen, J. R. Sedell, G. W. Lienkaemper, K. Cromack, Jr., and K. W. Cummins. 1986. Ecology of coarse woody debris in temperate ecosystems. *Advances in Ecological Research* 15:133-302.

Haslam, S.M. 1991. *The Historic River: Rivers and Culture Down the Ages.* Cobden of Cambridge Press, Cambridge.

Herget, J. 2000. Holocene development of the River Lippe valley, Germany: A case study of anthropogenic influence. *Earth Surface Processes and Landforms* 25:293-305.

Hoblitt, R.P., J.S. Walder, C.L. Driedger, K.M. Scott, P.T. Pringle, and J.W. Vallance. 1998. Volcano hazards from Mount Rainier, Washington, Revised 1998. U.S. Geological Survey Open-File Report 98-428.

Interstate Publishing Company. 1906. *An Illustrated History of Skagit and Snohomish Counties; Their People, Their Commerce and Their Resources, with an Outline of the Early History of the State of Washington.* Interstate Publishing Company, Chicago, Illinois.

Kern, K. 1992. Restoration of lowland rivers: the German experience. In P.A. Carling and G. E. Petts (eds.) *Lowland Floodplain Rivers: Geomorphological Perspectives*. John Wiley & Sons, Chichester and New York, New York, pp. 279-297.

Knox, J.C. 1977. Human impacts on Wisconsin stream channels. *Annals of the Association of American Geographers* 67:323-342.

Leopold, E.B., R. Nickman, J.I. Hedges, and J.R. Ertel. 1982. Pollen and lignin records of late Quaternary vegetation, Lake Washington. *Science* 218:1305-1307.

Mangum, A.W. and Party. 1909. *Reconnaissance Soil Survey of the Eastern Part of Puget Sound.* U. S. Soils Bureau, Government Printing Office, Washington, D.C.

McNeil, J.R. 2000. *Something New Under the Sun, an Environmental History of the Twentieth-Century World.* Norton, New York, New York.

Montgomery, D.R., B.D. Collins, J.M. Buffington, and T.B. Abbe. In press. Geomorphic effects of wood in rivers. In S.V. Gregory (ed.) *Wood in World Rivers*. American Fisheries Society, Bethesda, Maryland.

Nelson, J.C., R.E. Sparks, L. DeHaan, and L. Robinson. 1998. Presettlement and contemporary vegetation patterns along two navigation reaches of the upper Mississippi River. In T.D. Sisk (ed.) Perspectives on the land use history of North America: a context for understanding our changing environment. U.S. Geological Survey, Biological Resources Division, Biological Sciences Report USGS/BRD/BSR-1998-0003, pp. 51-60.

Nesbit, D.M. with Contributions from U.S. Coast Survey, S.L. Boardman, Eldridge Morse, and others. 1885. Tide marshes of the United States. USDA

Miscellaneous Special Report No. 7. Government Printing Office, Washington, D. C.

North, M.E.A. and J.M. Teversham. 1984. The vegetation of the floodplains of the Lower Fraser, Serpentine and Nicomekl Rivers, 1859 to 1890. *Syesis* 17:47-66.

Planty-Tabacchi, A.M., E. Tabacchi, R.J. Naiman, C. Deferrari, and H. Decamps. 1996. Invasibility of species-rich communities in riparian zones. *Conservation Biology* 10:596-607.

Plummer, G.H., F.G. Plummer, and J.H. Rankine. 1902. Map of Washington showing classification of lands. Plate 1. In Gannet, H., The forests of Washington, a revision of estimates. U.S. Geological Survey Professional Paper 5, Series H, Forestry, 2.

Radeloff, V.C., D.J. Mladenoff, H.S. He, and M.S. Boyce. 1999. Forest landscape change in the northwestern Wisconsin Pine Barrens from pre-European settlement to the present. *Canadian Journal of Forest Research* 29:1649-1659.

Reeves, G.H., J.D. Hall, T.D. Roelofs, T.L. Hickman, and C.O Baker. 1991. Rehabilitating and modifying stream habitats. In W.R. Meehan (ed.) Influences of forest and rangeland management on salmonid fishes and their habitats. *American Fisheries Society Special Publication* 19: 519-557.

Sears, P.B. 1925. The natural vegetation of Ohio. *Ohio Journal of Science* 25:139-149.

Sedell, J.R. and K.J. Luchessa. 1981. Using the historical record as an aid to salmonid habitat enhancement. In N.B. Armentrout (ed.) *Acquisition and Utilization of Aquatic Habitat Inventory Information*. American Fisheries Society, pp. 210-223.

Sedell, J.R. and Froggatt, J.L. 1984. Importance of streamside forests to large rivers: The isolation of the Willamette River, Oregon, U.S.A., from its floodplain by snagging and streamside forest removal. *Verh. Internat. Verein. Limnol. (International Association of Theoretical and Applied Limnology)* 22:1828-1834.

Simenstad, C.A. and Thom, R.M. 1992. Restoring wetland habitats in urbanized Pacific Northwest estuaries. In: G.W. Thayer (ed.) *Restoring the Nation's Environment*. Maryland Sea Grant, College Park, Maryland. pp. 423-472.

Thorson, R.M. 1989. Glacio-isostatic response of the Puget Sound area, Washington, *Geological Society of America Bulletin* 101:1163-1174.

Triska, F.J. 1984. Role of wood debris in modifying channel geomorphology and riparian areas of a large lowland river under pristine conditions: A historical case study. *Verh. Internat. Verein. Limnol. (International Association of Theoretical and Applied Limnology)* 22:1876-1892.

Turner, N.J. 1995. *Food Plants of Coastal First Peoples*. University of British Columbia Press, Vancouver, British Columbia.

U.S. War Department. *Annual Reports of the Chief of Engineers, U. S. Army, to the Secretary of War* (continues 1907-1944 as: *Report of the Chief of Engineers, U. S. Army*; 1945-1953 as: *Annual Report of the Chief of Engineers, U. S. Army*; 1954-current year as: *Annual Report of the Chief of Engineers, U. S. Army, on Civil Works Activities*), 1876-1906. Government Printing Office, Washington, D.C.

Vileisis, A. 1997. *Discovering the Unknown Landscape: A History of America's Wetlands*. Island Press, Washington, D.C.

White, R. 1992. *Land Use, Environment, and Social Change: the Shaping of Island County, Washington.* University of Washington Press, Seattle, Washington.

White, C.A. 1991. *A History of the Rectangular Survey System.* U.S. Government Printing Office, Washington, D.C.

Whitney, G.G. 1996. *From Coastal Wilderness to Fruited Plains: A History of Environmental Change in Temperate North America from 1500 to the Present.* Cambridge University Press, Cambridge.

Whitney, G.G. and J. DeCant. 2001. Government Land Office survey and other early land surveys. In D. Egan and E. A. Howell (eds.) *The Historical Ecology Handbook: A Restorationist's Guide to Reference Ecosystems*. Island Press, Washington, D.C. pp. 147-172.

Williams, M. 2000. Dark ages and dark areas: global deforestation in the deep past. *Journal of Historical Geography* 26:28-46.

Zedler, J.B., Principal Author. 1996. Tidal wetland restoration: A scientific perspective and southern California focus. California Sea Grant College System Report No. T-038, University of California, La Jolla, California.

5. Anthropogenic Alterations to the Biogeography of Puget Sound Salmon

George Pess, David R. Montgomery, Timothy J. Beechie, and Lisa Holsinger

Abstract

Geologic and geomorphic history and anthropogenic influences have altered the biogeography of Puget Sound salmon, by which we mean their morphological and genetic relationships, abundance, and distribution, from individual habitat units (e.g., pools and riffles) to the regional scale (e.g., North Sound). We focus on habitat isolation and degradation but also discuss the effects of harvest practices, hatchery practices, and the introduction of non-native aquatic species. Habitat isolation and degradation at the habitat and watershed scales have altered the biogeography of Puget Sound salmon at the regional scale because different juvenile Pacific salmon species spatially segregate into different habitats in the same watershed or biogeographic range. Harvest practices have had an impact on overall abundance at the regional scale, and at the even larger Pacific Northwest scale, while hatchery practices have altered Puget Sound salmon abundance and distribution patterns at the watershed and regional scales. Alterations due to non-native species are not as pronounced; however, these do occur at the habitat-unit scale with effects that integrate up to the watershed scale. Biogeographical information can help regional and local salmon recovery efforts by identifying and characterizing large-scale spatial and temporal variability that influence salmon abundance and distribution.

Introduction

Biogeography is the spatial and temporal distribution of organisms as explained by past and present events. In the Pacific Northwest, phylogenetic (morphological and genetic) relationships among salmonid species and the geologic, geomorphic, and anthropogenic history of an area are the two primary influences on salmon biogeography (Wiley 1983). Understanding biogeographic patterns of salmon can lead to insights about historic and current life history and can help identify differences among and variability within species occupying different areas at both large and small spatial and temporal scales. In this chapter, we examine controls on biogeographic patterns of salmon in Puget Sound at several temporal scales—millions of years, thousands of years, hundreds of years, and tens of years, and at several spatial scales—oceanic (Atlantic vs. Pacific), coastal (Pacific Northwest), regional (e.g., Puget Sound), watershed (e.g., Skagit River), reach (e.g., channel and valley segment), and specific habitat units (e.g., pools and riffles). We then use these relationships to address the question of how anthropogenic activities affect(ed) the biogeography of salmon in Puget Sound. To answer this question, we first examine how geologic and geomorphic processes help define the biogeography of salmon in the Pacific Northwest and Puget Sound. We then examine how the distribution and abundance of salmon in the Pacific Northwest and Puget Sound has been altered over the last 150 years. Lastly, we discuss how specific anthropogenic activities within Puget Sound watersheds, such as habitat isolation and degradation, harvest, hatcheries, and the introduction of non-native species, affect salmonid abundance and distribution. We focus our analysis on habitat isolation and degradation and discuss the other three human effects more generally. Our analysis leads us to hypothesize that anthropogenic effects at the habitat and watershed scales cause changes in salmon abundance and distribution at the regional scale, which in turn have altered the biogeography of salmon in Puget Sound.

Large-Scale Geological and Geomorphic Effects on Salmon

Atlantic and Pacific Salmon

Atlantic (*Salmo* spp) and Pacific (*Oncorhynchus* spp) salmon evolved from a common ancestor and are thought to have diverged 15 to 20 million years ago (Ma) during the early Miocene (Figure 1) (Stearley 1992). Pacific salmon differentiated into several species between 20 and 6 Ma (McPhail 1997), and there is strong evidence for sympatric speciation of the Pacific salmon

(Dimmick et al. 1999). Both Pleistocene glaciation and Tertiary marine cooling have been hypothesized as mechanisms for Pacific salmon speciation, but the timing of these events does not correspond well with their Miocene-Pliocene radiation (Thomas et al. 1986). Pacific salmon radiation does coincide with regional mountain building that started during the Miocene to early Pliocene (Montgomery 2000). Many of the major northwestern rivers (e.g., the Columbia, Fraser, and Skeena) that flow east to west are thought to predate this period of uplift (McPhail and Lindsey 1986), and topographic changes would have strongly influenced streams during this uplift. Hence, physiographic changes around the Pacific Rim due to tectonic activity led to topographic change and a diversity of stream types, which in turn may have triggered Pacific salmon evolution (Montgomery 2000).

Continental Glaciation and Salmonid Recolonization of Puget Sound

Continental glaciation during the Pleistocene had a major influence on potential salmon abundance and distribution in the Puget Sound region. Almost three-quarters of the landmass that currently is Puget Sound was covered with ice for approximately 1,000 years (Porter and Swanson 1998). During this

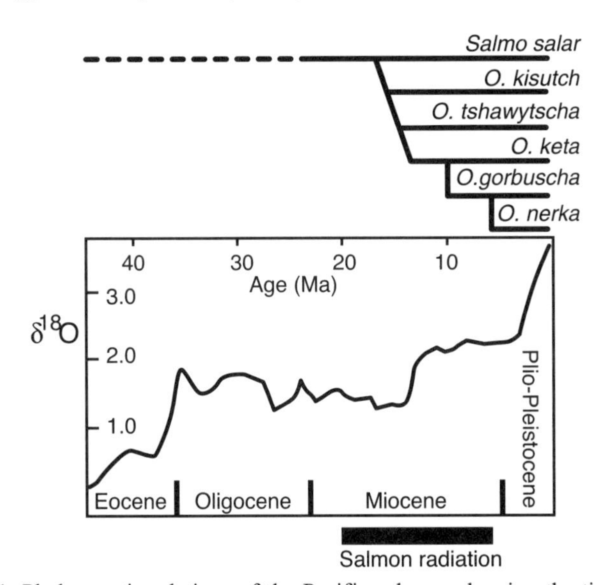

Figure 1. Phylogenetic relations of the Pacific salmon, showing the timing of their radiation 20 to 6 Ma in relation to global marine cooling as revealed from the oxygen isotope record. Based on Stearley (1992) and Montgomery (2000).

time period, pre-glacial fauna in Puget Sound were probably forced into ice-free refuges or died (McPhail and Lindsay 1986), and numerous channel networks were carved by subglacial water, creating distinct valleys or "channelways" (Booth and Hallet 1993). Much of the water from the Puget Sound lobe drained in a southerly direction through the Chehalis River basin (McPhail and Lindsey 1986). Continental ice sheet retreat also led to several processes that shaped the landscape of Puget Sound river basins, including isostatic uplift, valley erosion into glacial sediments, changes in sea level, and valley-burying mudflows (lahars) (Chapter 2). In short, salmon habitat quantity and quality during and after ice sheet retreat varied considerably.

Several hypotheses have been offered regarding the recolonization of Pacific Salmon in Puget Sound. The most well known describes dispersal of salmon from ice-free refuges such as the Columbia River, Chehalis River, and Coastal areas (Figure 2) (McPhail and Lindsey 1986). McPhail and Lindsey (1986) believe the majority of fish fauna dispersing into the Puget Sound region came from the marine environment because most of the fish fauna presently found in Puget Sound are high salinity tolerant species, while only one-third are low salinity tolerant species. Most salmonids are hypothesized to have recolonized Puget Sound rivers through the Chehalis River due to the large drainage connection through the glacial outflow (McPhail and Lindsey 1986).

Many of the Puget Sound salmonid populations, such as chinook (*Oncorhynchus tshawytscha*), steelhead (*O. mykiss*), coho (*O. kisutch*), and chum (*O. keta*) appear to constitute a genetically distinct group (Utter et al. 1989; Marshall et al. 1995; Weitkamp et al. 1995; Busby et al. 1996; Johnson et al. 1997; Myers et al. 1998). Specifically, Puget Sound chinook are genetically distinct from the coastal populations of chinook from California, Oregon, Washington (including the Chehalis), coastal British Columbia, and inland populations including the lower Fraser and Columbia rivers (Marshall et al. 1995; Meyers et al. 1998). One exception is Elwha River chinook, which is more transitional and lies between Puget Sound chinook and coastal populations (Marshall personal communication 2001). Puget Sound steelhead show a similar genetic distinction as chinook (Busby et al. 1996). Puget Sound coho show less genetic distinction than chinook and steelhead with some northern populations such as that from the Strait of Georgia (Weitkamp et al. 1995). However, the data are insufficient to conclude any relationship to Puget Sound coho (Weitkamp et al. 1995). Puget Sound chum also are genetically similar to populations in the Strait of Georgia (Johnson et al. 1997). Hood Canal and Strait of Juan de Fuca summer-run chum salmon are considered distinct from Puget Sound chum due to both genetic and life history traits (Johnson et al. 1997).

Several scientists have hypothesized that Puget Sound salmonids are genetically distinct because they were reproductively isolated from coastal and inland populations, and they subsequently remained within Puget Sound and did not migrate to the Pacific Ocean (Healey 1980; Marshall personal communication 2001). However, the time period for this isolation is unknown. It appears that some of the presently observed genetic diversity for several species already existed during the Pleistocene (Utter et al. 1989), meaning that the most recent glaciation was not the main cause for the genetic distinction.

Other factors predating the most recent glaciation may have had a large effect on genetic diversity. For example, Puget Sound populations of river-type sockeye salmon (*O. nerka*) are genetically more similar to river-type sockeye salmon in rivers more than 2,000 km away (e.g., Chilkat River in northern Southeast Alaska) than to lake-type sockeye salmon within Puget Sound (Gustafson and Winans 1999). This suggests that the common ancestry of river-type sockeye salmon is a greater influence on present-day genotypes than geographic proximity (Gustafson and Winans 1999).

Historic and Current Abundance and Distribution of Salmon

Gresh et al. (2000) estimated that salmon abundance (defined as the number of salmon returning to spawn) in the Pacific Northwest (from Alaska to northern California) has declined 20% to 40% since European settlement. The distribution of salmon during this time period has also changed (Figure 3).

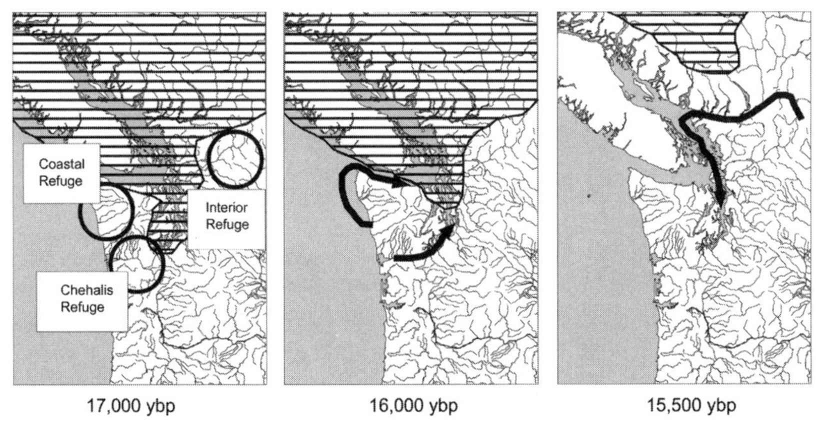

Figure 2. Potential pathways for post-glacial recolonization of salmonids in Puget Sound. Adapted from McPhail and Lindsey (1986). The lined patterns denotes the extent of continental glaciation at different points in time.

Historically, 84% of wild salmon returned to rivers in Alaska and British Columbia, and 16% returned to rivers in Washington, Oregon, Idaho, and California. Currently, 99% of wild salmon return to Alaska and British Columbia, while 1% return to Washington, Oregon, Idaho, and California. In Puget Sound, returns have declined an estimated 92% since the 1850s (Gresh et al. 2000).

This dramatic shift in the distribution and abundance of salmon in the Pacific Northwest, including Puget Sound, has ecological implications that may be proportionally greater than the species decline when the contribution of salmon to the nutrient cycle and stream productivity is considered. Salmon returning to spawn and die are an important source of nutrients, such as nitrogen and phosphorus, which enhance the growth and survival of young salmon and overall stream productivity (Bilby et al. 1998; Wipfli et al. 1998; Gresh et al. 2000). The historic decrease in salmon abundance may represent a larger ecological loss than originally suspected because many of the streams in the Pacific Northwest, including Puget Sound, are naturally nutrient-poor systems (Li et al. 1987).

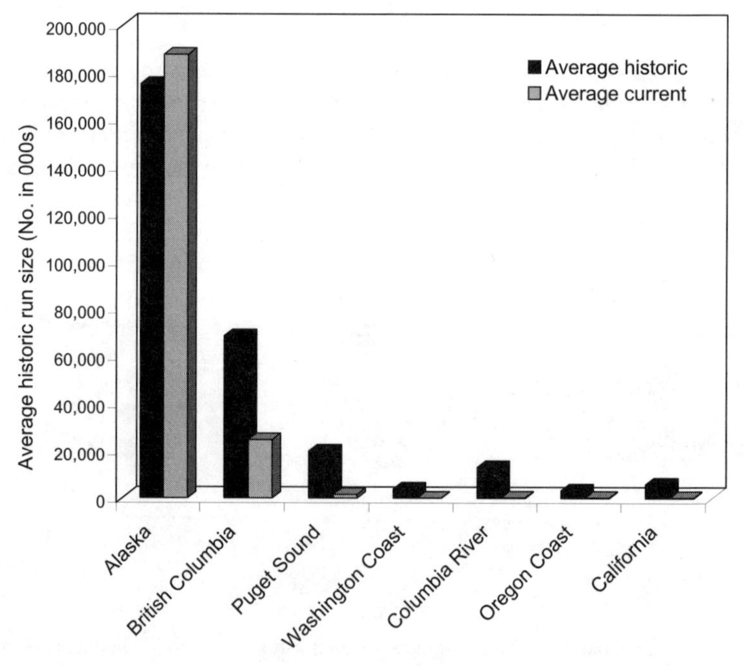

Figure 3. Estimated historic (~1800s) and current (~2000) Pacific salmon run sizes in the Pacific Northwest. Adapted from Gresh et al. (2000).

ANTHROPOGENIC AFFECTS ON THE BIOGEOGRAPHY OF SALMONIDS

Habitat isolation and degradation, the harvest of salmonids, hatchery practices, and the introduction of non-native species over the last 150 years have affected the fish assemblages throughout Pacific Northwest streams (Li et al. 1987). Habitat isolation and degradation in Puget Sound has resulted in the virtual eradication of certain habitat types, such as large freshwater wetland and forest floodplain habitats in the lower portion of river basins (Chapter 4). Other habitats in both the freshwater and estuarine environment, including blind-tidal channels, side-channel sloughs, and beaver ponds, have been reduced to less than 20% of their historic occurrence (Beechie et al. 1994, 2001; Chapter 4). In addition, mainstem and tributary habitats have been lost due to hydropower or fish passage blockages such as culverts, or degraded due to forest practices, land conversion, and stream-cleaning practices that have led to the loss of instream wood and preferred stream-channel types (Beechie et al. 1994; Montgomery et al. 1995, 1999).

Harvest of salmonids and hatchery practices have also altered the abundance, distribution, and phylogenetics of salmonids in Puget Sound (Weitkamp et al. 1995; Busby et al. 1996; Meyers et al. 1998; Gresh et al. 2000). Late nineteenth and early twentieth century harvest practices that focused on the selective removal of older and larger salmonids in gillnet fisheries resulted in an increase in the proportion of small precocious males, or jacks, among the fish that were not caught and returned to spawn (Rutter 1904; Hamley 1975). More recently, trolling gear that captures more fish early in the season when they grow rapidly could also lead to the loss of larger salmonids and could be one reason why the average size of salmonids, such as chinook, has decreased over the last several decades (Ricker 1980, 1981). A decrease in the size of adult salmonids could lead to a shift in the size and age structure of spawning salmonids, which could lead to a greater proportion of smaller, younger spawners (Rutter 1904). Smaller parental size has been shown to result in a significant decrease in offspring survivorship (Helle 1989).

Hatchery practices have altered salmonid populations to the point where the majority of coho and chinook returning to Puget Sound streams are of hatchery origin (Weitkamp et al. 1995; Meyers et al. 1998). This has an effect on the distribution and abundance of these salmonids at the reach, watershed, and regional scale. Hatchery populations have also altered the life history of specific salmonids such as coho (Weitkamp et al. 1995). The spawn timing of several coho hatchery populations has contracted from 10 weeks to 3 weeks, a decrease by more than 50% over a 25-year period of record (Flagg et al. 1995). These population-level changes in hatchery coho may affect the distribution and abundance of naturally spawning populations by hatching earlier, growing faster, and displacing the fry of later, naturally spawning coho (Nickelson et al. 1986). Large outplantings from hatcheries can also force naturally spawned coho fry into less desirable habitats, resulting in lower survival potential (Solazzi et al. 1990).

Introduced non-native species comprise 25% to 50% of the fish species currently found in Washington State (Rahel 2000). Many of these species that have been introduced to the Pacific Northwest and Puget Sound are warmwater species from the eastern United States (Li et al. 1987; Rahel 2000). In particular, the increase in piscivorous (fish-eating) fishes greatly increases the risk of predation to the native fauna and primarily affects the young-of-the-year salmonids and other species, such as sculpins, cyprinds, and catostomids (Li et al. 1987). Non-native species also compete with indigenous species for food and habitat, and they hybridize with existing species such as bull trout (*Salvelinus confluentus*) (Li et al. 1987; Markle 1992).

The degree to which non-native species have affected the biogeography of salmonids varies throughout the Pacific Northwest. For example, Li et al. (1987) found that dams in the Columbia River have created conditions that favor warmwater fishes and have considerably altered fish assemblages and the food web to the point where the fish community now resembles that found in Midwest rivers. In Puget Sound there is less documentation of such a large-scale effect, although there are some smaller scale changes. Fayram and Sibley (2000) found that introduced non-native smallmouth bass (*Micropterus dolomieu*) prey on out-migrating juvenile sockeye salmon from Lake Washington, where 28% to 38% of the bass diet consists of juvenile sockeye. However, they concluded that smallmouth bass predation had little impact on the observed decline in sockeye salmon survival in recent years, because of the relatively small amount of time these two species overlap in areas where smallmouth bass can prey upon out-migrating sockeye.

Habitat Loss and Degradation

How have habitat loss and degradation affected salmonid distribution and abundance? The spatial scales that we focus on include regional (e.g., Puget Sound), watershed (e.g., Skagit River), reach (e.g., channel and valley segment), and habitat unit (e.g., pool, riffle, glide). We believe that anthropogenic effects at the habitat unit scale have caused changes in salmonid abundance and distribution that have effects up to the regional scale.

Biogeographic patterns of salmonids in Puget Sound have been substantially altered by changes in land use since 1850. The most obvious changes are due to removal or isolation of habitats, but changes in physical characteristics of accessible habitats have also changed local abundance of some salmonids. We illustrate these changes with examples documenting habitat change in three Puget Sound river basins (the Skagit River, the Stillaguamish River, and the Snohomish River). By linking these changes to habitat preferences for different species, we show how shifts in availability of habitat types affect distribution and abundance of chinook salmon, coho salmon, and steelhead trout.

There is significant overlap in the ranges of the three species, and life history patterns vary considerably (Groot and Margolis 1991). Coho salmon and steelhead trout generally spawn in smaller and steeper streams, although steelhead trout also spawn in smaller mainstem rivers of the three basins. Juveniles of both species spend their first year or two in freshwater. Juvenile coho salmon show strong preferences for pools and woody debris cover in summer and for off-channel slough or pond habitats in the winter. Juvenile steelhead are more generalized in their selection of rearing habitats, but they are less reliant on woody debris cover and generally avoid ponds or sloughs as winter habitat (Williams et al. 1975). Chinook salmon typically spawn in the larger rivers, and most juveniles migrate to sea soon after they emerge from the gravel (Williams et al. 1975). Estuaries provide an important rearing area during smoltification for all species. Figure 4 illustrates habitat preferences for juveniles of each species.

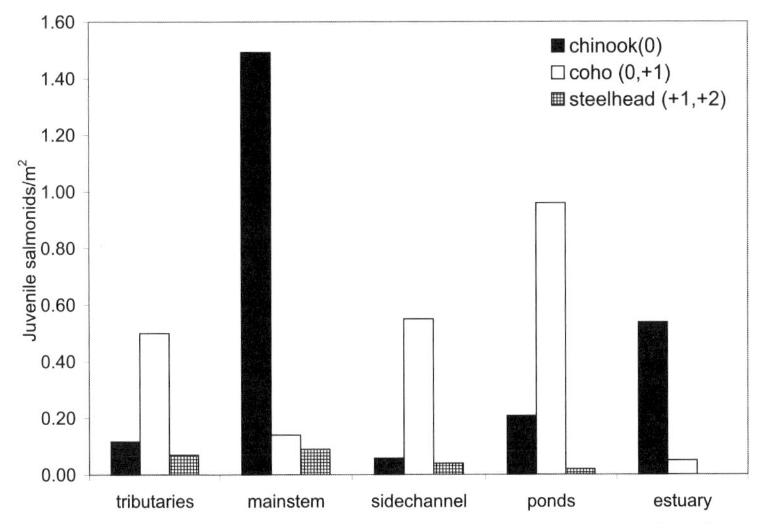

Figure 4. General juvenile salmonid use at the habitat scale. Compilation of over 60 references. Age class 0 denotes juvenile salmonids that have recently emerged or are less than one year of age. Age +1 denotes juvenile salmonids that are greater than one year of age.

Skagit and Stillaguamish Rivers

Historically, the freshwater range of coho salmon and steelhead trout in the Skagit and Stillaguamish Rivers was limited mainly by natural barriers to upstream migration such as bedrock falls. Chinook salmon have shorter migrations and were limited in range by availability of preferred habitats. Upstream migration to the Baker River system (tributary to the Skagit River) has been blocked by the installation of two hydroelectric dams, but access is presently maintained through trapping and hauling operations (Figure 5). Migration into the upper South Fork of the Stillaguamish River was naturally blocked at Granite Falls until 1954, when a fish ladder was constructed to allow anadromous fish into the upper basin.

Most of the salmonid habitat in the Skagit and Stillaguamish River watersheds were historically located in the floodplains and deltas (Beechie et al. 2001). However, intensive land uses such as agriculture and urbanization have also concentrated in these landforms, leading to large losses in off-channel habitats (Figure 6). The majority of distributary slough habitat in the Skagit delta has been removed or made inaccessible by levees, resulting in the loss of more than 75% of the delta habitat area (Beechie et al. 1994). Side-channel slough areas on the floodplain have been reduced by nearly half, and roughly 90% of beaver ponds have been isolated (Beechie et al. 1994, 2001). By contrast, in the Stillaguamish River basin the loss of beaver ponds is by far the greatest habitat loss. These two examples illustrate variation in the types of habitat losses from basin to basin within Puget Sound and also suggest variation in causes of habitat loss.

Habitat changes do not affect all species equally because different species prefer different habitat types. For example, detailed analyses of losses in coho salmon smolt production potential, based on the model of Reeves et al. (1989) show that declines in potential production of juvenile coho smolts are greater than 50% in both basins (Beechie et al. 1994, 2001); however, contractions in coho salmon distribution are focused mainly in the delta and floodplain of the Skagit basin and in floodplain and tributaries (the location of most ponds) in the Stillaguamish basin. These same habitat losses would have different effects on chinook salmon and steelhead trout distributions. The range of chinook salmon in the Skagit river basin may be most compressed in the delta because losses there are large compared to those in mainstem rivers. Compression of the chinook range in the Stillaguamish basin may also be focused in the delta, although the magnitude of effect is likely much lower than in the Skagit due to the smaller amount of historic habitat there (Chapter 4). Steelhead trout appear to depend less on delta habitats than either chinook or coho, so alterations of estuary habitats are less likely to impact their distribu-

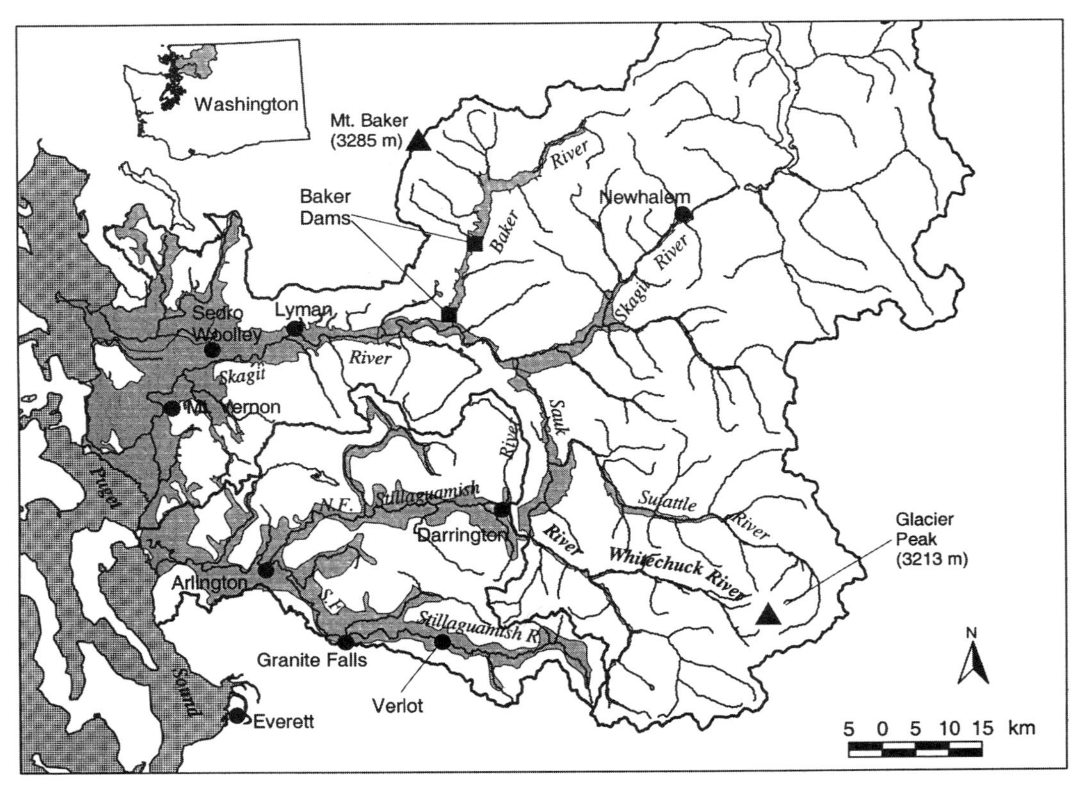

Figure 5. Location map of Skagit and Stillaguamish Rivers.

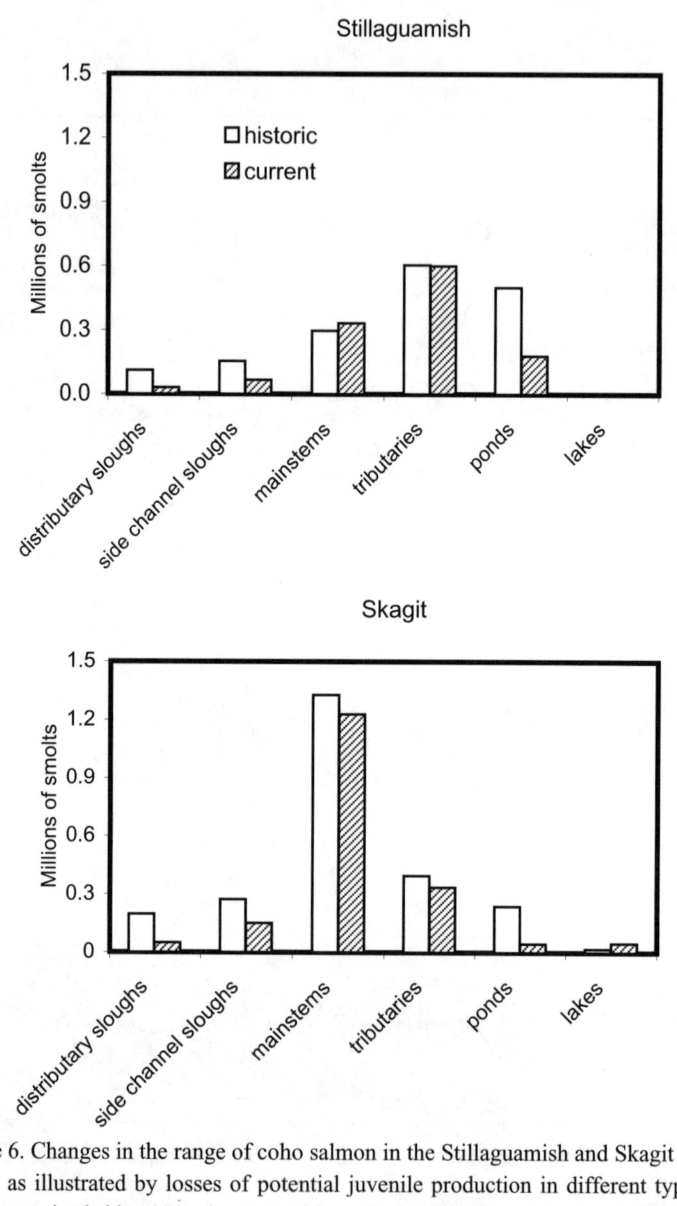

Figure 6. Changes in the range of coho salmon in the Stillaguamish and Skagit River basins as illustrated by losses of potential juvenile production in different types of summer rearing habitat. Note that most of the range compression, and subsequent juvenile production, in the Stillaguamish results in the loss of ponds, which are located in the tirbutaries. The range of Skagit coho rearing has been compressed by both loss of ponds and losses in the river delta and floodplains (adapted from Beechie et al. 2001).

tion. Rather, they are most likely affected by the loss or degradation of tributary and mainstem habitats. Because historic availability of different habitat types and the relative losses of each type vary by river basin, we anticipate that each basin in Puget Sound will exhibit different changes in the distribution and potential production of each species.

Snohomish River

Similar to the Skagit and Stillaguamish Rivers, the majority of salmonid habitat in the Snohomish River basin was historically located in the floodplains and deltas. In the Snohomish River basin, the spread of agriculture has coincided with the destruction and isolation of off-channel and wetland areas over the last 70 to 120 years (Chapter 4). Removal of riparian and floodplain vegetation began in the 1860s, and agriculture was well established in the basin by 1900 (Chapter 4).

Today, the Snohomish River floodplain and neighboring foothills along the channel are located predominantly in rural residential, agricultural, and urbanized areas. Throughout much of the lower portion of the Snohomish River basin, forested and agricultural areas are zoned to allow future development to rural residential, suburban, and urban land use. Even with this extensive change and loss of lowland habitat, more than 75% of the Snohomish River basin is still forested, with the vast majority in upland areas. The combination of extensive habitat loss in the lowlands and relatively intact-forested areas in the uplands provides an interesting and typical description of land-use patterns and habitat conditions in Puget Sound. These land-use patterns may have also led to changes in fish distribution and abundance due to changes in habitat- and reach-scale characteristics, such as the loss of off-channel over-wintering habitat due to floodplain isolation, or the loss of in-channel wood and subsequent reduction in pool habitat quantity and quality (Beechie et al. 1994; Montgomery et al. 1995). Both impacts reduce either the production potential or survival of salmonids (Beechie et al. 1994; Kruzic et al. 2001).

For example, adult Snohomish River coho show a definite trend in salmon abundance by land-use category (Figure 7). Stream reaches in forests support more than 2.5 times more coho than rural, urban, or agricultural streams. Streams in agricultural areas support the fewest salmon, where average weighted fish-days were 4 times lower than the other land-use categories. Differences in relative salmon abundance are observed between hydrologically altered and unaltered wetlands (Figure 8). Hydrologically altered wetlands included those ditched or separated from the stream channel by bank

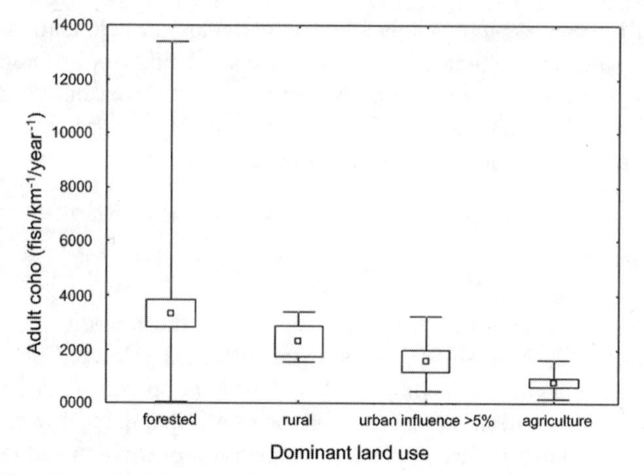

Figure 7. Plot of annual adult coho salmon returns per kilometer of stream length in fish-days by dominant land-use type within 100 m of the stream channel in the Snohomish River basin in northern Washington from 1984 to 1998. Clear squares are mean values, top and bottom of the clear rectangles are the mean value plus one standard error, and horizontal lines at the top and bottom are minima and maxima. Data from Pess et al. (1999). Fish-days were calculated by multiplying the number of live fish observed on each survey date by the number of days between surveys.

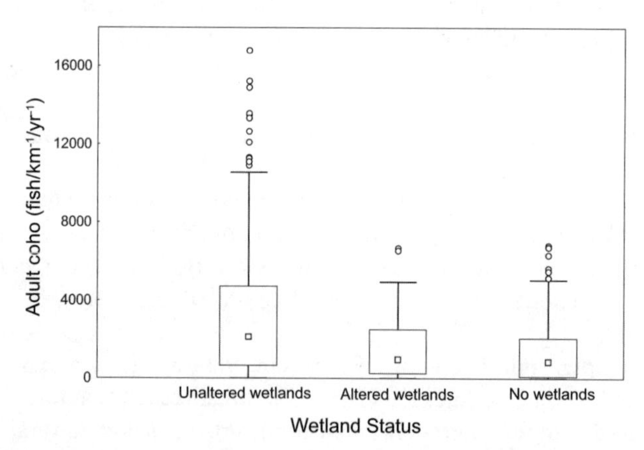

Figure 8. Relationship between wetland status and adult coho salmon per kilometer of stream length per year in fish-days in the Snohomish River basin from 1984 to 1998. Clear squares are median values, top and bottom of the clear rectangles are the 25th and 75th percentile, horizontal lines at the top and bottom are 5th and 95th percentile, and clear circles are outliers. Data from Pess et al. (1999).

armoring or diking. Stream reaches with unaltered wetlands associated with the stream channel (i.e., wetlands present at the reach scale) had adult coho salmon densities 2 to 3 times greater than spawner survey reaches with altered wetlands. Stream reaches with altered wetlands exhibited salmon densities comparable to stream reaches without wetlands. Changing habitat characteristics at the habitat-unit and reach-scale due to land-use effects, such as channel modification, can alter abundance patterns and redistribute salmonids, such as coho, to preferred habitats on a watershed-scale. If the relationships identified above are correct, then an increase in specific land-uses, such as urban or agricultural areas, or a decrease in specific reach-scale habitat characteristics, such as wetlands, can further alter coho abundance and future distribution.

Harvest, Hatcheries, and Non-Native Species

Harvest

Direct mortality from harvest (fishing) has reduced the overall abundance of salmonids at the entire Puget Sound scale. Total harvest rates for Puget Sound chinook, coho, and steelhead, including ocean and terminal fisheries from the 1970s to the 1990s, ranged between 60% and 90% (Weitkamp et al. 1995; Busby et al. 1996; Meyers et al. 1998). During the 1990s, decreased harvest rates for chinook ranged between 15% and 50% and averaged approximately 30% (PSSSRG 1997; Grayum personal communication 2001). Some stocks are targeted for a greater proportion of the total harvest, particularly if they have a large hatchery component, and such targeting practices could result in a change in the distribution and abundance patterns over time, although we did not find evidence for this in the most recent records at the regional scale (Figure 9).

Hatcheries

Hatchery practices have an effect at the Puget Sound, regional, and watershed-scale. Hatchery practices have altered salmonid populations to the point where the majority of chinook and coho returning to Puget Sound streams are of hatchery origin (Figure 10) (Weitkamp et al. 1995; Busby et al. 1996; Meyers et al. 1998). Chinook returning to hatcheries account for 57% of the adult fish returning to spawn (spawning escapement) (Meyers et al. 1998), while the average hatchery contribution rate for monitored coho stocks with

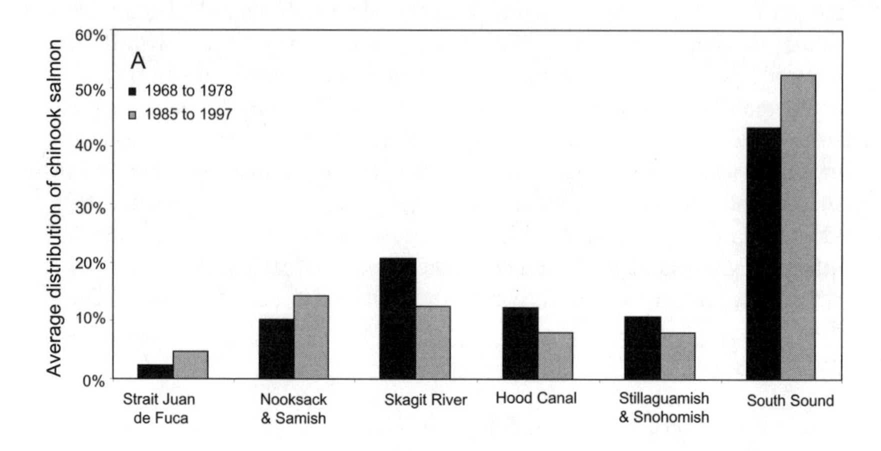

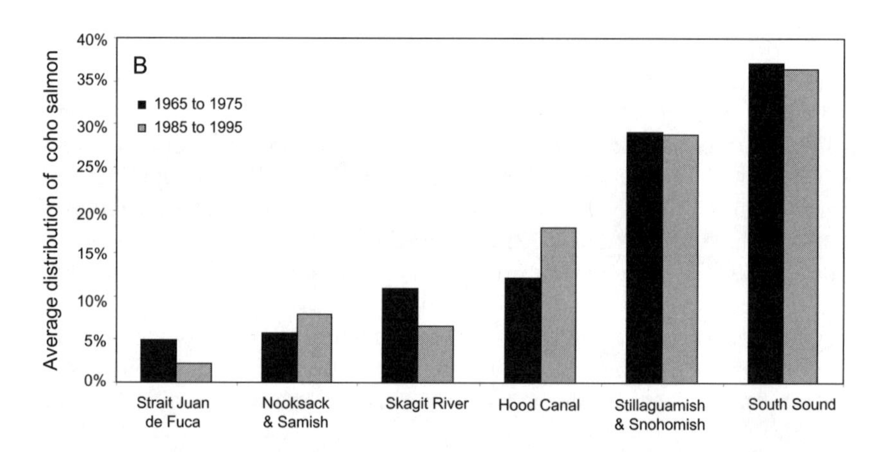

Figure 9. Distribution of harvested chinook (A) and coho (B) by region within Puget Sound from 1965 and 1975 and 1985 and 1997. Run-reconstruction data are for salmon landings and were developed by the Northwest Indian Fisheries Commission. Data is from http://www.nwifc.wa.gov/fisheriesdata/runreconstruction.asp. Regions include the following main rivers: Strait of Juan de Fuca—Dungeness and Elwha Rivers; Nooksack and Samish—Nooksack and Samish Rivers; Skagit River— Skagit River; Hood Canal—Dosewallips, Duckabush, Hamma-Hamma, Quilcene and Skokomish Rivers; Stillaguamish and Snohomish basins—Stillaguamish and Snohomish Rivers; South Sound—Lake Washington, Green-Duwamish, Puyallup, Nisqually, Deschutes Rivers, and Minters and Chambers Creeks.

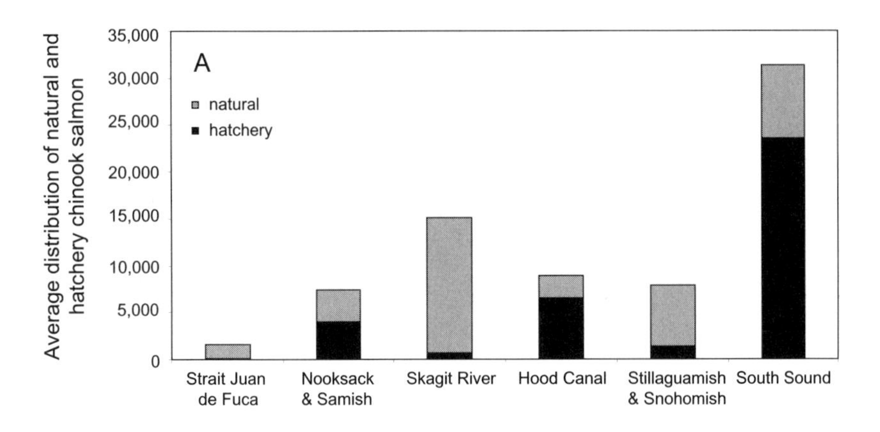

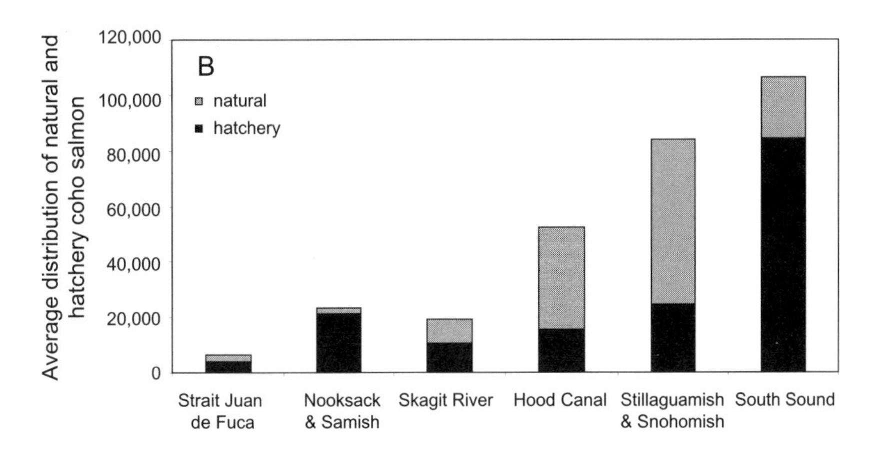

Figure 10. Distribution of natural and hatchery chinook (A) and coho (B) by region within Puget Sound from 1985 to 1995. Run-reconstruction data is for salmon landings and was developed by the Northwest Indian Fisheries Commission. Data can be accessed at http://www.nwifc.wa.gov/fisheriesdata/runreconstruction.asp. Regions include the following main rivers: Strait Juan de Fuca—Dungeness and Elwha Rivers; Nooksack and Samish—Nooksack and Samish Rivers; Skagit River—Skagit River; Hood Canal—Dosewallips, Duckabush, Hamma-Hamma, Quilcene and Skokomish Rivers; Stillaguamish and Snohomish basins—Stillaguamish and Snohomish Rivers; South Sound—Lake Washington, Green-Duwamish, Puyallup, Nisqually, Deschutes Rivers, and Minters and Chambers Creeks.

Puget Sound was 62% from 1981 to 1992 (Weitkamp et al. 1995). The proportion of steelhead spawning escapement derived from hatchery fish ranges between 1% and 51%, with higher proportions of hatchery steelhead occurring in Hood Canal and the Strait of Juan de Fuca (Busby et al. 1996).

Distribution of natural versus hatchery origin salmonids varies by region within Puget Sound. Hatchery influence is greatest for coho and chinook in the most northernly (e.g., Nooksack and Samish) and southerly (e.g., Green-Duwamish, Puyallup, and Nisqually) regions of Puget Sound, while north-central Puget Sound (e.g., Skagit, Stillaguamish, and Snohomish) has less of a hatchery influence. The Strait of Juan de Fuca has more hatchery influence with respect to coho than chinook, while Hood Canal is the opposite. Chinook salmon abundance appears to be closely correlated with hatchery effort, with greater abundance for stocks that have a large hatchery component than for those which have little influence from hatchery programs (Compare Figures 9 and 10) (Meyers et al. 1998).

The overall influence of hatchery salmonids on genetic diversity to Puget Sound has been similar for chinook, coho, and steelhead, because most of the hatchery-introduced species have been a within-Puget Sound transfer of stocks between river basins. Many of the chinook hatchery fish within Puget Sound are inter-watershed stock transfers from the Green River Hatchery, which has been ongoing since the early 1900s (Meyers et al. 1998). As late as 1995, 20 hatcheries and 10 net-pen programs were regularly releasing Green River Hatchery chinook in locations throughout Puget Sound (Marshall et al. 1995). The extensive use of Green River fall chinook has had an impact on among-stock diversity with the South Puget Sound, Hood Canal, and Snohomish stocks, and may have also impacted genetic diversity elsewhere in Puget Sound and the Strait of Juan de Fuca (Marshall et al. 1995) by reducing the genetic diversity and fitness of the naturally spawning chinook populations (Meyers et al. 1998). Only recently have stock integrity and genetic diversity become part of chinook management objectives (Meyers et al. 1998).

Similar to chinook, most coho salmon stock transfers have been derived from within the Puget Sound, although some stocks have been imported from the Columbia River and Olympic Peninsula (Weitkamp et al. 1995). Artificial propagation of coho salmon appears to have had a substantial impact on native, natural coho salmon populations, to the point that the NMFS Biological Review Team had difficulty identifying self-sustaining, native stocks in the Puget Sound (Weitkamp et al. 1995). Most steelhead hatchery transfers to Puget Sound Rivers come from two stocks—one within Puget Sound and the other from the Columbia River (Crawford 1979). Hatchery fish are considered a major threat to the genetic integrity of Puget Sound steelhead (Busby et al. 1996).

Hatcheries have also been identified as a potential factor in the change of the size of Puget Sound salmonids. Between 1972 and 1993, the average size of coho caught in the final fishery before entering their natal streams (terminal landings) has decreased from 4 kg to almost 2 kg, which could result from several causes including hatchery influence (Weitkamp et al. 1995). This size reduction can seriously reduce fecundity and fitness of naturally spawning fish (Weitkamp et al. 1995).

Non-native Species

Non-native species interaction first occurs at the habitat-unit scale and can alter the habitat and trophic interactions of different ecological communities. For example, *Spartina* spp., a cordgrass indigenous to the Eastern United States and the Gulf Coast, can alter mudflats in the Pacific Northwest into salt marshes and subsequently alter primary productivity, food-web dynamics, and overall community composition by out-competing native flora (Feist and Simenstad 2000). Another example of non-native species interaction at the habitat-scale is that between brook trout (*Salvelinus fontinalis*) and juvenile chinook salmon in the Salmon River basin, Idaho. The survival of juvenile chinook salmon was two times greater when non-native brook trout were absent and was positively associated with habitat quality; conversely, when brook trout were present, there was no relationship between habitat quality and juvenile chinook survival (Levin et al. in press). Thus the effects of non-native species interaction with native salmonids masked any effect of habitat quality on juvenile chinook survival. Little is known with respect to the impact of non-native species on salmonids in Puget Sound, but completed studies suggest that the habitat scale is where most of the interaction occurs (Fayram and Sibley 2000).

An example of habitat-scale interactions between non-native and native species that can scale up to the reach or watershed level in Puget Sound is cutthroat trout (*O. clarki*) and coho salmon competition in Huckleberry Creek, a tributary to the Deschutes River in south Puget Sound. Coho salmon are non-native to the Deschutes River, but they now occupy areas throughout the entire watershed (Haring and Konovsky 1999). Cutthroat trout occur in high densities only in those portions of Huckleberry Creek where coho salmon densities are low, which may be due to several reasons, including competition for food or space resources (Figure 11A) (Fransen et al. 1993). Time series data from Huckleberry Creek show that after adult coho salmon passage was blocked by a landslide in February of 1990, the age-0 cutthroat trout population in-

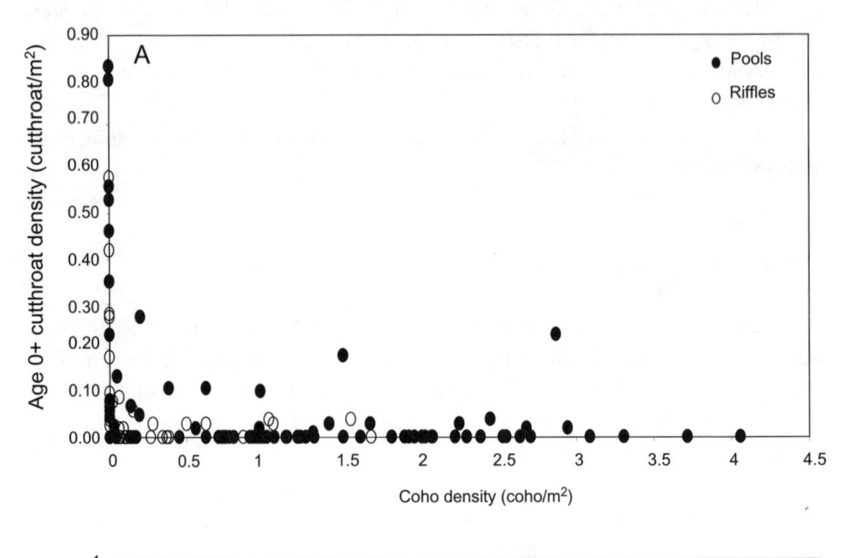

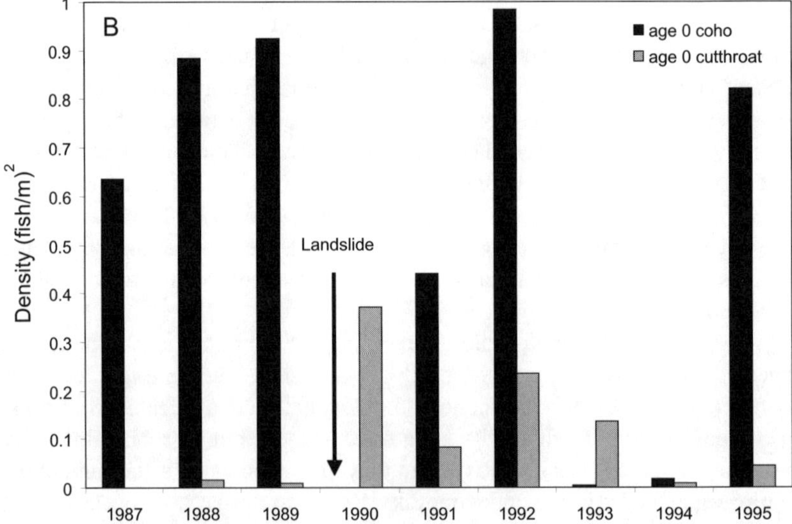

Figure 11. (A) Density of age 0+ cutthroat and juvenile coho salmon for 100 habitat units in Huckleberry Creek, Deschutes River, South Puget Sound. (B) Average density of age 0+ cutthroat and juvenile coho salmon from 1987 to 1995 in Huckleberry Creek, Dechutes River, South Puget Sound. Data from Fransen et al. (1993).

creased dramatically the following spring and remained a more conspicuous component of the fish community for several years afterwards, even though adult coho were moved around the landslide deposit by fall of 1990 (Figure 11B) (Fransen et al. 1993).

Conclusions

Habitat isolation and degradation, the harvest of salmonids, hatchery practices, and the introduction of non-native species over the last 150 years have affected the distribution and abundance patterns of Puget Sound salmonids. Many of these alterations, particularly habitat changes, begin at the habitat scale but are eventually seen at the reach, watershed, and regional scales. Historic availability of different habitat types such as forested floodplain, freshwater wetlands and off-channel habitats, and estuarine habitats, and the relative losses of each type, varies by river basin in Puget Sound. Species-specific preferences on a habitat and reach scale will also mean differences in salmonid distribution and production potential. Together, the effect of habitat-scale loss will depend upon which habitats were available historically and how salmonids evolved to utilize them. Changes in physical characteristics, such as the quality of current instream and off-channel conditions, or the hydrologic alteration of wetlands due to land-use practices, have also changed the abundance of some salmonids.

Large-scale shifts in the biogeography of salmon in the Pacific Northwest seem to correlate with salmon harvest practices. Today a larger percentage of salmonids are found farther north in their range than 150 years ago. Harvest has had more of an effect on abundance in the Puget Sound, while hatcheries seems to have had an effect on abundance, distribution, and phylogenetics. Hatchery fish now comprise the majority of chinook, coho, and steelhead produced in Puget Sound, and they have altered the distribution of salmonids towards the south Sound. Less is known about the effects of non-native species on salmon biogeography, although the documented interactions also seem to occur at the habitat scale.

In combination, the wide variety of natural and anthropogenic influences on salmonid biogeography indicate that a broad level of understanding should underlie salmon recovery efforts. In particular, we need to understand the historic types, abundance, and distribution of both salmon and their habitats, how such distributions have changed to the present, and the nature of the natural processes or human actions that have triggered such changes. Most importantly, the wide range of influences on salmonid biogeography mean that development of restoration (or rehabilitation) goals requires such understanding for the paticular river system of interest and for the species of interest.

References

Beechie, T.J., B.D. Collins, and G.R. Pess. 2001. Holocene and recent geomorphic processes, land use, and fish habitat in two Puget Sound watersheds. In J.M. Dorava, D.R. Montgomery, B. Palcsak, and F. Fitzpatrick (eds.) *Geomorphic Processes and Riverine Habitat.* American Geophysical Union, Washington, D.C. pp 37-54.

Beechie, T., E. Beamer, L. Wasserman. 1994. Estimating coho salmon rearing habitat and smolt production losses in a large river basin, and implications for restoration. *North American Journal of Fisheries Management* 14:797-811.

Bilby, R.E., R.R. French, and P.A. Bisson. 1998. Response of juvenile coho salmon (*Oncorhynchus kisutch*) and steelhead (*O. mykiss*) to the addition of salmon carcasses to two stream in southwestern Washington, USA. *Canadian Journal of Fisheries and Aquatic Sciences* 55:908-918.

Booth, D.B., and B. Hallet. 1993. Channel networks carved by subglacial water: Observation and reconstruction in the eastern Puget Lowland of Washington. *Geological Society of America Bulletin.* 105:671-683.

Busby, P.J., T.C. Wainwright, G.J. Bryant, L.J. Lierheimer, R.S. Waples, F.W. Waknitz, and I.V. Lagomarsino. 1996. Status review of steelhead from Washington, Oregon, and California. NOAA Technical Memorandum NMFS-NWFSC-27. Seattle, Washington.

Crawford, B.A. 1979. The origin and history of trout brood stocks of the Washington Department of Game. Washington State Game Department, Fishery Research Report. pp. 76.

Dimmick, W.W., M.J. Ghedotti, M.J. Grose, A.M. Maglia, D.J. Meinhardt, and D.S. Pennock. 1999. The importance of systematic biology in defining units of conservation. *Conservation Biology* 13:653-660.

Fayram, A.H. and T.H. Sibley. 2000. Impact of predation by smallmouth bass on sockeye salmon in Lake Washington, Washington. *North American Journal of Fisheries Management* 20:81-89.

Flagg, T.A., F.W. Waknitz, D.J. Maynard, G.B. Milner, and C.V. Mahnken. 1995. The effect of hatcheries of native coho salmon populations in the lower Columbia River. In H. Schramm and B. Piper (eds.) *Proceedings of the American Fisheries Society Symposium on the uses and effect of cultured fishes in aquatic ecosystems, March 12-17, 1994. Albuquerque, NM.* American Fisheries Society Symposium 15:366-375.

Fransen, B.R., P.A. Bisson, J.W. Ward, and R.E. Bilby. 1993. Physical and biological constraints on summer rearing of juvenile coho salmon (*Oncorhynchus kisutch*) in small western Washington streams. *Proceedings of the 1992 coho salmon workshop.* Canadian Ministry of Fisheries and Oceans, Nanaimo, British Columbia. pp. 271-288.

Feist, B.E. and C.A. Simenstad. 2000. Expansion rates and recruitment frequency of exotic smooth cordgrass *Spartina alternflora* (Loisel), colonizing unvegetated littoral flats in Willapa Bay, Washington. *Estuaries* 23:267-274.

Gustafson, R.G. and G.A. Winans. 1999. Distribution and population genetic structure of river- and sea-type sockeye salmon in western North America. *Ecology of Freshwater Fish* 8:181-193.

Grayum, M. 2001. Shared Salmon Strategy presentation in Port Ludlow, Washington on January 21, 2001. http://www.sharedsalmonstrategy.org/pl2.htm

Gresh, T., J. Lichatowich, and P. Schoonmaker. 2000. An estimation of historic and current levels of salmon production in the Northeast Pacific Ecosystem: evidence of a nutrient deficit in the freshwater systems of the Pacific Northwest. *Fisheries* 25(1):15-21.

Groot, C. and L. Margolis. 1991 *Pacific Salmon Life Histories*, UBC Press, University of British Columbia, Vancouver, British Columbia.

Hamley, J.M. 1975. Review of gillnet selectivity. *Journal of the Fisheries Research Board of Canada* 32:1943-1969.

Haring, D. and J. Konovsky. 1999. Washington State Conservation Comission salmon habitat limiting factors final report, Water Resource Inventory Area 13. Olympia, Washington.

Healey, M.C. 1980. The ecology of juvenile salmon in Georgia Strait, British Columbia. In W.J. McNeil and D.C. Himsworth (eds). *Salmonid ecosystems of the North Pacific.* Oregon State University Press and Oregon State University Sea Grant College Program, Corvallis, Oregon. pp. 203-229.

Helle, J.H. 1989. Relation between size-at-maturity and survival of progeny in chum salmon (*Oncorhynchus keta*) (Walbaum). *Journal of Fish Biology (Supplement A)* 35:99-107.

Johnson, O.W., W.S. Grant, R.G. Kope, K. Neely, F.W. Wanitz, and R.S. Waples. 1997. Status review of chum salmon from Washington, Oregon, and California. NOAA Technical Memorandum NMFS-NWFSC-32. Seattle, Washington.

Kruzic, L.M., D.L. Scarnecchia, and B.B. Roper. 2001. Comparison of mid-summer survival and growth of age-0 hatchery coho salmon held in pools and riffles. *Transactions of the American Fisheries Society* 130:147-154.

Levin, P.S., S. Achord, B.E. Feist, and B.W. Zabel. In press. Non-indigenous brook trout and the demise of Snake River Salmon: A forgotten threat?

Proceedings of the Biological Sciences (The Royal Society).

Li, H.W., Schreck, C.E., Bond, C.E., and Rexstad, E. 1987. Factors influencing changes in fish assemblages of Pacific Northwest streams. In W.J. Mathews and D. C. Heins (eds.) *Community and Evolutionary Ecology of North American Stream Fishes*. University of Oklahoma Press, Norman and London. pp. 193- 202.

Markle, D.F. 1992. Evidence of bull trout brook trout hybrids in Oregon. In P.J. Howell and D.V. Buchanan (eds.) *Proceedings of the Gearhart Mountain Bull Trout Workshop*. Oregon Chapter of the American Fisheries Society, Corvallis, Oregon. pp. 58-67.

Marshall, A.R., C. Smith, R. Brix, W. Dammers, J. Hymer, and L. LaVoy. 1995. Genetic diversity units and major ancestral lineages for chinook salmon in Washington. In C. Busack and J.B. Shakelee (eds.) *Genetic Diversity Units and Major Ancestral Lineages of Salmonid Fishes in Washington*. Washington Department of Fish and Wildlife Technical Report. RAD 95-02. pp. 111-173.

McPhail, J.D. and C.C. Lindsey. 1986. Zoogeography of freshwater fishes of Cascadia. In C.H. Horcutt and E.O. Wiley (eds.) *The Zoogeography of North American Fishes*. Wiley Interscience, New York. pp. 615-637.

McPhail, J.D. 1997. The origin and speciation of *Oncorhynchus* revisited. In D. J. Stouder, P.A. Bisson, and R.J. Naiman (eds.) *Pacific Salmon and Their Ecosystems: Status and Future Options*. Chapman and Hall, New York. pp. 29-38.

Meyers, J.M., R.G. Kope, G.J. Bryant, D. Teel, L.J. Lierheimer, T.C. Wainwright, W.S. Grant, F.W. Waknitz, K. Neely, S.T. Lindley, and R.S. Waples. 1998. Status review of chinook salmon in Washington, Idaho, Oregon, and California. NOAA Technical Memorandum NMFS-NWFSC-35. Seattle, Washington.

Montgomery, D.R. 2000. Coevolution of the Pacific salmon and Pacific Rim topography. *Geology* 28:1107-1110.

Montgomery, D.R., G. Pess, E.M. Beamer, and T.P. Quinn. 1999. Channel type and salmonid spawning distributions and abundance. *Canadian Journal of Fisheries and Aquatic Sciences* 56:377-387.

Montgomery, D.R., J.M. Buffington, R.D. Smith, K.M. Schmidt, and G.R. Pess. 1995. Pool spacing in forest channels. *Water Resources Research* 31:1097-1105.

Nickelson, T.E., M.F. Solazzi, and S.L. Johnson. 1986. Use of hatchery coho salmon (*Oncorhynchus kisutch*) presmolts to rebuild wild populations in Oregon coastal streams. *Canadian Journal of Fisheries and Aquatic Sciences* 43:2443-2449.

Pess, G.R., R.E. Bilby, and D.R. Montgomery. 1999. Stream-reach and watershed-scale variables and salmonid spawning distribution and abundance in the Puget Sound Region. In R. Sakrison and P. Sturtevant (eds.) *Watershed Management to Protect Declining Species.* American Water Resources Association, Middleburg, Virginia. pp. 397-400.

Porter, S.C. and T.W. Swanson. 1998. Radiocarbon age constraint on rates of advance and retreat of the Puget lobe of the Cordilleran ice sheet during the last glaciation. *Quaternary Research* 50:205-213.

Puget Sound Salmon Stock Review Group. 1997. An assessment of the status of Puget Sound Chinook and Strait of Juan de Fuca coho stocks as required under the salmon fishery management plan. Pacific Fishery Management Council, Portland, Oregon.

Rahel, F.J. 2000. Homogenization of fish faunas across the United States. *Science* 288:854-856.

Reeves, G.H., F.H. Everest, and T.E. Nickelson. 1989. Identification of physical habitats limiting the production of coho salmon in western Oregon and Washington. U.S. Forest Service General Technical Report PNW-GTR-245.

Ricker, W.E. 1980. Causes of the decrease in age and size of chinook salmon (*Oncorhynchus tshawytscha*). Canadian Technical Report of Fisheries and Aquatic Sciences.

Ricker, W.E. 1981. Changes in the average size and average age of Pacific salmon. *Canadian Journal of Fisheries and Aquatic Sciences* 38:1636-1656.

Rutter, C. 1904. Natural history of the quinnat salmon: A report of investigation in the Sacramento River, 1896-1901. *Bulletin of the U.S. Fish Commission* 1902:65-141.

Solazzi, M.F., T.E. Nickelson, and S.L. Johnson. 1990. An evaluation of the use of coho salmon presmolts to supplement wild production in Oregon coastal streams. Oregon Department of Fish and Wildlife Research Report 10. Portland, Oregon. pp. 22.

Stearley, R.F. 1992. Historical ecology of Salmoninae with special reference of *Oncorhynchus*. In Mayden, R.L. (ed.) *Systematics, Historical Ecology, and North American Freshwater Fishes*. Stanford University Press, Stanford, California. pp. 622-658.

Thomas, W.K., R.E. Withler, and A.T. Beckenback. 1986. Mitochondrial DNA analysis of Pacific salmon evolution. *Canadian Journal of Zoology* 64:1058-1064.

Utter, F., G. Milner, G. Stahl, and D. Teel. 1989. Genetic population structure of chinook salmon (*Oncorhynchus tshawytscha*) in the Pacific Northwest. *U.S. Fishery Bulletin* 87:239-264.

Weitkamp, L.A., T.C. Wainwright, G.J. Bryant, G.B. Milner, D.J. Teel, R.G. Kope, and R.S. Waples. 1995. Status review of coho salmon from

Washington, Oregon, and California. NOAA Technical Memorandum NMFS-NWFSC-24. Seattle, Washington.

Wiley, E.O. 1983. *Phylogenetics: The Theory and Practice of Phylogenetic Systematics*. John Wiley & Sons, New York.

Williams, W.R., R.M. Laramie, and J.J. Ames. 1975. A catalog of Washington stream and salmon utilization. Washington Department of Fisheries, v. 1. Puget Sound. Olympia, Washington.

Wipfli, M.S., J. Hudson, and J. Caouette. 1998. Influence of salmon carcasses on stream productivity: response of biofilm and benthic marcoinvertebrates in southeastern Alaska, USA. *Canadian Journal of Fisheries and Aquatic Sciences* 55:503-511.

6. Scientific, Institutional, and Individual Constraints on Restoring Puget Sound Rivers

Clare M. Ryan and Sara M. Jensen

Abstract

The tasks of restoring rivers, watersheds, and other critical habitats are complex and represent some of the most difficult challenges faced by natural resource scientists and managers today. Blame for society's inability to adequately deal with restoration challenges is placed on both scientists and policy makers. Some people argue that the necessary information and levels of certainty fall far short of scientific standards for decision making; others argue that science is not the issue, and indecisiveness merely reflects a lack of political leadership and will. Regardless, the discussion ultimately focuses on the science-policy interface as a root cause of the inability to address such complex management issues. Both science and policy hold unique cultural positions, values, and norms. When the two spheres try to communicate, these differences can interfere with developing, selecting, and implementing management alternatives. This chapter addresses the topic of science and policy communications, offers a conceptual framework to assist in understanding how scientists and policy makers might forge a new dialogue, and discusses additional institutional and individual constraints to restoration efforts. Examples of approaches toward overcoming these constraints provide hope for addressing and ultimately realizing the restoration challenges that lie ahead.

Introduction

"We must manage through sound science, not politics."
"We don't have enough data to make a decision."
"We don't have jurisdiction over this issue."
"It's my property, and I'll do as I damn well please on it."

Each of these statements illustrates the vexing nature of some of the major constraints to the restoration of Puget Sound rivers. In addition to a number of well-recognized institutional and individual constraints, the vast gap between scientific and policy-making cultures and processes present a formidable challenge. This chapter discusses several of the more common scientific, institutional, and individual constraints and presents examples of efforts that are making strides towards overcoming these barriers. These examples illustrate approaches that emphasize the strategies of incentives, innovation, vision, and leadership, each of which is a necessary component to realizing the ultimate goal of restoring Puget Sound rivers. Because much of the blame for lack of progress in current restoration efforts points to the two-way failures in science-policy interactions and communication, the chapter begins with an examination of the differences between science and policy cultures and processes and presents a conceptual framework for a more comprehensive understanding of science-policy communications.

Understanding Scientific and Policy-Making Cultures and Processes

Both science and policy are human constructs, developed over millennia and shaped by our learned and innate behavior. In the United States, we gather knowledge and information through science and make public decisions through policy. Science can come from many sources (universities, government agencies, think tanks) and policy can be considered at many levels (elected officials versus bureaucrats; city councils versus Congress). However, science and policy as a whole hold unique cultural positions, values, and norms. When the two try to communicate, these cultural differences can interfere.

The need for effective communication is important in general for the field of restoration ecology, and in particular for Puget Sound river restoration efforts. Many Puget Sound rivers flow through urban centers, which are located in ecologically important and sensitive areas. As human population grows and urbanizes, its citizens must increasingly understand the role a healthy ecosystem plays and weigh the benefits against other urban needs and desires. "Urban restoration ecology" may be a new term to some scientists, many of whom have traditionally studied more pristine and controlled habitats. It may also be new to urban governments, who may not grasp the relevance of ecology to the urban core. For these reasons, there is a unique set of variables to consider in urban river restoration ecology. However, the underlying relationship between science and policy in this arena remains the same. To understand and address the differences between science and policy-making processes, it is useful to step back and understand how science and policy differ from and fail each other and recognize the limits to the current practice of both science and policy in communicating and solving environmental and ecological problems.

Science and Policy Cultures

To begin with, scientists and policy-makers work in distinct cultures. Science holds an honored but insulated place in American society. As an ideal, science operates in a sphere apart from political, religious, or other concerns, allowing scientists to pursue analytical truth.[1] This separation also largely frees scientists from time pressures (science can wait decades or even centuries for the "truth" to emerge). The system is dispersed and relatively non-hierarchical so one view cannot gain a monopoly of resources. The scientific community controls quality through peer review, by encouraging disputes on analytical grounds, and by limiting entry to those that can set aside years for intensive, specialized training. As a result, science is primarily produced and published for other scientists (Taylor 1984).

In contrast, policy-makers do not undergo a uniform training to enter the occupation. Few come from scientific backgrounds—many are lawyers. The policy community is hierarchical and varied, with participants ranging from elected officials to career bureaucrats, from city council members to state governors. Furthermore, policy-makers face great external pressures from a wide range of actors. They are accountable to voters, special interests, campaign contributors, and superiors. In sum, "policy is not science—it is broader and must accommodate societal concerns within a political arena, not a scien-

[1] This is generally true. However, for many citizens, science is not neutral territory. For every romantic view of science as progress and modern magic, there is a Dr. Frankenstein or Dr. Strangelove (Antypas and Meidinger 1996). The questions of whether scientists and society as a whole can be trusted with their discoveries have been raised with such issues as nuclear power, atomic weapons, and more recently, genetic engineering.

tific one" (Hanley 1994, p. 530). This is a culture of compromise and favors, where timing can make or break a policy.

The two cultures therefore have inherent communication problems. For science, the very culture that produces impressive data hampers its communication with other fields. The insulation that protects scientists can, in some cases, blind them to trends and events in the outside world. The system offers little incentive to publish findings for public consumption. Scientists are more comfortable with both uncertainty and long research time frames and are therefore often unable to translate results without sounding vague to policy-makers (who often equate uncertainty and vagueness). The fragmentation of efforts that protects against monopoly of resources or truth often leads to a narrow research focus, and the related narrow scope of most scientific journals discourages multi-disciplinary research efforts.

The policy culture also creates a number of problems for understanding and incorporating science. The general lack of basic scientific knowledge, lack of awareness of inherent scientific uncertainties, and the relatively short decision-making time frames make it difficult for many policy makers to include science as an input. Even when science enters the policy process, the extreme external pressures on policy, common instances of short-term over long-term rationality, and fragmentation of responsibilities, authorities, interests, and values all dilute and hamper its use (Yaffee 1997). In short, there are many reasons that science and policy have difficulty communicating. The gap between cultures is wide, and few on either side have the inclination, time, and skills to understand the other.

Science and Policy Communication

Despite the formidable barriers, scientists and policy makers do communicate. There are four primary models representing this dialogue (Figure 1).

The One-Way Information Flow Model

In the first model (Figure 1A), science and policy operate in separate spheres and may never directly interact. For example, a scientist researching the effects of global warming on stream temperatures might submit the results to a peer-reviewed journal and consider the project completed. A policy-maker may never see those results and is unlikely to be able to incorporate them into a policy decision affecting Puget Sound rivers. In the same way, a policy maker may look to the scientific realm for information or data on an issue without directly interacting with the scientists involved (e.g., by conducting a literature search, which would not be a direct interaction with a scientist but only an interaction with published studies).

The Trans-Science Model

This model (Figure 1B), developed by Alvin Weinberg in 1972, considers that there are some policy issues that depend on scientific answers, but which science cannot answer. The trans-science sphere in this model represents a hybrid realm of policy-relevant scientific questions, whose resolution requires non-scientific methods (Jager 1998). An example might be the ques-

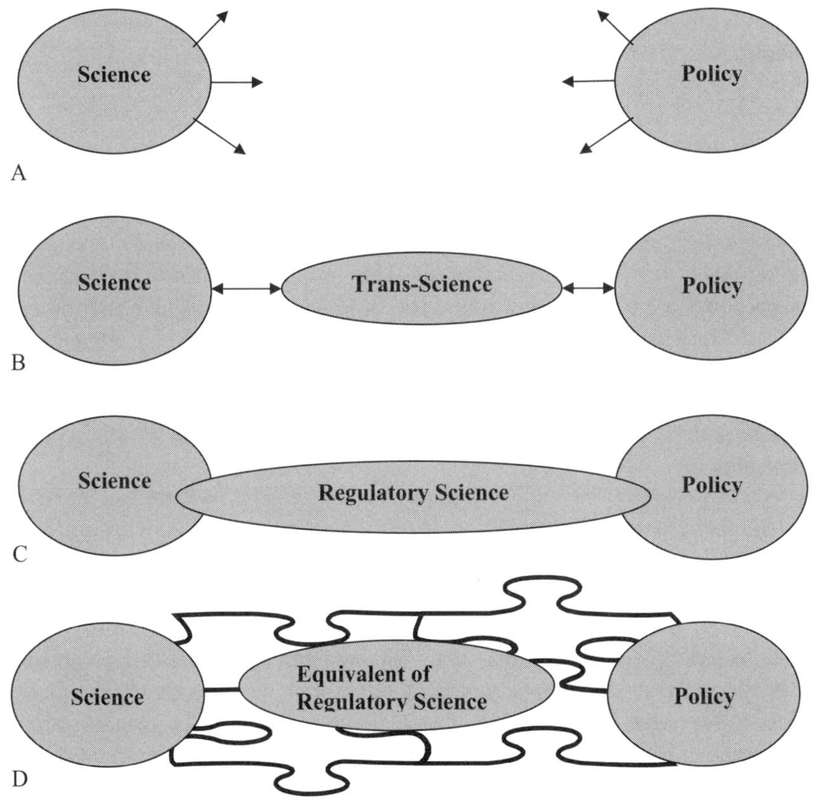

Figure 1. Models of Science-Policy Communications: (A) one-way information flow model, (B) trans-science model, (C) Regulatory Science model, and (D) Mutual construction model

tion of what the potential impacts are of removing one of the dams on the Columbia River to provide access to habitat for endangered salmon. Research in this area is likely to involve models and other estimates of impacts, which yield policy recommendations, but cannot provide a certain answer or prediction. This question would have to be resolved in a political arena, taking into account risks, costs, and other value trade-offs.

The Regulatory-Science Model

The regulatory science sphere in this model (Figure 1C) represents research in which the content and context are driven by specific policy or legally driven questions (Jager 1998). An example is research that supports the quantification and determination of Total Maximum Daily Loads (TMDLs) in water bodies that must comply with the Clean Water Act.

The Mutual Construction Model

In this model (Figure 1D), science and policy actively influence each other in constructing research questions and options. For example, there are many questions as well as various suggestions about which monitoring criteria and approaches are best able to capture the variables of concern in Puget Sound rivers. These suggestions are mutually influenced by the agencies that frame the question and fund such research, and by scientists who offer and respond with monitoring approaches that they hope will be incorporated into policy. Ralph and Poole (Chapter 9) provide an additional example of this type of interaction.

Towards a Conceptual Framework of the Science-Policy Dialogue

Science thus can and does influence environmental policy decisions. However, it is only *one* input among many that feed into policy-making processes. The weight that science plays in policy decisions depends on the issue, the decision-makers, and the stage in the policy process (i.e., agenda setting versus final legislation), among other things. In turn, policy has an influence on science. For example, federal, state, or local funding decisions may have a direct influence on the nature and content of scientific research.

What are some of the possibilities for future dialogue between science and policy? The post-normal science approach is often presented as an emerging field, one that has a chance to democratize science by extending the "peer" community to everyone with a stake in the issue. According to Hellstrom (1996), post-normal science differs from normal science by confronting underlying views, extending problems, including uncertainties, intersecting scientific and policy considerations, incorporating external pressure, and stressing trans-disciplinary research. The approach applies when difficult or complex policy choices must be made on the basis of uncertain scientific inputs (Hellstrom 1996), or when system uncertainties and decision stakes are high (Funtowicz and Ravetz 1993). One element of the approach is that the task of defining what is good and acceptable research moves from the scientific community to the policy and economic communities of interests. Although many argue that there is no difference between post-normal science and other applied scientific efforts, this is a potentially important approach for urban and agricultural ecosystems where the key variables (uncertainties and stakes) are high.

The conceptual framework presented in Figure 2 places scientific and policy processes along two gradients: external pressure and a science-policy continuum. This framework builds on the current literature and adds several additional elements. It purposely does not include the "trans-science" model discussed earlier because, short of a scientific dictatorship, no environmental problem can be solved solely with science. Therefore, *all* policy decisions are trans-scientific.

The Academic Science Sphere

Academic science is the realm of empirical inquiry, theory, and the pursuit of knowledge. Members of this sphere publish results under established systems of peer reviewers and referees to control quality. The scientific cultural characteristics discussed previously are present in this sphere. Within this general science culture, the conceptual framework recognizes three types of scientific endeavor. There is a core science realm made up of knowledge-based researchers pursuing basic science. This group receives the least external pressure on its work. Members of this scientific arena hold authority and status with policy makers. However, a diversity of expert opinion and high levels of uncertainty can diminish their influence on policy.

There is also an issue-based research realm. This group might include scientists researching the safety of new products at a chemical company, EPA toxicologists studying pesticides, and scientist-activists. These scientists subscribe to the importance of peer-review and other attributes of core science, but they differ in that they are working towards a pre-determined agenda

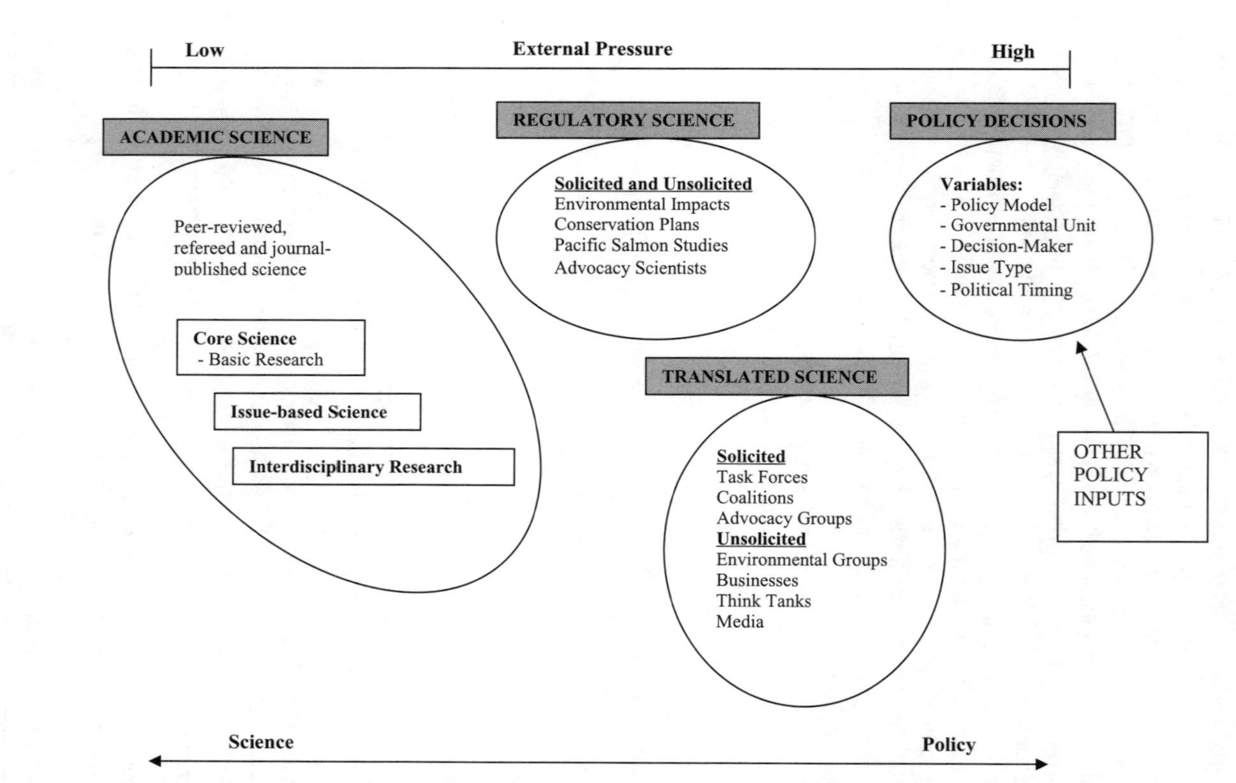

Figure 2. Science and Policy Conceptual Framework. There are two gradients included in the framework: external pressure and a science-policy continuum. The four process ovals are situated in the framework along these two gradients.

(perhaps stated, perhaps not). These scientists may hold authority and legitimacy in some policy arenas but are regarded suspiciously in others. This type of science may be more relevant to policy makers and as a result is exposed to slightly greater external pressure.

The third type of science in this sphere is interdisciplinary research. A large consortium of scientists addresses a particular research question (e.g., social scientists, economists, and ecologists), which has a broader potential audience. Of the three types, this research has the largest scope and most relevance to policy and therefore experiences the most external pressure.

Scientists can and do transfer between these three types of science, and they may fall in two or all three categories at once depending on current research projects. Importantly, information flows not only from these science realms into the policy realm but also from the policy realm into the science realm. Policy can influence all three research types, although basic research tends to be more insulated from outside policy influences. Influences can stem from grants, fellowships, tax codes, legislation on human and animal subjects, and definitions of long-term needs by agencies.

The Regulatory Science Sphere

The concept of a regulatory science sphere is adapted from current policy modeling literature. This sphere is one step away from science and one step closer to policy along the continuum; thus it is subject to greater external pressures than the academic science sphere. Original research prevails, although results may not be peer-reviewed, refereed, or published in scientific journals. Research or information gathering is generally conducted in direct response to a policy need or law, such as environmental impact statements (EIS) and habitat conservation plans. Research in this sphere can be solicited (usually the case for an EIS), or unsolicited (for example, a Sierra Club study of urban sprawl). Data are often directly relevant to policy-makers, but there is a potential for bias and lack of scientific rigor. Information flows in a two-way relationship between this sphere and the other spheres in the model.

The Translated Science Sphere

No original research is conducted in this sphere; rather existing science is translated to policy-makers. As shown by its location on the continuum, science and policy directly influence one another here, as various groups interpret scientific data to compose policy recommendations. This sphere includes task forces and scientific coalitions charged with making a coherent policy recommendation (solicited science). It also includes business and environmental groups that selectively use scientific information to support an agenda (unsolicited science). Members of this sphere essentially interpret science for policy makers. They thus have some of the authority of science as well as the ability to distill a message for policy. This power is diminished, however, by the strong likelihood or suspicion of bias.

The Policy Decisions Sphere

Policy decisions are made in complex ways in a variety of arenas and are subject to the greatest external pressure. The conceptual framework focuses on some of the scientific inputs to policy decisions rather than the mechanisms of those decisions. However, the framework does incorporate several of the important variables that affect these decisions. These variables are listed in the policy sphere and include assumptions about how policy is formulated and implemented (the policy model or theoretical frameworks), the governmental unit, the background and sentiments of the decision maker, the issue type, and political timing. These last two variables are discussed in more detail below.

Issue Type: The particular issue under consideration determines the weight that science will have in the decision and the scientific arena to which the policy makers will turn. For example, Congress could conceivably trust EPA scientists on an issue of toxicology but turn to the core science arena to learn about acid rain. Similarly, scientific considerations might predominate in setting drinking water standards, while politics and business interests drive superfund legislation. Thus the flows in the policy model change with every issue under consideration.

Political Timing: The point along a decision-making process can greatly affect the influence of science on policy and vice-versa. For example, science may strongly influence problem identification and agenda setting, but it may get diluted at later stages of the policy process. Adams and Hairston (1996, p. 28) noted that "By the time a policymaker says, 'Let's call in the experts,' an issue typically has become widely visible and controversial. At that point, resolution of the issue may rely as much or more on politics as on science." On the flip side, before an issue is on the political agenda, scientific processes may be able to proceed relatively free from direct policy influence.

The framework illustrated in Figure 2 attempts to capture information and influence flows rather than physical processes, patterns, or changes. Policy decisions (or non-decisions) are drivers in urban systems. Policy can directly affect land-use patterns, hydrology, climate and atmosphere, and the viability of wildlife. Land-use patterns and hydrologic systems act as secondary drivers for other physical processes. These physical processes in turn provide subject matter for scientists to study, which may ultimately lead to changes in policy. Importantly, policy and science are not simply drivers but are themselves processes, driven by human behavior.

The framework also illustrates that the science sphere is difficult for policy-makers to access, and its answers are usually perceived as vague. What fills the gap between science and policy (translated and regulatory science) can help, but there will likely be concerns about bias, hidden agendas, and scientific rigor. The post-normal science approach discussed previously is promising, but without a set of rules and norms is open to the same problems as the translated and regulatory science spheres. Policy makers are left without clear avenues for solid scientific information.

To have the greatest potential to affect policy, ecologists and other scientists should consider current policy questions when designing their research. Working in interdisciplinary groups may allow for a more comprehensive understanding of the problems and provide unique opportunities for integration of results. If these first two steps are followed, adding humans to the framework should not be a difficult stretch because many ecologists and other scientists are already considering economic and social values as part of their work. Szaro et al. (1998) recommend that future natural resource management include human motivation and responses as part of the system to be studied and managed.

Finally, there is often a lack of avenues for scientists to participate in policy-making processes, with a clear bias from both sides against scientist-activists. Two recent articles discuss this phenomenon within ecology and find similar attitudes, even though they also find a large segment of scientists who feel restricted by the boundaries of the academic science sphere (Brown 2000; Kaiser 2000). Scientists have the same concerns about translated and regulatory science as policy makers, with the additional problems that participating in these spheres can be time consuming, contribute little to or even damage a scientific or academic career, and be potentially frustrating for holders of a minority view.

The conceptual framework raises a number of questions regarding how we as a society ask scientific questions and what might be done to better inform policy-making processes. In the area of restoration, for example, do we have a clear idea of what we are trying to restore? Can we do a better job of framing or anticipating scientific questions in a policy-relevant way?

Reframing research questions and approaches may get us further towards bridging the gap between science and policy. Carefully crafted, applied research can be relevant to policy and still remain in the science sphere. Many issues fall in the realm *between* science and policy, however. For these issues, it is important to create rules, norms, and quality controls for the region between science and policy. Finally, policy makers can help by better educating themselves in the realm of science. If policy makers can grasp what science can and cannot answer, and if government regulatory agencies make hiring scientists a priority, we will also progress along the path towards bridging this gap. Technically trained policy makers can serve as their own scientific translators and thus avoid the potential bias in the inter-science spheres.

While the gaps in science-policy communication seem quite wide, there is hope that they can be bridged. Awareness of these gaps between science and policy cultures and processes provides an important context for examining and understanding the numerous on-the-ground restoration efforts currently underway in the Puget Sound region. Let us now turn to a brief discussion of some of the institutional and individual constraints to restoration of Puget Sound rivers.

Institutional Constraints on Restoration

A number of institutional behaviors constrain our ability to conduct restoration efforts (Yaffee 1997). The following institutional behavioral biases lead to policy impasses and poor choices. Restoration activity in the Puget Sound region exhibits many of the constraints that Yaffee identifies: short-term rationality, competitive behavior, and fragmentation.

Short-Term Rationality

Institutions have a tendency to make decisions that are rational and effective in the short term, yet counterproductive and ineffective over the long term. These usually end up being poor choices and reflect the "tragedy of the commons" idea so prevalent in natural resource management today.

Competitive, Not Cooperative, Behavior

Institutions exhibit a tendency to promote competitive behavior at the expense of cooperative actions, yet often cooperation is needed to identify and implement good solutions. Our institutions operate within a "win-lose" model of decision making. This is changing slowly, but we need to think about incentives for cooperation and integration. It is clear that institutions will soon be forced to begin thinking about performing the "unnatural" acts of cooperating and sharing information and resources to a much greater extent than they have in the past.

Fragmentation of Interests and Values

Interest group politics have become dominant, and restoration involves many groups who do not agree on goals, values, or data. There is a proclivity to split the different elements of society, avoiding the integration of interests and values necessary to craft and sustain effective courses of action on contentious issues.

Fragmentation of Responsibility and Authority

In the Puget Sound region, there are an inordinately large number of institutions that have responsibility and authority for restoration, but very few (perhaps no) incentives to cooperate. Again, we see the tendency to separate the institutions responsible for resource management, there by diminishing accountability and resulting in management strategies that are piecemeal solutions.

Fragmented Information and Knowledge

There is a tendency to fragment what is known about a situation and its context, so that decision makers are hampered and make poor choices because they are operating with inadequate information. This problem was addressed to some extent in our earlier discussion of science-policy dialogues.

To Yaffee's (1997) list, two additional constraints should be added: *Leadership and risk-taking*. Perhaps the lack of leadership and reluctance to take risks are merely logical results of the preceding behavioral biases, but many of the more successful efforts boil down to the simple (yet incredibly difficult-to-implement) ideas of leadership and risk-taking. Certainly in the Puget Sound area we have seen a leadership vacuum reflected in many key restoration questions, such as: Who is in charge? Where are we going? What is the "vision" of restoration that we want and how will we get there (Currens et al. 2000)?

These institutional biases operate at all levels of organizations, and they are extremely difficult to confront and solve. Institutions have a tendency to break things apart and hold them apart, when integration is what is needed. Several of the examples discussed later in the chapter illustrate approaches to changing or intervening in these tendencies.

Individual Constraints on Restoration

Because individuals are an integral component of our institutions, many of the individual constraints faced in restoration efforts are directly related to, and in fact are likely causes of, the institutional constraints. Although there are numerous other societal issues that contribute to our inability to conduct restoration efforts, this section briefly discusses only three individual constraints: private property, litigation, and individual behavior.

Private Property Ethic

The society of the United States exhibits and encourages a strong individual private property ethic, which often does not coincide with the goals and requirements of restoration efforts. Although a system of regulation and enforcement is in place that limits some behaviors and actions on private property (e.g., dredging and filling requirements, building standards), many of the regulations are not directly related to restoration goals. In addition, legal challenges arise in association with attempts to compel individuals to conduct activities on their property that will aid in restoration when such actions may preclude their use of the property for other uses (the "takings" issue). If in fact private landowners are conducting activities on their property that provide or contribute to a larger "public good" (such as restoration), management institutions and policy makers would be wise to explore compensation and incentive schemes such that individual landowners are not bearing the sole burden for contributing to this larger societal benefit. In short, restoration is expensive and society should be prepared to pay for it.

Litigation

The United States is a highly litigious society. In spite of energetic, costly, and lengthy courtroom efforts, litigation rarely resolves the real issues underlying the conflict. We must seek and use alternatives to litigation that will meet the many interests and goals in our restoration efforts. Again, leadership and creative risk-taking in developing solutions has the potential to get us further towards restoration than contentious legal battles.

Individual Behavior

Government institutions cannot do all that is necessary for restoration, and all levels of society will have to participate in some way if we are to achieve this goal. A difficult challenge is the question of how to educate and motivate individual citizens to change their behavior in ways that will benefit restoration efforts. For example, more comprehensive information is needed regarding what land and resource-use practices are appropriate to aid restoration, and how to encourage individual landowners to conduct certain activities on their land or as part of their daily lives. Much work is needed in the area of individual behavior change and on the incentives that can be used to influence individual behaviors.

Realizing Restoration: Incentives, Innovation, Vision, and Leadership

Although it is important to be aware of the constraints, it is also useful to provide examples where many of these barriers are beginning to be addressed (Chapter 7). Below are highlighted just a few examples of the literally hundreds of restoration and enhancement projects going on today in the Puget Sound area. Although it is difficult to identify clear "success stories," the following examples embrace the use of key strategies for realizing restoration—incentives, innovation, vision, and leadership. Several of the examples illustrate promising approaches to overcoming some of the institutional and individual barriers discussed earlier in this chapter.

Incentives in Bear Creek

Bear Creek, in the Lake Washington watershed, is one of the most productive streams for its size in Puget Sound, supporting chinook (*Oncorhynchus tshawytscha*), coho (*O. kisutch*), sockeye (*O. nerka*), and other fish species. Bear Creek continues to support salmon due to 15 years of effort by citizen activists, along with staff from King and Snohomish Counties, the city of Redmond, and involved landowners. All parties have cooperated to protect near-pristine stream banks, extensive wetlands, and major forested upland areas. Much of the stream system is protected because city and county staff worked directly with landowners, offering a variety of tax incentives, money for easements, and regulations. In addition, the Salmon Recovery Funding Board (SRFB) provided $250,000 to match more than $4 million in public and private funds to acquire land and timber rights for one of the largest wetlands in the headwaters of Bear Creek (Ruckleshaus 2000).

Innovation in Chimacum Creek

Chimacum Creek is located on the Olympic Peninsula south of Port Townsend, Washington. Hood Canal summer chum salmon (*O. keta*) have been listed as an endangered species and have been missing from Chimacum Creek for several years. Through a creative and innovative approach of community involvement, landowners in the Chimacum Valley and other Jefferson County citizens have been working to restore the health of the creek and reintroduce chum salmon. The local community has an annual salmon celebration and, some years back, started a treasure hunt to inform people about the area's natural and cultural history (Ruckleshaus 2000). Farmers, artists, tribal members, volunteers, and government officials have been fencing areas to protect the stream, creating small wetlands and other features once part of this natural system. In fall 2000, chum salmon were seen in Chimacum Creek for the first time in more than 10 years (Ruckleshaus 2000).

A Vision in Oregon—River Renaissance

The Willamette River, which runs through downtown Portland, Oregon, was recently designated as an American Heritage River. However, the health of the Willamette watershed has been severely compromised by urban growth and development. The listing of steelhead trout (*O. mykiss*) and chinook salmon, a likely listing of the harbor as a Superfund site, and numerous combined sewer overflows are only a few of its challenges. The city of Portland's Bureau of Planning has as its goal the creation of a community-wide vision and effort to revitalize the Willamette River, its banks, and its tributaries throughout Portland. At the core is a vision that attempts to integrate the natural, economic, urban, and recreational roles that make the Willamette important to the region. The plan attempts to encompass new initiatives and efforts already underway as a way of aligning city work plans and generating opportunities to coordinate efforts and benefits.

Leadership in Puget Sound—The "Shared Strategy"

Puget Sound is currently struggling with institutional and political gridlock and paralysis, which will likely prove to be untenable for the long term. Recently, leaders have emerged in the Puget Sound region in an attempt to overcome the political and jurisdictional barriers associated with restoring Puget Sound rivers, and ultimately, Puget Sound salmon. Specifically, Daniel J. Evans (former Washington State Governor and United States Senator) and William D. Ruckleshaus (currently Chair of the Washington Salmon Recovery Funding Board) gathered together many of the key players involved in restoration to discuss, develop, and initiate a solution to the restoration challenges faced by the region. Attendees included Governor Gary Locke and other state and local elected officials, representatives of federal, state, tribal, and local governments, as well as businesses, environmental and conservation organizations, and watershed groups. Convened in early 2001, the group has the goals of developing a "Shared Strategy" and implementing a proposed regional recovery plan for salmon in Puget Sound. Implementation of the strategy is yet to be seen, but the results from this meeting, which all key parties supported, attempted to create a new institution to assist in reaching the goal of restoration. State, federal, and tribal leaders with significant statutory authority for salmon recovery made a commitment to implementing the strategy, and there appears to be significant political and agency support. However, many people are skeptical of this approach, which may result in the creation of yet another institution to add to the list of numerous existing institutions responsible for the management and rehabilitation of Puget Sound rivers.

Realizing Restoration—Can We Get There From Here?

While rehabilitation and improvement of Puget Sound rivers is certainly possible, these efforts will require overcoming significant scientific, institutional, and individual constraints. In addition, to accomplish our restoration goals, we must have a clear vision of what those goals are along with an understanding and acceptance of the costs and trade-offs required to achieve them.

Scientists, institutions, and other individuals will play critical roles in the effort to achieve the goal of restoration. A major, but not insurmountable, barrier is the difficulty with science and policy communication. The current situation of political and institutional gridlock is not sustainable, and political leadership, innovation, and risk-taking are needed to break the impasse. Together with scientists and other citizens, we must think about going beyond the common view of restoration as restoring only the natural ecosystem and think about how we can "restore" the political and social systems that are necessary to support a comprehensive restoration effort. Finally, political and governmental institutions are not solely responsible for achieving restoration goals. Individual members of society, through their behavior and policy choices, will also be key players in restoration efforts.

In addition, we must accept the premise that restoration is expensive, and we should prepare to pay for it. Re-framing our current thinking about ecosystems, marketing, and education will help us make this leap. If we can begin to think of ecosystems as infrastructure and invest in ecosystems the same way that we do highways, for example, we may be able to make significant progress towards our restoration goals.

Finally, for many ecological restoration questions and policy choices, there are no clear answers. What restoration goals are appropriate and feasible? How much should we spend and who should pay? Should we prioritize drinking water, salmon, or energy? Because various segments of society often value competing priorities differently, in the end decision-makers are faced with extremely difficult policy choices, not scientific decisions, about where to put our limited resources (e.g., transportation, homelessness, restoration, education). These choices require an understanding of what values are traded off in order to achieve a particular policy goal, such as restoration. Science can only take us so far towards realizing the goal of restoration. Our political and social systems, fueled by imagination, ingenuity, and a willingness to take risks must carry us the rest of the way.

References

Adams, P.W. and A.B. Hairston. 1996. Calling all experts: Using science to direct policy. *Journal of Forestry* 94(4):27-30.

Antypas, A. and E.E. Meidinger. 1996. Science-intensive policy disputes: An analytic overview of the literature. People and Natural Resources Program, USFS, PNW Research Station.

Brown, K. 2000. A new breed of scientist-advocate emerges. News Focus. *Science* 287:1192-1195.

Currens, K.P., H.W. Li, J.D. McIntyre, D.R. Montgomery, and D.W. Reiser. 2000. Review of "Statewide Strategy to Recover Salmon: Extinction is Not an Option." Report 2000-1, Independent Science Panel, Olympia, Washington.

Funtowicz, S. and J. Ravetz. 1993. Science for the post-normal age. *Futures* 25:739-52.

Hanley, T.A. 1994. Interaction of wildlife research and forest management: The need for maturation of science and policy. *Forestry Chronicle* 70:527-532.

Hellstrom, T. 1996. The science-policy dialogue in transformation: Model-uncertainty and environmental policy. *Science and Public Policy* 23:91-97.

Jager, J. 1998. Current thinking on using scientific findings in environmental policy making. *Environmental Modeling and Assessment* 3:143-153.

Kaiser, J. 2000. Ecologists on a mission to save the world. News Focus. *Science* 287:1188-1192.

Ruckleshaus, W.D. 2000. Turning on the money spigot to save the salmon. Seattle Post-Intelligencer, July 23, 2000. Page G1.

Szaro, R.C., J. Berc, S. Cameron, S. Cordle, M. Crosby, L. Martin, D. Norton, R. O'Mallay, and G. Ruark. 1998. The ecosystem approach: Science and information management issues, gaps and needs. *Landscape and Urban Planning* 40:89-101.

Taylor, S. 1984. *Making Bureaucracies Think: The Environmental Impact Statement Strategy of Administrative Reform*. Stanford University Press, Stanford, California.

Yaffee, S. 1997. Why environmental policy nightmares recur. *Conservation Biology* 11:328-337.

7. The Politics of Salmon Recovery in Lake Washington

John Lombard

Abstract

How much and where humans consume land and water mean life or death for salmon. For this reason, expanding our approach to growth management is crucial to salmon recovery in Puget Sound in general and the Lake Washington watershed in particular. Growth management as we have known it provides a start toward limiting our consumption of land, but it has not been structured to limit our consumption of water, nor to account for human effects on ecological processes across broad landscapes. Ultimately, it must do all three for salmon recovery in Puget Sound to have a chance of success.

Over the next 50 years, the Puget Sound area's population is projected to double. This will profoundly affect its ecological processes and salmon runs, on a scale comparable to constructing multiple new dams on the Columbia River. To accommodate growth of that magnitude and still provide for salmon recovery, some areas will have to accept increased human impacts to ensure that ecological processes in other areas can continue to support sustainable runs of wild salmon. To begin toward this, a variety of specific actions should be taken soon, involving new approaches to habitat acquisition, land-use regulations, and funding for salmon recovery. They have received little discussion so far, in part because most parties involved have not openly acknowledged the long-term conflicts between salmon recovery and the growth of our population and economy. If long-term salmon recovery is to be possible in Puget Sound, its real requirements must not remain hidden from public debate.

INTRODUCTION

> I thought of the influence of this most impressive scenery upon its future pupils among men... No tameness of thought is possible here....Civilized mankind has never yet had a fresh chance of developing itself under grand and stirring influences so large as in the Northwest.
> —Theodore Winthrop, *The Canoe and the Saddle,* 1853 (Franklin-Ward Company, Portland, OR, 1913, p.210)

In considering efforts today to conserve and recover salmon in Puget Sound and the Lake Washington watershed, it is useful to imagine an environmental historian writing on the subject 50 to 100 years from now. Looking back, what will likely have been the most important issues? At that time scale, the success or failure of individual habitat projects (which receive a great deal of attention in the short-term) pales in comparison to broader, systemic issues. The extent and form of human alterations to the Puget Sound landscape, the ongoing management of harvests and hatcheries, and, if projections are correct, the effects of climate change will dominate nearly everything else. This chapter focuses on the first issue—our alterations of the landscape. Global climate change may fundamentally alter the ecosystems supporting salmon, but those involved with salmon recovery in Puget Sound cannot adequately address it; they can only do their part, and hope that others will do theirs. Over the next 50 years, fishing and hatchery practices will unquestionably have an enormous influence on the fate of salmon. But they are more or less technocratic activities, with a relatively small number of managers responsible for them. No one can tell today how well these managers will fulfill that responsibility. But there is reason for some hope, if only because those responsible have grown increasingly self-conscious in recent years as they have been watched more closely than ever before.

The extent and form of human alterations to the landscape of Puget Sound will produce the local habitat conditions that limit the size and distribution of wild salmon populations. These conditions will largely determine which runs are extinct by 2050. By then, current projections are that Puget Sound's human population will have more than doubled. There are many scientific uncertainties in salmon recovery, but at a gross scale, we know what to expect from such population growth unless we dramatically change the way we use land and water. We know how clearing forests, constructing impervious surfaces, and withdrawing and impounding water affect the ecological processes of our streams and rivers, which in turn create and sustain salmon habitat. We know enough to say that how and where the human population grows will be decisive for salmon in Puget Sound by 2050, let alone 2100.

With that knowledge at its foundation, this chapter considers three subjects:

- Growth management in Washington and its current weaknesses for addressing salmon recovery;
- The organization of salmon recovery planning in the greater Lake Washington watershed, and its similar weaknesses; and
- Strategies for protecting and recovering salmon habitat that can address the ecological processes at risk from growth in the Lake Washington watershed and elsewhere in Puget Sound.

Salmon recovery in the face of projected growth over the next 50 to 100 years requires very difficult choices for our society. The most important of these choices, concerning the large-scale, systemic effects of altering the landscape, are unacknowledged in public debates today. Continuing to ignore them will mean the end of self-sustaining wild salmon runs and numerous extinctions throughout much of Puget Sound by 2050.

The Greater Lake Washington Watershed

For planning purposes, the State of Washington recognizes 62 "Water Resource Inventory Areas" (WRIAs), which mostly follow watershed boundaries. WRIA 8, largely consisting of the greater Lake Washington watershed, may be the most politically and ecologically complex of them all (Figure 1). It is home to thirty general purpose local governments and 1.4 million people—about twice as many on both counts as any other WRIA. Its largest river, the Cedar, provides potable water serving nearly a million people. Lake Washington and Lake Sammamish, the second and sixth largest natural lakes in the state, respectively, break the watershed into distinct ecological areas and contribute complex ecological dynamics that distinguish the watershed from the large majority of WRIAs based on mainstem rivers. The watershed's "plumbing" has been radically altered: Lake Washington was lowered in 1916 and redirected to flow out the new Lake Washington Ship Canal; at about the same time, the Cedar River was redirected to flow into the lake rather than into the Black River (the historic outlet of Lake Washington, which flowed into the Duwamish before entering the Sound). Lowering Lake Washington also lowered Lake Sammamish and drained much of the vast complex of wetlands through which the other large river in the watershed, the Sammamish, meandered between the two lakes. This encouraged local residents to straighten and channelize the Sammamish River for farming, beginning in the 1920s. The U.S. Army Corps of Engineers deepened the river and systematically completed this channelization in the 1960s.

WRIA 8 is home to two salmonids listed under the Endangered Species Act (ESA): Puget Sound chinook (*Oncorhynchus tshawytscha)* and Puget Sound/ Coastal bull trout (*Salvelinus confluentus)*. As is true for much of Washington State, WRIA 8 is also subject to court findings concerning treaty rights held by Native American tribes. *U.S. v. Washington* (506 F. Supp. 187 [1980]) requires compensation to affected tribes if the habitat supporting their fisheries is degraded to the point that fishing can no longer meet their "moderate living needs."

Though salmon recovery in the Lake Washington watershed obviously faces major challenges, optimists can focus on a number of strengths. Current escapement targets (the minimum number of fish that must be projected to return to spawn before harvest is allowed) set by the Washington Department of Fish and Wildlife assume the watershed has the potential to produce more salmon per square mile than any other watershed in the state, even with all of the existing alterations. This is mostly because Lake Washington supports a large introduced sockeye salmon run, planted in the 1930s with stock from the Skagit River basin, and sockeye are a mass-spawning fish that need a lake for rearing. Large contiguous areas of good habitat remain in the Cedar River and Bear and Issaquah Creeks. There is a tremendous amount of local expertise regarding salmon recovery and watershed planning in the area, particularly at the University of Washington, King County, and the City of Seattle but also in the private sector. The electorate of the watershed leans toward making the sacrifices needed to save salmon, although with little knowledge of what the long-term sacrifices may be. The Lake Washington watershed is also the wealthiest in the state and can raise substantial funds for salmon recovery, if its voters and political leaders so choose. Lastly, most of the local governments in the watershed have already been working together to improve salmon habitat for many years. Key elected officials and staff have developed good working relationships and a general grasp of the issues related to salmon recovery, providing a better foundation for discussing difficult political and interjurisdictional choices than most other parts of the state.

Planning for Land and Water Use in the Watershed

Growth management is still relatively new in Washington. The Growth Management Act, originally passed in 1990, mandates and establishes the rules for comprehensive planning in parts of the state expected to grow rapidly, including much of the Puget Sound basin. The first King County Comprehensive Plan that was developed under the act's provisions was adopted in 1994; Snohomish County's plan followed in 1995. Most cities in the WRIA also adopted comprehensive plans in the mid-1990s. For long-term salmon recov-

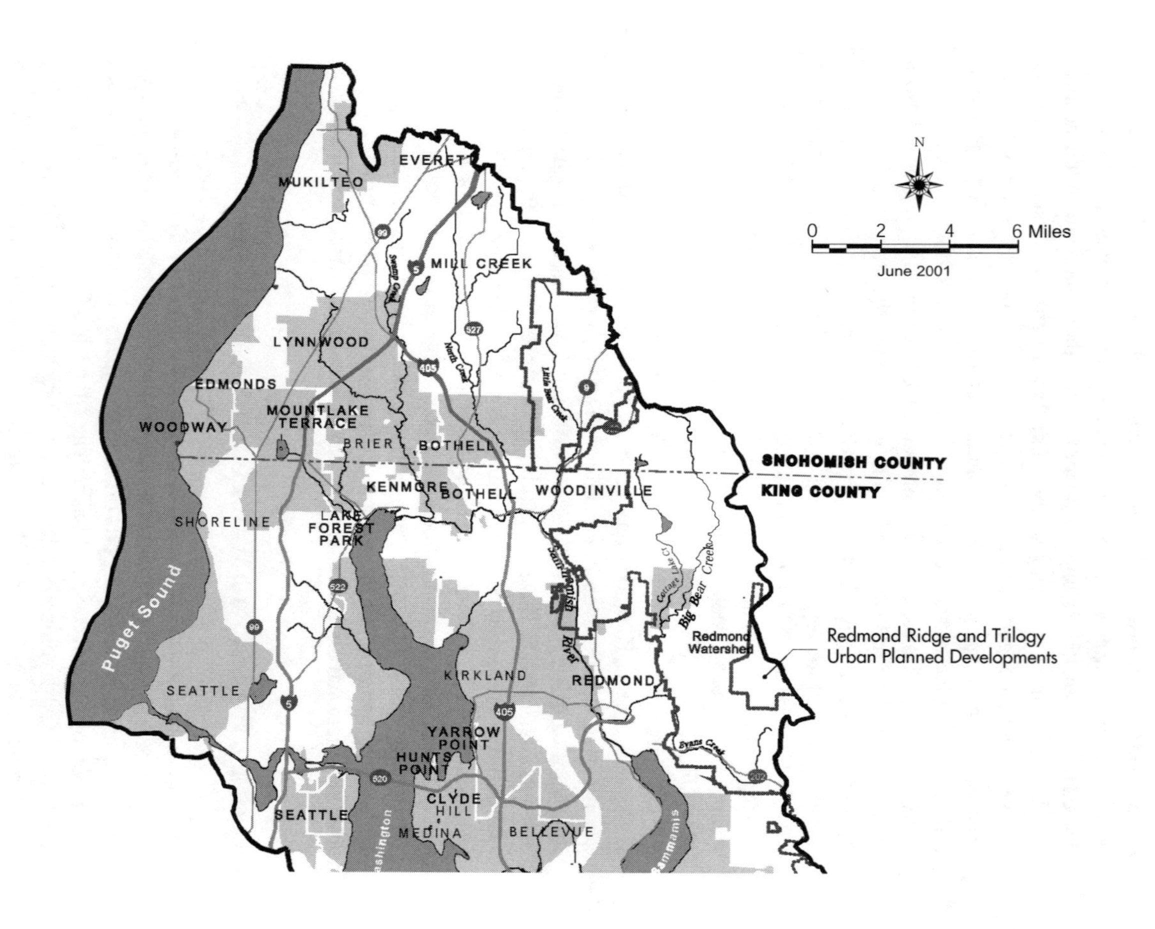
N
0
2
4
6 Miles
June 2001
SNOHOMISH COUNTY
KING COUNTY
Redmond Ridge and Trilogy
Urban Planned Developments
MUKILTEO
EVERETT
MILL CREEK
LYNNWOOD
EDMONDS
MOUNTLAKE TERRACE
WOODWAY
BRIER
BOTHELL
KENMORE
BOTHELL
WOODINVILLE
SHORELINE
LAKE FOREST PARK
Puget Sound
SEATTLE
KIRKLAND
REDMOND
Redmond Watershed
YARROW POINT
HUNTS POINT
CLYDE HILL
MEDINA
SEATTLE
BELLEVUE
Swamp Creek
North Creek
Little Bear Creek
Big Bear Creek
Cottage Lake Cr
Sammamish River
Evans Creek

Figure 1. Map of the Lake Washington Watershed (WRIA 8). Courtesy of King County Water and Land Resources Division.

ery, perhaps the most important provision in these plans sets the boundary for urban growth. As required by the act, this boundary was agreed upon jointly by local governments within each county to last for 20 years. It was set with salmon only partly in mind, but it resulted in the large majority of the three basins that provide core salmon habitat in the Lake Washington watershed (the Cedar River and Bear and Issaquah Creeks) being designated for rural use. The Urban Growth Boundary has received minor alterations already and is expected to be under significant pressure to move outward before its scheduled expiration. From the start, school districts and large churches have sought to build in the rural area, where they can purchase large parcels of undeveloped and relatively inexpensive land. Placing limits on the size of these churches and schools has long been a contentious political issue in King County government. Meanwhile, some cities, such as Redmond (much of which is in the lower Bear Creek basin), have grown much faster than projected, and are already approaching their 20-year targets. Across King County, new residential construction from 1995 through 1999 was 23% greater than planned. Unincorporated urban areas—often located at the edge of the Urban Growth Boundary—have grown especially fast. Overall, the population of the greater Lake Washington watershed is expected to grow from 1.4 million people in 2000 to 1.7 million in 2020, based on projections from the Puget Sound Regional Council.

In summary, though implementation of the Growth Management Act led to a substantial set of laws in the Lake Washington watershed intended to limit the amount of land developed to urban and suburban densities, the current limit was not set to support long-term salmon recovery. It is intended to last only to the middle of the next decade, and it is under increasing pressure already.

In contrast to land use, the Growth Management Act did not require that planning for water be coordinated across jurisdictions. Regional water planning for human consumption has proceeded under the state's law for coordinated water supply planning and through the City of Seattle, which supplies much of the Lake Washington watershed from its reservoirs on the Cedar River and the South Fork of the Tolt River (in the Snoqualmie River basin). Ecological aspects of water withdrawals in the state have largely been governed by the Washington Department of Ecology, which in 1979 stopped granting new rights for water withdrawals in WRIA 8.

The only existing state law that authorizes planning for water across ecological and human demands and across jurisdictions is RCW 90.82, the Watershed Planning Act. Planning under its guidelines is not mandatory, but the state provides incentives to participate, including some funding and automatic state recognition of some of its outcomes. The act defines the scale for watershed planning to be at least one WRIA. Though it has provisions for salmon recovery and habitat planning, the act's primary focus is on establishing water supplies for minimum instream flows and human consumption. It calls for a multi-stakeholder oversight body similar to the one that is overseeing salmon recovery planning in the Lake Washington watershed. However, the act establishes specific processes for how that new body must be formed and how its actions are to be approved, which differ from the existing Lake Washington body. A number of Indian tribes in the state (including the Muckleshoot Tribe, which has treaty rights to fish in the Lake Washington watershed) have objected to various provisions of the act. Because of tribal opposition, the focus on water supplies, and restrictions on how the oversight body operates, the leading governments in the Lake Washington watershed—King and Snohomish Counties, Seattle and Bellevue—have decided against using the Watershed Planning Act in WRIA 8, despite the incentives offered by the state.

Planning for regional water use has always tended to be long-term, because creating major new municipal water supplies is extremely difficult and expensive. Seattle's system has dominated past planning for water supply in the watershed, but as the region has grown the parties involved have also grown. The Central Puget Sound Water Suppliers Forum, which includes large water utilities, utility associations, and county governments in Snohomish, King, and Pierce Counties, was created in 1998 to coordinate long-term planning for water supply across the three-county area. From the start, it has been aware that it must take the Endangered Species Act and the needs of fish into account in planning for future supplies. However, without use of the Watershed Planning Act, there is no accepted process in the region to prioritize where withdrawals should take place or to guide the conservation, re-use, and distribution of water to meet the needs of the human population and the environment. Recognition is growing, however, that such a process is necessary. The Governor's Office has developed a four-year plan to make substantial reforms in water law, which may in time provide a new legal framework for such a process (either as an amendment of, or an alternative to, the Watershed Planning Act). However, even if these reforms are successful, the state would be authorizing a process; it would be up to local governments to carry it out. In the Lake Washington watershed and the Central Puget Sound area, most likely the state would attempt to recognize some process already taking shape by the time a new law is passed.

In summary, the region has shown a greater recognition of the long-term conflicts between salmon recovery and human consumption for water than it has for land, but it currently lacks a mechanism to put that recognition into action.

Growth Management to Become Ecosystem Management?

Nowhere in the Growth Management Act does the word "ecosystem" appear. The goals of the act include conserving fish and wildlife habitat, but only within a larger goal of encouraging retention of open space, developing new parks, and increasing access to natural resource lands and water. The act also includes a goal of protecting the environment but relates this only to "the state's high quality of life, including air and water quality, and the availability of water." While forward-thinking for its time, the act is insufficient to support long-term salmon recovery, largely because it does not address ecological processes and how human activities affect them. However, its goals of reducing sprawl, conserving fish and wildlife habitat, and protecting the environment are critical to long-term salmon recovery. Moreover, the act places these goals in a larger context of societal goals, including affordable housing, efficient transportation, cost-effective public services, and a healthy economy. Salmon recovery will not be politically sustainable if it is seen to be in fundamental conflict with these goals. Salmon recovery also will not be sustainable if it is viewed as distinct from the larger benefits (economic, aesthetic, spiritual, and otherwise) provided by functional watersheds, even in heavily urbanized areas. The Growth Management Act is a satisfactory beginning, but it needs to be amended for salmon recovery. It needs to incorporate "ecosystem management," a phrase that was just being defined when the Growth Management Act was adopted.

Today, ecosystem management goes by many definitions. Most, however, incorporate the following core concepts, which form the basis for how the Growth Management Act must be amended for salmon recovery. Ecosystem management acknowledges humans as part of the ecosystem. It focuses on managing human interactions with ecological processes over broad landscapes, at least at the scale of major watersheds, such as Lake Washington's. Its approach is underlain by two fundamental beliefs: (1) ecosystems are resilient, within limits; and (2) if natural processes can at least be approximated within an ecosystem, the structure and function of habitats within it will generally be maintained over time and will support a diversity and distribution of species adapted to it.

Science informs ecosystem management by identifying the limits of an ecosystem's resiliency—the degree to which its natural processes can be altered before they no longer can support some approximation of a natural diversity and distribution of habitats and species. Just what approximation of "natural" is acceptable, or whether it is worth taking the actions necessary to provide it, are value choices. They are not for science to decide. However, it is critically important for the success of ecosystem management that scientists make the consequences of different choices clear. From a societal standpoint, salmon recovery is not necessarily more important than economic development or affordable housing. But given the strength of political interests tied to economic development and affordable housing, salmon recovery will almost certainly lose by default if it is not built into debates over those issues with a full appreciation of what is at stake.

If the Growth Management Act is eventually amended to incorporate ecosystem management, one of the most important questions for the state legislature to consider would be the scale at which ecosystem management should be applied. A major determinant of this would be the scale of ESA listings. Based on the listings of chinook and bull trout, the Lake Washington watershed should be viewed as part of the Puget Sound ecosystem. Ecosystem management would probably be applied by the state at that level, with direction to local governments to apply it further at the larger level of major watersheds, and probably to smaller areas as well.

Salmon Recovery Planning in the Lake Washington Watershed

Given past history, if ecosystem management is to be integrated into the Growth Management Act, it will likely be after King County has begun applying it. King County articulated growth management concepts in its first comprehensive plan, which it adopted 26 years before passage of the Growth Management Act, in 1964. Elected officials in King, Snohomish, and Pierce Counties have been far in front of the governor and other state elected officials in developing a response to ESA listings. Therefore, in considering the prospects for ecosystem management in Puget Sound and the State of Washington, it is worth looking closely at how salmon recovery planning is currently proceeding in King County and the Lake Washington watershed.

In WRIA 8, local governments developed an agreement in 2000 that provides the decision-making structure and funding base for salmon recovery planning (referred to as "the Agreement" below). A similar agreement was approved in WRIA 9, the Green/ Duwamish watershed, which covers most of the rest of the heavily populated parts of King County. The goal of the Agreements is to develop, by 2005, a customized, multi-species plan that, together with related implementation agreements, can protect and enhance the habitat necessary to support salmon recovery. Local governments want implementation of this plan to be the basis for the federal government to grant them long-term exemptions from "take" prohibitions under the ESA.

The Agreement establishes the following four key bodies for governance and decision-making in each WRIA:

- *WRIA Forum*: The ultimate approval authority within the watershed for the salmon recovery plan. Its membership consists of one elected official from each local government that signs the Agreement. The Forum also adopts the annual scope of work and budget for watershed planning and coordination and appoints the following three bodies: WRIA Steering Committee, WRIA Management Committee, and the Service Provider.
- *WRIA Steering Committee*: The multi-stakeholder, multi-jurisdictional body responsible for developing the salmon recovery plan to be sent to the WRIA Forum for approval. Its membership includes representatives of environmental and business groups, state and federal agencies and key elected officials. Under the Agreement, all participating governments have now authorized this committee to develop the WRIA salmon recovery plan. The membership of the Steering Committee complies with the state Salmon Recovery Act, which allows it to apply for state and federal funds for habitat projects and research on behalf of the WRIA.
- *WRIA Management Committee*: A body that oversees the work of the Service Provider, made up of five delegates from the governments on the WRIA Forum.
- *Service Provider*: The organization hired to provide staff support to the Steering Committee and the Forum. At the direction of the Forum, Steering Committee, and Management Committee, the service provider is responsible for facilitating production of the salmon recovery plan and related documents. The current provider is King County, working with Snohomish County in that part of WRIA 8 within its overall jurisdiction.

In evaluating whether the Agreement provides the institutional structure to support ecosystem management, the following questions are especially important. Does the Agreement:

- Allow decisions affecting large-scale ecological processes to be made across the appropriate landscape?
- Integrate scientific knowledge of ecological processes into those decisions, using an appropriate time scale?
- Integrate regional decisions on related issues (e.g., land use, water resources management, and transportation) so that they can support ecological goals (including salmon recovery)?

The answer to the first question is a qualified yes. The Agreement allows decisions affecting large-scale ecological processes to be made across the entire WRIA. This is not a sufficient scale for salmon recovery, but it is as large a scale as is practically possible for the governments involved, and it can be built on. The Agreement accepts the practical necessity that no one government can be in charge if the salmon recovery plan is to be implemented widely. Within their jurisdictions, the 28 cities and two counties in the WRIA determine land-use regulations and zoning, fund and oversee construction of most habitat projects, and deliver most education and public involvement programs. The Agreement provides a mechanism for the governments to coordinate their actions for common goals. The large number of autonomous local governments in the WRIA is, without question, an obstacle to getting a strong salmon recovery plan approved. Every significant step requires an extraordinary level of coordination and the likelihood of some compromise. Even when good proposals are brought forward with skillful diplomacy, they can be blocked by poor political relations between affected governments that often have little or nothing to do with the substance of salmon recovery. Still, these problems are inherent in the WRIA's circumstances; no single agreement could be expected to overcome them.

With respect to the second question—how well does the Agreement bring scientific knowledge of ecological processes to bear on its decisions—the Agreement is weak, but could be strengthened. Currently, it is silent on the scientific and technical resources needed for the plan. It provides essentially no funding for technical work and never addresses how science should relate to the WRIA salmon recovery plan. The Agreement does not even mention the WRIA Technical Committee, though the committee (comprised of scientists participating from key jurisdictions and stakeholders) has existed since mid-1999 and is presumed by all parties involved to be providing the technical foundation for the 2005 salmon recovery plan. Negotiations for the Agreement eliminated funding for technical staff, based partly on objections to cost. This had the unintended consequence of keeping the Technical Committee from being accountable to the WRIA-level bodies established by the Agreement. All technical work for the WRIA salmon recovery plan is proceeding through the voluntary actions of individual governments.

In its initial products, the Technical Committee has emphasized the importance of preserving or rehabilitating natural ecological processes as the best means of protecting or enhancing salmon habitat. It has, therefore, supported one of the foundations of ecosystem management. But the Technical Committee has not yet detailed the actions needed to accomplish this, particularly in the face of the projected growth of the human population. Ultimately, these actions will require value choices for the WRIA Forum to consider. Integrating scientific understanding into these choices will require clarifying the Technical Committee's role in salmon recovery planning, which in turn will require amendment of the Agreement.

Lastly, with respect to the third question—how well does the Agreement integrate regional choices on issues relating to salmon recovery—the Agreement is a start, but it has much further to go if it is to provide the basis for ecosystem management. The Growth Management Act led to the establishment of two separate bodies in the WRIA, one county-wide Growth Management Planning Council in each of King and Snohomish Counties, to decide interjurisdictional issues for regional land-use planning. Interjurisdictional decisions for transportation planning across King, Snohomish, Pierce, and Kitsap Counties are made by the Puget Sound Regional Council. Thus, to implement major decisions on land use and transportation, the WRIA Forum would need the agreement of these existing regional bodies. As discussed above, there is no existing body in King or Snohomish Counties to make interjurisdictional decisions for regional water planning in the WRIA. The WRIA Forum may therefore find it even more difficult to implement decisions on water than on land use and transportation.

In summary, the institutional structure to apply ecosystem management in King County and WRIA 8 does not now exist, but the Agreement for salmon recovery planning is a move in that direction.

Allocating Scarce Resources: Money, Land, and Water

In considering the prospects for salmon recovery in Puget Sound over the next 50 to 100 years, the likely ecological effects of the area's population and economic growth are comparable to building multiple new dams on the Columbia River. Just as is true of eastern Washington, the economy and social organization of Puget Sound are based on the disruption of ecological processes that support salmon. One important difference between the two is that, in contrast to new dams on the Columbia, there really is no choice for Puget Sound about accepting growth. Perhaps growth in the region can be slowed, but the question is not *whether* the Puget Sound area will have analogs of new dams, but how many, and where? As long as the area continues to use ever greater amounts of water and land for its growing population and economy, improvements in how it uses them are likely to be on the order of improvements in the artificial passage of salmon around new dams. By and large, no one talks about removing Puget Sound's existing counterparts to Columbia River dams. No one seriously proposes eliminating the Lake Washington Ship Canal, raising Lake Washington to its historic level, and re-routing its outlet through downtown Renton.

Speaking practically, planning for salmon recovery must focus on where the greatest ecological improvements can be made at the least economic and social cost. This involves considering where natural ecological processes are the most intact, or where they can be approximated or restored most cost-effectively. Watershed planning groups, such as the WRIA 8 Steering Committee, are already engaging in these debates. But these debates must also be held across Puget Sound. They are most contentious, of course, when considering where resources should *not* be directed. Local advocates are quick to charge that their areas are being "written off"—which in some cases they are, for some purposes. As a practical matter, large investments of funding, severe restrictions on land uses, and attempts to re-create historic habitat processes cannot be done everywhere, either within or across watersheds. This would be true even without the doubling of Puget Sound's population in 50 years. But to say an area should not be targeted for major protective or restorative actions is not to say that it should be abandoned. Some problems in highly altered or urbanized parts of a watershed may need to be addressed systemically to ensure the long-term survival of salmon in the watershed as a whole. Moreover, all parts of an ecosystem have some effects on the whole. Areas not targeted for major salmon recovery actions may still be important locally and be supported as such. In the Lake Washington watershed, nearly all urban streams have been receiving local funds and attention, largely at the urging of local residents. This benefits the WRIA as a whole, so long as the improvements are not at the expense of areas of greater ecological importance.

The Lake Washington watershed itself may someday be deemed unworthy of limited resources for salmon recovery across the state or Puget Sound. It will never be as important as the Skagit, Stillaguamish, and Snohomish watersheds in any serious attempt to conserve and recover wild Puget Sound salmon. The Lake Washington watershed is not unimportant to Puget Sound salmon recovery, but with the human population of the Puget Sound basin projected to more than double over the next 50 years, difficult choices concerning the allocation not only of public funds but also of land and water will be necessary to ensure that ecological processes in key places within Puget Sound can support sustainable runs of wild salmon. Given that so much political and economic power is concentrated in the Lake Washington watershed, it is uncertain whether deciding that Lake Washington is not a priority for salmon recovery across Puget Sound is politically tenable. Under the ESA or tribal treaties, it is also unclear how such a decision would bear on the area's legal responsibilities. Practically, however, the fate of salmon in the Lake Washington watershed depends above all on whether the people who live and work there are willing to accept locally greater density of development, significant restrictions on water use, and the higher costs of living necessary to protect the habitat for sustainable runs of wild salmon in the face of continued growth. If they are not, then as a practical matter the area should not be a priority for the limited resources available for salmon recovery.

Near-Term Actions and Long-Term Strategies for Lake Washington

In Lake Washington, the WRIA Steering Committee approved a "Near-Term Action Agenda for Salmon Habitat Conservation" in spring 2002 to guide actions until the full WRIA Salmon Recovery Plan is completed in 2005. It focuses on habitat preservation and restoration projects, but also provides some broad, relatively uncontroversial guidance regarding regulatory, programmatic, and policy issues. The very fact that the Agenda was "Near-Term" made it easier to delay difficult recommendations—the difficult work was seen as still ahead. But there is an irony in this. Recommendations based on the best science generally would not have focused on restoration projects, which many people in the watershed see as uncontroversial and the essence of near-term action. As discussed in other chapters of this book, restoration projects are often of questionable ecological value. Instead, recommendations based on science would generally have focused on the larger issues that affect the watershed's ecological processes: where will growth go, how much will there be, and how will it use land and water? The consequences of these larger issues are fairly well-established scientifically; politically, though, these are the most controversial issues for salmon recovery planning.

In prioritizing among near-term actions, the Steering Committee has generally taken a politically and ecologically safe course, concentrating on conservation of the best remaining habitat in the watershed, largely through acquisition. This has also generally been the recommendation of the state Salmon Recovery Funding Board (SRFB, commonly referred to as the "Surf Board"), which allocates state and federal funds for habitat projects. It is an intelligent interim strategy. But its long-term success relies on protecting the ecological processes that support this publicly acquired habitat, which is mostly riparian. This means protecting the landscapes of the basins in which the habitat is located. If the basins are cleared and developed, public ownership of riparian areas will be of vastly reduced benefit.

If the Steering Committee and the WRIA Forum wanted to pursue the most important near-term actions to support long-term salmon recovery in the watershed, several efforts could have been initiated that do not require detailed scientific study. These include new approaches to habitat acquisition, land-use regulations, and regional funding to support salmon recovery. These approaches are important wherever population and economic growth come into serious conflict with the needs of salmon, and so they will be relevant at some point throughout most of Puget Sound.

New Approaches to Habitat Acquisition

Neither the SRFB nor the WRIA Steering Committee has yet strategically considered the long-term effects of population and economic growth in their evaluation of habitat projects. The Steering Committee has considered the threat of development or timber harvest on specific parcels proposed for acquisition, and therefore has prioritized some higher than merited by their habitat value alone. However, even the Steering Committee has not yet considered acquisition of properties that are not used by fish at all, which have important strategic effects on how and where future development takes place, thus affecting the ecological processes that create and sustain good salmon habitat. Such acquisitions might be the smartest long-term investment available in the Lake Washington watershed.

These acquisitions might avoid the cost of outright land purchases, and need not bar all development. They might purchase only the right to develop rural properties beyond their existing zoning—i.e., they could keep a parcel that is currently zoned one house per five acres at that density for perpetuity. They might also include limits on forest clearing and impervious surfaces. Such acquisitions could be very effective in reducing pressures to move the Urban Growth Boundary in key places.

New Approaches to Land Use

The long-term importance for salmon of where and how growth occurs should also affect regulations for new development, particularly in our most urbanized areas. In contrast to the present approach of the National Marine Fisheries Service (NMFS), the effects of new development in urban basins should not be compared against "properly functioning conditions," which could not be achieved there without massive social dislocation (if at all). New regulations based on that approach will significantly increase the cost of development in urban areas, which will increase pressure to develop rural areas—the worst possible result for salmon. New regulations should consider not just the immediate effects of particular developments but also their substitution effects—i.e., what those developments might do to salmon habitat and ecological processes if located elsewhere in the region, particularly in rural areas. This does not mean eliminating protective regulations for urban streams; the condition of urban streams can and should be improved. But improvement does not require restoration of "properly functioning conditions." NMFS and local governments should be allied in wanting new development to occur in urban areas as much as possible.

Land-use regulations should encourage experimental developments to test drainage and forest retention practices in urban areas, which could dramatically reduce the ecological damage done by stormwater (Chapter 11). Such developments typically require exceptions to existing building code requirements for parking, street widths, sidewalks, roof types, and other issues. They do not provide conventional lot sizes or lawns, but they may provide large areas of shared, forested open space. Though they have been suggested for a few pilot projects, they have not been fully implemented anywhere in the region. If maintaining the existing Urban Growth Boundary for much beyond 2014 is not realistic, the ability of the region to maintain its salmon runs in the face of future growth may depend on the widespread implementation of these new development styles.

New Approaches to Funding Salmon Recovery Actions

Near-term recommendations must also address the fact that existing funding sources for the conservation and recovery of salmon habitat do not meet the need. It is not merely that they are unable to provide enough money; they simply are not structured to do the job. In King County government, the two main funding sources devoted to salmon recovery are utility fees for wastewater treatment and for surface water management. Legally, utility fees must be used to benefit ratepayers. Funds received by the wastewater system are, with only small exceptions, supposed to be spent on the transmission, treatment, and discharge of wastewater. The King County surface water management fee is supposed to serve unincorporated areas, which comprise only a relatively small portion of the 700-square-mile Lake Washington watershed. The only existing source of cross-jurisdictional funding for salmon recovery in the watershed is the assessment for the King Conservation District, which raises about $600,000 a year across the WRIA—a miniscule amount compared to the need. Worse, the District assessment is not collected and cannot be spent in the 15% of the watershed in Snohomish County.

The watershed's elected officials have debated creating a new charge for salmon recovery that could be collected and spent across the watershed or one county (including incorporated areas). This would probably require a new chapter in state law. Existing legal authorities for new taxes and fees to benefit salmon recovery are essentially limited to property taxes, either through a bond issue or by exceeding the normal maximum tax rate set by state law. Both actions require a public vote. Surcharges on water or wastewater utility rates, if legal, would need to be separately adopted by more than 40 utilities in the watershed and probably could not be spent outside of their individual boundaries, severely limiting their capacity to provide the most benefit to salmon recovery. Increased surface water utility rates would face similar difficulties. Worse, the state statute that authorizes surface water utilities may not even provide a sufficient basis to collect funds specifically devoted to salmon recovery rather than drainage. Given the systemic problems with existing legal authorities, and the dilemma that salmon are generally most abundant where local tax revenues are the smallest (i.e., the areas with the fewest people), it may be time for a new chapter in state law.

Should the region pursue such a course, it should consider taxes and fees that do more than just raise new revenues. Conceived strategically, taxes and fees can reward good stewardship of the environment and penalize poor stewardship. Surface water charges based on impervious surface could reward dense development and penalize sprawl by taking lot sizes into account. (Today, a downtown condominium and a five-acre rural mansion are both categorized simply as residential properties for purposes of surface water fees.) Additions to the gas tax and motor vehicle charges could recognize that the majority of our impervious surface is necessitated by cars, not people. Pesticides and herbicides that are not banned outright could be given large surcharges, as could the sale of turf. In contrast, new construction in built-out neighborhoods could qualify for tax breaks, as could new subdivisions that incorporate large-scale preservation of trees and topsoil. Such a system of taxes, fees and incentives could have great educational value, in addition to raising funds and influencing behavior.

Scientists Must Help Focus the Debate

Broadly speaking, scientists working in salmon recovery advise three types of "clients": regulating agencies, regulated entities, and the public. Regulating agencies (the federal services responsible for administering the ESA, other permitting agencies, and the more diffuse group of tribes and state and federal agencies concerned with the maintenance of tribal treaty rights for harvest) have sizable scientific bureaucracies, which are working to define what recovery must involve and what actions are needed to accomplish it. Regulated entities (the private sector, local governments, state and federal agencies in their non-permitting capacities, and the consultants that support all of them), also have sizable scientific bureaucracies, which are attempting to advise their employers how to respond to the regulations relating to salmon. All of the regulators and regulated might accurately say that they do not have sufficient scientific resources to do these jobs as well as they would like, but their collective demands have pushed existing resources to the limit. There are only so many qualified scientists to go around.

The public, though, has largely been left out. When a democracy makes clear value choices, the public can expect to be served by scientists and engineers in government, who can focus on accomplishing those choices. In the 1960s, there was no need for scientists and engineers outside of NASA to figure out how to get to the moon. But this was a broadly supported value choice. It was a natural extension of our technological society. According to opinion polls, salmon recovery is also a broadly supported value choice. But it is in conflict with much of our technological society. The polls do not reflect an understanding of what salmon recovery actually requires, especially in combination with the population and economic growth expected for our region. With some exceptions, the public's lack of understanding is shared by the media and our elected officials. As a consequence, scientists working on salmon recovery in government constantly receive mixed messages or are assigned irreconcilable objectives. They have difficulty serving the public, because the public is difficult to serve—its desires are in conflict, due to both denial and ignorance.

Under these conditions, scientists can provide the greatest public service by bringing these conflicts to light. Much of what is needed to describe the conflicts, such as the effects of development and land clearing on ecological processes and salmon habitat, is well-established in scientific literature. With sufficient attention and resources, scientific organizations, such as the Society for Ecological Restoration or the American Fisheries Society, can answer many critical questions that today no formal body is even asking. These include:

- What are the likely consequences of filling the existing Urban Growth Boundaries for salmon and the ecological processes that support them?
- What are the likely consequences of extending those boundaries to include population projections for 2025 or 2050, given current densities and mitigation techniques?
- What are the likely consequences of such growth for the investments in salmon habitat we are making now, or may make in the near future?

THINKING AT THE RIGHT TIME AND SPACE SCALES

> The social structures and institutions that have been operating in the Pacific Northwest have proved incapable of ensuring a long-term future for salmon, in large part because they do not operate at the right time and space scales. — National Research Council, *Upstream: Salmon and Society in the Pacific Northwest*, 1996 (National Academy Press, Washington, D.C., p. 4)

Surely the right time scale for salmon recovery planning is at least 50 years—a short period ecologically, but probably near the practical maximum allowed by our fast-changing human society. Regularly thinking at least 50 years ahead would force us to ask essential questions about larger historical forces, such as the relation of ecological processes to habitat structures and functions, and, in turn, the effects of human population and economic growth on those processes. The time scale of our thinking should also influence its spatial scale. Local governments naturally think mainly about problems within their own jurisdictional boundaries. Regional thinking for issues such as transportation rarely goes beyond a 10- to 15-mile radius from a jurisdiction's boundaries. But thinking 50 years ahead, and considering the larger historical forces affecting salmon, the appropriate spatial scale for Puget Sound salmon recovery is the Sound's entire basin.

When the Growth Management Act was passed it represented a major advance for planning across time and spatial scales in Washington. But planning for urban and rural land uses on a 20-year time scale, county-by-county, is not enough to save our wild salmon runs. We must move toward ecosystem management, integrating planning for water and land use, at the appropriate time and spatial scales. One of the special challenges this poses is that the traditional place for leading such change in the state, King County, may ultimately need to play a lesser role in this process. Its heavily urbanized ecosystems likely will not be the most important for the success of long-term salmon recovery across Puget Sound. Its people, however, will have to decide what vision of success they want for their immediate area. Given their size and wealth, they will also play a large role in determining the vision for all of Puget Sound. In addition, they may pioneer new strategies to protect ecosystem processes against the effects of population and economic growth. Given the intense pressures of growth at the center of Puget Sound's population and economy, it is possible that those strategies will fail in their own area. But the strategies discussed in this chapter are relevant to any area facing major conflicts between human development and functioning ecosystems. Within 50 years, this will likely describe most of Puget Sound below the Cascade and Olympic mountains. The strategies may have a better chance elsewhere. They must, however, be started somewhere if wild salmon are to remain part of this region's natural heritage.

8. Role of Watershed Assessments in Recovery Planning for Salmon

Timothy J. Beechie, George Pess, Eric Beamer, Gino Lucchetti, and Robert E. Bilby

Abstract

The term "watershed assessment" describes a general approach for understanding watershed processes, their effects on riverine ecosystems, and the role people play in their modification and restoration. Identifying the goals of a watershed assessment focuses effort on specific questions important to achieving the goal and limits extraneous effort and expense in the planning phase of restoration. We propose two roles for watershed assessments in restoration planning: (1) estimating historical and current smolt production potentials based on available habitat quantity and quality, and (2) identifying causes of habitat loss and restoration actions needed to recover salmon. In the first role, watershed assessments can help identify which habitat losses have had the greatest impact on populations. In the second role, they can elucidate causes of habitat change in stream networks (both human and natural) and can identify needed habitat restoration actions. We discuss methods for both roles, including inventory and assessment of watershed processes, reach-level processes, and aquatic habitat.

Introduction

The term "watershed assessment" connotes a general methodology for evaluating landscape and land-use factors that affect riverine ecosystems. Similar to the term "environmental impact assessment," the term watershed assessment and its variants (e.g., watershed analysis, sub-basin assessments, basin planning) include many different methods and assessment techniques. The diversity of assessment techniques has left ample room for differing perceptions of the purposes and products of watershed assessment. These differing perceptions have led to some confusion in salmon recovery planning, where watershed assessments have been touted as a primary tool for identifying and prioritizing habitat restoration and protection actions.

Individual watershed assessments should clearly spell out the questions they are intended to answer, collect data relevant to those questions, and then elucidate answers clearly and simply. However, many assessments are overly broad and lack sufficient direction to provide focused answers. In this chapter, we assert that clearly defined goals and objectives will help define the assessment needs and limit extraneous effort and expense in the planning phase of restoration. We then propose two roles of watershed assessment in recovery planning for Puget Sound salmon listed under the Endangered Species Act (ESA). First, watershed assessments can be used to estimate changes in habitat conditions and smolt production capacity from the time prior to non-Native American settlement to the present day (Beechie et al. 1994). These results can be used to identify loss of important habitats, and they provide an independent means of checking population goals derived from other methods (e.g., population viability analysis). Second, watershed assessments can identify causes (human and natural) of habitat change or loss, and identify needed habitat protection and restoration actions. We give examples of assessment methods that are useful in each of the two roles and describe the products that each type of assessment delivers.

A Conceptual Framework for Watershed Assessments

Development of any watershed assessment methodology should begin only after the central goals for habitat restoration have been defined. These goals help define the tasks required in watershed assessment and identify the products required for recovery planning efforts. We propose that watershed assessments can help answer two important questions in salmon recovery planning:

1. What are historic and current habitat abundance for watersheds containing one or more genetically distinct salmon populations?
2. What are the causes of habitat degradation in each watershed?

The first question is driven by the need to estimate population goals for salmon stocks within an Evolutionarily Significant Unit (ESU). An ESU is "a population or group of populations that are (1) substantially reproductively isolated from other populations, and (2) contribute substantially to the ecological or genetic diversity of the biological species" (NOAA 1998). In Puget Sound, the Technical Recovery Team (TRT) for threatened chinook salmon estimates population goals by population viability analyses, which do not account for availability of habitat. The habitat-based method described here provides an independent estimate of the historical spawner abundance or smolt production, which the TRT can use for comparison with estimates made by population viability analysis. Assessing both the historical and current conditions within a watershed help in addressing these two questions. Contrasting the historical and current conditions provides an indication of the types of habitats most impacted and where those impacts have been greatest, which is useful for identifying and prioritizing habitat restoration actions. Information on the current and historic distribution of human activities in the watershed is essential to address the second question: identifying the causes of habitat change in each watershed. This information provides the basis for identifying the specific restoration actions needed to recover habitats.

A simple diagram of linkages between watershed processes, habitat, and fish abundance illustrates how watershed assessments can be used to address these two questions (Figure 1). The underlying principles in this approach are that aquatic habitats are sensitive to landscape process inputs (e.g., water, sediment, wood, nutrients) and that different types of aquatic habitats have differing capacities to support salmon. Assessing changes in the absolute abundance of each habitat type allows estimation of changes in fish abundance. Assessing changes in habitat-forming processes (e.g., hydrology, sediment supply, or wood recruitment) then indicates how land uses have altered habitat. Once the important habitat-altering processes have been identified, inventories of specific land-use actions that cause these changes can be used to identify habitat recovery options, which may include protection of good quality habitats as well as restoration of degraded habitats. Finally, recovery actions can be targeted to recoup the largest incremental increases in biological response at least cost (e.g., Beechie et al. 1996).

The outcome of these assessments is a science-based identification of potential recovery actions, which can be subsequently combined with social and economic considerations to arrive at a recovery plan. Recovery actions identified through the combination of both types of assessments have several attributes that make them desirable in a management setting. First, the assessments and inventories are systematic and consistent across a basin, so the technical justification for each project is clear. Second, because actions target recovery of natural processes that create the natural array of habitat types and qualities in a basin, there is low risk that any project will harm the ecosystem it is intended to recover. Finally, projects can be prioritized based on costs and expected benefit for recovery (Beechie et al. 1996). Other factors that may influence priorities are identification of refugia, land-owner willingness to participate, and availability of funds for specific types of projects (Beechie and Bolton 1999).

Salmon recovery planning efforts based on this approach should initially focus on collecting historic and current data of three types: (1) condition and distribution of aquatic habitats, (2) salmon abundance and distribution, and (3) status of habitat-forming processes, such as sediment supply, hydrologic regime, or wood recruitment. Such efforts should focus on the full range of potential habitats in the planning area, including freshwater, estuary, and nearshore marine areas. The conceptual model itself does not limit the spatial

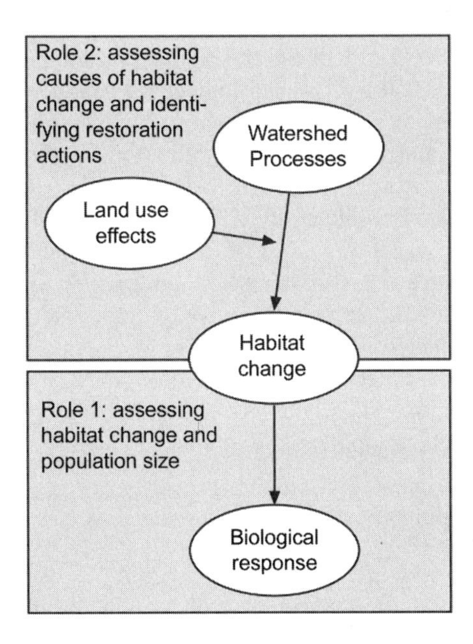

Figure 1. Linkages between the two roles of watershed assessment.

or temporal resolution of the results. Rather, specificity of results is driven by the resolution of the data, which in turn is driven by the assessment goals. For example, coarse resolution remote sensing data might be used to investigate broad regional patterns in relationships among geologic or climatic variables and ecosystem conditions or salmon populations (Lunetta et al. 1997; Pess et al. 1999). However, detailed field data must be used to identify specific recovery actions. As a general rule, larger scale assessments help managers understand where landscape processes have been most altered, and site-specific inventories identify specific actions that are needed to restore those processes to their natural rates.

Role 1. Estimating Historic and Current Habitat and Fish Production Potential

Watershed assessments for estimating habitat availability and production potential first require quantifying areas of different habitat types and second require associating seasonal fish use and survival with each habitat type (Reeves et al. 1989; Beechie et al. 1994). Quantifying fish use of different habitat types requires specific studies designed to identify these relationships (e.g., Bisson et al. 1988). However, it may not be necessary to develop new relationships between fish use and habitat types in each watershed if it can be demonstrated that fish use and comparable habitat types are similar to those in previous studies. By contrast, the relative proportions of different habitat types and land uses vary from watershed to watershed, so habitat inventory data are generally not transferrable from one watershed to another (Lunetta et al. 1997; Beechie et al. 2001; Collins and Montgomery 2001). Therefore, habitat inventories must be conducted separately in each river basin or planning area.

Assessments for estimating historic and current habitat and production potentials are conducted in three steps:

1. identify habitat types;
2. estimate historic and current habitat abundance by type; and
3. estimate production potential based on habitat-fish relationships.

Identify Habitat Types

A habitat classification system suitable for estimating historic and current habitat and potential fish production must have two main attributes. First, analysts must be able to associate fish abundance and survival with each habitat type in order to estimate total fish abundance in a watershed. Second, it must be possible to quantify historical and current habitat areas in order to estimate changes in potential production over time. We recommend a suite of habitat types at two hierarchical scales (Table 1). The coarse resolution habitat classes can be mapped from remotely sensed data at the reach scale (e.g., topographic maps, aerial photography, or satellite information), whereas the finer resolution of habitat types must be identified in the field at the habitat-unit scale (sometimes with the aid of aerial photography). Because these typing systems are nested, all reaches within a watershed can be stratified by landscape and land-use factors using the remotely sensed coarse-resolution data, and reaches within each strata can be sub-sampled to develop an understanding of habitat types within each reach type. This hierarchical relationship enables extrapolation of habitat conditions for unsampled reaches within the watershed.

Stratification of reach types may include several different landscape and land-use factors, although a relatively small number of strata are desirable to reduce the complexity and number of assumptions and calculations. For example, tributary reaches may be stratified simply by slope and land use in order to identify changes in pool area as result of land uses (Beechie et al. 2001). This stratification is useful because reaches of different slope have different natural potentials for pool formation (e.g., Montgomery and Buffington 1997), and reduced wood abundance (a common impact of land use) has a more pronounced effect on pool area in certain slope classes (e.g., Montgomery et al. 1995; Beechie and Sibley 1997). The same slope classes are not particularly relevant for large rivers, where some combination of slope and discharge may be more useful in predicting natural channel patterns (e.g., Leopold et al. 1964). Stratification by potential natural channel pattern and land use can be used to help identify where river-floodplain interactions have been altered, and therefore where changes in the finer resolution habitat types have occurred.

Estimate Habitat Abundance

Methods for estimating current and historic habitat abundance differ among habitat types. For example, estimated changes in tributary pool areas are based primarily on data from reference sites within the study area (Beechie et al. 1994), whereas estimated changes in channel and wetland areas on deltas are primarily from maps and survey notes made prior to extensive settlement by non-Native Americans in the mid 1800s (Collins and Montgomery 2001). Therefore, it is not possible to describe a single methodology for assessing changes for all reach types. Instead, we provide an overview of different meth-

Table 1. Habitat types used for the two types of watershed assessments described in this paper. Coarser-scale habitat types are mapped from topographic maps, aerial photography, and satellite imagery. Finer-scale habitat types are mapped using a combination of aerial photography (for larger units) and field measurements.

Habitat Type (Coarser Scale)		*Habitat Type (Finer Scale)*

Large mainstems (>50 m bfw*) by channel type based on gradient and confinement	Edge
	Mid-channel
		Pool
		Glide
		Riffle
		   Boulder/Cobble
		   Cobble/Gravel
		Bar edge
		Bank edge
		   Natural
		   Hardened
		Backwater (alcove)

Small mainstems (10–50 m bfw) and tributaries (<10 m bfw) by channel type based on gradient and confinement	Pools
	Riffles
		Pool
		   Scour
		   Plunge
		   Trench
		   Backwater
		Glide
		Run
		Rapid
		Riffle

Off-channel habitat within large main channel floodplains	Channel-like Pond-like	Ponds <500 m^2 Ponds >500 m^2 and <5 ha Lakes >5 ha
Impoundments		
Palustrine Wetland	Forested Scrub/shrub	Openwater area by season
Riverine Tidal Wetland	Forested Scrub/shrub	Openwater area by season and tidal stage

* bfw = bankfull width

Table 1 (continued).

Habitat Type (Coarser Scale)		*Habitat Type (Finer Scale)*
Estuarine Wetland	Scrub/shrub Emergent	Openwater area by season and tidal stage
Estuarine Channel	Mainstem Blind Distributary	

ods that one might use for assessing habitat conditions historically and at present. More detail on individual methods can be found in Beechie et al. (1994), WDFW (1998), and Collins and Montgomery (2001).

Reduction of pool areas in tributary habitats can be estimated by comparing pool areas in streams impacted by land use to pool areas in reference streams relatively unimpacted by human activities (Beechie et al. 1994). Even in watersheds unaffected by logging or other land use, factors such as wildfire, debris torrents, or floods introduce variability in habitat conditions among reaches (Bisson et al. 1997). Therefore, enough reach types in each stratum should be sampled in both impacted and reference watersheds to assess the variability among reaches in the relative abundance of different habitat types. This variability can also be reduced by ensuring that reference streams are as similar as possible to those in impacted areas with respect to geomorphic setting and potential natural vegetation. That is, stratification of reach types for sampling will help limit variability within reach types and increase chances of detecting differences between impacted and unimpacted reaches. A similar approach can be used for larger rivers (e.g., >30 m wide), although processes of habitat formation and the range of habitat types in these rivers differ from those in tributaries and the availability of appropriate reference reaches may be limited (Collins and Montgomery 2001).

A second approach for assessing habitat changes in larger channels uses historical records and maps to reconstruct historical habitat conditions. Historical areas of slough habitats (both side channels and distributaries) can be estimated from historical maps, notes, and photos and often can be field verified by residual evidence of their prior locations (Beechie et al. 1994; Collins and Montgomery 2001) (Figure 2). Present-day areas can be measured from aerial photographs and in the field. Comparisons of the two inventories identify areas of significant habitat loss and suggest how land uses have changed habitat-forming processes. Similar methods are used to identify losses of off-channel habitats on floodplains.

Changes to lake areas are measured directly from historical and current maps (Beechie et al. 1994), and typically indicate where rivers have been dammed for hydropower or water supplies. Lake areas tend to increase, whereas mainstem and tributary habitats are inundated by reservoirs and decrease in extent. Pre-settlement beaver pond areas can be estimated based on frequencies of beaver ponds in relatively pristine areas (e.g., Naiman et al. 1988), and present-day pond areas within the study area can be measured using field surveys and aerial photography. Losses of beaver pond habitat are then estimated by comparing estimates of historical and current areas.

Portions of tributaries that are no longer accessible to salmon can be mapped using inventories of habitat upstream of migration barriers. Natural barriers to salmon migration must first be identified to delineate the assessment area. All structures crossing streams within the assessment area (culverted crossings, bridges, small dams) should then be inventoried to determine if they meet passage criteria for salmon (Figure 3). Finally, habitat areas upstream of each

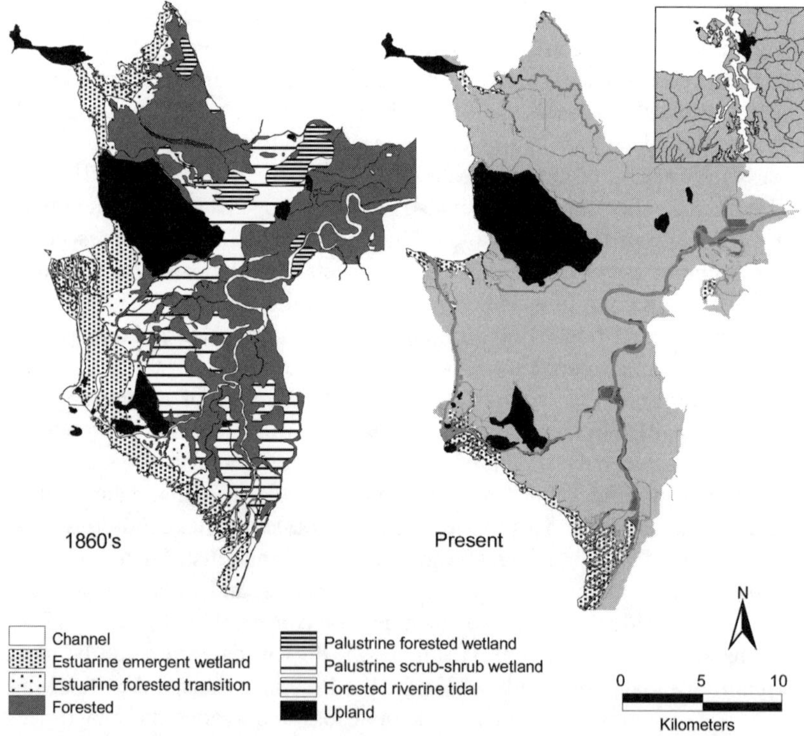

Figure 2. Comparison of historic delta conditions in the Skagit River basin (ca. 1860) from Collins and Montgomery (2001) and current conditions (Beamer, unpublished data).

man-made barrier must be surveyed to determine how much habitat is inaccessible (i.e., that area upstream of the identified barrier and downstream of the natural barrier to salmon migration). Use of a standardized method for determining blockages (e.g., WDFW 1998) will streamline identification and prioritization of isolated habitats and provide a standard set of criteria for monitoring progress toward re-opening these habitats.

Linking Fish Production and Habitat

At its simplest, smolt production potential from a given habitat type or area is calculated as:

$$\text{habitat area} \times \text{average fish density} \times \text{survival to smolt.} \qquad (1)$$

However, comparing the impact of different habitat alterations on smolt production potential requires making separate estimates for each habitat type. Thus, the production potential of a habitat for each life stage (e.g., spawning, egg to fry, summer rearing, winter rearing, smolt migration) can be expressed mathematically as

$$N = \left(\sum_{i=1}^{n} \left(\left(\sum_{j=1}^{n} A_{ij} \right) \times d_i \right) \right) \qquad (2)$$

where A_{ij} is the sum of areas of all habitat units (j =1 through n) of type i, and d_i is the density of fish in habitat type i. To compare capacities among life stages and identify which habitats may be limiting smolt production, the population estimate (N) for each life stage in a given habitat is multiplied by density independent survival to smolt stage so the capacities can be compared in terms of number of smolts ultimately produced (Reeves et al. 1989). Equation 2 can also be used to estimate historical spawner capacity based on estimates of historical habitat availability. Both spawning and rearing capacities then can be incorporated into assessments of factors that limit population size.

Application

Two examples of this assessment procedure are a historical assessment of coho salmon rearing habitat and smolt production losses in the Skagit River basin (Beechie et al. 1994), and a similar unpublished study for the Stillaguamish basin. In these assessments, inventories of current and historical habitat types

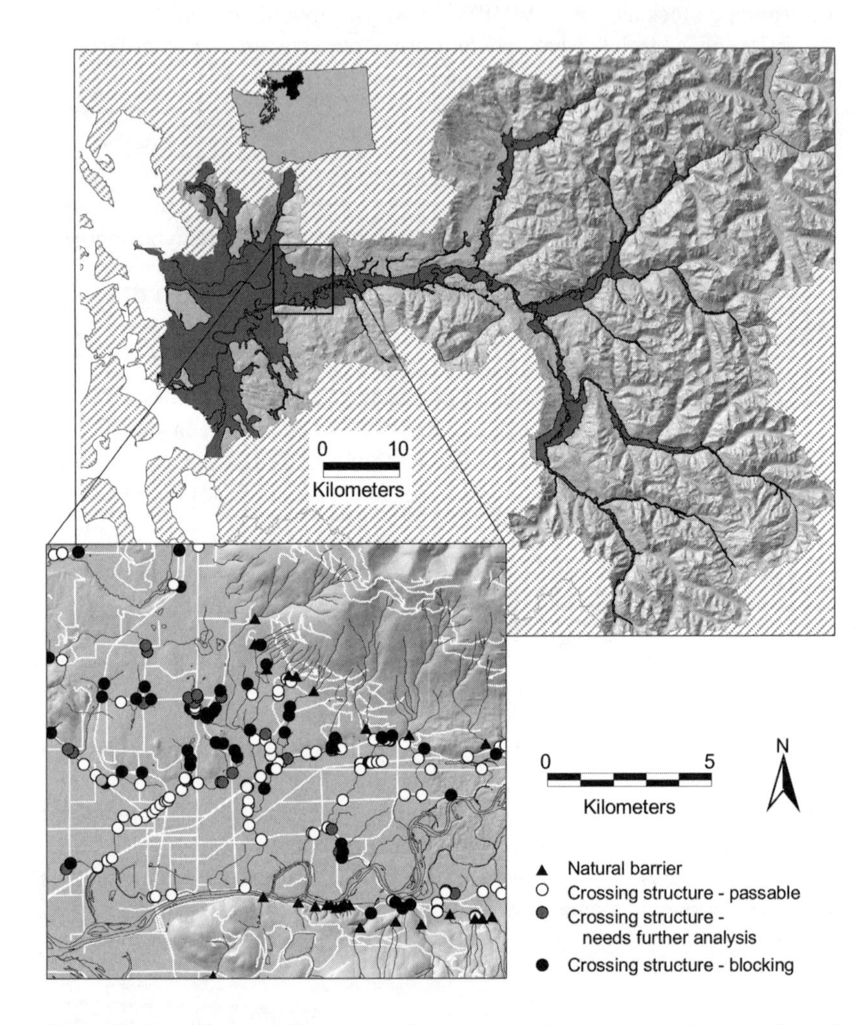

Figure 3. Example map of inventoried stream crossing structures for a portion of the Skagit River basin (enlarged area), showing structures that are blocking, passable, or needing further analysis. Dark gray area in watershed map indicates the historical extent of anadromous fish access based on mapping of 480 natural barriers. All stream crossing structures (1,587 total) were inventoried within the historical anadromous zone.

and abundance provide the basis for estimating historical and current smolt production. Both studies used habitat-specific rearing densities and average seasonal survivals from Reeves et al. (1989) to estimate historical and current coho salmon smolt production, evaluate the effects of different land uses on coho production, and identify recovery of lost off-channel habitats such as sloughs and ponds as a necessary focus for any recovery effort intended to significantly increase coho smolt production (Figure 4) (Beechie et al. 2001). A similar effort to estimate population responses of ocean-type chinook salmon to habitat change is currently underway in the Skagit River basin. Because of the importance of estuary rearing to ocean-type chinook, the assessment will also include relationships developed for estuarine rearing habitat.

As illustrated by Beechie et al. (1994), it may not be necessary to develop new relationships between habitat and fish use for each study area. However, use of any model requires corroboration within the application area to verify that the assumptions and parameters included in the model accurately reflect the conditions within the watershed. That is, data on fish use of different habitats should be collected and used in all watersheds where the model is to be applied, and the model should be modified as necessary based on local data.

Transferability of relationships between fish population parameters and habitat condition does not imply that the habitat factors controlling salmon abundance are also transferable. For example, the type of freshwater habitats that have been most compromised differ between the Skagit and Stillaguamish (Figure 4). More beaver ponds have been lost in the Stillaguamish, whereas

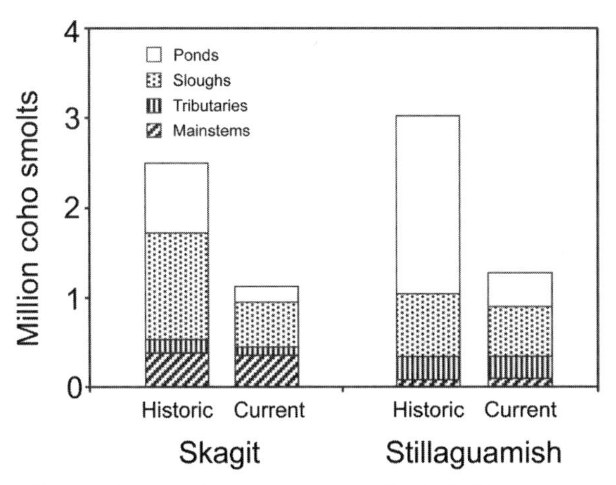

Figure 4. Distribution of potential coho smolt production by habitat type in the Skagit and Stillaguamish basins, both currently and historically.

more freshwater slough habitat has been lost in the Skagit. This is partly due to differences in the underlying physical templates of the watersheds, which are a function of landscape evolution over the past 16,000 years (Beechie et al. 2001). Similarly, each of the remaining watersheds in Puget Sound has different natural histories as well as differences in the type and intensity of land uses that alter habitat-forming processes and habitat conditions. Therefore, historical and current abundance of habitat will differ from watershed to watershed, and assessments must be conducted separately in all locations.

Regardless of the tools or models selected for estimating historic and current production potential, outputs must be interpreted in light of assumptions made in choosing parameters for the model. A good rule of thumb is "*All models are wrong, some are useful*" (Deming 1996). In this case, simplifying assumptions and lack of data limit the ability of any model to accurately represent population response to habitat change, but some models may provide useful insights into land-use impacts on salmonid habitats and production. The tools discussed above can indicate which types of habitat losses are most likely responsible for changes in production potential, and where those losses have occurred. However, the ability of these methods to accurately predict fish abundance, survival, productivity, or distribution is limited by the quality and length of the data record, as well as year-to-year variations in freshwater or marine habitat conditions. Therefore, these models provide only a relative indication of fish population responses to restoration or protection actions (i.e., action x implemented at site y is likely to generate a better fish population response than other options).

Interim Restoration Actions

Implementation of the above procedures will require a number of years of inventory and research in most watersheds. However, even without predicting the relative impacts of different habitat changes on fish populations, identification of habitat types or watershed processes that are important to the species of interest and that have been disproportionately disrupted can provide a basis for prioritizing initial restoration actions. A variety of watershed analysis methods have been developed over the last decade and can be used to determine current conditions. Historical reconstructions of habitat conditions are perhaps more difficult, but available methods provide valuable information on the natural function of watersheds and aquatic habitats. Comparing historical conditions with current conditions provides a solid basis for identifying the range of habitat types to be restored, the processes that form and maintain them, and the degree to which those processes have been affected by modern development. Using this information in conjunction with general knowledge of the habitat needs of the species of interest provides a starting point for developing restoration strategies. However, population responses to restoration action cannot be predicted without information on the relationships between fish and habitat for that watershed.

To date, watershed assessments and historical reconstructions on Puget Sound watersheds have identified a number of common priority habitat features for protection and restoration. The measures can be categorized by actions that protect or restore: (1) access to habitat, (2) the quality of existing habitat, and (3) watershed processes. These measures include:

1. Protecting remaining high quality and highly functioning habitats;
2. Opening disconnected habitat in floodplains and estuaries (which is where we often see the greatest loss of habitat, as well as the greatest amount of potential use by multiple species);
3. Removing unnatural bank protection materials, which enhances interactions between the channel and its floodplain and can immediately increase fish use in a reach;
4. Eliminating or reducing sediment inputs from roads and other development-related activity, especially where they have historically been major sources of sediment (which can increase survival at several life stages for all salmonids); and
5. Accelerating the restoration of riparian vegetation as a source of large wood and shade.

While it is not yet possible to systematically prioritize these measures with respect to their benefits for chinook recovery, each is a necessary component of restoring riverine ecosystems that support listed species. Thus an important near-term action for restoration planning is to inventory blocked access to habitats and disruptions to landscape processes that form habitats. These inventories identify immediate opportunities for habitat protection or restoration while more detailed and complex watershed assessments are underway.

Role 2: Assessing Causes of Habitat Change

Watershed assessments in their second role identify: (1) the natural landscape processes active in a watershed, (2) the effects of land use on natural processes, and (3) the causal relationships between land use and habitat conditions. Natural rates of landscape processes are defined here as those that existed prior to European settlement and development activities, mainly forestry, agriculture, and rural and urban residential and commercial development. Habitat restoration and protection actions resulting from these assessments are directed at protecting and restoring beneficial habitat-forming processes instead of attempting to build specific habitat conditions (Beechie and Bolton 1999). These assessments systematically identify land-use disruptions to habitat-forming processes at two levels of resolution. Coarser level assessments locate disturbed habitat-forming processes using a combination of Geographic Information System (GIS) data and field-based inventories to identify alterations to peak flows, sediment supply, riparian conditions, blockages to salmon migration, water quality, and channel and floodplain interactions. These assessments provide broad-brush screening tools for understanding where processes are disrupted, and in some cases, for estimating total costs of restoration actions. The finer level of resolution relies solely on field-based inventories and identifies specific restoration or protection actions that are required for recovery.

Overview of Assessment

Figure 5 is a conceptual diagram illustrating how watershed controls (ultimate and proximate) and natural landscape processes combine to form various habitat conditions. Ultimate controls are independent of land management over the long term (centuries to millennia), act over large areas (>1 km^2), and shape the range of possible habitat conditions in a watershed. Proximate controls are affected by land management over the short term (≤ decades), act over smaller areas than ultimate controls, and are partly a function of ultimate controls. Landscape processes are typically measured as rates and characterize what ecosystems or components of ecosystems do. For example, sediment or hydrologic processes in a watershed may be characterized by the rates (volume/area/time) at which sediment or water is supplied to and transported through specific locations of a watershed. Certain riparian functions can be viewed similarly. For example, wood recruitment to streams from riparian forests and wood depletion from the channel are both rate functions.

Landscape processes assessed should include (at a minimum) hydrology, sediment supply, riparian functions, channel-floodplain interactions, habitat isolated from salmon access, and water quality (Table 2). This suite of assessments is based on current scientific knowledge of their effects on salmonid habitat and survival of salmon in freshwater as well as knowledge of how various land-use practices affect the processes. The list may not include all impacts to salmon in a watershed, but it includes those that are clearly supported by scientific literature and that are responsible for a significant proportion of the total loss in salmon production from Puget Sound river basins.

For each process, a series of diagnostics can be developed based on rates from scientific literature and local studies. We describe two examples of these assessments to illustrate how a distributed watershed process (sediment supply) and reach-level processes (riparian functions) are assessed. We also revisit inventories of isolated habitats (discussed under Role 1) to explain their utility in the second role of watershed assessments in recovery planning.

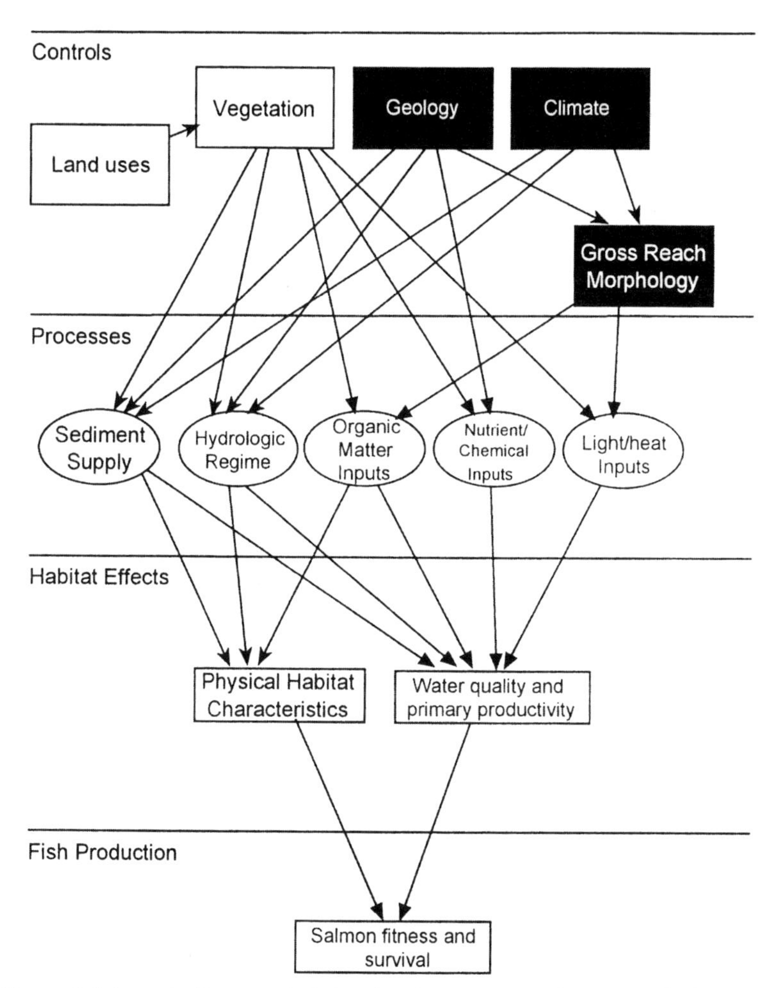

Figure 5. Schematic diagram of relationships between controls on watershed processes, effects on habitat conditions, and salmon survival and fitness (adapted from Beechie and Bolton 1999). Dark boxes in upper row are ultimate controls; light boxes are proximate controls.

Table 2. Summary of methods used for rating individual landscape processes in the Skagit River basin. Details of methods and supporting references can be obtained from the Skagit Watershed Council (www.skagitwatershed.org).

Hydrology - Peak flow in lowland basins

Hydrologic impairment in lowland basins rated based on planned effective impervious area (EIA), which is the weighted average EIA upstream of the stream reach under fully developed conditions. EIA <3% is considered "functioning," EIA between 3% and 10% is "moderately impaired," and EIA >10% is "impaired" (based on Booth and Jackson 1997).

Hydrology - Peak flow in mountain basins

Peak flow ratings for mountain sub-basins in the Skagit were developed based on an empirical correlation between land use and elevated peak flow in an adjacent basin because sub-basin flow data are limited in the Skagit. Sub-basins with more than 50% watershed area in hydrologically immature vegetation due to land use and more than 2 km of road length per km^2 of watershed area are rated "very likely impaired." Sub-basins exceeding one or the other of the criteria are considered "likely impaired." Sub-basins that do not exceed either criterion are considered "functioning."

Sediment supply

Estimating impairment of sediment supply: Average sediment supply for each sub-basin estimated based on average sediment supply rates for 13 combinations of geology and vegetation cover, which were derived from ten sediment budgets conducted within the basin (Paulson 1997). Using GIS, we calculated average current sediment supply for each sub-basin and the average increase over the natural sediment supply for each sub-basin (current/natural). Sediment supply process is considered "functioning" where average sediment supply is <100 $m^3/km^2/yr$, or where average sediment supply is >100 $m^3/km^2/yr$ but <1.5 times the natural rate. Sediment supply is "impaired" where average sediment supply is >100 $m^3/km^2/yr$ and >1.5 times the natural rate.

Forest road inventory—identify sediment reduction projects: The inventory rates factors that influence road-related landslides and the consequences of landslides. All ratings concerning the likelihood of landsliding are summed and then multiplied by a rating of the likelihood that significant stream resources will be

Table 2 (continued).

impacted (Renison 1998). The final value, called the risk rating, ranks roads with respect to the threat that they pose to salmon habitat. Higher risk ratings indicate greater chance that a road will fail and impact salmon habitat. Final ratings were grouped into three categories of risk. A rating >30 is high, 16 to 30 is moderate, and <15 is low.

Riparian Function

Remote sensing assessment: Riparian forests that are > 40 m wide are considered "functioning." Forested buffers 20 to 40 m wide are considered "moderately impaired." Forested buffers <20 m wide are considered "impaired." The proportion of impaired, moderately impaired, and functioning riparian forests can be estimated using LANDSAT classifications of vegetation (see Table 4).

Field inventory: Ratings are the same as above. In addition to documenting forested buffer width, field inventories also classify stand types by species mix and seral stage, which gives sufficient information to prescribe generalized management regimes for each segment of riparian forest. Inventories also identify areas of livestock access and potential fencing projects.

Channel-floodplain

Floodplain areas were delineated where the 100-year floodplain was greater than two channel widths using Federal Emergency Management Agency maps or U. S. Geological Survey 7.5-minute quadrangles and aerial photographs. Reach breaks were based on differences in floodplain width and changes in channel pattern. Hydromodified areas were delineated on aerial photos by rafting or jetboating each main channel within floodplain reaches.

Isolated habitat

Human-made barriers to anadromous fish habitat are identified through a systematic field inventory of channel crossing structures (culverts, tidegates, bridges, dams, and other human-made structures). The inventory identifies the type and physical dimensions of structures as well as physical attributes necessary for modeling water flow conditions and comparing results to passage criteria for salmonids (WDFW 1998).

Examples of Application

Sediment Supply

Changes in the sediment supply to stream channels affect both salmonid habitat characteristics and survival. Large increases in coarse sediment supply tend to fill pools and aggrade channels (e.g., Lisle 1982; Madej and Ozaki 1996), resulting in reduced habitat complexity and reduced rearing capacity for some salmonids. Large increases in total sediment supply to a channel also tend to increase the proportion of fine sediments in the bed (Dietrich et al. 1989), which may reduce the survival of incubating eggs in the gravel and change benthic invertebrate production (Everest et al. 1987).

As mentioned earlier, the assessment of altered sediment supply to stream channels has two levels of resolution. The coarser resolution identifies areas within a watershed where land uses have increased sediment supply above natural background levels. The detailed analysis identifies specific areas where restoration actions may be most effective. One example of these assessments in the Skagit River basin is based on partial sediment budgets for 10 sub-basins covering nearly 10% of the total watershed area (Paulson 1997). These sediment budgets quantified sediment produced by landslides since the 1940s using the aerial photograph record. The size of each visible landslide in each photo series was measured, and volumes of a sub-sample of recent landslides were measured in the field. Field-measured volumes were regressed on photo-measured areas to develop an equation for estimating landslide volume from photo-measured landslide areas on older photos.

From these data we compiled average sediment supply rates by lithologic group and land use in order to estimate average sediment supplies for the remaining watershed area (Table 3). We then mapped these average sediment supplies using GIS and calculated average sediment supplies for each subwatershed area in the basin (Figure 6). This coarse screen indicates where average sediment supply from landsliding has likely increased more than 50% over natural rates and where restoration or rehabilitation may be necessary. However, it does not identify specific restoration actions needed to reduce sediment supplies in impaired sub-basins.

Two types of inventory can be used to identify specific restoration actions. First, mapping of landslide hazard areas identifies locations that are particularly prone to landsliding and sensitive to land uses, such as clearcut logging or road building (Figure 6). Such maps are tools for passive restoration, which allows recovery of sediment supply rates by preventing or modifying land uses within hazard areas. Second, inventory of road landslide hazards identifies specific areas for active restoration. Road inventories should identify

Table 3. Average sediment supply rates from landsliding ($m^3/km^2/yr$) by lithology and land-use class, based on landslide inventories in ten sub-basins of the Skagit River (Paulson 1997).

Lithology	*Forest*	*Clearcut*	*Road*
Low-grade metamorphic rocks	130	520	1040
High-grade metamorphic rocks	53	318	4346
Glacial sediments	33	99	1485

segments of road that are at risk of failure (e.g., Renison 1998) as well as specific stream crossings, cross drains, or fills that are likely to fail. Each potential failure site can be itemized on project lists for restoration action. The restoration actions can then be prioritized based on potential impact to stream habitat and smolt production and cost.

Riparian Function

An extensive body of literature describes linkages between riparian forest functions and stream habitat (Chapter 10), which in turn affect the productivity and abundance of salmonids. Riparian functions include supply of wood and leaf litter to streams, shading, and root reinforcement of stream banks and floodplain soils. For this example, we focus on recruitment of wood to streams and its function in channels, which are among the most studied of riparian functions (e.g., Murphy and Koski 1989; Bilby and Ward 1991; Montgomery et al. 1995; Abbe and Montgomery 1996; Beechie and Sibley 1997). The level of wood input or other riparian functions increases with widening forest buffers on streams (Figure 7), and the proportion of the function occurring within a given distance of the channel edge varies by function (Sedell et al. 1997). These relationships can be used to evaluate the current status of functional interaction between a stream reach and riparian area and indicate whether existing levels of riparian protection are sufficient to ensure continued function.

The Skagit Watershed Council conducted such an assessment for the Skagit River watershed, choosing to classify forested buffers > 40 m wide as functioning because they achieve more than 80% of the wood recruitment and shading functions, as well as all of the root strength and litter fall functions.

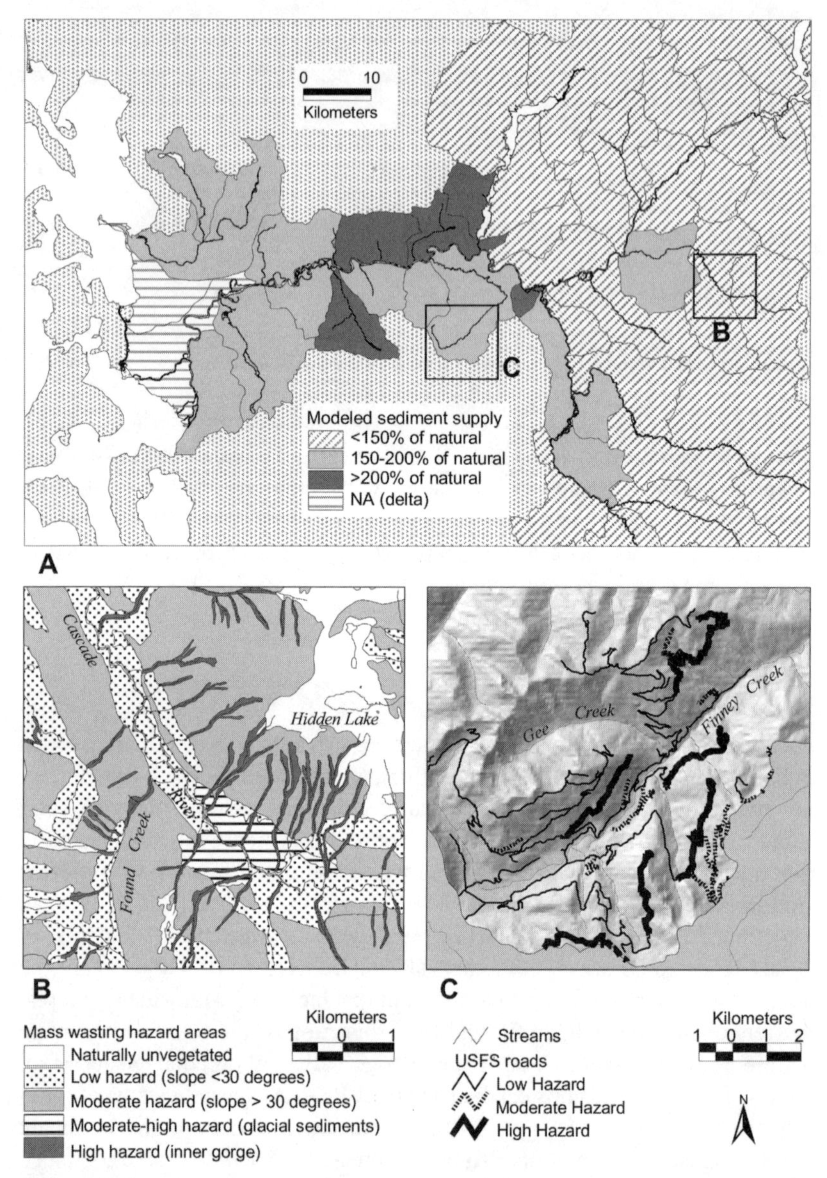

Figure 6. (A) Map of areas in the Skagit basin where sediment supply has likely increased due to land use, based on extrapolation of data from sediment budgets (described in text). (B) Landslide hazard map for a portion of the upper Cascade River basin. (C) Hazard map of U.S. Forest Service roads classified as high risk of failure, moderate risk, or low risk.

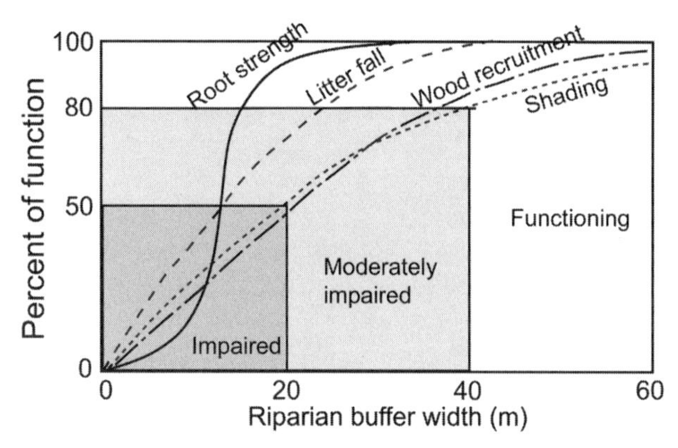

Figure 7. Illustration of change in riparian function with distance from channel, and classification of impaired, moderately impaired, and functioning (curves adapted from Sedell et al. 1997).

Buffers < 20 m wide were classified as impaired because they achieve only 50% of the wood recruitment obtained in a mature conifer forest and 50% to 90% of the other three functions. Areas that achieve 50% to 80% of wood recruitment (20 m to 40 m wide) were classified as moderately impaired.

As with the preceding example for sediment supply, riparian condition was mapped at two levels of resolution. The coarser level, used for watershed screening and basin-level planning of riparian restoration, was based on LANDSAT classification of forest conditions along streams (Figure 8) (Lunetta et al. 1997). These LANDSAT classifications were ground-truthed to help interpret the meaning of forest cover classes derived from satellite data. Field inventory results from over 200 riparian sites describe the distribution of riparian conditions within each LANDSAT forest class (Table 4). All of the sampled "late-seral conifer forest" sites and between 88% and 92% of the "mid-seral conifer forest" and "early-seral conifer forest" sites met the definition of functioning buffer (>40 m wide riparian buffer). Conversely, 90% of the areas mapped as "non-forest" had < 20 m wide riparian buffers, matching the impaired designation. Areas mapped as "other forest" (ranging from clearcuts to mature hardwoods) were found to be 43% functioning, 15% moderately impaired, and 42% impaired. Based on this analysis, we estimate that 29% of the non-mainstem channels in the anadromous zone (by length) are in the non-forest land cover category, and therefore have a very high likelihood of being impaired. By contrast, 19% of the non-mainstem channels in the anadromous zone are in the mid- to late-seral forest cover category, and therefore have a high likelihood of being functioning.

Table 4. Distribution of 234 field-sampled widths of forested riparian buffers (row headings) by GIS-based land cover type (column headings, cover types described in text). Bold numbers highlight dominant buffer condition in each LANDSAT-based class.

	Late-seral Forest (n=24)	*Mid-seral Forest (n=13)*	*Early-seral Forest (n=24)*	*Other Forest (n=96)*	*Non-Forest (n=77)*
<20 m forested buffer (impaired)	0%	8%	8%	**42%**	**90%**
20-40 m forested buffer (moderately impaired)	0%	0%	4%	15%	6%
>40 m forested buffer (functioning)	**100%**	**92%**	**88%**	**43%**	4%

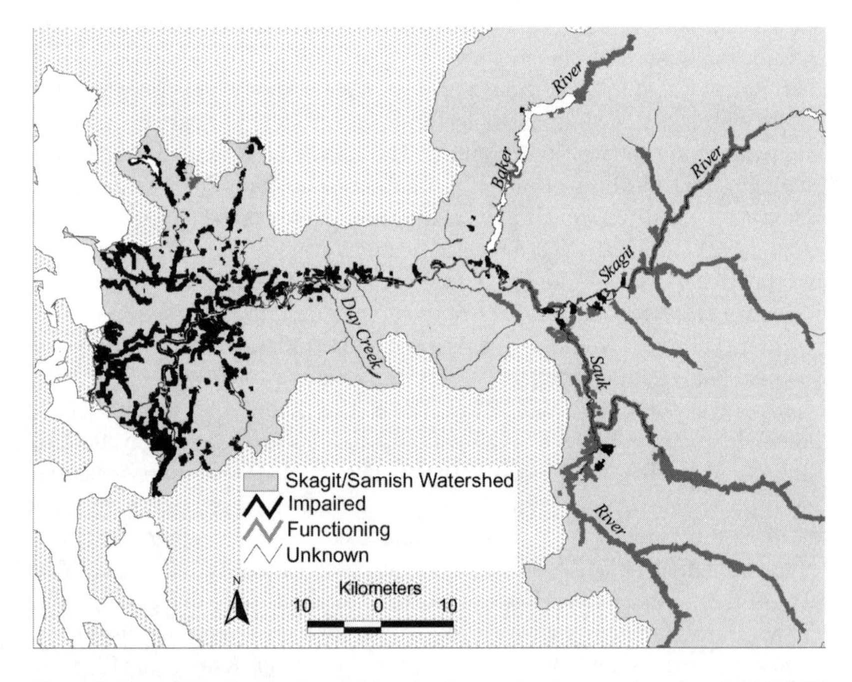

Figure 8. Map of functioning, impaired, and unknown riparian areas based on LANDSAT thematic mapper data (1993).

Regardless of current condition of riparian areas, establishing protected areas along the channel where natural riparian vegetation can develop through time and interact with the stream is a necessary component of riparian restoration. Active restoration efforts may be appropriate at currently impaired sites. Riparian restoration may include the planting of desired riparian plant species or manipulation of the existing vegetation to accelerate tree growth and the development of desired stand structural characteristics (Berg et al. 1996; Beechie et al. 2000).

The distribution of riparian conditions at this larger spatial scale can provide a general sense of the change in riparian function from historic conditions (e.g., Lunetta et al. 1997). Sub-watersheds where the current distribution of riparian conditions deviates markedly from that expected under a natural disturbance regime are locations where riparian restoration efforts may be appropriate. The same data can also help managers understand how different land-use practices differ in their degree of impact on riparian functions (Figure 9). These relationships can then help assess the potential impacts of large-

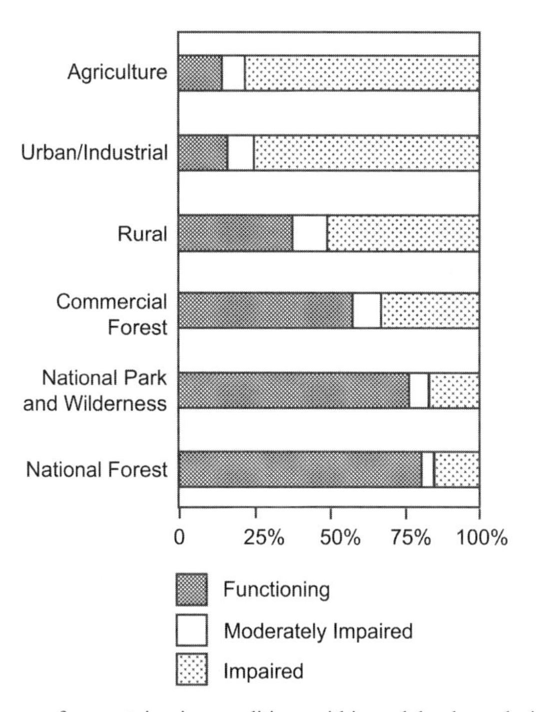

Figure 9. Summary of current riparian conditions within each land-use designation along non-mainstem channels in the anadromous zone of the Skagit River basin.

scale land-use policies on salmon habitat recovery (e.g., evaluating potential effects of growth management legislation).

Field inventories of riparian sites must be used to identify specific restoration actions because of limitations in the satellite classification of riparian forests. Field inventories may consist of initial measurements and classification from aerial photography, combined with field confirmation of the riparian vegetation conditions for each stream reach. At a minimum, the inventory should classify riparian conditions by buffer width, stand type, and age of vegetation. From these inventory data, managers can identify impaired or moderately impaired stream segments (i.e., those with forest buffers < 40 m wide), determine the likely cause of that impairment, and identify required restoration actions.

Inventory of Blocked Tributary Habitat

Stream crossing structures that block fish access to usable habitats can account for as much as 50% of lost smolt production from tributaries in Puget Sound river basins (Beechie et al. 1994). Assessing such isolation of habitats is one of the simplest inventories that can be conducted because criteria for fish migration blockages are relatively clear, and identifying the amount of habitat affected involves little subjectivity. Moreover, combining these inventory results with cost estimates for restoration actions allows managers to rank the cost-effectiveness of individual projects in order to more effectively direct the expenditure of limited restoration funds.

The Skagit System Cooperative used such an inventory to identify isolated habitat projects in the Skagit River basin (Figure 3), following the Fish Passage Barrier and Prioritization Manual of Washington Department of Fish and Wildlife (WDFW 1998). Based on inventory of 1,755 structures as of 2001, they identified 688 structures that do not meet passage criteria on tributaries, 29 that require a more in-depth hydraulic analysis to determine whether or not they meet the criteria, and 794 that meet passage criteria (Klochak and Olis, Skagit System Cooperative, personal communication). The remaining 244 structures, mainly those on side channels or distributary channels that are currently disconnected from the river, were labeled 'unknown' because passage cannot be evaluated until designs for reconnecting river flows to these channels have been developed.

Using this inventory the Skagit Watershed Council evaluated the cost-effectiveness of projects based on the habitat area upstream of the project, multiplied by the average life span of a blockage (approximately 50 years) and divided by the cost of the project. These results allowed the Council to identify the most cost effective projects for reconnecting blocked tributary habitats based on benefits to multiple salmonid species as well as costs of reconstructing individual stream crossings (Table 5). This simple example illustrates that, in some cases, projects opening relatively small areas of habitat are more cost-effective than projects opening large areas because project costs are much lower.

Use of Assessment Results

Evaluating disruptions to habitat-forming processes recognizes that land use and resource management activities influence natural landscape processes, which result in altered habitat conditions. Therefore, restoration and protection actions identified by these assessments should be directed at the habitat-forming processes instead of attempting to build specific habitat conditions. Focusing actions on "building" habitat for specific species may be to the detriment of other species and may not be sustainable due to potential conflicts with natural processes (Frissell and Nawa 1992; Kauffman et al. 1997; Beechie and Bolton 1999). Actions implemented by this strategy aim to re-establish natural landscape processes at levels similar to those that existed historically, which should (1) result in a high likelihood of long-term project success, (2) protect and restore habitat for all salmonid species as well as other native aquatic and riparian dependent species, and (3) ensure the effective use of public and private restoration funds.

These assessments provide a consistent process and set of principles that guide restoration actions, enabling the systematic identification and prioritization of restoration and protection projects. A complete river basin overview of landscape processes and resulting habitat conditions illustrates the extent to which different processes have been altered, and helps in estimating total effort and cost required for restoring this suite of watershed processes in Puget Sound river basins. This approach also allows priorities to be based on locally defined objectives, such as recovery of a certain species or completion of certain types of restoration (Lichatowich et al. 1995; Beechie et al. 1996). However, prioritization does not alter the types of projects enacted but only alters the sequence in which projects are completed. Currently the Skagit Watershed Council prioritizes projects based on the relative cost-effectiveness of different projects, which means that projects protecting or restoring the greatest proportion of anadromous fish habitat function per dollar cost are considered higher priority. Additionally, individual restoration groups may choose projects from the list of possible projects in order to fulfill their respective missions.

Table 5. Example of cost-effectiveness calculations for stream crossing structures that do not meet WDFW (1998) criteria for fish passage (excerpted from an unpublished report by the Skagit Watershed Council). Juvenile salmon typically cannot pass upstream of these structures, and delays to adult migration may occur at some locations. Cost effectiveness is (wetted area)x (project life)/(estimated cost).

Stream Crossing	*Wetted Width (m)*	*Reach Length (m)*	*Wetted Area (m^2)*	*Estimated Project Cost*	*Project life (yrs)*	*Cost-effectiveness*
Careys Slough at Pettit Road	45	1354	60930	$100,000	50	30.5[a]
Red Cabin Cr. at Hamilton Cemetery Rd.	5.8	5345	31001	$100,000	50	15.5
Careys Cr.	4	1000	4000	$30,000	50	6.7[a]
unnamed	6	620	3720	$30,000	50	6.2
Davis Slough at S. Skagit Highway	4	3200	12800	$250,000	50	2.6
Careys Cr. at Maple St.	4	890	3560	$100,000	50	1.8[a]
unnamed	1.9	495	941	$30,000	50	1.6
Gilligan Cr. at S. Skagit Highway	6	750	4500	$250,000	50	0.9
unnamed	2.5	170	425	$30,000	50	0.7
unnamed	2.2	173	381	$30,000	50	0.6[b]
Unnamed at S. Skagit Highway	dry	160	0	$250,000	50	0.0[b]

[a] Three blockages on same tributary. All blockages must be opened to realize full benefit for the stream.
[b] Two blockages on same tributary. All blockages must be opened to realize full benefit for the stream.

These inventories also provide the data necessary to begin tracking the progress of watershed restoration in different river basins. Each restoration project can be mapped on a GIS theme, and relevant data stored in the associated databases. These themes can be updated as new inventories are completed or as project status changes (e.g., design phase, construction, completed, monitoring). Additionally, one can develop related databases for monitoring the effectiveness and costs of different project types. Over time, the GIS maps and databases will help display progress made in restoring habitats in a river basin and will help modify our actions to more efficiently restore habitat in the basin.

Conclusions

Assessments of the historic and current condition of a watershed can greatly improve our efforts to plan, implement, and monitor habitat restoration for the recovery of Pacific salmon. Systematically collected habitat data and a more thorough understanding of fish responses to habitat change will allow refinement of the modeling tools used to predict fish response from application of different restoration strategies. These refinements will improve estimates of rates and pathways of recovery for many species in Puget Sound rivers, and assist in prioritizing restoration actions. However, many of these refinements are still several years from completion.

In the interim, systematic inventories of disrupted habitat-forming processes and blockages to salmon migration should be conducted to provide a complete river basin overview of necessary restoration actions that can be prioritized and sequenced logically. A minimum set of inventories for Puget Sound river basins should include barrier inventories, landslide inventories, floodplain and riparian characterization, channel and valley type classification, and road inventories. Some of these data are already available for parts of many watersheds in Puget Sound. These data provide the basis for identifying needed restoration actions, which can be prioritized by cost-effectiveness, influence on particular species, adjacency to existing centers of biological productivity or diversity (commonly referred to as: refugia, biological hot spots, source watersheds, core areas, key habitat), or other strategies.

There are many sources of uncertainty in these assessments as well as in the political landscape surrounding restoration of Puget Sound watersheds. Uncertainties in the two assessments stem from natural variability in habitat-forming processes, habitat characteristics, and fish populations as well as from errors in assumptions and limitations of data or knowledge. Our ability to characterize these types of uncertainty is limited by availability of data on watershed processes, habitat conditions, and fish populations over long periods of time. Lack of knowledge about current habitat conditions or responses of fish populations to changing habitat conditions introduce uncertainty into predictions of fish responses to watershed and habitat restoration. As with any model, improving the quality of the data reduces uncertainty related to knowledge gaps and improves ability to address the uncertainty related to natural variability in fish response to habitat conditions.

The changing nature of the political landscape also introduces uncertainty into recovery planning. Political uncertainties include potential changes in the species of concern (or listing of additional species in the future) and shifting public policy affecting regulatory protection of salmon habitat and funding for restoration. In the early 1990s, management of salmon habitat in Puget Sound was driven largely by interest in coho salmon, which limited fishing on many stocks under weak-stock management policies. The 1999 listing of Puget Sound chinook salmon as threatened under the ESA dramatically shifted the emphasis of habitat studies and restoration efforts. Potential future listings of other species may further complicate recovery planning because the habitat requirements of different species will not match those of chinook salmon. Therefore, prioritizing restoration actions for single species may create additional conflict in recovery planning.

Recovery plans designed to protect and recover processes that create and sustain riverine habitats in Puget Sound are more likely to recover salmon of all species and help avoid future conflicts among species. Using a comprehensive assessment process to develop restoration plans focused on the reestablishment of habitat forming-processes minimizes conflicts that can arise with species-centric restoration approaches. Restoration of habitat-forming processes targets restoration of the natural array of habitat types and conditions within a watershed, which is consistent with the concepts of watershed and ecosystem management supported by the scientific community. Moreover this approach focuses on the natural potential of each watershed and therefore is most likely to restore the diversity and abundance of stocks appropriate to each watershed in Puget Sound.

Acknowledgments

Environmental Systems Research Institute provided a substantial grant enabling Skagit System Cooperative to acquire GIS software for this project. The U.S. Forest Service provided road inventory data for the identification of landslide hazard areas. We thank Blake Feist for his assistance in GIS-based estimation of sediment supplies.

References

Abbe, T.B. and D.R. Montgomery. 1996. Large woody debris jams, channel hydraulics and habitat formation in large rivers. *Regulated Rivers: Research & Management* 12:201-221.

Beechie, T., E. Beamer, B. Collins, and L. Benda. 1996. Restoration of habitat-forming processes in Pacific Northwest watersheds: A locally adaptable approach to salmonid habitat restoration. In D.L. Peterson and C.V. Klimas (eds.) *The Role of Restoration in Ecosystem Management.* Society for Ecological Restoration, Madison, Wisconsin. pp. 48-67.

Beechie, T., E. Beamer, and L. Wasserman. 1994. Estimating coho salmon rearing habitat and smolt production losses in a large river basin, and implications for restoration. *North American Journal of Fisheries Management* 14:797-811.

Beechie, T.J. and S. Bolton. 1999. An approach to restoring salmonid habitat-forming processes in Pacific Northwest watersheds. *Fisheries* 24(4):6-15.

Beechie, T.J., B.D. Collins, and G.R. Pess. 2001. Holocene and recent geomorphic processes, land use, and fish habitat in two Puget Sound watersheds. In J. M. Dorava, D.R. Montgomery, B. Palcsak, and F. Fitzpatrick (eds.) *Geomorphic Processes and Riverine Habitat.* American Geophysical Union, Washington, D.C. pp. 37-54.

Beechie, T.J, G. Pess, P. Kennard, R.E. Bilby, and S. Bolton. 2000. Modeling recovery rates and pathways for woody debris recruitment in northwestern Washington streams. *North American Journal of Fisheries Management* 20:436–452.

Beechie, T.J. and T.H. Sibley. 1997. Relationships between channel characteristics, woody debris, and fish habitat in northwestern Washington streams. *Transactions of the American Fisheries Society* 126:217-229.

Berg, D.R., T.K. Brown, and B. Blessing. 1996. Silvicultural systems design with emphasis on the forest canopy. *Northwest Science* 70:31-36.

Bilby, R.E. and J.W. Ward. 1991. Characteristics and function of large woody debris in streams draining old-growth, clear-cut, and 2nd-growth forest in southwestern Washington. *Canadian Journal of Fisheries and Aquatic Sciences* 48:2499-2508.

Bisson, P.A., K. Sullivan, and J.L. Nielsen. 1988. Channel hydraulics, habitat use, and body form of juvenile coho salmon, steelhead, and cutthroat trout in streams. *Transactions of the American Fisheries Society* 117:262-273.

Bisson, P.A., G.H. Reeves, R.E. Bilby, and R.J. Naiman. 1997. Watershed management and Pacific salmon: Desired future conditions. In D.J. Stouder, P. A. Bisson, and R. J. Naiman (eds.) *Pacific Salmon and Their Ecosystems: Status and Future Options*. Chapman and Hall, New York. pp. 447-474.

Booth, D.B. and C.R. Jackson. 1997. Urbanization of aquatic systems: Degradation thresholds, stormwater detention, and the limits of mitigation. *Journal of the American Water Resources Association* 33:1077-1090.

Collins, B.D. and D.R. Montgomery. 2001. Importance of archival and process studies to characterizing pre-settlement riverine geomorphic processes and habitat in the Puget Lowland. In J.M. Dorava, D.R. Montgomery, B. Palcsak, and F. Fitzpatrick (eds.) *Geomorphic Processes and Riverine Habitat*. American Geophysical Union, Washington, D.C. pp. 227-243.

Deming, W. E. 1996. *Out of the Crisis*. MIT Center for Advanced Engineering Study, Cambridge, Massachusetts.

Dietrich, W.E., J.W. Kirchner, H. Ikeda, and F. Iseya. 1989. Sediment supply and the development of the coarse surface layer in gravel bed rivers. *Nature* 340:215-217.

Everest, F.H., R.L. Beschta, J.C. Scrivner, K.V. Koski, J.R. Sedell and C.J. Cederholm. 1987. Fine sediment and salmonid production: A paradox. In E.O. Salo and T.W. Cundy (eds.) *Streamside Management: Forestry and Fisheries Interactions*. Contribution No. 59, Institute of Forest Resources, University of Washington, Seattle, Washington. pp. 98-142.

Frissell, C.A. and R.K. Nawa. 1992. Incidence and causes of physical failure of artificial habitat structures in streams of western Oregon and Washington. *North American Journal of Fisheries Management* 12:182-197.

Kauffman, J.B., R.L. Beschta, N. Otting, and D. Lytjen. 1997. An ecological perspective of riparian and stream restoration in the western United States. *Fisheries* 22(5):12-24.

Leopold, L.B., M.G. Wolman, and J.P. Miller. 1964. *Fluvial Processes in Geomorphology*. W. H. Freeman and Co., San Francisco, California.

Lichatowich, J., L. Mobrand, L. Lestelle, and T. Vogel. 1995. An approach to the diagnosis and treatment of depleted Pacific salmon populations in Pacific Northwest watersheds. *Fisheries* 20(1):10-18.

Lisle, T.E. 1982. Effects of aggradation and degradation on pool-riffle morphology in natural gravel channels, northwestern California. *Water Resources Research* 18:1643-1651.

Lunetta R.S., B. Cosentino, D.R. Montgomery, E.M. Beamer, and T.J. Beechie. 1997. GIS-based evaluation of salmon habitat in the Pacific Northwest. *Photogrammetric Engineering and Remote Sensing* 63:1219-1229.

Madej, M.A. and V. Ozaki. 1996. Channel response to sediment wave propagation and movement, Redwood Creek, California, USA. *Earth Surface Processes and Landforms* 21:911-927.

Montgomery, D.R. and J.M. Buffington. 1997. Channel-reach morphology in mountain drainage basins. *Geological Society of America Bulletin* 109:596-611.

Montgomery, D.R., J.M. Buffington, R.D. Smith, K.M. Schmidt, and G. Pess. 1995. Pool spacing in forest channels. *Water Resources Research* 31:1097-1105.

Murphy, M.L. and K.V. Koski. 1989. Input and depletion of woody debris in Alaska streams and implications for streamside management. *North American Journal of Fisheries Management* 9:427-436.

Naiman, R.J., C.A. Johnston, and J.C. Kelly. 1988. Alteration of North American streams by beaver. *BioScience* 38:753-762.

NOAA. 1998. Status review of chinook salmon from Washington, Idaho, Oregon, and California. NOAA Technical Memorandum NMFS-NWFSC-35, Northwest Fisheries Science Center, Seattle, Washington.

Paulson, K. 1997. Estimating changes in sediment supply due to forest practices: A sediment budget approach applied to the Skagit River Basin in northwestern Washington. Master's thesis. University of Washington. Seattle, Washington.

Pess, G.R., R.E. Bilby, and D.R. Montgomery. 1999. Stream-reach and watershed-scale variables and salmonid spawning distribution and abundance in the Puget Sound Region. In R. Sakrison and P. Sturtevant (eds.) *Watershed Management to Protect Declining Species*. American Water Resources Association, Middleburg, Virginia. pp. 397-400.

Reeves, G.H., F.H. Everest, and T.E. Nickelson. 1989. Identification of physical habitats limiting the production of coho salmon in western Oregon and Washington. U.S. Forest Service General Technical Report PNW-GTR-245.

Renison, B. 1998. Risky Business. *Engineering Field Notes* 30:7-20.

Sedell, J.R., G.H. Reeves, and P.A. Bisson. 1997. Habitat policy for salmon in the Pacific Northwest. In D.J. Stouder, P.A. Bisson, and R.J. Naiman (eds.) *Pacific Salmon and Their Ecosystems: Status and Future Options*. Chapman and Hall, New York. pp. 375-388.

WDFW (Washington Department of Fish and Wildlife). 1998. Fish Passage Barrier and Prioritization Manual of Washington Department of Fish and Wildlife. Washington Department of Fish and Wildlife, Olympia, Washington.

9. Putting Monitoring First: Designing Accountable Ecosystem Restoration and Management Plans

Stephen C. Ralph and Geoffrey C. Poole

Abstract

Recovery of Puget Sound rivers and their native fish fauna will depend upon carefully documenting the ultimate effectiveness of restoration actions. Yet, as currently designed and implemented, monitoring programs are predestined to fail in this task. Consequently, our attempts to implement iterative, adaptive restoration or management actions will also fail unless managers and researchers: (1) alter their current conceptual models about the relationship between monitoring and management/restoration; (2) design and implement monitoring programs before planning restoration/management actions; (3) recognize the need for hierarchical monitoring programs and learn how to implement them; and (4) eliminate myths about monitoring, including the assumption that we can generate reliable new information about management and restoration actions simply by observing their outcomes. In order for monitoring programs to provide reliable and timely information required by iterative and adaptive approaches to ecosystem restoration and management, monitoring programs must serve as a scientifically rigorous framework for "Empirical Management" of natural resources. To accomplish this, managers and researchers must work together first to design hierarchically-structured monitoring experiments and then to plan on-the-ground management and restoration actions that serve as experimental manipulations within the context of the monitoring experiment. Unlike current approaches, this empirical approach has the potential to generate rigorous new scientific information about the efficacy of implemented actions and therefore could support adaptive, iterative improvement in management and restoration plans.

Introduction

> "[A] functional long-term monitoring program can become the key component for bringing together the efforts of management organizations, decision makers, and researchers that intend to improve and protect natural ecosystems." (Wissmar 1993, p. 219)

The widespread decline of native salmon populations in Puget Sound watersheds and the quality and quantity of their aquatic habitats is indicative of the cumulative effects and unintended consequences of past and present land-use and water-use decisions over the last 150 years (Chapters 4 and 5). Current and future listings of native salmon, trout, and char under the Endangered Species Act will require explicit recovery plans to be designed and implemented throughout the Pacific Northwest. In addition, as a result of court-sanctioned settlement agreements, water quality management plans (i.e., Total Maximum Daily Loads) are now legally required to be developed for the literally hundreds of locations within Puget Sound waterways that fail to meet current criteria associated with water quality standards. These recovery and management plans will affect land- and water-use decisions at all levels of government, and potentially, society at large.

Recovery of Pacific salmonid habitats should involve a two-pronged strategy that emphasizes protection of the remaining intact aquatic systems while making intelligent, strategic decisions on restoring important ecological processes and functions of riparian and nearshore habitats. Development of effective and timely salmon recovery strategies requires innumerable decisions regarding future land and water use that are ideally based on adequate scientific understanding of the ecology of freshwater and marine ecosystems in Puget Sound and its catchment. Unfortunately, such decisions are routinely made with an imperfect or even wholly inadequate understanding of ecosystem response to protection and restoration actions. While in some cases decisions are made without considering information that already exists (see Chapter 6); in other cases management decisions are uninformed because the information necessary to *fully* inform the decisions does not (and may never) exist. Ecosystems are simply too complex to expect perfect understanding of the dynamics, structures, and feedback loops occurring therein.

Adaptive management—incorporating management activities into scientific experiments and modifying future management actions based on experimental results—is a widely embraced mechanism to make management decisions in light of uncertainty while learning from these decisions. Although adaptive management is generally applied to resource management decisions having to do with extraction or exploitation, we assert that restoration is as much a form of management as resource extraction, equally as fraught with uncertainty, and thus equally as reliant upon our ability to learn from our mistakes. Thus, if ecosystem management efforts (including salmon recovery) are to succeed, monitoring the outcomes of protection, restoration, and resource-extraction actions needs to be factored into the mix of planning and implementation to form a truly effective and integrated strategy (Currens et al. 2000).

In this chapter, we examine the broad role of monitoring as an applied science, which helps guide salmon recovery planning and other forms of management, particularly by providing a means to reduce uncertainties associated with past, present, and future land-use decisions controlling aquatic habitats. We argue that: (1) monitoring the outcome of actions is a fundamental underpinning of an iterative and adaptive process designed to manage resources in the face of uncertainty; (2) widespread myths about monitoring currently ensure that monitoring programs will not succeed; and therefore, (3) iterative, adaptive approaches to resource management cannot succeed without fundamental changes in the design, implementation, and integration of monitoring programs.

Many millions of dollars have already been spent in the PNW on river enhancement projects aimed to aid recovery of native fish, and less often, the processes responsible for shaping rivers and riparian areas. Although very few systematic evaluations have been made of the success or failure of such projects, published accounts suggest a significant disconnect between on-the-ground implementation of such projects and any subsequent, explicit attempt to evaluate the outcome or success (Frissell and Nawa 1992; Beschta et al. 1994; Kondolf 1995; Frissell and Ralph 1998; and Chapter 12). Management actions—even those taken in the name of restoration—should be subjected to rigorous scientific scrutiny to ensure that we gain a better understanding of their ultimate and proximate contribution to recovery.

Two Flavors of Adaptive Management

> "There is abundant evidence of poor or unsuccessful management of ecosystems, but little evidence of successful management." (Ludwig 1996, p. 16)

When resource management or restoration decisions are based on imperfect knowledge, there will always be risks associated with these decisions. Conceptually, "adaptive management" (Holling 1978) has been widely embraced as a means of dealing with these risks. Yet critical assessments (e.g., Halbert

Table 1. Three fundamental conclusions of a critical assessment of adaptive management (from Lee 1999).

Adaptive management has been more influential as an idea than as a practical means of gaining insight into the response of ecosystems inhabited and used by humans.
Adaptive management should be used only after all parties to the dispute have agreed on a list of key questions that are to be answered by the approach.
Efficient and effective social learning and consequent change in behaviors, of the kind that could be facilitated by adaptive management, are likely to be of strategic importance in determining the fate of ecosystems as humanity searches for a sustainable economy.

1993; Walters 1997; Johnson 1999; Lee 1999) have concluded that adaptive management is difficult to initiate and maintain over periods of time sufficient to show success (Table 1). We believe that adaptive management has failed largely because many processes implemented under the label "adaptive management" have only superficial similarities to the concept outlined by Holling (1978). To illustrate, we contrast Holling's Adaptive Management (HAM), a science-based process, with the more commonly initiated process, which we term "socio-political adaptive management" (SPAM).

Holling's Adaptive Management is a complete resource-management paradigm designed to provide a means of addressing the uncertain ecological risks associated with land-use and water-use decisions. In theory, Holling's Adaptive Management builds a credible scientific foundation by envisioning land-use activities (e.g., laying out timber sales, setting prescribed fire, building roads, stream restoration, and so on) as experimental manipulations that are implemented within the context of well-designed monitoring experiments. This strategy seeks to simultaneously generate economic value *and* scientific understanding of ecosystem response to human activities (see also Holling and Meffe 1996; Walters 1997).

Socio-political adaptive management concepts emerge from socio-political decision-making processes (Chapter 6). Socio-political adaptive management concepts generally assume that an *independent* monitoring effort will be able to document any negative ecological impacts associated with continued land use, even though monitoring is not typically viewed as a series of well-designed experiments. In part because of their genesis in the policy-making realm, socio-political adaptive management concepts often are scientifically incomplete and ineffective. Often, they are based on only casual or uninformed interpretations of Holling's Adaptive Management.

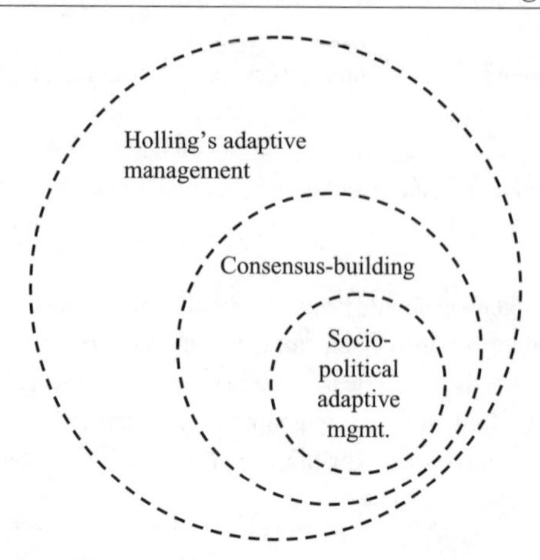

Figure 1. Venn Diagram of relationship between Holling's Adaptive Management, consensus-building, and socio-political adaptive management.

To understand the difference between Holling's Adaptive Management and socio-political adaptive management, it is useful to understand the relationship between these concepts and other socio-political processes such as consensus building. Consensus building is an interpersonal and political process designed to facilitate decision-making in the divisive and contentious political environment that surrounds the development of natural resource management policies. Thus, where implemented within the context of a pre-existing science-based process such as Holling's Adaptive Management, consensus building is apt to be a valuable tool for implementing adaptive management (Figure 1). Yet, as in any complex decisions-making process based on both inadequate information and political compromise between parties with different views and objectives, there are situations where participants simply cannot reach consensus. Lack of consensus typically arises when human land-use activities that can create economic value (e.g., resource extraction) might degrade ecological values (e.g., degradation of habitat for salmon and other native biota). Often, these impasses arise when one or more participants in the consensus group can successfully characterize ecological risks as uncertain. These friction points can overshadow and potentially derail other decisions where consensus *is* possible unless there is a means of addressing fundamental points of disagreement. In the face of "uncertain" ecological risk and the "assured" economic benefits, the impasse is typically resolved by allowing land-use actions to proceed while enduring the ecological risks, but with assurances that the actions will be monitored to determine whether ecosystem values are harmed. The results of monitoring, then, are intended to catalyze any necessary future "adaptive" improvements in management action. Thus, arises socio-political adaptive management—a tool for facilitating consensus-building (Figure 1).

Consensus-building processes convened to design restoration strategies may suffer from similar tensions. For instance, political pressure to "do something positive" can overshadow the more deliberate and careful design and implementation of a restoration project done as part of an experimental evaluation program. Similarly, political pressure to implement piecemeal restoration strategies compatible with status-quo resource extraction (e.g., placing large wood to create artificial pools in streams) may preempt the more comprehensive but politically difficult task of restoring a balance between stream flow, sediment sources, and riparian vegetation at a watershed scale.

Weaknesses of Socio-Political Adaptive Management

> "[L]ong-term monitoring and planning are often considered to be more a philosophical exercise than one of practical value." (Ziemer 1998, p. 131)

Explicitly recognizing the role of the socio-political adaptive management concept in consensus-building processes underscores Lee's (1999) first conclusion (Table 1) by revealing that socio-political adaptive management has little utility beyond facilitating consensus-building processes. Any resulting consensus-based management/restoration plan is unlikely to induce adaptive social learning and changes in behavior.

There are two reasons for this failing. First, consensus-building processes typically focus first and foremost on the nuts-and-bolts of determining allowable or acceptable management actions (e.g., defining best management practices, determining when they should apply, and deciding which should be mandatory and which should be voluntary). Therefore, the consensus process results in a relatively complete blueprint for management actions, but no more than a statement of need for a monitoring plan and a requirement that it be developed in the future. Although management actions and monitoring programs are originally envisioned as interdependent activities (Figure 2a), management actions typically are designed to proceed prior to implementation of the monitoring program (Figure 2b). The process may be well-intentioned and earnest, but the substance and schedule of the monitoring plan is often poorly defined. Thus there is little economic or political impetus to carry through on the monitoring component of the agreement. Given that adequate monitoring is both time-consuming and expensive, planned moni-

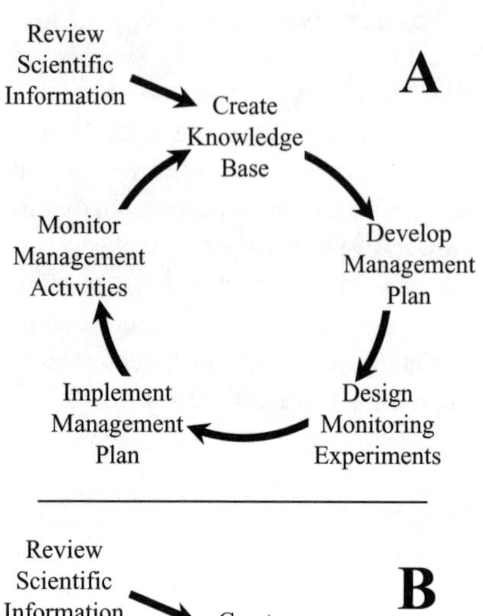

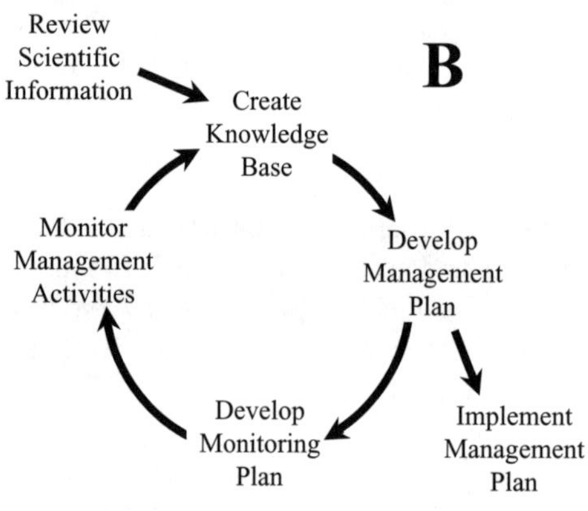

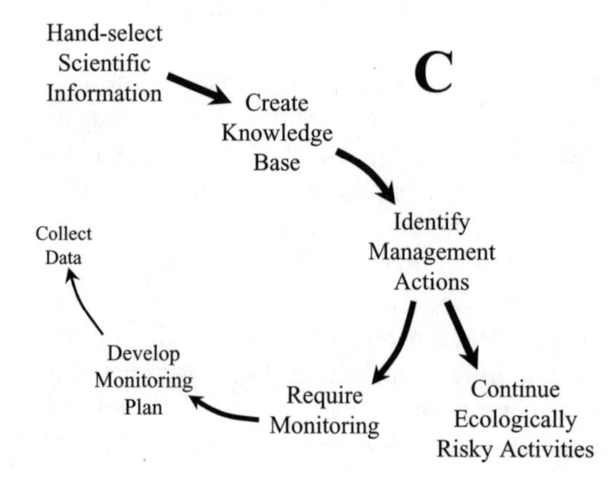

Figure 2. Schematics of socio-political adaptive management: A) as perceived by participants in consensus-building processes; B) as typically designed during consensus-building; and C) as generally implemented, often as the direct result of "myths" associated with monitoring (see text).

toring programs are sometimes not implemented; even when implemented, they may be short-lived. This results ultimately in the failure of the planned adaptive process and the loss of the opportunity to collectively explore the efficacy of agreed-to management decisions. Monitoring programs that do not last long enough to generate new information result in a linear rather than iterative process (Figure 2c). The burden of proof to show the harmful effects of management decisions thus remains with the ecological system at risk, with no real prospect for lessening that burden through learning.

Second, monitoring programs that accompany socio-political adaptive management plans typically fail to recognize that reliable new information can only be generated by conducting well-planned scientific experiments. This requires generating credible hypotheses and designing monitoring experiments to adequately test these hypotheses. Although some have argued that monitoring must be approached as an experiment with testable hypotheses (Walters 1986; Conquest and Ralph 1998; Currens et al. 2000), contemporary socio-political adaptive management plans tend to result in scientifically ineffectual monitoring programs (Walters 1997).

We illustrate this point by outlining several commonly held and deeply entrenched "myths" about monitoring and argue that most contemporary monitoring programs are built upon one or more of these myths, each of which can eliminate necessary scientific rigor from monitoring programs.

Myth 1: We can monitor anything, it's just a matter of figuring out how.

Because of real-world limitations arising from political, technical, and budget realities, some ecosystem responses are more easily measured over time than others. Yet managers often set management benchmarks without considering our ability to accurately and repeatedly determine the status and trend of the benchmark (e.g., Poole et al. 1997). Natural resource management goals, such as salmon recovery, need to be framed in terms of what we can (and will) measure so that we can determine success or failure. In contrast with contemporary management planning, management goals (in the form of benchmarks) should be set *after* determining what we are politically, technically, and financially able to measure.

Myth 2: We can learn from our management actions alone.

Landscapes and watershed processes that control the expression of salmon habitat can vary substantially in how they respond to disturbances (Reeves et al. 1995). For example, the frequency and magnitude of sediment inputs from steep unstable hillslope terrain will increase in proportion to logging and road building in comparison to similar timber harvest activities conducted in flat terrain with few erodible features. In part because of this variability, management actions conducted outside of the context of a rigorous experimental design do not generate new knowledge that is broadly applicable. In the absence of an experimental control, there is no way to determine whether the effect of the management action or the effects of other events and processes are linked to observed changes. Traditionally, land managers have taken a trial-and-error approach, where future decisions may be made based upon implementing a management action to "see what happens" and figuring they "would not do it again if the desired outcome is not achieved." If the outcome "looks good" based on limited, informal observation over a short period of time, the activity is assumed to have succeeded. This approach can lead to innumerable problems, such as the increasing frequency of perceived "acts of God" which result from delayed or cumulative effects of management activities.

Myth 3: Monitoring can be a separate activity from management; i.e., an adequate monitoring program can be developed in response to proposed management or restoration actions.

If monitoring is to generate new information, it has to be approached as an experiment that tests hypotheses about the effects of management actions. If monitoring represents such an experiment, management activities (whether intended to restore watersheds or extract resources) must be planned as experimental manipulations associated with the monitoring experiment. Thus, for monitoring to fulfill its requisite role in a rigorous, iterative and adaptive strategy for natural resource management, on-the-ground actions must be planned within the context of a monitoring experiment, not after-the-fact.

Interestingly, debunking any of these myths results in the same conclusion—monitoring programs must be designed *before* agreeing on management benchmarks, *before* determining what management actions are appropriate, and *before* laying out management or restoration activities across the landscape. In other words, for adaptive management to succeed, on-the-ground activities must be designed *within the context of* rigorous monitoring programs. Therefore, monitoring programs must be designed first.

Hierarchical Monitoring Design

> "There is a critical need to begin multi-scaled monitoring—not just for point-source pollution but monitoring of key features of normal ecosystem function and indicators of the demands imposed by human society." (O'Neill et al. 1996, p. 24)

Designing a comprehensive and integrated monitoring program that will meet the needs of a salmon recovery strategy for Puget Sound rivers is a daunting task. In concept, such a monitoring program should address specific questions and identify meaningful variables that reflect the consequences of both protection and restoration actions on important components of aquatic environments. If properly framed, monitoring the outcomes of management decisions could increase our understanding of the variety of factors that either contribute to or pose impediments to recovery of river processes and the ecological functions they provide to native salmonids. Monitoring could act as an accounting system to establish an understanding of restoration actions and ecosystem response, elucidate the role of the past in shaping the present, and anticipate the added challenges of future expansion of human settlements throughout Puget Sound.

There is a hierarchy of ecosystem responses to human actions. Local conditions respond immediately to local actions, but the cumulative effects of multiple localized actions manifest themselves later in time and at progressively coarser spatial scales (hillslope, catchment, basin, and so on). Therefore, monitoring experiments must be similarly hierarchical to capture these multi-scaled responses. Although useful and requisite for improving site-specific management techniques, site-specific monitoring of individual management activities documents neither the cumulative watershed scale effects of site-specific actions *nor* the effects of site context on monitoring results. One cannot legitimately extrapolate local-scale results to a larger scale without understanding (1) synergistic interactions between multiple disturbances, (2) the influence of context on local results, and (3) the variation in context at coarser scales. For monitoring experiments to successfully document the array of potential management outcomes, the experimental framework must address patterns and process across spatial scales and link to the scale at which outcomes of management decisions are expressed (Naiman et al. 1992; Conquest and Ralph 1998; Bauer and Ralph 1999).

Variability across large land areas influences the results of monitoring efforts and confounds our ability to interpret resulting information. There are several schemes to stratify landscapes by determinant features (geology, climate, vegetation, elevation) that drive the expressions of habitat forming processes operating at large spatial and temporal scales (Frissell et al. 1986; Omernik and Bailey 1997; Bryce et al. 1999; Montgomery 1999, Montgomery et al. 1999; Chapter 8). Monitoring programs that incorporate a hierarchy of nested monitoring designs with spatially explicit experiments can address multiple objectives in an integrated fashion. We recommend a program that is designed at four distinct spatial scales: (1) the *basin* scale, incorporating major drainages (such as the Puget Sound or Snake River drainage basins); (2) the *watershed* scale, which focuses on watersheds of major tributaries within a given basin; (3) the *segment* scale, encompassing specific stream/riparian, floodplain, and hillslope complexes (for example, a discrete stream segment and its associated hillslopes); and (4) the *site* scale, encompassing a single management or restoration action (Chapter 12). Selection of sampling locations by scale would be further refined by identification of appropriate stratification schemes to minimize confounding influences of inherent variations in landscape characteristics. Table 2 is a hypothetical illustration of how this framework might be applied to a spatially integrated monitoring system to evaluate riparian zone management prescriptions for forest lands. For each spatial scale, it defines a purpose, identifies monitoring questions and objectives, suggests appropriate monitoring variables, and gives guidance on specific design criteria to aid selection of individual sampling sites (see also MacDonald et al. 1991; Conquest and Ralph 1998).

Monitoring applied at the *basin* scale would provide information on the status and trends of key indicators across the larger landscape. This provides information on spatial variability and therefore provides context to help with interpretation of related information gathered at the watershed scale. Similarly, information at the *watershed* scale provides context for *segment*-scale information, which in turn provides context for experiments at the *site* scale. An extensive network of monitoring locations, if properly designed, would provide information on the range of variability in key indicators, while reference sites would provide information on the potential range of expression and system potential that a given watershed may have. This would provide a basis for comparison to landscapes where intensive land uses such as forestry, agriculture, or urbanization occur.

At the watershed scale, multiple factors can be evaluated in an integrated monitoring network over multiple years. Examples of where this has been successful include Coweeta Watershed (Webster et al. 1992) and Carnation Creek (Hartman and Scrivener 1990). A number of segment-scale units should be developed in different areas to better support our desire to extrapolate findings from one area to other areas. It is at this level where *cumulative*

Table 2. The hypothetical application of a nested hierarchy framework for monitoring the effect of alternative riparian zone management configurations on stream temperatures. The FFR refers to Washington State's Forest Practice Rules and are used here only to illustrate the concept of how such a system might be structured.

Monitoring Spatial Scale	*Characteristics*	*Design Criteria*
Basin	*Purpose:* Provide estimates of the status and trends in riparian stand characteristics, riparian shading, and basin temperature regimes across drainage basins (e.g., Puget Sound Basin, Snake River Basin). • Allow stratification of status and trends by dominant land use and ecoregions. • Evaluate whole-basin trends in riparian stand characteristics, shade, and stream temperature in the context of land use history and the application of riparian best management practices. *Example Question*: What is the current status of riparian shade and water temperature across Washington's commercial forests? Are changes in this status occurring over time? *Objective of Question*: Estimate landscape patterns of response of riparian shade and daily water temperatures to the application of FFR riparian management prescriptions. *Example Monitoring Variables*: Seasonal and daily air temperature, seasonal and daily stream temperature, riparian stand characteristics, stream flow.	• Stratification and site selection criteria must allow for extrapolation of results to the majority of the commercial forest lands in Washington State, within relevant ecoregions. • Iterative sampling should continue over an extended timeframe to allow for changes to occur. • Data collection and analysis must include probable covariates (linkages) to differentiate between changes due to FFR and other sources of variability (stream flow, weather, ecoregions, etc.). • Data analysis methods should be specified in the study design (power analysis) along with the time needed (years) for positive changes to occur. • A searchable database to store and provide ready access must be developed and its maintenance provided for.

Table 2 (continued).

Monitoring Spatial Scale	*Characteristics*	*Design Criteria*
Watershed	*Purpose:* Identify spatial and temporal distribution of water temperature within a watershed and its proximate response to adjacent land use and riparian management prescriptions. • Examine outcomes of adjacent clearing on riparian stand characteristics, shade, microclimate, and water temperature. *Example Question*: What are the cumulative effects of FFR riparian prescriptions for small streams on downstream water temperature characteristics? *Objectives of Question*: Determine if and how non-fish bearing streams help maintain cool temperatures in downstream fish bearing streams; evaluate the effectiveness of riparian prescriptions for non-fish streams in maintaining any downstream contribution. *Example Monitoring Variables*: Seasonal and daily water and air temperature through the riparian zone, and upstream/downstream of units; riparian stand conditions; stream flow; groundwater temperature; current and historic land use.	• Watersheds stratified by physiographic regions. • Criteria for selecting watershed must allow for extrapolation to a substantial portion of Washington's commercial forests. • Treatment and control (reference) watershed design, if possible. • Integrate with BMP effectiveness monitoring and with other studies within the basin. • Long-term monitoring time-frame • Studies must be designed to determine cause and effect. • Study design should include probable covariates so that data analysis may differentiate between natural variation (e.g., weather, streamflow) and effects of management activities.

Table 2 (continued).

Monitoring Spatial Scale	*Characteristics*	*Design Criteria*
Stream Segment	*Purpose:* Evaluate the effectiveness of riparian management prescriptions in meeting water quality standards and providing cool water habitat needs of native fish and amphibians. • Quantify how riparian stand characteristics (species composition, site class, structure, aspect, elevation, and buffer width) change in response to harvest prescriptions in terms of percent shade, groundwater, microclimate, and stream temperature. *Example Questions*: Are the FFR shade targets adequate for protecting the temperatures of aquatic habitats in stream segments? *Objective of Questions*: Test the effect of the various FFR regulations in maintaining cool water temperatures locally; identify variability in local water temperature response to FFR prescriptions due to riparian stand characteristics. *Example Monitoring Variables*: Seasonal and daily water and air temperatures throughout the riparian zone; riparian stand characteristics; stream flow and channel characteristics; groundwater temperature; upstream land-use history.	• Treatment control and/or pre- and post-treatment experimental design to isolate the effects of forest practices. • Sampling sites stratified by key physical variables that exert strong influence on riparian stand conditions. • Active monitoring approach to test effect of specific prescriptions. • Use power analysis to optimize sample size, magnitude of minimum detectable effect, and probability of Type I and II errors. • Study design must have unbiased site selection process.

Table 2 (continued).

Monitoring Spatial Scale	*Characteristics*	*Design Criteria*
Site	*Purpose:* Determine if individual and collective management actions associated with timber harvest have a discernable effect on aquatic systems, including channel or bank stability, water quality, or fish habitat. *Example Questions*: Is bank stability disrupted within yarding corridors across streams? Do new or existing culverts associated with haul roads discharge sediment to the adjacent stream? Is blowdown of remaining riparian corridor trees excessive (e.g., > 15% of stand density) following adjacent timber harvest? *Objective of Questions*: Determine how specific aspects of a land-use activities cause proximal disruption to streams; determine how specific aspects of a land-use activities can be modified to mitigate or eliminate stream disruption. *Example Monitoring Variables*: Bank stability where bank has been disrupted; sediment yield associated with road crossings; blowdown rates associated with various riparian prescriptions in different settings.	• Site characteristics should be described relative to context. • Link should be established to site scale cause-effect and consequence on biota. • For some questions, evaluation may be more empirical than subject to long-term monitoring. • Should help to provide the basis for monitoring at coarser scales that track cumulative net effects of distributed "site" scale events.

effects of past, present, and future management actions could be evaluated with carefully designed paired watershed studies.

The effectiveness of particular ordinances governing management practices (e.g., forest practice rules for protection of riparian zones and stream temperatures or local government sensitive-area ordinances to protect against sediment input into stream courses or maintain riparian zone functions) can be assessed at the *segment* scale. Examples include the cumulative effect of several proximal management activities on response variables such as water temperature, habitat diversity, and channel stability.

Evaluation of individual management actions is best suited to monitoring experiments at the *site* scale. Site-specific activities such as culvert replacement, road drainage structures, or placement of large wood in streams can be evaluated on a case-by-case basis to assess effectiveness. When multiple site-scale evaluations are clustered within a framework of intensive segment- and watershed-scale sampling units, more information is revealed about the outcome of management practices applied to sites with different susceptibilities to disturbance. Moreover, the cumulative contribution of multiple restoration actions within a watershed can be more readily assessed.

The Empirical Management Paradigm

> "[M]onitoring [must] be developed as a science in its own right, rather than be the uncritical application of convenient contemporary techniques." (Schindler, 1987, p. 14)

We have argued that monitoring programs must possess specific characteristics in order for "adaptive" natural resource management strategies, including ecosystem restoration, to succeed. Monitoring programs must be designed as scientific experiments wherein management actions serve as experimental manipulations that test well-defined hypotheses. Additionally, monitoring programs must be designed in a manner that mimics the natural hierarchy of both individual and cumulative ecosystem responses to management actions (e.g., at the site, segment, watershed, and basin scales). These requirements emphasize empiricism as the basis for iterative management strategies that facilitate changes to institutional behaviors, responding to new information generated as the result of management actions. They call for a fundamental shift in the way natural resource institutions view monitoring: from a "follow-up" activity that responds to management actions to an organizational framework that provides guidance to designing management or restoration activities. Fundamentally, they underscore the need for monitoring programs that are designed *prior to* management and restoration planning, highlighting the importance of a proactive approach to iterative, self-correcting management actions.

For several reasons, we are reticent to refer to our proposal as a new means of implementing "adaptive management." First, the phrase "adaptive management" has been used so broadly that it is now virtually meaningless. It has been applied to nearly every form of proposed iterative management strategy, from Holling's Adaptive Management to the ill-defined and freewheeling form we term socio-political adaptive management (which is to say that the term refers to everything from HAM to SPAM). Second, the term "adaptive" connotes a *reactive* approach to management, perhaps contributing to the bastardization of the phrase "adaptive management" away from Holling's original (proactive) vision. We therefore use "Empirical Management" as a term to describe our proposed approach (Figure 3). The phrase "Empirical Management" emphasizes the need for up-front scientific experimental design in the form of a well-planned monitoring experiment that should apply equally well to traditional resource management or ecosystem restoration activities.

As originally conceived, adaptive management requires the development of "contingency plans" (Holling 1978; Walters 1997), that define ahead of time the change you will make if your implemented strategy fails to produce the desired results. This remains an important element of any iterative, self-correcting management strategy including Empirical Management. With its focus on up-front experimental design, however, Empirical Management provides a means of developing contingencies by allowing the simultaneous

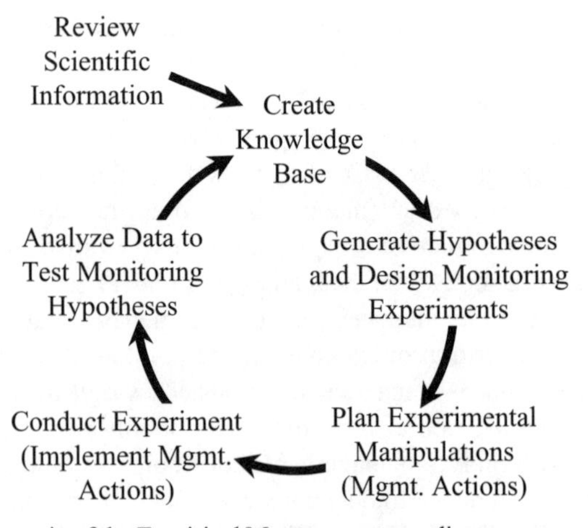

Figure 3. Schematic of the Empirical Management paradigm.

testing of multiple approaches within the context of a single, rigorous experimental design. This is especially important in cases where "new" land management guidelines are being used in hopes of increasing the level of protection to aquatic and riparian habitats associated with river systems. It is equally true where multiple stream restoration projects are advanced as part of an overall recovery strategy.

It makes little difference whether Empirical Management is truly a new approach, a modification of adaptive management, or simply a new name for Holling's original vision. Regardless, the important characteristics of the Empirical Management process are: (1) developing the monitoring plan as the first step in the process of defining the management plan; (2) developing the monitoring plan as a statistically sound scientific experiment; (3) designing the monitoring experiment to capture ecosystem responses across spatial scales; and (4) using the experimental design of the monitoring strategy to guide management activities so that on-the-ground actions will serve as effective experimental manipulations at multiple spatial and temporal scales. Failing to consider any of these characteristics will substantially reduce the rate at which new information is generated and its overall quality.

Conclusions

> "Many have begun to understand that you can't possibly manage what you don't measure." (Law Professor Deborah Ramirez, Northeastern University, speaking on the need for police departments to monitor racial profiling by their officers. All Things Considered, National Public Radio, July 12, 2001)

In order to implement Empirical Management or any similar strategy successfully, managers must broaden their expectations for management actions to include the need to generate new information. This is true for management actions and restoration activities alike. These actions must be implemented as experimental manipulations that support well-planned monitoring experiments designed to generate new information. This blurs the line between research and management/restoration, and it will likely require close collaboration between university research scientists, who have the requisite skills to design effective monitoring experiments, and land management agencies with the budgets and mandate to perform large-scale manipulations of ecosystems. Although this task is daunting, successful ecosystem management and restoration depends on learning from our mistakes and adapting our practices accordingly (McLain and Lee 1996; Lee 1999).

For a variety of reasons, contemporary approaches to adaptive management preclude iterative, self-correcting management approaches by promising but failing to implement adequate and integrated monitoring programs. In contrast, Empirical Management provides a framework for implementing management and restoration activities as part of an integrated monitoring experiment, thereby improving our ability to generate new knowledge about ecosystem response to resource management/restoration. If paired with an improved means of encouraging public acceptance of reliable scientific information, Empirical Management may provide a means to facilitate an iterative, self-correcting management or restoration strategy. The importance of adopting an Empirical Management approach is illustrated by reconsidering Lee's (1999) three conclusions regarding adaptive management (Table 1). By putting monitoring first, Empirical Management could: (1) avoid the pitfalls of contemporary approaches to adaptive management (Figure 2c) that preempt development of new insights, (2) force all parties to agree on the list of key questions to be answered, and (3) provide reliable scientific information as the basis for social learning by integrating management/restoration actions into a well-designed monitoring experiment.

Acknowledgment

Partial funding for this work was provided by Region 10 of the U.S. Environmental Protection Agency. In the preface of their 1996 report, Dale McCullough and Al Espinosa provided the quotes by Ludwig (1996) and O'Neill et al. (1996) used in this chapter.

References

Bauer, S.B and S.C. Ralph. 1999. Aquatic habitat indicators and their application to water quality objectives within the Clean Water Act. US Environmental Protection Agency USEPA-910-R-99-014, Seattle, Washington.

Beschta, R.L., W.S. Platts, J.B. Kauffman, and M.T. Hill. 1994. Artificial stream restoration: Money well spent or an expensive failure? In *Proceedings of A Symposium on Environmental Restoration*. The Universities Council on Water Resources, Montana State University, Bozeman, Montana. pp 76-102.

Bryce, S.A., J.M. Omernik, and D.P. Larsen. 1999. Ecoregions: a geographic framework to guide risk characterization and ecosystem management. *Environmental Practice* 1(3):141-155.

Conquest, L.L. and S.C. Ralph. 1998. Statistical design and analysis considerations for monitoring and assessment. In R.J. Naiman and R.E. Bilby (eds.) *River Ecology and Management: Lessons from the Pacific Coastal Ecoregion.* Springer–Verlag, New York. pp. 455-475.

Currens, K.P., H.W. Li, J.D. McIntyre, D.R. Montgomery, and D.W. Reiser. 2000. Recommendations for monitoring salmonid recovery in Washington State. Independent Science Panel, Report 2000-2, Governor's Salmon Recovery Office, Olympia, Washington.

Frissell, C.A., W.J. Liss, C.E. Warren, and M.D. Hurley. 1986. A hierarchical framework for stream habitat classification: Viewing streams in a watershed context. *Environmental Management* 10:199-214.

Frissell, C.A. and R.K. Nawa. 1992. Incidence and causes of physical failure of artificial habitat structures in streams of western Oregon and Washington. *North American Journal of Fisheries Management* 12:182-197.

Frissell, C.A. and S.C. Ralph. 1998. Stream and Watershed Restoration. In R.J. Naiman and R.E. Bilby (eds.) *River Ecology and Management: Lessons from the Pacific Coastal Ecoregion.* Springer–Verlag, New York. pp. 599-624.

Halbert, C.L. 1993. How adaptive is adaptive management? Implementing adaptive management in Washington state and British Columbia. *Reviews in Fisheries Science* 1:261-283.

Hartman, G.F. and J.C. Scrivener. 1990. Impacts of forestry practices on a coastal stream ecosystem, Carnation Creek, British Columbia. Department of Fisheries and Oceans, Canada, Ottawa. Contribution No. 150, Canadian Bulletin of Fisheries and Aquatic Sciences, 223.

Holling, C.S. (ed.). 1978. *Adaptive Environmental Assessment and Management.* John Wiley, New York.

Holling, C.S. and G.K Meffe. 1996. Command and control and the pathology of natural resource management. *Conservation Biology* 10:328-337.

Johnson, B.L. 1999. Introduction to the special feature: adaptive management —scientifically sound, socially challenged? *Conservation Ecology* 3(1):10.

Kondolf, G.M. 1995. Geomorphological stream channel classification in aquatic habitat restoration: Uses and limitations. *Aquatic Conservation: Marine and Freshwater Ecosystems* 5:127-141.

Lee, K.N. 1999. Appraising Adaptive Management. *Conservation Ecology* 3(2):3.

Ludwig, D. 1996. The end of the beginning. *Ecological Applications* 6:16-17.

McCullough, D.A. and F.A. Espinosa, Jr. 1996. A monitoring strategy for application to salmon-bearing watersheds. Technical Report 96-5. Columbia River Inter-Tribal Fish Commission, June, 4, Portland, Oregon.

McLain, R.J. and R.G. Lee. 1996. Adaptive management: promises and pitfalls. *Environmental Management* 20:437-448.

MacDonald, L.H., A.W. Smart, and R.C. Wissmar. 1991. Monitoring guidelines to evaluate effects of forestry activities on streams in the Pacific Northwest and Alaska. Center for Streamside Studies, University of Washington and U.S. Environmental Protection Agency, EPA/910/9-91-001, Seattle, Washington.

Montgomery, D.R. 1999. Process domains and the river continuum. *Journal of the American Water Resources Association* 35:397-410.

Montgomery, D.R., E.M. Beamer, G.R. Pess, and T.P. Quinn. 1999. Channel type and salmonid spawning distribution and abundance. *Canadian Journal of Fisheries and Aquatic Sciences* 56:377-387.

Naiman, R.J., D.G. Lonzarich, T.J. Beechie, and S.C. Ralph. 1992. General principles of classification and the assessment of conservation potential in rivers. In B.J. Boon, P. Calow, and G.E. Petts (eds.) *River Conservation and Management.* John Wiley & Sons, New York. pp. 93-123.

Omernik, J.M. and R.B. Bailey. 1997. Distinguishing between watersheds and ecoregions. *Journal of the American Water Resources Association* 33:935-949.

O'Neill, R.V., J.R. Kahn, J.R. Duncan, S. Elliott, R. Efroymson, H. Cardwell, and D.W. Jones. 1996. Economic growth and sustainability: A new challenge. *Ecological Applications* 6:23–24.

Poole, G.C., C.A. Frissell, and S.C. Ralph. 1997. Instream habitat unit classification: Inadequacies for monitoring and some consequences for management. *Journal of the American Water Resources Association* 33:879-896.

Reeves, G.H., L.E. Benda, K.M. Burnett, P.A. Bisson, and J.R. Sedell. 1995. A disturbance-based ecosystem approach to maintaining and restoring freshwater habitats of evolutionarily significant units of anadromous salmonids in the Pacific Northwest. In J.L. Nielsen (ed.) Evolution and the Aquatic Ecosystem: Defining Unique Units in Population Conservation. *American Fisheries Society Symposium* 17:334-349.

Schindler, D.W. 1987. Detecting ecosystem responses to anthropogenic stress. *Canadian Journal of Fisheries and Aquatic Sciences* 44:6-25.

Walters, C. 1986. *Adaptive Management of Renewable Resources*. MacMillan, New York.

Walters, C. 1997. Challenges in adaptive management of riparian and coastal ecosystems. *Conservation Ecology* 1(2):1.

Webster, J.R., S.W. Golladay, E.F. Benfield, J.L. Meyer, W.T. Swank, and J.B. Wallace. 1992. Catchment disturbance and stream response: an overview of stream research at Coweeta Hydrologic Laboratory. In P.J. Boon, P. Calow, and G.E. Petts (eds.) *River Conservation and Management*. John Wiley & Sons, New York. pp. 231-253.

Wissmar, R.C. 1993. The need for long-term stream monitoring programs in forest ecosystems of the Pacific Northwest. *Environmental Monitoring and Assessment.* 26:219-234.

Ziemer, R.R. 1998. Monitoring watersheds and streams. USDA Forest Service General Technical Report PSW-GTR-168. Arcata, California.

10. Restoring Floodplain Forests

Dean Rae Berg, Arthur McKee, and Michael J. Maki

Abstract

Floodplain riparian forests influence aquatic ecosystems by providing shade, allochthonous input of organic matter to the stream food chain, filtration of nutrients and fine sediments from overland and shallow groundwater flows, and a source of large wood critical to the health of aquatic ecosystems. When it comes to restoration of floodplain forests one must understand the spatial watershed context and stage of forest stand development. Returning to pre-European settlement conditions is nearly impossible in urban and industrial floodplains, extremely difficult in agricultural systems, and unlikely over large portions of forested regions. Therefore, we believe that efforts to attain desired future conditions realistically refer to the conservation and enhancement of intact riparian forest components and restoration of processes through changes in management of forested floodplains. Floodplain forest recovery strategies are beginning to foster a return to riparian stand conditions and ecological linkages analogous to pre-European conditions, complemented with management activities based on understanding the effects of physical and biological processes between terrestrial and aquatic ecosystems. Active management practices (e.g., thinning, planting, and shrub and herb control) may accelerate achievement of desired conditions in severely degraded riparian forest systems and can result in an ecologically healthy river if done with due consideration to both local processes and the position in the watershed.

Introduction

This chapter discusses the restoration of floodplain forests, the associated terrestrial-aquatic linkages and functions, and their relative roles and contributions along a stream network from headwaters to estuaries. The floodplain, as the terrestrial context for riverine processes, varies both physically and biologically through a watershed. We discuss the effects of human activities on the riparian landscape and the potential roles active management can play in stream restoration in forested, agricultural, and urban/industrial reaches. Natural riparian forests tend to be both structurally diverse and species rich compared to adjacent upland stands (Gregory et al. 1991). Landscape position within the watershed and local geomorphology combine with the disturbance regime to create a mosaic of complex, multi-story riparian forests (Pabst and Spies 1998).

The last century has seen the most dramatic changes at the watershed level in the Puget Sound region since the last Ice Age (Chapter 4). Human activities, such as intensive harvest of salmon and timber and conversion of riparian and upland forested areas to farmland and settlement, have led to extensive habitat degradation and placed salmon productivity and survival at risk (Nehlson et al. 1991). Historically, snags and large wood were removed from rivers to facilitate transportation (Sedell and Luchessa 1981). "Cleaning" the rivers of wood and the construction of dikes and levees for flood protection structurally simplified formerly complex lowland river reaches. The dramatic reduction in the volume of semi-stationary wood in the system (e.g., log jams and snags), combined with hydro-modification (e.g., riprap, dikes, and channelization), occurred synchronously with substantial increases in sediments and logging debris moving through the system.

Pre-European settlement streamside forests have been largely replaced with early seral riparian forest strips or non-forested land uses, which has decreased currently available and projected supplies of in-channel large wood (e.g., Andrus et al. 1988). Silvicultural strategies to replenish desirable forms and volumes of wood can greatly enhance long-term natural recovery processes, especially when patterned on natural processes and enhanced by deliberate and well-monitored management objectives (e.g., Beschta 1991, Chapter 9). The numerous functions that large wood provides vary across the landscape, as does the ability of the local forest to provide shade and trees of various sizes.

The relative importance of different riparian functions varies along the channel network (White 1991) depending on stream type and adjacent land uses (e.g., Cooper et al. 1987; Dillaha et al. 1989). Functional vegetation cover types in these areas may consist of trees (e.g., Peterjohn and Correll

1984; Lowrance et al. 1984b), managed herbaceous filter strips (e.g., Magette et al. 1989; Van Dijk et al. 1996), or a combination of vegetation types (e.g., Haycock and Pinay 1993). Other land uses, such as residential and industrial development, can integrate buffers of varying width and structure to enhance riparian functions and benefit aquatic resources (Budd et al. 1987).

Silviculture is the art of cultivating forests to meet landowners' objectives (Smith 1986) and derives from the latin roots of *Silvae*—the forest trees of a certain area—and the production, development, or improvement of a particular plant, animal, or commodity. Silviculture commonly refers to manipulating forest species assemblages toward some desired state or production level. Forest landowners and stewards have a variety of objectives in addition to timber production, including wildlife habitat, water quality concerns, recreation, and recovery of riparian floodplain functions. Reforestation is the placement of trees in areas to direct stand development toward a desired vegetation type. Establishing trees in riparian areas, many of which have not had trees for many decades, can be a difficult and expensive process, drawing on methods from both forestry and agriculture.

Agriculture is the production of annual or perennial crops, using early successional stages as models for output productivity with rapid organic matter turnover rates and microbial soil processes (Griffith et al. 1997). While soil management strategies for forest and field crops have some significant differences, all have as a goal soil carbon accretion and nutrient retention for conversion into desired outputs (e.g., crops, trees, and habitat). Because of available soil nutrients, moisture, and tilth, agricultural development has largely taken place in alluvial floodplain areas. Intensive forms and levels of agriculture have resulted in varying degrees of riparian habitat degradation, including nutrient and pesticide runoff, riparian vegetation removal or degradation, and consequent sediment delivery to streams in addition to altered hydrologic, soil, and plant community conditions.

In Washington, forested buffers are mandated to protect streams and rivers from the effects of timber harvest, agriculture, and other development. Strategies for riparian forest creation, maintenance, and enhancement draw upon combined knowledge from the fields of silviculture, agriculture, aquatic ecology, geomorphology, and hydrology. Managed natural processes can direct and accelerate desired riparian plant community development. Floodplain processes comprise the terrestrial component of complex, dynamic aquatic systems, which support salmonid species of Pacific Northwest ecosystems.

Floodplain riparian forests provide a variety of ecological services or functions (e.g., stream bank protection, shade, allochthonous nutrient input, nutrient and fine sediment filtration, and the source of large wood) that influence the stream channel to varying degrees (NRC 1996). Riparian forests provide the large diameter wood from mature trees that partly control geomorphic process on the valley floor (Triska et al. 1982; Bisson et al. 1987; Abbe and Montgomery 1996, Chapter 3). Large woody debris creates complex channel morphologies, alters flow direction and turbulence, as well as influences sediment storage and acts as organic substrate for periphyton and other aquatic organisms. Large wood acts as a physical structure within and adjacent to the stream channel, providing hydraulic diversity similar to bedrock and boulder features, but with some unique characteristics. Large wood can move during flood events and reconfigure in jams and as single key pieces, forming the basis for deep pools and other channel structures such as islands (Fetherston et al. 1995; Abbe and Montgomery 1996). Wood further benefits the aquatic system as it eventually breaks down into organic sediments in estuarine and other downstream areas (Maser et al. 1989; Sedell et al. 1989).

As a riparian management goal, return to pre-European settlement conditions is constrained along more intensively developed reaches, but management of floodplain forests even in these areas can improve riparian functions. Reference to late seral forest conditions and watershed assessment (Chapters 4 and 8) provide management direction to achieve specific functions. We hold that the measure of stream restoration should be actual, desired instream changes, while recognizing that the long-term solution includes the broader recovery of riparian forest and other landscape functions. Past restoration efforts targeted at salmonids have had mixed results (House and Boehne 1985, 1986; Frissell and Nawa 1992), and while much has been learned, longer term analysis and subsequent data-driven adaptive management is needed throughout the Pacific Northwest (Hollings 1978; Chapter 9).

One restoration philosophy promotes a passive restoration strategy (e.g., Knutson and Naef 1997), relying on the resilience of natural systems to re-establish themselves. However, given the introduction and establishment of exotic species within aquatic and riparian systems, a future equilibrium will not be the same as pre-European settlement conditions. Where riparian forests are relatively intact, there should be a strong effort to preserve those stands. Such stands have a range of seral stages with a variety of species and an area large enough to include multi-species habitat conditions. The urgency of Pacific salmon recovery in the face of fishery overharvest and ineffective management of freshwater habitat over the last century, however, argues for using riparian silvicultural strategies to assist in accelerated stream habitat rehabilitation.

Silvicultural activities can be used in concert with natural processes to significantly improve short- and long-term salmon and aquatic habitat enhancement (Franklin et al. 1996). Current forestry regulations and policies in the western United States are supposed to protect and conserve existing riparian forests. However, tens of thousands of stream miles have already been modified by logging and development and do not resemble historic riparian forests—the model for "fully functional" conditions. Simply setting aside riparian areas may not return a full suite of desired riparian functions for many decades, especially within lower sections of rivers where urban, agriculture, and industrial land use are well established, and where structures such as dikes and levees have been in place for many decades. Silvicultural and engineering approaches offer a broad range of tools that can provide interim analog components and natural process functions.

Floodplain size varies along the channel network and floodplain functions can be altered by land use. In the Pacific Northwest, land use tends to be tightly coupled to landscape position and geomorphic conditions. For example, present day forests typically occupy the steep headwaters, agriculture is located on the alluvial plains of larger rivers, and industrial/urban development occurs in the lower basin.

Overall recovery of riparian functions can be accelerated through active management of the streamside forests and the channels themselves. The "taming" of river systems that has resulted through engineered bank stabilization and erosion control can be mitigated partially by off-channel habitat creation and maintenance, woody debris supplementation, and riparian forest recovery. Presently, most riparian forests in the Pacific Northwest are a poor source of large wood. Numerous watershed analysis recommendations include retention and production of large dimension conifer trees as a future source of large wood. While some riparian forest functions (e.g., shade/stream temperature, sediment storage and transport, bank stability, and allochthonous inputs) directly impact stream conditions, other impacts have indirect results, such as over time from genetic diversity conservation, stochastic processes, and the production of large-dimension decay-resistant logs recruited typically from coniferous trees.

Functions of Floodplain Riparian Forests

Floodplain riparian forests provide a variety of ecological functions (Swanson et al. 1982). We have grouped these functions into three categories—Energy, Nutrients, and Habitat—which act in concert to create and structure habitat. Riparian forests play an integral role in watershed processes, connecting laterally and longitudinally the wide variety of aquatic and terrestrial habitats (Figure 1). Re-establishment or enhancement of riparian forests requires preserving existing native forested riparian plant communities and directing stand establishment and development where these communities are absent. However, attempting to provide the same complements and levels of stream and riparian function throughout all reaches is neither realistic nor cost-effec-

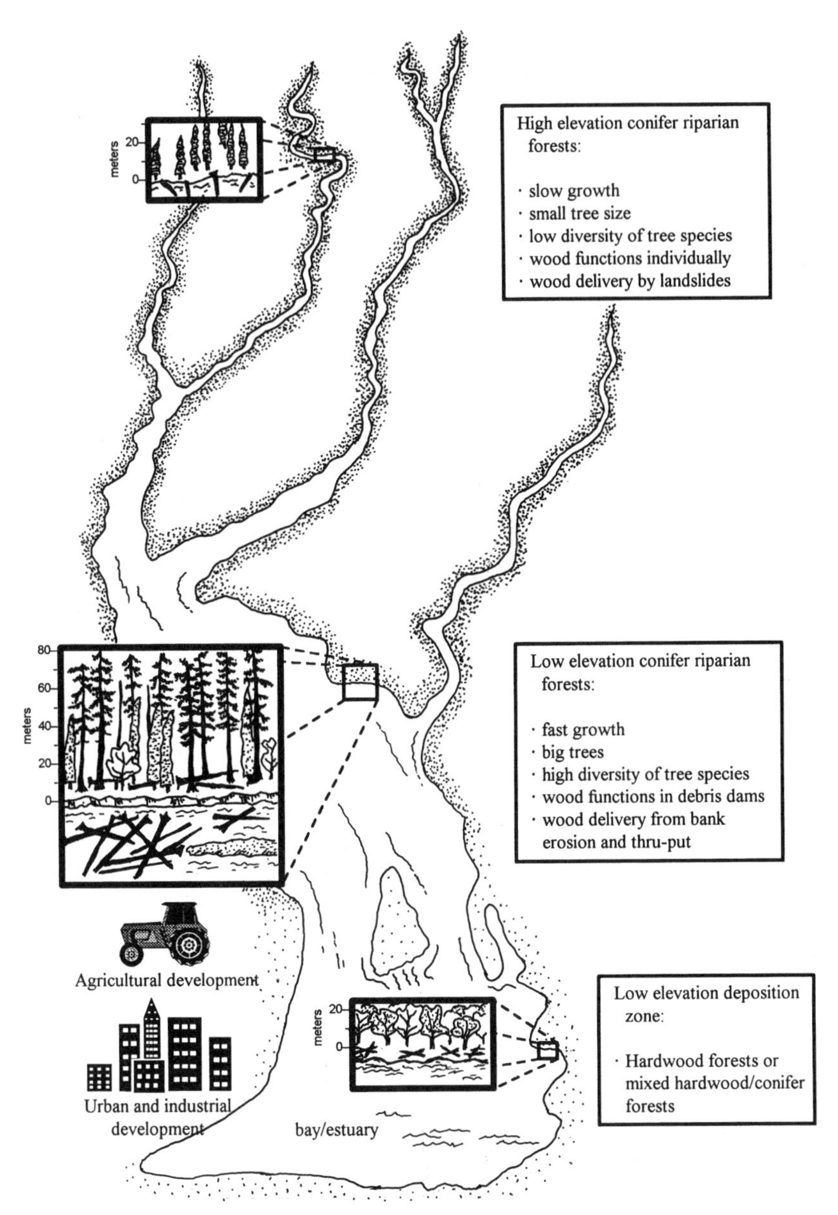

Figure 1. A simplified view of a watershed applied to riparian floodplain forests. Development typically increases in intensity from the headwaters to the estuary. Many of the functions of riparian forest floodplains and their structure change along a continuum as the river flows downstream (modified from Vannote et al. 1980).

tive. Instead, inventory and assessment are needed to integrate functional components longitudinally along the river from headwaters to mouth (Osborne and Kovacic 1993; Chapter 8).

Energy

Aquatic systems do not thrive by allochthonous inputs alone (Minshall 1978), and while conventional wisdom for riparian rehabilitation holds that "more shade is better," sunlight can have a beneficial role in stream autotrophic production. In general, light penetration, which drives photosynthetic and autotrophic production, increases with increasing cahnnel width even in streams with forested buffers, up to a point where stream depth and turbidity limit light penetration (Minshall 1978; Naiman and Sedell 1980). Light penetration increases again in alluvial fans and estuarine areas (Vannote et al. 1980; Triska et al. 1982). Concurrently, stream temperature generally increases downstream due to changes in elevation, air temperature, and stream width. In combination with changes in light interception and instream photosynthesis, this downstream temperature increase alters the aquatic community structure.

Shade from riparian vegetation provides salmonid habitat needs, including temperature maintenance and protection from predators. Smaller streams can be more effectively shaded by riparian vegetation (depending on solar aspect), and this is where shading can have a relatively greater effect on stream heating (Beschta et al. 1987). Stream shading is most important for temperature moderation during peak summer high temperatures; in winter, riparian canopy shade is insignificant and perhaps detrimental to autotrophic production, except for providing thermal cover during extreme cold (Beschta et al. 1987). Microclimate effects as measured by riparian air temperature, humidity, and soil temperature show little transference to stream temperatures (Brosofske et al. 1997). Stream temperatures are largely a function of combined temperatures of confluent streams, based on the relationship of temperature and volume, but variably influenced by groundwater and hyporheic inflow, geomorphic channel conditions, and larger climatic factors (Beschta et al. 1987).

Removal of streamside vegetation in upper reaches through harvest and other mechanisms can increase diurnal temperatures (ranging from less than 0.5° C to over 10° C) from incoming solar energy directly striking the water's surface (Murphy et al. 1986; Holtby 1988). Forest practices, such as large-scale clearcutting and road building that influence riparian cover, mass erosion, slope failure, and stream reconfiguration can have cumulative effects on stream temperatures (Beschta and Taylor 1988). Shade by itself does not decrease temperatures; rather, it moderates incremental temperature increases from direct solar exposure. When overall temperature increases are produced by complete or partial exposure of a stream to sun, they will subsequently return to ambient temperature over time as riparian vegetation regrows. Sites in moist coastal ecosystems of the Pacific Northwest recover more rapidly than those in more arid ecosystems; sites at lower elevations recover more rapidly than those at higher elevations (Beschta et al. 1987).

Changes in aquatic communities brought about by canopy openings are variable and dynamic in nature (Chapman and Knudsen 1980, Murphy et al. 1981), but they generally follow trends of increased productivity for a range of species (Hawkins et al. 1982; Wilzbach and Cummins 1986). Hawkins et al. (1983) studied production of invertebrates and vertebrates (including salmonids) in shaded and unshaded stream reaches. They showed a consistent increase in density and biomass of all groups when open canopy conditions are present. Even the negative effects of fine particle sedimentation were overcome by increased solar radiation.

Several key subjects merit additional investigation. The relationship between channel orientation and temperature is a refinement that would improve site-specific silvicultural prescriptions. The geometric relationship between tree height, solar angle, streamside terrain slope, and stream temperature has a direct influence on the impact of management on stream temperature (Figure 2). While late summer is the most limiting season in terms of thermal effects, the influence of the time of year (angle of incidence) and canopy density (attenuation) on solar radiation are still important and poorly described. An understanding of the magnitude of the changes in stream temperature throughout stream networks is essential for prioritizing restoration efforts targeting stream temperature.

Nutrients

Input of Organic Matter

The effects of organic debris on anadromous fish vary among salmonid species, with different types of features being preferred by one species or life stage over another (Bryant 1983). Deciduous trees (e.g., black cottonwood [*Populus trichocarpa*], red alder [*Alnus rubra*], maples [*Acer macrophylum, A. circinatum*], and willows [*Salix* spp.]) supply leaves, branches, and twigs to the aquatic and riparian environment. These allochthonous inputs provide structure, substrate, and nutrient values to the stream ecosystem (Bilby and Likens 1980; Bilby 1981, Harmon et al. 1986). Seasonal input of deciduous litter and herbaceous streamside plant materials rapidly decompose in warmer fall water temperatures and provide basic nutrients to aquatic organisms (Suberkropp 1998). Conifer needles and twigs, which break down much more slowly, provide a less nutritious but longer lasting source of energy and nutrients for stream invertebrates (Cummins 1974; Bird and Kaushik 1981). Decomposition of organic matter results in smaller and less nutritious particles, which causes the relative metabolic contribution of the detritus community to decrease with increasing distance downstream (Bisson and Bilby 1998). Allochthonous supplies of organic nutrients spiral through the system, increasing in amplitude as they move downstream when adequate wood and hydraulic diversity is present (Bilby 1981; Bilby and Likens 1989; Bilby and Bisson 1992).

Wood in all its forms provides a variety of functions to aquatic systems and ultimately to salmon (Maser et al. 1989). Large wood retains leaves and other

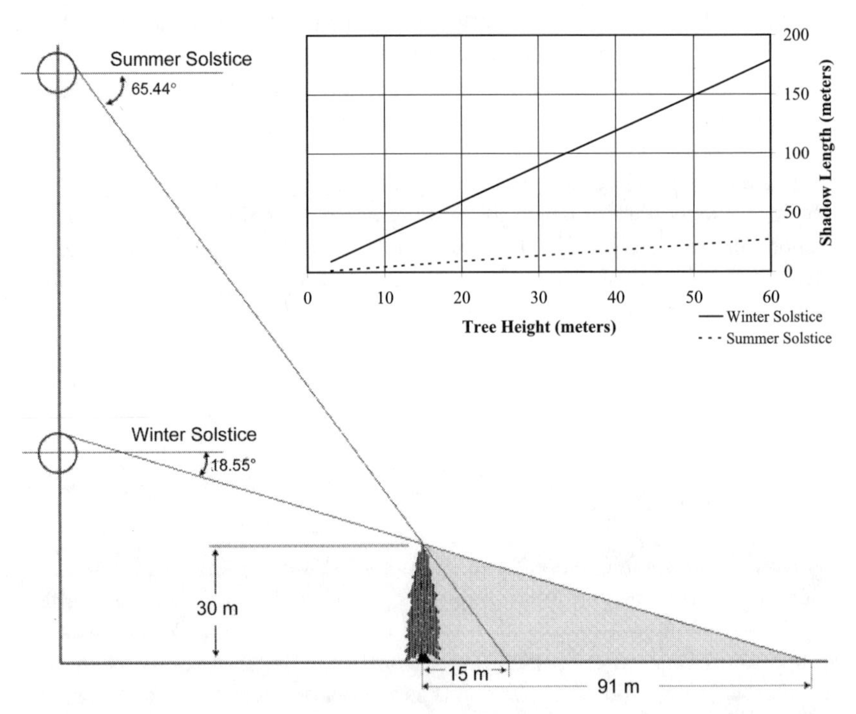

Figure 2. Actual shading dimensions are dependent on not only canopy height but also geographic location and solar aspect. Shadow length is behind 30 m tree at 48° north latitude for summer and winter solstices. Inset shows shadow length related to tree height at different times of the year at 48° north latitude.

allochthonous inputs, including salmon carcasses, by anchoring debris dams and creating pools of varying sizes (Boling et al. 1975; Sedell et al. 1975; Bilby and Likens 1980; Cederholm and Peterson 1985). Instream allochthonous nutrient processing rates decrease significantly in winter months (Bilby and Bisson 1992), due to lower temperatures, lower biotic respiration rates, and shorter retention times, exacerbated where debris dams are not in place to retain litter. Long-term studies of large-wood budgets of watersheds (analogous to sediment budgets) have not been done, but climate, geomorphology, forest community type, and human management practices will most likely influence wood budgets (Harmon et al. 1986).

Nutrient and Chemical Regulation

Riparian vegetation and various biological and physical filtration processes accumulate and use essential elements such as nitrogen, sulphur, and phosphorus and incorporate them into the forest ecosystem biomass. Aquatic systems in shaded headwater streams depend almost exclusively on allochthonous inputs (Bilby and Bisson 1992; McClain et al. 1998). Nitrogen and phosphorus, as well as trace elements, are limiting factors in productivity for aquatic and terrestrial species in forest ecosystems (Bilby and Bisson 1992; Naiman et al. 1992). However, in lower gradient alluvial floodplain areas where agriculture and other intensive human development occur, increased temperatures, decreased dissolved oxygen (DO) levels, cumulative nutrient inputs, and increased autotrophic production may negatively affect salmonid populations (Lowrance et al. 1984a). Riparian areas stabilize sedimentation and erosion processes while they uptake nutrients and moderate stream temperatures with shade. Over the long term, concerns have been raised about phosphorus and nitrogen saturation of established agricultural riparian buffers (Karr and Schlosser 1978; Omernik et al. 1981; Osborne and Koviacic 1993; Cooper et al. 1995).

In riverine ecosystems in the Pacific Northwest and elsewhere, anadromous salmon return nutrients from the marine environment on an annual basis. Marine-derived nutrients are recycled through the food webs of the aquatic and riparian ecosystems. Spawned-out carcasses are the main source of returning nutrients and, where salmon remain abundant, are distributed extensively throughout the aquatic system where salmon are able to spawn (e.g., Cederholm and Petersen 1985; Bilby et al. 1996; Larkin and Slaney 1997; Cederholm et al. 1999). Red alder stands also provide nitrogen, influence the short-term mobilization of phosphorus (Robert Bilby, personal communication), and are a valuable component of any functional riparian forest in the maritime Pacific Northwest.

The nutrient status of low-gradient streams has fluctuated over the last century. Prior to European settlement, recurrent salmon runs returned large quantities of marine-derived nutrients to aquatic systems. Intensive harvest of salmon drastically reduced these renewable nutrient sources. As floodplain forests were cleared and developed for farming, agriculturally-derived nutrient levels increased. Improved agricultural practices have reduced nutrient input to streams, and contemporary fisheries management does not consider the returning marine-derived nutrients in setting fishery harvest levels (Michael 1998, 1999; Gresh et al. 2000). Nutrient levels needed to support aquatic biota and salmonid recovery have not been empirically determined.

Input of large wood

Historically, conifers dominated the headwaters of forested watersheds of the Pacific Northwest (Franklin and Dyrness 1969). The lower floodplain areas where residential, agriculture, and commercial development are located today once contained extensive lowland forests of black cottonwood, red alder, and bigleaf maple as well as conifers including western redcedar (*Thuja plicata*), Sitka spruce (*Picea sitchensis*), western hemlock (*Tsuga heterophylla*), grand fir (*Abies grandis*), and Douglas fir (*Pseudotsuga menzesii*) (Chapter 4). In diverse riparian zones all along Puget Sound watercourses, early successional species such as red alder are interspersed with patches of flood tolerant willows and riparian shrub species. Such patchiness and consequent structural heterogeneity was maintained by floods and channel migration (Bragg 2000). Tree species specially adapted to riparian conditions such as seasonal flooding and high water tables (e.g., Oregon ash [*Fraxinus latifolia*]) or new alluvial deposits (e.g., black cottonwood) grow in uniform or mixed stands. More stable riparian settings are able to support coniferous species such as Sitka spruce and western redcedar.

Large wood is integral to aquatic ecosystems in the Pacific Northwest because it provides key structural elements that create and enhance salmonid habitat (Murphy et al. 1986; Lienkaemper and Swanson 1987; Murphy and Koski 1989). A diversity of wood sizes creates debris dams and pools, log jams, and bank features that support a range of salmonid life stages. Large wood decays and disintegrates in a stream over time and needs replacement from the source. Relative persistence, mobility, and aquatic functioning of large wood varies along the length of a river system (Bilby and Likens 1989). Physical abrasion, species characteristics, exposure, and ambient temperature are all factors that affect the rate of decay (Harmon et al. 1986; Beechie et al. 2000). Large conifers provide the most desirable structural elements because of their form and slow decomposition rates; hardwoods decompose more quickly than conifers unless they are completely submerged (Hodkinson 1974). At an average decomposition rate of 2% per year (Murphy and Koski 1989), the full replacement of the standing crop should occur over 50 years. Random recruitment delivers a fraction of the potential large wood through episodic disturbances such as windstorms, fires, landslides, and stream bank erosion caused by floods. Trees recruited to the stream can force creation of pools and bars—thereby increasing habitat complexity (Lienkaemper and Swanson 1987). Large wood may immediately affect channel structure and thus structural diversity or require an indeterminate time to become functional instream components. Montgomery et al. (1995a) note that less than one quarter of in-channel large wood contributes to pool formation. This implies riparian silvicultural systems should consider such inefficiency in setting goals for recruiting large wood.

In old growth (>200 year old) forest channels surveyed by Murphy and Koski (1989), more than 70% of the large wood of small headwater (first through third order) streams originated within 20 m of the channel, and 11% originated within 1 m of the bank. Of the large wood pieces in mature (80–200 year old) stands, McDade et al. (1990) found that 83% of hardwood large wood pieces originated within 10 m of the stream channel, compared to 53% of the conifer pieces in the same stand. Under pre-European settlement conditions in mature stands, nearly 50% of large wood could be recruited from within 20 m of the channel bank (McDade et al. 1990). Wind-throw combined with bank undercutting is the principal method of recruitment. In a study of first through fifth order streams, Lienkaemper and Swanson (1987) found that about one-third (34%) of the total large wood pieces originated within 1 m of the stream over a 9-year study period, and report wind-throw as the most common delivery mechanism.

Volumes and sizes of large wood in streams have declined since the initial logging of old growth forests. Legacy wood (wood that is embedded in the channel from decades or centuries past) continues to be the most significant part of the structurally functional large wood (Lisle 1986). This includes wood exhumed through erosional processes. In a study of Oregon streams by Andrus et al. (1988), fifty years after initial logging and subsequent forest re-establishment, only 14% of the large wood in study streams came from new forest wood. Grette (1985) found that second growth debris from 50–60 year old forests only contributed 20–30% of the total large wood, much of it in the less long-lasting form of hardwood debris. Depending on native processes of riparian forest re-establishment, large wood recruitment approaching old growth levels would take centuries to produce equivalent levels and forms. Second-growth forests have not developed to the point of delivering wood of sufficient size to act as geomorphic elements (Grette 1985; Andrus et al.

1988), and short rotation forestry holds little prospect for production of key large wood pieces. The need for mid- to long-term recruitment of key pieces calls for a supplementation strategy or engineered substitution if levels of large wood similar to old growth reference conditions are to be achieved under short rotation forestry.

In heavily managed areas with sufficiently large standing wood (snags and green trees), random recruitment will need to be augmented with a more deliberate conscription and directed placement of wood pieces to maximize the likelihood of providing functional value in the aquatic ecosystem. In the short term, hardwoods provide utility and may be grown and delivered streamside as part of riparian silvicultural strategies and schedules. The large basal diameter of cottonwoods in particular may provide key pieces in a short time frame (Collins and Montgomery 2002), which offer the added benefit of regenerating colonades of live trees along fallen stems.

Habitat

Forests provide large woody debris, which serve a number of functions and settings as they are transported downstream (Bryant 1983; Heede 1985). Downstream reaches are sink areas, using material derived from upstream to retain sediments, increase instream structural complexity, and eventually decompose in estuarine or nearshore ecotones. Floodplain vegetation provides bank stability, reduces hydraulic force, captures sediments during flood events, and provides microclimatic amelioration to streamside areas.

The continuum of riparian forest species, from conifer-dominated uplands to more deciduous lower floodplain areas, is determined by climatic and geomorphic conditions, as well as stochastic flood events that reset riparian forest plant communities. Also, the physical interactions of the multiple process domains along a river system (Montgomery 1999) affect the development and stability of riparian forests and influence the type and characteristics of instream habitat, forest development, and subsequently the time required to recover. Riparian management in lowland floodplains can help buffer other important sediment and temperature functions (e.g., sorted clean gravel for spawning beds and cool, clear water).

Streambank Stability and Cover

Floodplain forests affect channel bank erosion and therefore channel migration. Root strength of streamside vegetation can contribute significantly to bank stability. Conversely, fallen trees can direct flow into and destabilize channel banks. In large rivers, floodplain forests stabilize mid-channel bars and islands (Fetherston et al. 1995; Abbe and Montgomery 1996). Logs and other debris in or above the water level, stream bank overhangs (bank cover) (Bustard and Narver 1975), and water turbulence provide the most effective protective cover for salmonids. The rootwads of large trees contribute an important ephemeral structure to overhanging stream banks and offer refuge habitat. Boulders and other geomorphic features play an equally important role in aquatic infrastructure as they anchor large wood and generate hydraulic diversity. Under natural conditions flood events create these cover features, and many restoration projects build or place such features to increase salmonid habitat diversity (Nickelson et al. 1992). Stability of human-placed large wood and other structural elements depends on placement techniques and intensity of erosional events (Frissell and Nawa 1992). Central to many stream and riparian functions are log jams (Maser et al. 1989), many of which have been removed over the past century to aid navigation, log transport, and salvage.

Hyporheic Flow

Input from small tributaries and springs (Bilby 1984) and from subsurface exchange of hyporheic flow provide a significant source of cool water (Boulton et al. 1998). The hyporheic zone is not only a source of cool water to support base flow, it is a biologically active subsurface section of the riparian zone, containing a moving medium that includes bacteria and invertebrates (Hynes 1983, Stanford and Ward 1988). Disconnection from hyporheic exchange due to excessive sedimentation can increase stream temperature, reduce dissolved oxygen (Ringler and Hall 1975), and decrease beneficial microbiological activity.

Floodplain Forests and River Size

Floodplains, forests, and woody debris have different functions in different positions in a watershed (Figure 3). Headwater streams benefit from buffers that maintain adjacent slope stability, shade, and a source of large wood. Slightly larger streams with greater flow and velocity capable of floating large wood often serve as source areas for wood recruitment (Van Sickle and Gregory 1990). Freely migrating floodplain streams regularly recruit large wood that replenishes that removed by floods, decomposition, and burial.

Wood loading at any one point represents a portion of the cumulative flow of material transport from other points upstream. Large wood delivery to the channel is slow in some forest types that are composed of physically stable, long-lived conifers. The accumulation of wood in jams has different impacts along the river.

(A) (B) (C) (D) (E) (F) (G) (H)

(I) (J)

(K) (L)

(M) (N)

Figure 3. Typical floodplain forests along a continuum of channel sizes in the Pacific Northwest. The left column is an aerial view of the riparian forest and the right column shows more detail of riparian stand structure relative to stream size. (A) Valhalla Wilderness, B.C.; (B) West Twin Creek, Olympic National Park, WA; (C) Finney Creek, Mt. Baker-Snoqualimie NF, WA; (D) Slocan Valley, B.C.; (E) Deer Creek, Snohomish County, WA; (F) North Fork Calawah River, Olympic Experimental Forest, WA; (G) South Fork Hoh River, Olympic National Park, WA; (H) Naches River, Wenatchee NF, WA; (I) Bear Creek, King County, WA; (J) Samish River, Skagit County, WA; (K) Skagit River at the mouth of the Sauk River, WA; (L) Lower Columbia River, near Kalama, WA; (M) Mouth of the Duwamish River, Seattle, WA; (N) Mountlake cut, Lake Washington, WA.

Headwater streams tend toward high accumulations of wood relative to the channel width because of the limited ability to transport wood. These headwater stream systems are able to contribute large wood downstream only during episodic flooding, debris torrents, and mass wasting events. Sediment retention around large wood combined with geomorphic features increases nutrient retention and biological processing time (Bilby and Likens 1980; Bilby 1981).

Typically, mid-sized streams contain a variety of habitats including spawning, rearing, foraging, and refuge sites. Here, large wood creates and maintains pools and bars along the longitudinal profile of the stream channel.

As streams increase in size, large wood plays a different role in shaping channels and storing sediments as floods regularly redistribute wood downstream and can reset forest stand development. Meandering mainstem channels create oxbows, backwaters, and sloughs that provide shelter for salmonids and other species during high flow periods and follow a geomorphic successional process of sediment infilling over time. Cottonwood forests develop on low terraces created by episodic floods (Braatne et al. 1996). Middle and lower reach habitat is highly influenced by log accumulations (jams), creating a complex channel structure (Bisson et al. 1987). Maintaining the complexity of these habitats demands a continual source of large wood. Conifer logs are most persistent under these circumstances of exposure and mobility. Although only large trees can provide key pieces, wood of all sizes can play an important role in shaping aquatic habitat, and sustained supplies are necessary to maintain structure to aquatic systems.

The lower portions of Puget Sound watersheds are affected by a wide variety of land uses. Floodplains are often well developed and streams constrained by structures (e.g., riprap and diking). Transportation corridors (e.g., railroads, highways, and bridges) control high water flow as dramatically as any landforms. Historically, rivers meandered across the valleys (Sedell and Luchessa 1981; Sedell and Froggatt 1984) with regular bank overflow events spilling sediment-rich waters onto the floodplain, periodically recharging marshes. This occasional exchange of sediments, nutrients, propagules, and larger organic materials created valley soils and greatly influenced riparian forest communities.

Active Management of Floodplains

Effective design of recovery strategies for key riparian functions demands expertise in forest ecology, fisheries biology, geomorphology, hydrology, silviculture and forest stand dynamics, forest engineering, and other scientific disciplines. Forests, although dynamic in nature over time, are basically stationary; fish are essentially mobile, both through anadromy and through foraging behavior at various life stages. The union of these two ecological systems into a single management prescription is largely undeveloped and needs to have substantial monitoring associated with a number of performance parameters. A key component of any aquatic conservation strategy is riparian silviculture. Silvicultural prescriptions within a watershed will range from no treatment to very active management (Table 1). The degree of intervention in the stand development process depends on the forest stand structure (e.g., size, age, density, and composition), present land use and ownership, and acceptable level of risk. Young stands are quite malleable and are more responsive to designed changes in growth and development trajectory (Oliver and Larsen 1990). Older stands can be managed to contribute important ecological features, such as large wood through direct conscription rather than random recruitment.

Riparian forest types have different influences along the longitudinal profile of a river (Vannote et al. 1980; Triska et al. 1982). Management strategies should be coupled with the functional requirements of the system at the reach level, using measurable criteria and following outcomes through time with an adaptive management strategy (Berg et al. 1996). Active management techniques address particular functions and limiting factors. A landscape or watershed level view is essential for prioritizing where recovery activities should take place and identifying the functional requirements of the systems being rehabilitated or enhanced (Chapter 8). Diagnosis requires measurable criteria that describe the system and monitor progress or deviation from the system design (Ishikawa 1982). Prescriptions are designed to provide the forest structure and composition necessary to support one or more specific ecological functions. These prescriptions must be embedded, if possible, in a larger watershed and landscape-level strategy to engage positively with habitat-forming processes (Montgomery et al. 1995b).

Silviculture of Floodplain Forests

Because large wood is so important to the ecological properties of streams, the source of this material—riparian forests—requires strong measures to protect high quality sources and provisions to recover the composition and structure of the forest necessary to provide large woody debris to streams and rivers. Present-day lower elevation floodplain forests do not produce logs large enough to provide many aquatic and geomorphic functions. Hardwood-dominated floodplains offer a valuable short-term supply (McDade et al. 1990; Beechie et al. 2000) but can never achieve the size and decay resistance of many conifer species. For this reason, where riparian forest composi-

Table 1. Hypothesized and demonstrated effects of various forestry practices on selected functional relationships between forests and streams. Symbols indicate direction and strength of effect: ++ strongly positive; + positive; # uncertain; +/- or -/+ case-by-case, or species-specific basis, with most probable short-term effect shown first; - negative; and -- strongly-negative.

Functional Relationship	*Logging*	*Slash Removal*	*Planting*	*Thinning*	*Fertilize*	*Herbicide*	*Insecticide*	*Fencing*
Shading	--	+/-	+	+/-	+	-	+	+/++
Light	++	+/-	-	+/-	-	+/-	-	-
Fine debris input	--	-/+	+	-	+	-	#	++
Large debris input	--	-	+	+	+	+	#	++
Retention	+/-	-	+	+/-	+/-	-	+	+
Nutrient uptake	--/-	-/+	+	-	+/-	-	#	+/++
Peak flow	+	#	-	+	#	+	#	-/--
Summer low flow	+/-	#	-/+	+/-	#	+/-	#	+/-
Bank stabilization	-/--	-	+	-/+	#	-/+	#	++
Terrestrial habitat	+/-	-	+/-	+/-	#	#	#	+/-

tion has shifted to hardwoods exclusively over large portions of a watershed, some measures to recolonize these sites with conifers would be prudent. Restrictions on silvicultural hardwood conversion of riparian areas are justified because poorly designed recovery and regeneration operations can deplete short-term supplies of functional wood from hardwood stands. Hardwood-to-conifer conversion redirects forest stand dynamics. We would not suggest the extirpation of hardwoods from riparian areas, however, because it is neither desirable nor possible. Cottonwood, for example, is an essential element of the lower reach floodplain forest (Braatne et al. 1996), and red alder provides numerous ecological functions that conifers cannot (e.g., seasonal litter input, nitrogen fixation, and phosphorus mobilization). Active management of alder and shrub-dominated riparian forests will accelerate the process of coniferous riparian forest development. For example, maple, willow, and poplar can be managed by coppice methods to maintain root systems that enhance soil stability while freeing up growing space for conifers.

A range of approaches from passive (no human intervention) to active can maintain or recover riparian conifer or mixed conifer-hardwood forests. Future forests are subject to the stochastic disturbances that have historically shaped riparian forests. A passive approach will follow successional pathways; however, the stand may develop in an uncontrolled fashion toward some less desired structure or composition. For example, many of the pure red alder stands in riparian areas will naturally thin as they age due to mortality, and eventually some conifers will establish themselves in the gap. Dense hardwood thickets may develop tenacious understory communities that will remain patchy, sparsely stocked conifer stands for well over a century (Tappenier et al. 1991), especially in the absence of appropriate shade tolerant conifer seed sources (Beach and Halpern 2001). Hardwoods and shrubs dominate many riparian bottomlands (Hibbs and Giordano 1996) as a result of various disturbances. Selective thinning of hardwoods accelerates conifer dominance by directing where and how much conifer regeneration can be established. Ultimately, as the conifer stands mature, thinning and natural mortality will restore the processes of large wood recruitment.

Silvicultural system design uses prescriptions based on key diagnosis for a high degree of stand manipulation, as opposed to the more passive approaches of allowing successional process to occur unimpeded over subsequent decades or centuries. The goals for stand structure are derived from the decades of research that have detailed the role of riparian forests in salmonid habitat (e.g., large wood, canopy cover). Riparian set-asides (e.g., fixed width, non-managed buffers) in stream reaches with degraded forest composition and structure are an uncertain management strategy and may take a very long time to recover functions.

Active management of riparian forest zones for large wood recruitment is currently being developed for Pacific Northwest watersheds. Although there are numerous sources of information on the yield and dynamics of various tree species (e.g., Johnson 1955; Nystrom et al. 1984; Long et al. 1988; Curtis et al. 1997), the process of tree death and recruitment to streams is poorly understood. Foresters have only recently begun to follow-up modeling results with research about relative source distances and recruitment rates (McDade et al. 1990). Growth and yield models (e.g., Curtis et al. 1981) have been used in a variety of ways to predict the effect of a variety of silvicultural scenarios (e.g., Berg 1995, 1997). Modeling based on these assumptions should be tempered with an understanding of the limitations of forest growth and yield models to accurately predict mortality, let alone the delivery process (Van Sickle and Gregory 1990; Hairston-Strang and Adams 1998; Kennard et al. 1998). While informative for some decision making, modeling outputs are relative and should not be construed as absolute values or used to write resource management laws.

Tree death in riparian forests is poorly understood and will require more intensive long-term research. Mortality is predicted by simulation models, but most often the estimate is based on the least robust equations in a particular model. A new set of tools is being developed to explore the wood recruitment implications of a variety of management scenarios (e.g., Kennard et al. 1998; Bragg 2000). This approach is useful for comparing management scenarios but has limitations in the reliability of the absolute numerical values of the projected wood loading in streams. While interesting, the emphasis for long-term silvicultural prescriptions should be grounded in regional empirical relationships, some of which are still being developed.

Wind-throw requires careful attention, as riparian buffers are often wind-prone following logging (Steinblums 1984). Prevailing winds and episodic windstorms have the potential to become the determining factors in timing and availability of naturally recruited large wood (Grizzel and Wolf 1998). Concerns about windthrow are difficult to address, but slow adaptation to decreased stand density is best achieved through light thinning and multiple entries over many years.

Thinning

Thinning is largely an effort to increase the overall value of a stand by concentrating growth on fewer stems (Zeide 2001). After centuries of study (e.g., Brown 1882; Curtis et al. 1997), the results of thinning forests on standing volume still remain clouded (Zeide 2001), although there is little question that over-dense, young stands will likely stagnate, causing massive declines in productivity and increased mortality. Application of thinning prescriptions to floodplain forests can have a great impact. Total biomass is rarely the parameter of ecological function most sought in riparian ecosystem management. Rather, thinning of less desirable species is intended to establish or accelerate the development of species with greater ecological utility. An active approach of tightly controlled, prescribed thinning can shift the forest towards the desired composition. Where the conifer component is absent, a system of planting beneath a thinned canopy may be necessary.

Basic riparian forest stand dynamics and riparian silviculture research are in early stages of inquiry. For recovery of riparian habitat functions, the goal of thinning is first and foremost a functional forest structure and composition. While thinning has obvious benefits, there is still uncertainty about the long-term effects. Riparian thinning methods are largely untested in floodplain forest systems (Beechie et al. 2000), although there are ongoing trials in Oregon and Washington. Where the environmental risk is unacceptably high, guidelines must be developed to limit thinning based on the projected impact to habitat and stream channels.

As forest stands mature, there is increasing competition for limited light and nutrients. The crowding that occurs in unthinned stands leads to competition-induced mortality in a predictable way. There are a number of guides to predict the density and size of trees at a particular stage of stand development (Hibbs 1987; Smith 1987; Long et al. 1988). This type of thinning guide is generally referred to as a density management diagram (DMD) and graphically relates the number of trees per hectare that a forest is capable of carrying at a point in time measured by various size criteria (e.g., height, diameter, volume, and canopy closure). As stand density increases, growth slows to the point where competition for limited resources becomes so great that mortality occurs. This point can be expanded to a line (the empirical line of imminent competition induced mortality) if it is projected on two axes: tree density on the independent axis (X axis) and the size parameter on the dependent axis (Y axis). By following the empirical line of imminent competition induced mortality, silviculturists control losses from tree death, and the growth rate of the stand can be controlled to a degree. While tree selection is never perfect, the application of thinning guidelines enhances the chances that the stand will arrive at a specified size (e.g., key piece diameter and tree height). Not only can the stand development be directed to generate a specific size, but silviculturists often can accelerate the time to a target through thinning.

There may be some optimal stocking that produces stems of sufficient ecological requirements that also have high merchantable value. Once forests are established, tight, dense stands will likely stagnate and cease growth while those stands that are open grown may develop rapidly in diameter relative to height. Both situations produce very different crown and stem dimensions. Tight, closed forests are likely to produce tall, thin trees with height/diameter ratios far in excess of 100, well past the range for stable wind firm stems (Oliver and Larsen 1990). Open grown trees are much more stable and have full deep crowns often with large diameter branches.

The DMD approach is designed for even-aged, single species stands, which are not characteristic of riparian forests. Empirical yield tables (e.g., Meyer 1939; McArdle et al. 1949) for a variety of species can help interpret the uneven age and incomplete stocking data from actual riparian forests. The mixed species problem is not well understood, but some simulation models predict growth of these stand types, provided the models include the species that are interacting. A deficiency of a number of public domain models, such as ORGANON (e.g., Hester at al. 1989), is that they do not include important Northwest riparian trees (e.g., western redcedar and red alder). Growth information from other species has been used as a surrogate for missing species; however, no validation has been done.

Maintenance of existing riparian vegetation is critical for recovery of the shade function. The width or design of the protection zone from the streambank should reflect an analytical process that considers potential tree height, solar angle, stream orientation, and factors in management opportunities and costs. For thinning, a uniform removal with continuous forest cover has very different implications for productivity than a series of openings (gaps) interspersed in a matrix of forest patches. As the available light increases, the stability of the saplings increases, as measured by the ratio of the height to diameter at breast height (Ht/dbh). At around 30% available light, growth results in Ht/dbh that exceeds 100, roughly the limit of stability for conifers.

An approach for the development of management strategies for riparian areas is to utilize habitat diversity and the resilience and adaptability of species to take advantage of a range of available ecological niches. The role of sunlight is universal, directly driving autotrophic systems and indirectly supporting heterotrophic systems. Dent and Walsh (1997) found that a series of small openings along the stream (approximately 90 m long) had less impact on stream temperature than one large opening (approximately 180 m). Perhaps a series of streamside forest openings could be combined with other prescriptions, such as preservation of high quality riparian forest and an intermediate level of thinning and planting (Oliver and Hinckley 1987). In riparian forests with moderate wood recruitment potential, a program of variable density thinning combined with underplanting of select forest trees would enhance suitable future recruitment and shade functions (Franklin et al. 1996).

Planting

Composition of a riparian forest stand changes with time as ecological succession proceeds along a chronological pathway that depends on the level and timing of disturbance and competition (Oliver and Larson 1990). Plant associations are indicators of the complex environmental gradients of elevation (e.g., temperature and moisture), edaphic conditions (e.g., soil nutrients, base geology, and soil moisture), disturbance regime (e.g., wind, fire, and flood), and distance from stream (e.g., terrace level and microclimate).

Planting is one of the ways to rapidly re-establish or shift stand composition. There are several planting approaches that influence stand structure and survival, including the planting pattern and use of elevated planting sites (Figure 4). Dispersed planting, a convention transferred to riparian recovery from terrestrial tree farms, creates an evenly spaced, uniform plantation. Aggregated planting may have utility if rapid early growth is necessary to outcompete brush and recover large size trees as soon as possible. The added advantage of aggregated planting is that it allows for a heterogenous forest structure with many different species between the aggregates and may exhibit rapid early stem growth (Scott et al. 1998). Elevated planting sites have been successful in the regeneration of conifer seedlings in a number of spots in the Pacific Northwest. Whereas existing forest cover present at a site provides numerous ecological services, planting and tending a stand of another species within or adjacent to these conditions may diversify forest functions. Red alder forests fix nitrogen and provide shade, seasonal detritus, and small dimension wood. However, when planted with redcedar, for example, they add large decay-resistant wood over time. Cottonwoods could play a much larger part of the silvicultural design in the re-colonization of low elevation fluvial channels (Braatne et al. 1996).

Elevated planting sites are preferable where available. On river terraces on the South Fork of the Hoh River, downed log substrate for seedling establishment exceeded by an order of magnitude the regeneration directly on the forest floor (~35,000 stems per hectare vs. 700 stems per hectare) (McKee et al. 1982). Harmon and Franklin (1989) found this to be a regional phenomena that has a definable number of causes that include pathogens, predation, competition, and hydric soils (Beach and Halpern 2001).

Planting density affects early stem growth. Scott et al. (1998) found the growth of seedlings planted at about 3000 stems per hectare to be greater than all other lower densities that were tested. This counter-intuitive result has several as of yet untested explanations. The most probable is that the seedlings were able to more completely occupy the growing space and through

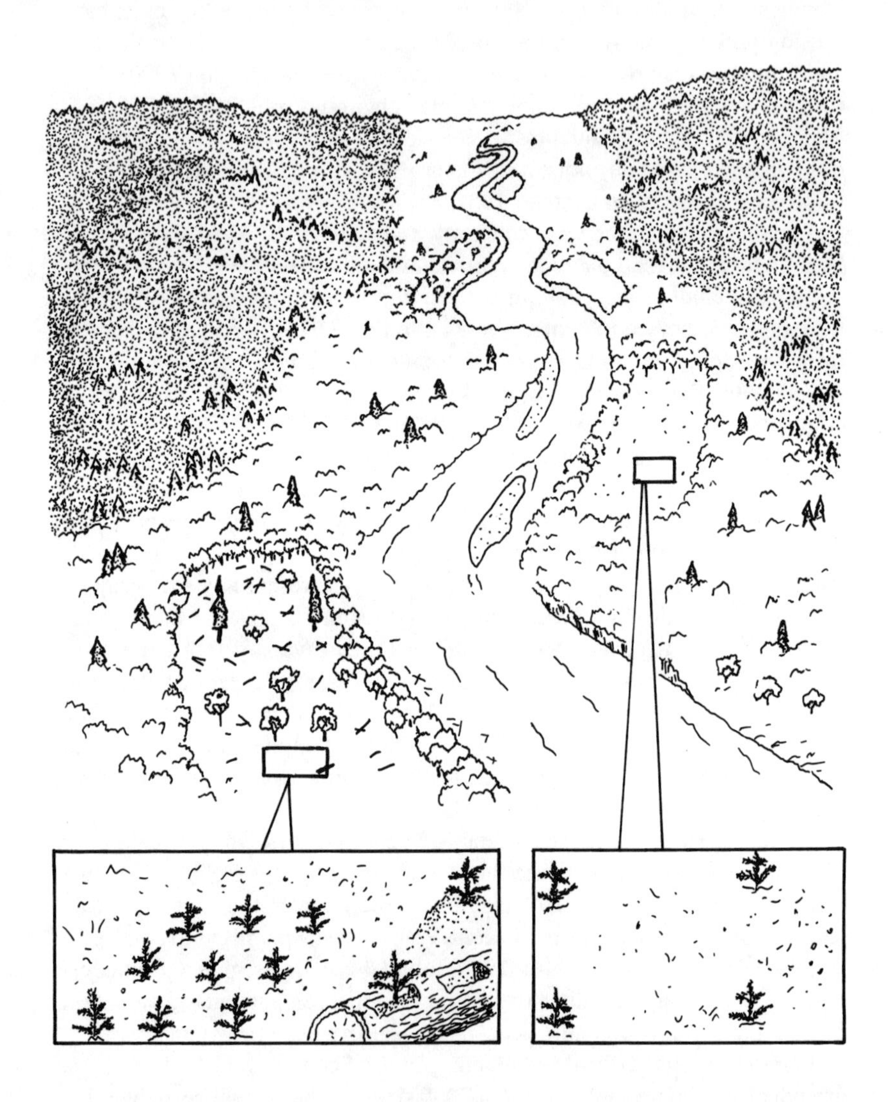

Figure 4. Riparian planting techniques used in demonstrations and experiments throughout the Pacific Northwest. Dispersed, aggregated, and elevated planting sites have had success in the recovery of conifer seedlings in a number of spots in the Pacific Northwest.

root grafting were able to exclude any competition for resources. Tight clumps also have the advantage of being more frost hardy, a problem in some riparian areas, while preventing soil moisture loss during late summer dryspells. The implication is that rapid early growth will more fully utilize growing space and lead to more complete occupation of a site. Thinning following this high-density treatment may be required to prevent early stagnation from over-crowded growing conditions.

The benefits of using planted nursery stock are also debated. While areas planted with appropriate tree species have clear advantages in terms of the establishment time, a forest will develop dense conifer composition on most sites in Pacific Northwest floodplains given enough time, reasonable regeneration conditions, and a seed source (Beach and Halpern 2001). The genetic diversity of the latter is an argument for advance and natural regeneration. However, Nehlson et al. (1991) describe a type of condition that dictates that at least in some areas an active approach of planting seedlings is appropriate.

Hazards to planted stock are numerous and include animal browsing, plant competition, and harsh growing conditions. Competition can be particularly detrimental, and therefore some sort of brush control and more intensive early manipulation may increase survival. Environmental stress is documented near streams, which can concentrate cold and create frost pockets. The other extreme, hot summer dryspells, can de-water some coarse-grained floodplain soils and lead to wilting of seedlings.

Perennial riparian vegetation, whose stems bend during floods, is suitable for stream bank stabilization, stream force reduction, and sediment accumulation during floods. Planting willows, red osier dogwood (*Cornus stolonifera*), or other streamside-adapted species through deep planting of cuttings or "live stakes" can give immediate structural stability while the new plant develops a reinforced root system. These streamside plants can also reduce the force of floodwaters and collect organic debris and sediments, effectively developing a bank of increasing stability and serving as a filter, fixing subsurface nutrients.

Recovery Period

Recovery of riparian sources of large wood can take decades or centuries. Managing the large wood recruitment area (Berg 1995; Beechie and Sibley 1997) can shorten this recovery period. In channels less than 10 m (33 feet) wide, small diameter wood (<20 cm; 8 in) can potentially form pools after 25 years (Beechie and Sibley 1997), a short timeframe relative to forest stand development. Recovery of a specific forest type will depend on the recent stand history as well as many of the factors mentioned above. There are multiple pathways or trajectories that can lead a forest stand to where it is at any point in time. These trajectories are important to understand if one is going to effectively direct forest development. The direction the stand will develop is strongly influenced by the trees planted. It is ecologically more appropriate to plant species that are historically represented or at least native to the region.

Techniques for replenishment of large wood include conservation of existing large wood pieces and combinations of cabling (including use of available hardwoods), engineered supplementation, placement of large wood pieces, and use of large wood analogs such as boulders and manufactured structures to increase hydraulic diversity (Chapters 15, 16, and 17). While some researchers see such engineered solutions as only marginally effective (Frissell and Nawa 1992) and call for wide (e.g., site potential tree height) unmanaged buffers to deliver old-growth equivalent large wood supplies over long periods of time, the desire to improve instream conditions and levels of large wood within shorter time frames motivates active management for the near term.

Thousands of kilometers of floodplain have become dominated by early seral, tenacious plant communities (e.g., red alder and salmonberry) and will proceed only slowly toward a mixed conifer forest (Tappenier et al. 1991). Accelerating the succession and the composition of these forests will take a combination of silvicultural activities over time—a silvicultural system—to recover desired ecological function (Berg et al. 1996). Where a dearth of conifer species limits recruitment of large woody debris, some combination of planting and thinning may be useful. In particular, brush competition can be severe and measures are often taken to allow seedling establishment. While mixed hardwood-conifer stands have the potential to provide sufficient regeneration (e.g., seed source and elevated platforms for establishment), augmentation of planted stock and thinning can direct stand development (Beach and Halpern 2001).

Agricultural Floodplains

Agricultural floodplains were historically a diverse patchwork of forest types and plant communities with an equally diverse set of alluvial soil and hydrologic conditions. Today these areas are broken up into an even more complex set of conditions with a variety of land uses. Agricultural development required the removal of preexisting forest cover, and in some cases also required installation of drainage systems to facilitate crop production. Once converted to agriculture and other uses, floodplains were protected from flood damage by dikes and levees. Networks of smaller streams and seasonal drainages throughout lowland floodplain areas, many of them historically occupied by beavers, were channelized, straightened, and in some cases culverted. Nevertheless, many riparian lowland areas "too wet to plow" were neglected or used only for grazing. In some areas, riparian trees such as cottonwoods, ash, and alders were deemed unmerchantable and left to line stream margins. At cultivated edges, non-native species of shrubs and forbs have invaded fence lines, field margins, roadsides, and unmanaged pastures. All these vegetative components contribute to riparian function but do not resemble the original floodplain forest.

The legacy of human agricultural development today provides one setting in which riparian recovery efforts must take place. In agriculturally developed floodplains, issues associated with riparian forest strip establishment are related to reforestation on sites heavily colonized by competing, aggressive, largely non-native vegetation. Agro-forestry strategies may therefore be the best way to first address the long-term process of functional riparian forest development. Agro-forestry strategies include effective site preparation before planting, use of appropriate planting stock and effective establishment techniques (e.g., fallow cycles, cover or smother crops, and soil building rotation crops), protection from herbivores, fertilization, timely irrigation, thinning, pruning, soil and nutrient supplementation, and conservation methods (Doyle et al. 1977).

The abundant resources of the riparian zone (moisture, nutrients, and colonization opportunities) make establishment of desired vegetation a competitive struggle that can only succeed with protracted and diligent management. Once established, riparian trees can be given progressively less attention, and over time they develop into self-sustaining woodland ecosystems.

At a larger landscape level, agricultural riparian restoration strategies must look at processes that are part of the natural regime, those that are active parts of current land uses, and legacies of past land uses. Watershed-level analyses (Chapter 8) suggest trajectories of current land-use patterns and provide direction based on ecological (intertwined natural and anthropogenic) trends, including disturbance regimes and a species mix that takes into account native and exotic species serving combined and analogous functions. In any case, it should be recognized that we are not returning to the past. In all likelihood, rarely again will old growth riparian forests dominate active agricultural floodplains. It may be impossible to return to past conditions, but efforts of floodplain recovery can re-establish some of the natural processes and functions.

Agricultural lands have been largely cleared of forest and are maintained as open areas except where there is a lower terrace that is difficult terrain and is not tillable. Agricultural and urban floodplains that have extant riparian forest are worthy of protection to maintain ecological function. In this case, thinning and planting could be used to shift or encourage a species shift toward conifer (e.g., redcedar and Douglas fir), perhaps in concert with fencing where livestock are abundant. Riparian floodplain silvicultural systems, such as those based on mixed rotation age hybrid poplar cultivation could provide some ecological functions while bringing a modest return to farmers.

Urban

The first objective in urban areas is to ensure no further degradation of riparian and aquatic habitat by protecting the remaining floodplain areas and stream reaches that are in good condition. Second, focus rehabilitation efforts on reaches that are able to respond to treatment, and finally, work toward a system that minimizes or ameliorates input of toxins, maintains water quality, and minimizes hydrologic disturbance (Chapter 11). In some urban areas, there may need to be less emphasis on restoring riparian habitat and more focus on regulating pollution so that water quality is good for fish health (Chapter 11).

Urban streams should not be written off in an overall program to improve aquatic and riparian habitat. Many salmon stocks traverse urban streams enroute to spawning habitat. Even small streams within urban areas can and do support local stocks, despite degraded conditions. Portland and Seattle are both working to improve streams such as these through rehabilitation projects. City parks and reserves provide opportunities to improve riparian habitat along streams. Often park management directives are purely recreation-driven, but could be easily modified to include environmental goals. Trails can be relocated a short distance away from streams; downed logs on the floodplain and in the streams need not be removed. Management should allow shrubs and herbs to grow in more natural conditions along streams and plant native plants where landscaping is required (Young-Berg and Berg 1994). These "natural" reaches could play very important roles in comprehensive rehabilitation plans.

The biggest technical and conceptual challenge facing floodplain restoration is how to build a comprehensive, equitable network of recovery efforts along the entire river system. Even when marginally effective in fisheries recovery, floodplain forest rehabilitation serves other valuable ecological and educational benefits. Active management of riparian areas has been applied mainly to the forested sections of watersheds, ignoring many of the urban and agricultural settings. These lower reaches are connected ecologically and should be part of any broad-based recovery effort targeted to return rivers to a more functional level with historical conditions as an ideal but very long-term goal. Ecological constraints in urban settings are often immovable and economically unfeasible to rehabilitate (e.g., roads, buildings, and ship canals). Agricultural barriers are less hardened physically but equally difficult. Farmers face declining financial margins and increasing pressure to convert to suburban development. In either case, careful consideration of the balance between ecologically possible outcomes and financial feasibility should yield viable recovery of a desired suite of floodplain function.

Monitoring and Adaptive Management

Using the principles of adaptive management, a flexible, scientifically-based approach to riparian habitat enhancement can restore functional conditions for support of salmonid recovery (Chapter 9). This approach uses quantitative tools to assess monitoring results to correct or adjust the progress of a system. For example, rigorous forest sampling can assess the degree of conifer establishment beneath a canopy of mixed species and hardwood stands. The level of conifer stocking defines the long-term large wood recruitment potential and indicates where stand composition deficiencies exist. This information is easily compiled to determine if regeneration goals are achieved, and if not, what could be prescribed to move toward projected targets.

Published Literature on Large Wood Loading and Delivery

Several key studies on Pacific Northwest watersheds establish target size, quantity and volume of large wood (Bilby and Ward 1991; Beechie and Sibley 1997; Chapter 14). The riparian forest varies along the longitudinal profile in the ability to provide structural elements of the channel, from higher elevations with narrow streams to lower elevations with generally wider streams (e.g. Vannote et al. 1980; Triska et al. 1982). At high elevations tree growth is slow and time to achieve logs of key size may be considerably longer than at lower, more productive elevations. Application of empirically derived density management concepts can help direct the type and timing of silvicultural systems that supply wood of suitable dimension. Because of the stochastic nature of disturbance and response of forests, merely setting the target to achieve some minimum level of tree size and distribution does not allow for the random events that may reset the riparian system to a less desirable state of forest development. Redundancy in biological systems is a key to ecological resiliency and, ideally, the silvicultural system will do more than merely strive for mediocrity.

Riparian ecosystem functions of shade, streambank stabilization, allochthonous inputs, large wood, wildlife cover, nutrient filtration, and microclimate buffering all vary with distance from the stream and are most effective in close proximity to streambanks. Ensuring future delivery of large wood is important to floodplain restoration (McDade et al. 1990), and site potential tree height in forested settings is commonly used as a determinant for estimating maximum effective buffer width. Source-area relationships can be used to evaluate the buffer width from which to expect reasonable recruitment (Table 2 and Figure 5), and management within this zone will vary depending on local processes and the relative balance desired between financial and ecological optimization.

Experimental Designs

Experimental designs test methods that integrate a long-term systemic approach to watershed recovery. Once broad-scale criteria and goals are established, ecological restoration still occurs at the site level, where actual recovery, guided by landscape and ecosystem management guidelines, are implemented. Adaptive management calls for all work to be seen and carried out as experimental (Chapter 9) with the results altering future management decisions and implementation strategies. Doing applied research requires taking some risks when operating in the environment where recovery is intended to

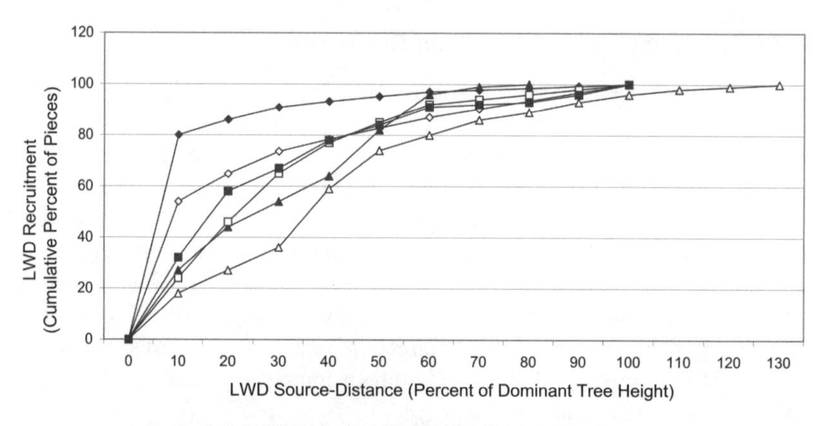

Figure 5. Large wood source curves for Coastal Pacific Northwest streams. (Courtesy of John Peters, US Fish and Wildlife Service, Arcata, California). For more information about the studies, see Table 2.

Table 2. Source and additional informaiton of data used in Figure 5. This table was generated by and is used with permission of John Peters, U.S. Fish and Wildlife Service, Arcata, California.

Authors	*Codes*	*Empirical or Model?*	*Forest Type*	*Forest Condition*	*Adjacent Forest*	*Tree Ht. (m)*	*Tree Fall Tendency*
Reid & Hilton (1998)	RB55	Empirical	Redwood	Age 90-130	Buffer	55	Downhill
Reid & Hilton (1998)	RT55	Empirical	Redwood	Age 90-130	Buffer	55	Downhill
Reid & Hilton (1998)	RP55	Empirical	Redwood	Age 90-130	Forest	55	Downhill
Murphy & Koski (1989)	KO30	Empirical	Spruce/Hemlock	Old-Growth	Forest	30	—
McKinley (1997)	Mc20	Empirical	Red alder	2nd Growth	Forest	20	—
McKinley (1997)	Mc30	Empirical	Hemlock	2nd Growth	Forest	30	—
McDade et al. (1990)	MH25	Empirical	Hardwood	Mature	Forest	25	Random
McDade et al. (1990)	MD48	Empirical	Douglas Fir	Mature	Forest	48	Random
McDade et al. (1990)	MD58	Empirical	Douglas Fir	Old-Growth	Forest	58	Random
McDade et al. (1990)	MM40	Model	Douglas Fir	—	Forest	40	Random
McDade et al. (1990)	MM50	Model	Douglas Fir	—	Forest	50	Random
VanSickle & Gregory (1990)	VPU50	Model	Douglas Fir	Uniform	—	50	Variable
VanSickle & Gregory (1990)	VPM50	Model	Douglas Fir	Mixed	—	50	Variable

occur. Hypotheses about riparian recovery and gaps in scientific knowledge can be addressed with a science-based approach and cooperation from stakeholders (Franklin et al. 1999). Riparian forest science is very much in need of empirical studies of the direct ecological impact of silvicultural activities, such as thinning hardwood and brush dominated stands to release or regenerate conifer (Hayes et al. 1996).

Recent reviews by Roni et al. (2002) have reported results of some of the last decades' active restoration efforts. Longer term analysis and subsequent data-driven adaptive management may now be possible throughout the Pacific Northwest, as the "grand experiment" in multiple site, reach, and watershed level process restoration moves forward over time and through the political and social process.

Riparian Restoration

Can we grow floodplain forests that supply the proper size and amount of wood to rivers and streams? This chapter illustrated some of the information that should be considered for prescribing design criteria based on diagnostic attributes of the functional requirements that describe riparian ecological systems. In many situations, active management could accelerate the recovery of riparian forest function by directing the delivery of wood in ways that are effective to both long-term and interim strategies. One strategy is to increase conifer composition or other desired species in forest types with low potential for woody debris recruitment to channels. Wherever recovery plans are implemented, provision for some means to satisfy economic objectives can offer land owners and managers incentives and flexibility in the amount of recovery measures undertaken.

As floodplain and riparian forest recovery projects are initiated, it is becoming more important to assess the effectiveness of silvicultural systems in the maintenance or rehabilitation of riparian and aquatic habitat. If we are going to manage the riparian floodplain ecosystem, we need to begin with simple conceptual models or examples of relationships between the multiple ecological and economic functions for different forest and stream types. Assessment of the physical setting for riparian habitat recovery projects can be quite extensive in large rivers. Design development must incorporate the system that is to be managed, variables to be measured as the elements of recovery, and some bounds to measure performance. A detailed floodplain restoration plan will include the technical details of fluvial geomorphology, the principles of forest stand dynamics (i.e., growing the source of future wood) and forest engineering (e.g., delivering large wood to the channel efficiently with minimal site impact), and the ecology of aquatic flora and fauna, which are essential for directing recovery efforts. Collaborative research and adaptive management can objectively describe the problems and craft technically sound, economically feasible solutions. A faint signal exists from the magnificent biological resources that our rivers and floodplain forests once held. Using a set of objectives based on ecosystem functions, a variety of riparian silvicultural strategies may be employed to help restore or rehabilitate aquatic-terrestrial ecosystem processes.

References

Abbe, T.E. and D.R. Montgomery. 1996. Large woody debris jams, channel hydraulics and habitat formation in large rivers. *Regulated Rivers: Research & Management* 12:201-221.

Andrus, C.W., B.A. Long, and H.A. Froehlich. 1988. Woody debris and its contribution to pool formation in a coastal stream 50 years after logging. *Canadian Journal of Fisheries and Aquatic Sciences* 45:2080-2086.

Beach, E.W. and C.B. Halpern. 2001. Controls on conifer regeneration in managed riparian forests: effects of seed source, substrate, and vegetation. *Canadian Journal of Forest Research* 31: 471-482.

Beechie, T.J. and T.H. Sibley. 1997. Relationships between channel characteristics, woody debris, and fish habitat in Northwestern Washington streams. *Transactions of the American Fisheries Society* 126:217-229.

Beechie, T., G. Pess, P. Kennard, R.E. Bilby, and S. Bolton. 2000. Modeling recovery rates and pathways for woody debris recruitment in Northwestern Washington streams. *North American Journal Fisheries Management* 20:436-452.

Berg, D.R. 1995. Riparian silvicultural system design and assessment in the Pacific Northwest Cascade mountains, USA. *Ecological Applications* 5:87-96.

Berg, D.R. 1997. Active management of riparian habitats. In K.B. McDonald and F Weinmann (eds.) Wetland and Riparian Restoration:Taking a Broader View. Publication EPA 910-R-97-007, US EPA, Region 10, Seattle, Washington. pp. 50-61.

Berg, D.R, T.K. Brown, and B. Blessing. 1996. Silvicultural systems design with emphasis on forest canopy. *Northwest Science* 70:31-36.

Beschta, R.L. 1991. Stream habitat management for fish in the Northwestern United States: the role of riparian vegetation. *American Fisheries Society Symposium* 10:53-58.

Beschta, R.L. and R.L. Taylor. 1988. Stream temperature increases and land use in a forested Oregon watershed. *Water Resources Bulletin* 24:19-25.

Beschta, R.L., R.E. Bilby, G.W. Brown, L.B. Holtby, and T.D. Hofstra. 1987. Stream temperature and aquatic habitat: fisheries and forestry interactions. In E.O. Salo and T.W. Cundy (eds.) *Streamside Management: Forestry and Fishery Interactions*. College of Forest Resources, University of Washington. Seattle, Washington. pp. 191-232.

Bilby, R.E. 1981. Role of organic debris dams in regulating the export of dissolved and particulate matter from a forested watershed. *Ecology* 62:1234-1243.

Bilby, R.E. 1984. Removal of woody debris may affect stream channel stability. *Journal of Forestry* 92:609-613.

Bilby, R.E. and G.E. Likens. 1980. Importance of organic debris dams in the structure and function of stream ecosystems. *Ecology* 61:1107-1113.

Bilby, R.E. and G.E. Likens. 1989. Changes in characteristics and function of woody debris with increasing size of streams in western Washington. *Transactions of the American Fisheries Society* 118:368-378.

Bilby, R.E. and J.W. Ward. 1991. Characteristics and function of large woody debris in streams draining old-growth, clear-cut, and second-growth forests in southwestern Washington. *Canadian Journal of Fisheries and Aquatic Sciences* 48:2499-2508.

Bilby, R.E. and P.A. Bisson. 1992. Allochthonous versus autochthonous organic matter contributions to the trophic support of fish populations in clear-cut and old-growth forested streams. *Canadian Journal of Fisheries and Aquatic Sciences* 49:540-551.

Bilby, R.E., B.R. Fransen, and P.A. Bisson. 1996. Incorporation of nitrogen and carbon from spwaning coho salmon into the trophic system of small streams: evidence from stable isotopes. *Canadian Journal of Fisheries and Aquatic Sciences* 53: 164-173.

Bird, G. A. and N.K. Kaushik. 1981. Coarse particulate organic matter in streams. In M.A. Lock and D.D. Williams (eds.) *Perspectives in Running Water Ecology*. Plenum Press. New York. pp. 41-68.

Bisson. P.A.and R.E. Bilby. 1998. Organic matter and trophic dynamics. In R.J. Naiman and R.E. Bilby (eds.) *River Ecology and Management: Lessons from the Pacific Coastal Ecoregion.* Springer-Verlag. New York. pp. 373-398.

Bisson, P.A., R.E. Bilby, M.D. Bryant, C.A. Dolloff, G.B. Grette, R.A. House, M.L. Murphy, K.V. Koski, and J.R. Sedell. 1987. Large woody debris in forested streams in the Pacific Northwest: Past, present, and future. In E.O. Salo and T.W. Cundy (eds.) *Streamside Management: Forestry and Fishery Interactions*. University of Washington. Seattle, Washington. pp. 143-190.

Boling, R.H., E.R. Goodman, and J.A. V. Sickle. 1975. Toward a model of detritus processing in a woodland stream. *Ecology* 56:141-151.

Boulton, A.J., S. Findlay; P. Marmonier, E.H. Stanley, and H.M. Valett. 1998. The functional signficance of the hyporheic zone in streams and rivers. *Annual Review of Ecological Systematics* 29:59-81.

Braatne, J.H., S.B. Rood, and P.E. Heilman. 1996. Life history, ecology, and conservation of riparian cottonwoods in North America. In R. Stettler, H. Bradshaw, P. Heilman, and T. Hinkley (eds.) *Biology of Populus*. NRC Research Press, National Research Council of Canada, Ottawa. pp. 57-85.

Bragg, D.C. 2000. Simulating catastrophic and individualistic large woody debris recruitment for a small riparian stream. *Ecology*. 81:1383-1394.

Brosofske, K.D., J. Chen, R.J. Naiman, and J.F. Franklin. 1997. Harvesting effects on microclimatic gradients from small streams to uplands in western Washington. *Ecological Applications* 7:1188-1200.

Brown, J. 1882. *The Forester or a Practical Treatise on the Planting, Rearing, and General Management of Forest Trees*. 5th edition. Blackwood, Edinburgh.

Bryant, M.D. 1983. The role and management of woody debris in west coast salmonid nursery streams. *North American Journal of Fisheries Management* 3:322-330.

Budd, W.W., P.L. Cohen, P.R. Saunders, and F.R. Steiner. 1987. Stream corridor management in the Pacific Northwest: I. Determination of stream-corridor widths. *Environmental Management* 11:587-597.

Bustard, D.R. and D.W. Narver. 1975. Preferences of juvenile coho salmon (*Onchorhynchus kisutch*) and cutthroat trout (*Salmo clarki*) relative to simulated alteration of winter habitat. *Journal of the Fisheries Research Board of Canada* 32:681-687.

Cederholm, C.J. and N.P. Peterson. 1985. The retention of coho salmon (*Oncorhynchus kisutch*) carcasses by organic debris in small streams. *Canadian Journal of Fisheries and Aquatic Sciences* 42:1222-1225.

Cederholm, C.J., M.D. Kunze, T. Murota, and A. Sibatani. 1999. Pacific salmon carcasses: essential contributions of nutrients and energy for aquatic and terrestrial ecosystems. *Fisheries* 24(10):6-15.

Chapman, D.W. and E. Knudsen. 1980. Channelization and livestock impacts on salmonid habitat and biomass in Western Washington. *Transactions of the American Fisheries Society* 109:357-363.

Collins, B.D. and D.R. Montgomery. 2002. Forest development, wood jams and restoration of floodplain rivers in the Puget Lowland, Washington. *Restoration Ecology* 10:237-247.

Cooper, A.B., C.M. Smith, and M.J. Smith. 1995. Effects of riparian set-aside on soil characteristics in an agricultural landscape: implication for nutrient transport and retention. *Agriculture, Ecosystems and Environment* 55:61-67.

Cooper, J.R., J.W. Gilliam, R.B. Daniels, and W.P. Robarge. 1987. Riparian areas as filters for agricultural sediment. *Soil Science Society of America Journal* 51:416-420.

Cummins, K.W. 1974. Structure and function of stream ecosystems. *Bioscience* 24:631-641.

Curtis, R.O., G.W. Clendenen, and D.J. DeMars. 1981. A new stand simulator for coastal Douglas-fir: DFSIM users guide. General Technical Report PNW-128 USFS Pacific Northwest Forest and Range Experiment Station, Portland, Oregon.

Curtis, R.O., D.D. Marshall, and J.F. Bell.1997. LOGS: A pioneering example of silvicultural research in coastal Douglas-fir. *Journal of Forestry* 95(7):19-25.

Dent, L.F. and J.B.S. Walsh. 1997. Effectiveness of riparian management areas and hardwood conversions in maintaining stream temperature. Forest Practices Technical Report No. 3. Oregon Department of Forestry, Salem, Oregon.

Dillaha, T.A., R.B. Reneau, S. Mostaghimi, and D. Lee. 1989. Vegetative filter strips for agricultural nonpoint source pollution control. *American Society of Agricultural Engineers* 32:513-519.

Doyle, R.C., G.C. Stanton, and D.C. Wolf. 1977. Effectiveness of forest and grass buffer strips in improving the water quality of manure polluted runoff. *Transactions American Society Agricultural Engineers*. Paper 77-2501. St. Joseph, Michigan.

Fetherston, K.L., R.J. Naiman, and R.E. Bilby. 1995. Large woody debris, physical process, and riparian forest development in montane river networks of the Pacific Northwest, USA. *Geomorphology* 13:133-144.

Franklin, J.F. and C.T. Dyrness. 1969. *Natural Vegetation of Oregon and Washington*. USDA-Forest Service Research Paper PNW 80. Pacific Northwest Forest and Range Experiment Station. Portland, Oregon.

Franklin, J.F., D.R. Berg, D. Thornburg, and J. Tappenier. 1996 Alternative Silvicultural approaches to timber harvest: Variable retention harvest system. In K. Kohm (ed.) *Forestry for the 21st Century*. Island Press. Washington D.C. pp. 111-140.

Franklin, J.F., L. Norris, D.R. Berg, and G. R.Smith. 1999. The History of DEMO (Demonstration of Ecological Management Operations): An experiment in regeneration harvest of Northwestern forest ecosystems. *Northwest Science* 73:3-11.

Frissell, C.A. and R.K. Nawa. 1992. Incidence and causes of physical failure of artificial habitat structures in streams of Western Oregon and Washington. *North American Journal of Fisheries Management* 12:182-197.

Gregory, S.V., F.J. Swanson, W.A. McKee, and K.W. Cummins. 1991. An ecosystem perspective of riparian zones: Focus on links between land and water. *Bioscience* 41:540-551.

Gresh, T., J. Lichatowich, and P. Schoonmaker. 2000. An estimation of historic and current levels of salmon production in the Northeast Pacific ecosystem: Evidence of a nutrient deficit in the freshwater systems of the Pacific Northwest. *Fisheries* 25(1):15-21.

Grette, G.B. 1985. The role of large organic debris in juvenile salmonid rearing habitat in small streams. Masters thesis. University of Seattle.

Griffith, S.M., J.S .Owen, W. R. Horwath, P.J. Wigington Jr., J.E. Bahan, and L.F. Elliot. 1997. Nitrogen movement and water quality at a poorly drained agricultural and riparian site in the Pacific Northwest. *Soil Science and Plant Nutrition* 43:1025-1030.

Grizzel, J.D. and N. Wolff. 1998. Occurrence of windthrow in forest buffer strips and its effect on small streams in northwest Washington. *Northwest Science* 72: 214-223.

Hairston-Strang, A.B. and P.W. Adams. 1998. Potential large woody debris sources in riparian buffers after harvesting in Oregon, USA. *Forestry Ecology Management* 112:67-77.

Harmon M.E. and J.F. Franklin. 1989. Age distribution of western hemlock and its relation to Roosevelt elk populations in the South Fork of the Hoh River valley, Washington. *Northwest Science* 57:249-255.

Harmon M.E., J.F. Franklin, F.J. Swanson, P. Sollins, S.P. Cline, N.G. Aumen, J.R. Sedell, G.W. Lienkaemper, K. Cromack, Jr, and K.W. Cummins. 1986. The ecology of coarse woody debris in temperate ecosystems. In A. MacFadyen and E. D. Ford (eds.) *Advances in Ecological Research.* Volume 15. Academic Press, New York, pp. 133-302.

Hawkins, C.P., M.L. Murphy, and N.H. Anderson. 1982. Effects of canopy, substrate composition, and gradient on the structure of macroinvertebrate communities in Cascade Range streams of Oregon. *Ecology* 63:1840-1856.

Hawkins, C.P., M.L. Murphy, N.H. Anderson, and M.A. Wilzbach. 1983. Density of fish and salamanders in relation to riparian canopy and physical habitat in streams of the northwestern United States. *Canadian Journal of Fisheries and Aquatic Sciences* 40:1173-1185.

Haycock, N.E. and G. Pinay. 1993. Groundwater nitrate dynamics in grass and poplar vegetated riparian buffer strips during the winter. *Journal of Environmental Quality* 22:273-278.

Hayes, J.P., M.D. Adams, D. Bateman, E. Dent, W.H. Emmingham, K.G. Maas, and A.E. Skaugset. 1996. Integrating research and forest management in riparian areas of the Oregon Coast Range. *Western Journal of Applied Forestry* 11:85-88.

Heede, B.H. 1985. Interactions between streamside vegetation and stream dynamics. In *Proceedings of Riparian Ecosystems and Their Management: Reconciling Conflicting Uses*. General Technical Report RM-120. U.S. Forest Service Rocky Mountain Forest and Range Experiment Station, Ft Collins, Colorado. pp 54-58.

Hester, A.S., D.W. Hann, and D.R. Larson. 1989. Organon: Southwest Oregon growth and yield model user manual, version 2.0. Forest Research Lab, College of Forestry, Oregon State University, Corvallis, Oregon.

Hibbs, D.E. and P.A. Giordano. 1996. Vegetation characteristics of alder-dominated riparian buffer strips in the Oregon coast range. *Northwest Science* 70:213-222.

Hibbs, D.E. 1987. The self-thinning rule and red alder management. *Forest Ecology and Management* 18:273-281.

Hodkinson, I.D. 1974. Dry weight loss and chemical changes in vascular plant litter of terrestrial origin, occurring in a beaver pond ecosystem. *Journal of Ecology* 63:131-142.

Hollings, C.S. 1978. *Adaptive Environmental Assessment and Management*. John Wiley and Sons, London.

Holtby, L.B. 1988. Effects of logging on stream temperature in Carnation Creek, British Columbia, and associated impacts on the Coho salmon (*Oncorhynchus kisutch*). *Canadian Jouranal of Fisheries and Aquatic Science* 45:502-515.

House, R.A. and P.L. Boehne. 1986. Effects of instream structures on salmonid habitat and populations in Tobe Creek, Oregon. *North American Journal of Fisheries Management* 6:38-46.

House, R.A. and P.L. Boehne. 1985. Evaluation of instream enhancement structures for salmonid spawning and rearing in a coastal Oregon stream. *North American Journal Fisheries Management* 5:283-295.

Hynes, H.B.N. 1983. Groundwater and stream ecology. *Hydrobiologia* 100:93-99.

Ishikawa, K. 1982. *Guide to Quality Control*. Asian Productivity Society. Tokyo, Japan.

Johnson, F.A. 1955. Volume tables for Pacific Northwest trees. Agriculture Handbook No.92. USDA Forest Service. Pacific Northwest Forest and Range Experiment Station. Portland, Oregon.

Karr, J.R. and I.J. Schlosser. 1978. Water resources and the land-water interface. *Science* 201:229-234.

Kennard, P., G. Pess, T. Beechie, R.Bilby, and D.R. Berg. 1998. Riparian-in-a-box: A manager's tool to predict the impacts of riparian management on fish habitat. In *Proceedings of Forest-Fish conference: Land management practices affecting aquatic ecosystems*. Natural Resources Canada, Canadian Forestry Service, North. Forestry Centre, Edmonton, Alberta. Information Report NOR-X-356.

Knutson, K.L. and V.L. Naef. 1997. Management recommendations for Washington's priority habitats: Riparian. Washington Department of Fish and Wildlife, Olympia, WA.

Larkin, G.A. and P.A. Slaney. 1997. Implications of trends in marine-derived nutrient influx to south coastal British Columbia salmonid production. *Fisheries* 22(11):16-24.

Lienkaemper, G.W. and F.J. Swanson. 1987. Dynamics of large woody debris in streams in old-growth Douglas-fir forests. *Canadian Journal of Forest Research* 17:150-156.

Lisle, T.E. 1986. Effects of woody debris on anadromous salmonid habitat, Prince of Wales Island, Southeast Alaska. *North American Journal of Fisheries Management* 6:538-550.

Long, J.N., J.B. McCarter, and S.B. Jack. 1988. A modified density management diagram for coastal Douglas-fir. *Western Journal Applied Forestry* 3:88-89.

Lowrance, R.R., R.L. Todd, and L.E. Asmussen. 1984a. Nutrient cycling in an agricutural watershed: II. Streamflow and artificial drainage. *Journal of Environmental Quality* 13:27-32.

Lowrance, R.R., R.L. Todd; J. Joseph Fail, J. Ole Hendrickson, R. Leonard and L. Asmussen. 1984b. Riparian forests as nutrient filters in agricultural watersheds. *Bioscience* 34:374-377.

Magette, W.L., R.B. Brinsfield, R.E. Palmer, and J.D. Wood. 1989. Nutrient and sediment removal by vegetated filter strips. *Transactions of the American Society of Agricultural Engineers* 32:663-667.

Maser, C., R.F. Tarrant, J.M. Trappe, and J.F. Franklin. 1989. From the forest to the sea: The story of a fallen tree. General Technical Report PNW-GTR-229 Pacific Northwest Forest and Range Experiment Station, Portland, Oregon.

McArdle, R.E., W.H. Meyer, and D. Bruce. 1949. Yield of Douglas-fir in the Pacific Northwest, Revised. USDA Technical Bulletin 201.

McClain, M.E., R.E. Bilby, and F.J.Triska. 1998. Nutrient cycles and responses to disturbance. In R.J. Naiman and R.E. Bilby (eds.) *River Ecology and Management: Lessons from the Pacific Coastal Ecoregion*. Springer-Verlag. New York. pp. 347-372.

McDade, M. H., F. J. Swanson, W. A. McKee, J. F. Franklin, and J. V. Sickle. 1990. Source distances for coarse woody debris entering small streams in western Oregon and Washington. *Canadian Journal of Forest Research* 20:326-330.

McKee, W. A., G. LaRoi, and J. F. Franklin. 1982. Structure, composition, and reproductive behavior of terrace forests, South Fork, Hoh River, Olympic National Park. In E. E. Starkey, J. F. Franklin, and J. W. Matthews (eds.) *Ecological Research in the National Parks of the Pacific Northwest.* Oregon State University. Corvallis, Oregon0. pp. 22-29.

McKinley, M. 1997. Large woody debris source distances for Western Washington Cascade Streams. Undergraduate thesis, University of Washington. Seattle, Washington.

Meyer, W. H.1939. Yield of even aged stands of Sitka spruce and western hemlock. Technical Bulletin 544. USDA, Washington D.C.

Michael, J. H. 1998. Pacific salmon spawner escapement goals for the Skagit River watershed as determined by nutrient cycling considerations. *Northwest Science* 72:239-248.

Michael. J. H. 1999. The future of Washington salmon: Extinction is not an option but may be the preferred alternative. *Northwest Science* 73:235-239.

Minshall, G. W. 1978. Autotrophy in stream ecosystems. *Bioscience* 28:767-771.

Montgomery, D. R.1999. Process domains and the river continuum. *Journal American Water Resources* 35:397-410.

Montgomery, D. R., J. M. Buffington, R. D. Smith, K. M. Schmidt, and G. Pess. 1995a. Pool spacing in forest channels. *Water Resources Research* 31:1097-1105.

Montgomery, D. R., G. E. Grant, and K. Sullivan. 1995b. Watershed analysis as a framework for implementing ecosystem management. *Water Resources Bulletin* 31:369-386.

Murphy, M. L. and K. V. Koski. 1989. Input and depletion of woody debris in Alaska streams and implications for streamside management. *North American Journal of Fisheries Management* 9:427-436.

Murphy, M. L., C. P. Hawkins, and N. H. Anderson. 1981. Effects of canopy modification and accumulated sedimentation on stream communities. *Transactions of the American Fisheries Society* 110:469-478.

Murphy, M. L., J. Heifetz, S. W. Johnson, K. V. Koski, and J. F. Thedinga. 1986. Effects of clear-cut logging with and without buffer strips on juvenile salmonids in Alaskan streams. *Canadian Journal of Fisheries and Aquatic Sciences* 43:1521-1533.

Naiman. R.J., T.J. Beechie, L.E. Benda, D.R. Berg, P.A. Bisson, L.H. MacDonald, M.D. O'Conner, P.L. Olsen, and E.A. Steele. 1992. Fundamentals of ecologically healthy watersheds in the Pacific Northwest coastal ecoregion. In R.J. Naiman (ed.) *Watershed Management*. Springer-Verlag. New York. pp. 127-188.

Naiman, R.J. and J.R. Sedell. 1980. Relationships between metabolic parameters and stream order in Oregon. *Canadian Journal of Fisheries and Aquatic Sciences* 37:834-847.

National Research Council (NRC). 1996. *Upstream.* National Academy Press,Washington D.C.

Nehlsen, W., J.E. Williams, and J.A. Lichatowich. 1991. Pacific salmon at the crossroads: Stocks at risk from California, Oregon, Idaho, and Washington.

Fisheries 16(2):4-21.

Nickelson, T.E., M. F. Solazzi, S.L. Johnson, and J.D. Rodgers. 1992. Effectiveness of selected stream improvement techniques to create suitable summer and winter rearing habitat for juvenile coho salmon (*Oncorhynchus kitsutch*) in Oregon coastal streams. *Canadian Journal of Fisheries and Aquatic Sciences* 49:790-794.

Nystrom, M.N., D.S. DeBell, and C.D. Oliver. 1984. Development of young growth western redcedar stands. USFS Research paper. PNW 324. Pacific Northwest Forest and Range Experiment Station, Portland, OR.

Oliver, C.D. and T.M. Hinckley. 1987. Species, stand structures, and silvicultural manipulation patterns for the streamside zone. In E.O. Salo and T.W. Cundy (eds.) *Streamside Management: Forestry and Fishery Interactions*. College of Forest Resources, University of Washington. Contribution no. 57. Seattle, Washington. pp. 259-276.

Oliver, C. D. and B. C. Larson. 1990. *Forest Stand Dynamics*. McGraw-Hill, Inc. New York.

Omernik, J.M., A.R. Abernathy, and L.M. Male. 1981. Stream nutrient levels and proximity of agricultural and forest lands to streams: some relationships. *Journal of Soil and Water Conservation* July-August:227-231.

Osborne, L.L. and D.A. Kovacic. 1993. Riparian vegetated buffer strips in water-quality restoration and stream management. *Freshwater Biology* 29:243-258.

Pabst, R.J. and T.A.Spies. 1998. Distribution of herbs and shrubs in relation to landform and canopy cover in riparian forests of coastal Oregon. *Canadian Journal of Botony* 76:298-315.

Peterjohn, W.T. and D.L. Correll. 1984. Nutrient dynamics in an agricultural watershed: observations on the role of a riparian forest. *Ecology* 65:1466-1475.

Reid, L.M. and S. Hilton. 1998. Buffering the buffer. In R.R. Ziemer, (ed.) *Proceedings of the Conference on Coastal Watersheds: The Caspar Creek Story*, USDA Forest Service Pacific Southwest Research Station General Technical Report PSW-GTR-168, Albany, California. pp. 71-80.

Ringler, N.H. and J.D. Hall. 1975. Effects of logging on water temperature and dissolved oxygen in spawning beds. *Transactions of the American Fisheries Society* 1:111-121.

Roni, P., T.J. Beechie, R.E. Bilby, F.E. Leonetti, M.M. Pollock, and G. Pess. 2002. A review of restoration techniques and a hierarchical strategy for prioritizing restoration in Pacific Northwest watersheds. *North American Journal of Fisheries Management 22:1-20*

Scott, W., R. Meade, R. Leon, D. Hyink, and R. Miller. 1998. Planting density and tree size relations in Coast Douglas–fir. *Canadian Journal of Forestry Research* 28:74-78.

Sedell, J.R. and J.L. Froggatt. 1984. Importance of streamside forests to large

rivers: the isolation of the Willamette River, Oregon, U.S.A., from its floodplain by snagging and streamside forest removal. *Verhandlungen-Internationale Vereinigung für Theorelifche und Angewandte Limnologie* 22:1828-1834.
Sedell, J.R. and K.J. Luchessa. 1981. Using the historical record as an aid to salmonid habitat enhancement. In N. Armantout (ed.) *Acquisition and Utilization of aquatic habitat inventory information.* Portland, Oregon. American Fisheries Society, Bethesda, Maryland. pp. 210-223.
Sedell, J.R., F.J. Triska, and N.S. Triska. 1975. The processing of conifer and hardwood leaves in two coniferous forest streams: Weight loss and associated invertebrates. *Verhandlungen-Internationale Vereinigung für Theorelifche und Angewandte Limnologie* 19:1617-1627.
Sedell, J.R. P.A. Bisson, F.J. Swanson, and S.V. Gregory. 1989. What we know about large trees that fall into streams and rivers. In C. Maser, R.F. Tarrant, J. M. Trappe, and J. F. Franklin. (eds.) *From the Forest to the Sea: A Story of Fallen Trees.* General Technical Report PNW-GTR-229 Pacific Northwest Forest and Range Experiment Station, Portland, Oregon.
Smith, D.M. 1986. *The Practice of Silviculture.* Wiley and Sons, New York
Smith, N.J. 1987. Stand density control diagram for western redcedar, *Thuja plicata. Forest Ecology and Management* 27:235-244.
Stanford, J.A. and J.V. Ward. 1988. The hyporheic habitat of river ecosystems. *Nature* 335:64-68.
Steinblums, I.J. 1984. Designing stable buffer strips for stream protection. *Journal of Forestry* 82:49-52.
Suberkropp, K.F. 1998. Microorganisms and organic matter decomposition. In Naiman, R.J. and R.E. Bilby (eds.) *River Ecology and Management: Lessons from the Pacific Coastal Ecoregion.* Springer-Verlag. New York. pp. 373-398.
Swanson, F.J., S.V. Gregory, J.R. Sedell, and A.G. Campbell. 1982. Land and water interactions: The riparian zone. In R.L. Edmonds (ed.) *Analysis of Coniferous Forest Ecosystems in the Western United States.* Hutchinson Ross Publishing Co., Stroudsburg, Pennsylvania. pp. 267-291.
Tappenier, J.C., J. Zasada, P. Ryan, and M. Newton. 1991. Salmonberry clonal and population structure in Oregon forests: The basis for persistent cover. *Ecology* 72:609-618.
Triska, F.J., J.R. Sedell, and S.V. Gregory. 1982. Coniferous forest streams. In R. L. Edmonds (ed.) *Analysis of Coniferous Forest Ecosystems in the Western United States.* Hutchinson Ross Publishing Co., Stroudsburg, Pennsylvania. pp. 293-332.
VanDijk, P.M., F.J.P.M. Kwaad, and M. Klapwijk. 1996. Retention of water and sediment by grass strips. *Hydrological Processes* 10:1069-1080.
Vannote, R.L., G.W.Minshall, K.W.Cummins, J.R.Sedell, and C.E.Cushing. 1980. The river continuum concept. *Canadian Journal of Fisheries and Aquatic*

Sciences 37:130-137.

Van Sickle, J. and S.V. Gregory. 1990. Modeling inputs of large woody debris to streams from falling trees. *Journal of Forest Research* 20:1593-1601.

White, R.J. 1991. Objectives should dictate methods in managing stream habitat for fish. *American Fisheries Society Symposium* 10:44-52.

Wilzbach, M.A. and K.W. Cummins. 1986. Influence of habitat manipulations on interactions between cutthroat trout and invertebrate drift. *Ecology* 67:898-911.

Young-Berg, K.J. and D.R. Berg. 1994. Native plants to attract wildlife. *Douglasia* 18:17-19.

Zeide B. 2001. Thinning and growth. *Journal of Forestry* 99:20-25.

11. Opportunities and Constraints for Urban Stream Rehabilitation

Christopher P. Konrad

Abstract

The profound and pervasive influences of people on urban streams create many opportunities for stream rehabilitation but ultimately constrain the restoration of stream ecosystems. Dense human settlements, commercial and industrial centers, and transportation corridors in urban areas influence every attribute of stream ecosystems: the paths, timing, and volume of runoff that generates streamflow; the supply and size of sediment and organic material delivered to stream channels; the thermal and chemical characteristics of stream water; the structure, form, and materials of stream channels and aquatic habitats; and the demographics of the biologic populations forming stream and riparian communities. Ecological restoration may not be feasible for most urban streams as some elements of stream ecosystems cannot be re-established in urban areas, while others can only be re-established to a limited degree or geographic extent. Even though ecological restoration may not be feasible, ecological concepts can inform urban stream rehabilitation. Important concepts include the spatial scale of ecological processes, the mosaic of habitats within a stream, and the linkages between a stream and terrestrial ecosystems. Urban stream rehabilitation is largely experimental given the complexity of ecological controls on streams, the multitude of anthropogenic constraints in urban areas, and the uncertainty of techniques that aim to compensate for the constraints. Effective urban stream rehabilitation depends on identification of the specific causes of degradation in a stream, development of techniques that address these causes, and monitoring to demonstrate success.

INTRODUCTION

People have altered much of the Pacific Northwest, clearing forests for farms, building roads and homes, and developing cities. These land-uses influence stream ecosystems, both directly by reducing the habitat area available for other species and indirectly by modifying the biogeophysical processes that form and maintain stream ecosystems. Public agencies and private organizations in the region advocate, fund, and implement projects to recover attributes of stream ecosystems, but the rehabilitation of urban streams is particularly challenging because of the types and extent of changes in stream ecosystems and the pervasive influence of people. The National Research Council recognized that restoration of aquatic ecosystems may be infeasible or undesirable in areas where it would conflict with existing human uses including urban development (NRC 1992). Although urban streams in the Pacific Northwest cannot be restored to their condition prior to the extensive conversion of forests (Chapter 4) to agricultural, commercial, and residential developments (ca. 1850), even the proposition of stream rehabilitation must be examined to identify feasible objectives and effective techniques for urban areas.

Urban stream rehabilitation stands apart from ecosystem restoration. Rehabilitation includes almost any socially-valued change in the condition of an ecosystem, whereas restoration is the "return of an ecosystem to a close approximation of its condition prior to (anthropogenic) disturbance" (NRC 1992, p. 523). Stream restoration may be either passive, where anthropogenic influences are removed, or active, where an element of a stream ecosystem is re-created or introduced (Kauffman et al. 1997). The difference between rehabilitation and restoration of streams is more profound than their endpoints. Stream restoration is limited to activities that re-establish attributes of stream ecosystems, whereas rehabilitation also includes activities intended to compensate for anthropogenic influences on streams and may include features that were not part of the stream ecosystem prior to urban development, such as high flow bypass channels, engineered channel features, and control of exotic species. Compensatory techniques are a necessary component for urban stream rehabilitation because of the many constraints on re-establishing stream ecosystems in urban areas. Although compensatory techniques do not represent part of the pre-development ecosystem, they still must be designed with ecological principles in mind and evaluated to determine their effectiveness.

Opportunities for Urban-Stream Rehabilitation

Urban development in the Pacific Northwest has transformed the lowland areas where development is concentrated. The Puget Lowland represents the largest expanse of urban development in the region (Matzke 1993). It is inhabited by nearly 4 million people and hosts the major cities in western Washington State. Many different measures have been used to distinguish urban from suburban or rural streams, such as total impervious area, road density, or urban land cover. The response of a stream ecosystem to a given level of land use, however, depends on other factors such as basin geology, topography, and the configuration of channel networks (Allan et al. 1997; Konrad 2000). In this chapter, "urban streams" refer to those streams in basins with extensive residential or commercial developments without using specific criteria, except when such criteria are cited in other studies, and is used frequently to refer to streams at one end of the continuum of urban development in comparison to those at the other end.

Human activities in urban areas modify many ecological processes and elements in ecosystems. The modifications affect every factor influencing the biologic conditions of streams: streamflow patterns, channel form and materials, physical characteristics of water, energy flux, and biotic conditions (Karr et al. 1986). Consequently, there are many opportunities for rehabilitating urban streams either by re-establishing a process or element or by compensating for a missing or modified process or element. The multitude of anthropogenic changes in urban streams, however, complicates their rehabilitation because the recovery of one process or element may not improve the condition of the stream where there are multiple causes of degradation. Thus urban stream rehabilitation requires the assessment of human influences on all attributes of streams.

Streamflow Patterns

Streamflow is a fundamental attribute of stream ecosystems that influences the composition and structure of biological communities (Shelford and Eddy 1929; Odum 1956; Poff and Ward 1989). In the forests of the Pacific Northwest, streamflow is generated by rain and snowmelt predominately through shallow subsurface flow and saturation-overland flow (Dinicola 1990; Burges et al. 1998). Urban development modifies hillslope hydrologic processes when trees and soils are cleared, the land surface is graded, and roads, drainage systems, and buildings are constructed. These development activities reduce the water-storage capacity of hillslopes provided by forest-canopy interception, wetlands and other topographic depressions, and the soil column.

Runoff from hillslopes reaches streams more rapidly in urban basins than in forest basins because of changes in runoff processes and drainage networks. Shallow soils generate more saturation overland flow while impervious surfaces generate infiltration-limited overland flow. Runoff from urban hillslopes generally has a short distance to travel to a drainage network, which includes roads and ditches in addition to stream channels.

The changes in hillslope storage, runoff processes, and drainage networks resulting from urban development in Puget Lowland streams produce characteristic changes in streamflow patterns that can generally be described by the redistribution of runoff from base flow to storm flow. During storms, flow in urban streams rises rapidly, attains high peak discharge rates, and recedes rapidly (Figure 1). In spite of the increase in storm-flow magnitude, the duration of increased storm flow is brief due to the rapid rise and recession of storm flow (Konrad 2000).

Urban development also affects base flows in streams. During the wet season, base flow normalized for drainage area is generally lower in urban streams, defined as those basins having a road density greater than 6 km/km^2, than in less developed suburban streams, defined as those basins having a road density less than 6 km/km^2, in the Puget Lowland (Konrad 2000). During the dry season, the effects of urban development are less clear. Area-normalized mean discharge for August 1994, a dry month in the region, did not vary systematically with urban development in 38 Puget Lowland streams (Konrad 2000). In contrast, hydrologic simulation modeling of Bear and Evans Creeks predicts that base flow decreases during early stages of urban development (Hartley 2001).

The different effects of urban development on base flow during the wet and dry seasons and, potentially, on the range of flow depths illustrate the complexity of the hydrologic effects of urban development. The period of surface flow (hydroperiod) in intermittent streams may lengthen in response to urban development, but the increase in the hydroperiod is a result of frequent but brief periods of flow (Figure 2). Thus, the longest period of continuous flow may be abbreviated in intermittent urban streams as the channels go dry earlier in the summer.

Many ecological processes and conditions in streams are influenced by streamflow patterns, including disturbance patterns, transport of organic material and nutrients, stream temperature, and the availability and condition of aquatic habitat. Given the broad dependence of sediment transport processes and habitat conditions on streamflow, the hydrologic effects of urban development could influence stream ecosystems in many ways, though changes in the disturbance regime of streams represent one of the most likely ways.

Increased storm flow from urban development has been implicated in lower salmonid populations (Moscrip and Montgomery 1997), lower salamander populations (Osher and Shure 1972), and lower benthic bacterial densities (Jancarkova et al. 1997). These biological changes may result from an increase in the frequency and extent of flood disturbance that are likely to occur when peak discharge rates increase relative to recessional and wet-season base-flow rates (Konrad 2000). The instability produced by a flashy hydro-

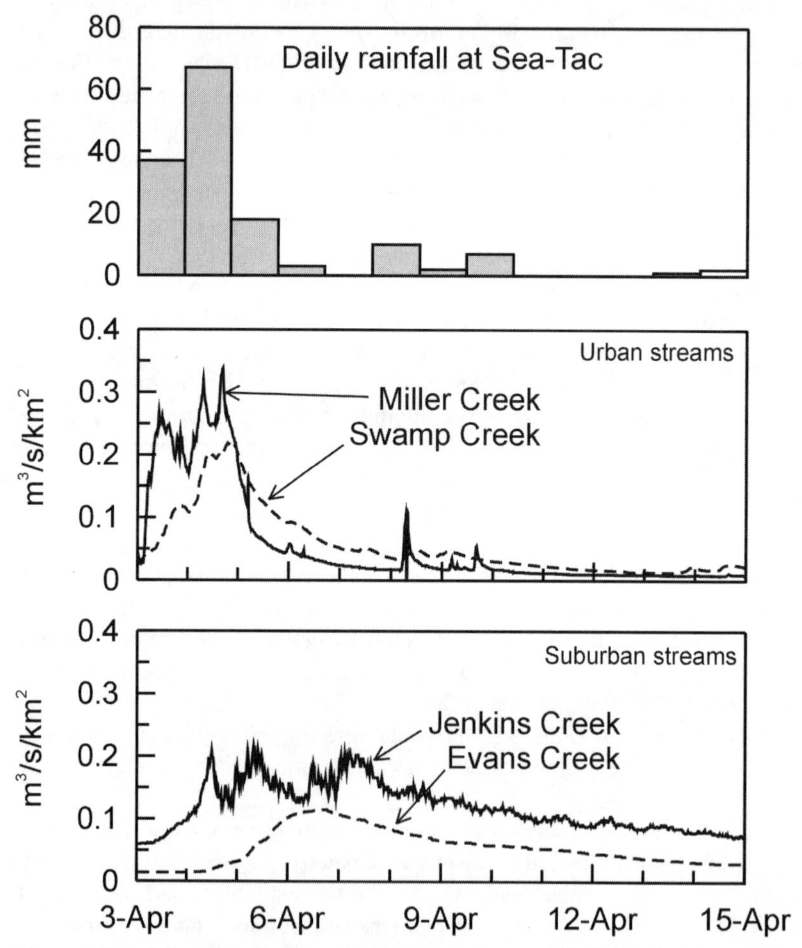

Figure 1. Rainfall and stream flow during a large storm in the Puget Lowland during April 1991 (Sources: National Weather Service, King County, and Snohomish County).

logic regime in urban streams is comparable to that of streams in arid regions, which have high rates of bedload transport during floods (Reid and Laronne 1995). The potential cessation of flow earlier in the spring or summer in intermittent streams also represents a form of hydrologic disturbance. An increase in the frequency and extent of disturbance generally can be expected to result in less diverse aquatic assemblages, less complex trophic structures in stream communities, and less complex age structures in populations of aquatic species (Allan 1996).

Poff and Allan (1995) found that the diversity and structure of fish assemblages were more closely linked to variation in daily discharge and base flow in streams in Wisconsin and Minnesota than extremely high or low flows. Thus, the reduction in wet-season base flow and increased variability between base flow and storm flow may also affect stream ecosystems in urban areas.

Restoration of hydrologic processes, let alone streamflow patterns, is daunting in urban areas because of the areal extent of human development. The hydrologic processes of lowland forests can be actively restored in urban areas by planting trees, re-creating wetlands, recontouring hillslopes, and amending soils. However, roads and buildings limit hydrologic restoration in urban areas to spatial scales smaller than hillslopes (which control runoff processes). Where hydrologic processes cannot be restored over the length of a hillslope to streams, properly sized on-site stormwater detention systems can increase the storage of stormwater on hillslopes (Konrad and Burges

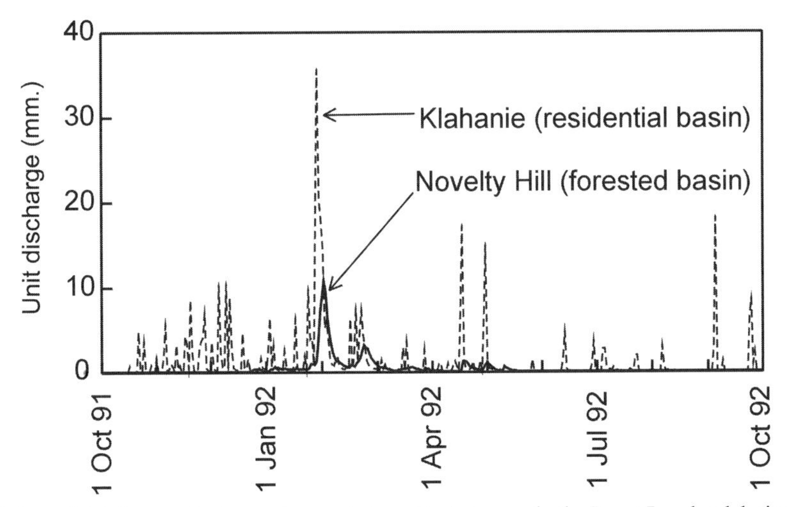

Figure 2. Daily unit discharge for two intermittent streams in the Puget Lowland during Water Year 1992 (Source: Wigmosta et al. 1994).

2001), and infiltration systems can divert runoff from surface pathways (roads, ditches, and saturated lawns) to subsurface pathways even in glacial till soils (Konrad et al. 1995).

A common dilemma in urban stream rehabilitation is illustrated by the hydrologic effects of urban development. An initial change, such as the increased production of runoff from hillslopes, produces a cascade of other changes. Although properly sized regional stormwater detention systems and high-flow bypass channels may reduce the magnitude and duration of peak storm flow, they are unlikely to increase recessional and base flow.

Because any number of streamflow patterns may have ecological effects (Poff et al. 1997), the ecologically relevant changes in streamflow patterns must be identified before effective rehabilitation strategies can be designed. Moreover, the partial recovery of streamflow patterns in lieu of their complete restoration is largely an experimental approach, which must be tested to identify which streamflow patterns are most important to the ecosystem (Chapter 9).

Channel Form and Materials

The interaction of streamflow with channel forms and materials establishes the basic physical template of stream habitat for aquatic organisms, including fish (Bisson and Sedell 1984), invertebrates (Smock et al. 1992), periphyton (Poff et al. 1990) and bacteria (Vervier et al. 1993). Urban stream channels have been modified directly by human activities and indirectly as the consequence of changes in storm runoff, sediment delivery, and riparian vegetation. Many of the morphologic changes were initiated by earlier land-use activities such as timber harvesting (Ralph et al. 1994). Stream channels in the Puget Lowland have been straightened, dredged, constrained with levees, and re-routed to new channels (Rees 1957). In extreme though common cases, stream channels have been replaced by concrete, metal, or asphalt (Figure 3). Where an urban stream is not enclosed in a culvert, its banks are likely lined with large boulders, pilings, or other materials placed to prevent bank erosion (Figure 4). Urban streams retain their natural form and materials only for short reaches where bank erosion and channel migration are allowed to occur.

Deforestation of riparian corridors influences the morphology of Pacific Northwest streams through changes in bank strength and reduction of large woody debris (LWD) supplied to the stream. The cross-sectional shape of a stream channel reflects the vegetation on the stream banks. Although trees have strong roots that effectively increase the cohesive strength of soil (Sidle et al. 1985), grasses can produce high root densities near the soil surface, particularly at the edge of a stream bank, promoting lateral and vertical accretion of floodplains (Trimble 2000). As a result, stream channels in forested reaches are frequently wider with gradually sloping banks in comparison to channels where riparian forests have been cleared and replaced by grass on their banks (Figure 5). Streamflow in a channel with a narrow cross-section is likely to be deeper and generate higher shear stresses at a given discharge than in a broader channel assuming no differences in the energy gradient of the streamflow or texture of the stream bed.

LWD is less abundant in urban streams (total impervious area greater than 14%) than in rural streams (total impervious area less than 8%) in the Pacific Northwest (Finkenbine et al. 2000). As a result, the frequency of pools is likely to be lower, especially in steeper streams (gradient >2%) where pool-

Figure 3. Miller Creek, King County, Washington.

riffle morphology is forced by LWD (Montgomery et al. 1995). Keller (1978) found no evidence for changes in pool-to-pool spacing—with respect to channel width—as a result of land uses in the eastern U.S., but the study may have included channels that are insensitive to changes in wood loading or failed to consider undisturbed channels with high wood loads. Hydraulic conditions (flow velocity and depth) and bed material are more spatially uniform without the local variation in the direction and velocity of streamflow created by LWD. Compared to streams with LWD, flow resistance is lower and average velocities are higher without LWD in the channel (Buffington and Montgomery 1999). Thus, streams in basins where urban land cover is dominant are

Figure 4. Bank protection and grade control structures in the reconstructed channel of Dewatto Creek, Mason County, Washington.

likely to have higher and more uniform current velocities than streams in basins with more agricultural or forest land cover (Ebbert et al. 2000).

Urban development influences other geomorphic processes that create and maintain channel forms. Streams incise deeper or wider channels, particularly at the margins of plateaus throughout the Puget Sound region, as the additional runoff from urban areas increases a stream's sediment transport capacity (Booth 1990). Grading of building sites generally increases the supply of fine sediment to streams (Guy and Ferguson 1962), while bank and hillslope stabilization projects reduce the delivery of coarse sediment. The increased sediment supply from construction and channel incision decreases the size of bed material in streams (Wolman and Schick 1967; Dietrich et al. 1989; Ebbert et al. 2000), fills in pools (Lisle and Hilton 1992), and aggrades low-gradient reaches of stream channels (Ebisemiju 1989).

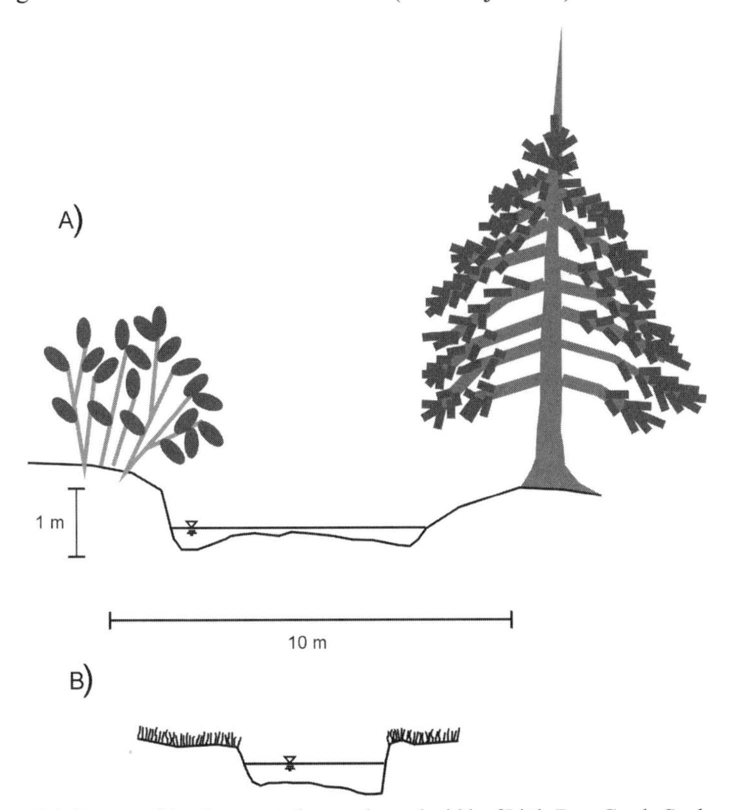

Figure 5. Influence of riparian vegetation on channel width of Little Bear Creek, Snohomish County. (A) Upstream reach with forested banks, and (B) downstream of reach with grass banks.

Over time, the supply of sediment from urban areas is likely to decrease as channel expansion slows and erosion control measures are implemented to prevent the loss of property. However, urban hillslopes continue to produce elevated quantities and rates of runoff relative to pre-development levels. As a result, urban channels may become sediment-supply limited, particularly in steep and confined reaches that transport rather than store sediment delivered from upstream sources. The size of bed material may increase in response to decreased sediment supply (Thoms 1987), and gravel bars may degrade (Lisle et al. 1993), which over time could transform a pool-riffle channel into a plane-bed channel. Reaches of some streams with the most extensive and oldest development in the Puget Lowland region (e.g., Thornton, Miller, and Des Moines Creeks) manifest signs of limited sediment supply: alluvium comprises only a thin (<1 m) veneer over the parent glacial and lacustrine deposits. The loss of coarse sediment from streams is likely to reduce the depth and area of pools as the bars forming the pools degrade. As the thickness of the alluvial deposit decreases, so does the thickness of the hyporheic zone supported by the alluvium. Thus a stream can more easily scour down to bedrock or other parent material.

A number of common anthropogenic activities directly modify the morphology of urban stream channels and thus the type and extent of aquatic habitats in streams (Black and Silkey 1998). Streams are often straightened in urban areas, reducing their length and area of aquatic habitat. Straightening channels also eliminates meander bends and the associated pools and riffles. Levees and filling of floodplains reduce side-channel habitat and low velocity backwater habitat during floods.

Morphologic and sedimentologic changes (e.g., channel straightening and increased sediment loads) associated with urban development typically reduces the roughness of stream channels. As a result, urban streams have more uniform spatial distributions of current velocities at the reach scale (Ebbert et al. 2000). Some changes are intentional efforts to reduce channel roughness, such as straightening channels and lining them with concrete to increase their flood conveyance. Other changes are unintentional consequences of urban development that reduce channel roughness. The increase in fine sediment supply associated with the initial stages of urban development will lower the grain roughness of a stream bed and the depth of the boundary layer of lower-velocity water above the stream bed. Likewise, form roughness is reduced in channels where gravel bars are eroded away by increased storm flow or where the supply of LWD is limited.

The form and material of urban stream channels can be rehabilitated in areas with a channel-migration zone, where riparian forests can recover, and geomorphic processes can proceed without human intervention. However, the changes in the supply of sediment and runoff that can be expected in urban areas are likely to produce changes in the form and materials of the channel. Thus the form and material of an urban channel shaped and maintained by current fluvial processes may be quite distinct from its pre-development form and material. Moreover, efforts to rehabilitate such an urban stream must resolve whether to re-establish the pre-development channel form or to promote a new equilibrium form. This dilemma exemplifies the difference between stream restoration, in which the pre-development channel form could be recovered and would be in "equilibrium" with fluvial processes, and stream rehabilitation, in which either the pre-development channel form or the equilibrium between channel form and fluvial process can be recovered—but not both. Although partial recovery may be the best that can be hoped for an urban stream, it may not be evident which elements or processes should be recovered. In this case, ecological concepts and an experimental approach to urban stream rehabilitation may help to identify activities that will have the broadest and longest lasting effects.

At a smaller scale, there are many opportunities for rehabilitating the form and materials of urban stream channels. Rehabilitation (e.g., eliminating some human influences) can be achieved by removing riprap or other materials placed to stabilize stream banks. Restoration of urban stream channels (re-establishing pre-development conditions) would require reforesting riparian corridors and other areas that supply wood to channels, adding LWD to channels, "daylighting" stream channels in culverts, excavating active floodplains along incised channels, controlling excess runoff, increasing infiltration, and where appropriate reconstructing streams with a meandering plan form.

As a general approach, existing geomorphic features in a stream channel that provide habitat can be used as a model for the design of new structures (Sear 1994; Chapters 13 and 15). The design of in-channel structures intended to create specific types of habitats or channel forms are described comprehensively in Wesche (1985); Reeves et al. (1991); Flosi et al. (1998); and Center for Watershed Protection (2000). Because in-channel structures typically increase channel roughness and flow depth, their hydraulic effects need to be evaluated carefully in an urban environment.

Projects that modify channel forms and materials may be able to compensate for some effects of urban development. For example, complex channel forms buffer organisms from flood disturbances in streams (Gorman and Karr 1978; Borchardt 1993) and, thus, may limit the effects of hydrologic changes in urban streams. In many cases, however, in-channel projects (e.g., bank protection) are primarily intended to achieve a social objective, such as protecting property. In some cases, bank protection may be warranted where land use has reduced bank strength. For example, lower groundwater levels led to a decline in bank vegetation on the Carmel River, California (Kondolf and Curry 1986). In this case, bank protection serves to recover the bank strength that existed prior to groundwater depletion.

The primary ecological claims for bank-protection projects in urban streams are that stream banks would have been more stable prior to development and such projects reduce sediment delivery to the stream. In many urban streams, however, extensive bank protection may exacerbate an imbalance between sediment supply and the sediment transport capacity of a stream (i.e., allowing bank erosion in response to increased streamflow). It may be useful to assess whether bank stability was a feature of the pre-development channel or if bank erosion and channel migration were natural processes that now threaten infrastructure. If storm flows have increased in an urban stream and bank erosion must be prevented, then rehabilitation options in addition to bank protection include upstream stormwater management, by-passing high flows, increasing the conveyance of the channel, and adding gravel to compensate for the increased sediment-transport capacity of the stream.

Physical and Chemical Characteristics of Water

Urban development affects many physical and chemical characteristics of water. Ebbert et al. (2000) noted higher mid-day water temperatures in streams where urban land cover is dominant than in streams where agricultural or forest land cover is dominant in the Puget Sound basin. Preliminary analysis of one-day, instantaneous summertime stream-temperatures measured in the Puget Lowland in August 1998, 1999, and 2000 indicated that local riparian and flow conditions may have a greater influence on summer maximum stream temperature than basin-scale levels of urban development (Booth et al. 2000). These findings suggest that re-forestation and hydrologic process restoration may be effective rehabilitation techniques in streams where temperatures are high.

The most obvious chemical change in urban streams is the introduction of synthetic chemicals not found naturally in stream water. Insecticides and herbicides have been detected in the surface water and stream-bed sediment of urban streams in the Puget Sound basin at levels that adversely affect aquatic organisms (Bortelson and Ebbert 2000; Ebbert et al. 2000). In addition to the chronic effects of synthetic chemicals in urban streams, occasional spills of chemicals such as gasoline, paints, or organic solvents may have acute effects on urban stream ecosystems.

Urban development also influences the concentration and yield of compounds that naturally occur in surface waters. Embrey and Inkpen (1998) found high nitrogen yields (greater than 350 $kg/km^2/yr$) from stream and river basins with urban land use in the Puget Sound basin. The yields from urban basins were the same order of magnitude as agricultural basins and an order of magnitude higher than the yield from less developed, forested basins. The greatest contributions of nitrogen were from atmospheric deposition and fertilizer applications.

Activities associated with urban development may increase the concentration of trace elements (e.g., cadmium, copper, lead, selenium) in streams. The National Urban Runoff Program found relatively high concentration of heavy metals in urban runoff, but concluded that pollutants were unlikely to have toxic effects in Mercer Creek, the Puget Lowland stream included in the study (Athayde et al. 1983). Long-lived organisms, however, can bioaccumulate trace elements. Indeed, the concentrations of trace elements in some fish tissue samples from Puget Sound basin streams were elevated compared with mean values from the U.S. Fish and Wildlife Service National Contaminant Biomonitoring Program (Black and Silkey 1998).

The chemical characteristics of water can be restored in urban areas by preventing chemical contamination of runoff and re-establishing biogeochemical processes that may have been altered, for example, by the increase in runoff production via overland flow relative to subsurface flow. Pollution prevention in urban areas includes reducing or eliminating the application of fertilizers, herbicides, and insecticides in landscaping, disconnecting combined sewer overflows that discharge untreated wastewater directly to streams, and preventing spills of hazardous materials in areas that drain to streams.

Water-treatment systems can also reduce the physical and chemical changes in stream water associated with urban development. Runoff can be treated in sedimentation ponds, vegetated swales, or vegetated buffer strips (Osborne and Kovacic 1993) before it enters streams. Nitrogen compounds and other nutrients may be treated *in situ* by benthic bacteria (Jancarkova et al. 1997). Other types of treatments are not commonly used (e.g., instream cooling of surface waters, addition of materials to adsorb/neutralize pollutants or re-establish a water quality parameter) but may help re-establish the physical and chemical characteristics of urban stream water.

Energy Flux

The flux of energy through a stream ecosystem is regulated by inputs from the sun and by trophic relationships among organisms in the ecosystem. Trophic interactions provide a primary control on the types and abundance of organisms in a stream community, with changes at one trophic level cascading through the community (Wooton et al. 1996). Urban development can influence stream communities by altering the energy flux at each trophic level.

Riparian deforestation increases the light available for primary production and, as a consequence, produces changes in the diversity and abundance of diatom assemblages in urban streams (Hession et al. 2000). The effect of deforestation on light availability is greatest for small streams where a mature forest canopy would have shaded the stream for much of the day. Discharge of organic wastes to urban streams increases the sources of energy and nutrients for microbial decomposers. Changes in vegetation from conifers to deciduous species in riparian areas along streams influence the type, quantity, quality, and timing of allochthonous inputs to a stream. Finally, fewer adult salmon returning to urban streams reduces the carcasses available to detritivores.

Rehabilitation of trophic interactions in urban streams could begin with reforestation of riparian areas, particularly with conifers, to reduce light available to periphyton. Tertiary treatment of wastewater would reduce availability of suspended organic compounds as a source of energy allowing the trophic base of stream communities to re-establish and, as a consequence, support the recovery of benthic macroinvertebrate assemblage diversity (Baker and Sharp 1998).

Although there are techniques to compensate for the availability of nutrients at different trophic levels in stream communities, their effects in urban streams have not been tested. For example, fertilization of streams with inorganic phosphorous and nitrogen has been a successful approach to increasing periphyton, invertebrate, and salmonid abundance in nutrient-poor systems (Reeves et al. 1991). Stream fertilization may be an option for increasing the productivity of urban stream ecosystems that have been shown to be nutrient deprived. Rehabilitating the energy fluxes of an urban stream by manipulating nutrient inputs rather than restoring them, for example, through an appropriate mix of point and non-point controls, re-forestation, and recovery of salmonid populations, requires careful design including the type, form, and quantity of nutrients and is by no means guaranteed to be successful. Fertilization could further alter trophic relations, for example, because of increased light availability in a stream. Moreover, fertilization represents an on-going intervention in an ecosystem, although it may promote the development of a more self-sustaining ecosystem as trophic relations recover.

Biological Conditions

The influence of urban development on the biological conditions of streams in the Puget Lowland is evident in the proliferation of exotic species, depressed fish stocks, lower diversity, less complex trophic structure, and increasing dominance of fewer taxa in benthic macroinvertebrate assemblages (Scott et al. 1986; Bledsoe et al. 1989; Nehlsen et al. 1991; Black and Silkey 1998; and Morley and Karr 2000). Although biological organisms may be viewed as an endpoint in an assessment of an urban stream, they also are an important mediator of stream ecosystems through biotic interactions (Allan 1996) and their influence on physical conditions of streams (Keller and Swanson 1979). Thus direct efforts to recover biological communities may be a means to rehabilitating urban stream ecosystems.

With the exception of riparian re-vegetation projects, rehabilitation projects typically focus on re-establishing physical conditions of streams with the expectation that biological recovery will result. However, biological conditions may not spontaneously recover in streams even after the physical conditions have been restored because of long distances to other streams harboring potential colonists (Fuchs and Statzner 1990). In this case, re-introduction of an extirpated species or re-stocking declining populations may be a necessary step in stream rehabilitation (Charbonneau and Resh 1992).

Conversely, exotic species may have to be actively removed from urban streams where natural succession cannot be relied on for biological recovery. For example, the prevention of flooding along Boulder Creek, Colorado, favored the establishment of invasive, exotic vegetation and limited the extent of indigenous plains cottonwood (*Populus deltoides*) in riparian areas (Auble et al. 1997). In this case, however, removing the exotic vegetation and planting cottonwoods may not be a successful rehabilitation strategy without re-establishing flooding.

Where pre-development biological conditions cannot be restored, rehabilitation projects in urban streams may be necessary to compensate for missing biological elements of streams. For example, riparian trees can be felled manually if beavers no longer inhabit a stream, or logs can be placed in a stream where stream banks lack trees. Although fish hatcheries do not restore the natural reproductive cycles of fish populations (Chapter 5), they can produce juvenile salmon for stocking in a stream and may increase the numbers of returning adults. As with other stream rehabilitation projects that do not restore ecological processes, projects that focus on recovery of specific species or attempt to compensate for missing elements may not promote broader recovery of stream ecosystems or be self-sustaining.

An Ecological Approach to Urban-Stream Rehabilitation

Individual components of an ecosystem depend on and, conversely, are necessary for a broad array of physical and biological processes and conditions. Because of the interdependency of elements in an ecosystem, it is generally more efficient and reliable to restore whole systems than to attempt to recover individual elements (NRC 1992). However, there are many constraints in urban areas that prevent the restoration of the processes and conditions necessary for self-regulating stream ecosystems.

Some constraints limit the degree or spatial extent to which attributes (e.g., runoff processes, flooding, bank erosion, riparian forests) of streams or wetlands can be recovered in urban areas. For example, roads and drainage networks reduce the length of hillslope flow paths traveled by water. Bridges and streamside property limit where streams can be allowed to erode their banks. Flooding is prevented to protect roads and structures from being damaged. Most people living in riparian areas actively landscape these areas rather than allowing natural succession of vegetation. In other cases, former elements of the ecosystem, such as beaver (*Castor canadensis*) or black bear (*Ursus americanus*), may be altogether missing from urban areas. At the most basic level, these constraints result from the lack of land available to provide natural habitat-forming processes.

The fundamental challenge for rehabilitating an urban stream is to recover attributes of an ecosystem limited to only a fraction of the surrounding watershed. Even where social values, political will, and financial resources are available to eliminate or compensate for some of the constraints on urban streams, effective approaches and techniques for urban stream rehabilitation are not known *a priori* and some constraints will persist. An understanding of ecological processes, such as hydrologic regime, channel forming-processes, energy flow, and biological interactions, are necessary for developing feasible objectives and successful techniques for recovering attributes of urban streams in spite of the limits on restoring these ecosystems. This is particularly important where people will be asked to accommodate streams and where ecological processes and conditions can only be partially recovered.

Ecological processes require a finite amount of space that varies from process-to-process. In urban areas, the space available for ecological processes in streams and the contributing watershed is limited by roads, structures, and people. The feasibility of recovering processes (Chapter 8) depends on their spatial scale relative to the available (unconstrained) space. For example, primary production from a colony of benthic macrophytes can occur over a scale of 1 m^2. The eddies and backwater formed behind a small obstruction in a streams may occupy a volume of 10 m^3. Subsurface flow may travel 100 m through a hillslope. Anadromous salmon migrate thousands of meters in a stream as smolts and returning adults. Processes requiring large spaces are likely to be highly constrained in urban areas, while processes occurring in small spaces may be recoverable in some places.

The theory of island biogeography (MacArthur and Wilson 1967) proposed that the diversity of an island was related fundamentally to its area and its distance to other islands. This theory has been cited as the basis for the increase in species diversity with the drainage area of a stream (Allan 1996). Similarly, the distance separating streams may be an important factor in the re-population of rehabilitated streams (Fuchs and Statzner 1990). Given the principles of island biogeography, urban stream rehabilitation projects are likely to be successful where they increase the amount, not just the quality, of habitat available to aquatic organisms. Additionally, efforts to rehabilitate large streams and streams adjacent to less-disturbed areas may best promote increased biological diversity. Likewise, rehabilitation projects that re-establish the connectivity between unimpaired reaches of a stream or between adjacent hyporheic and riparian zones may be most successful (Gore 1985).

A stream ecosystem is a mosaic of habitat patches with varying substrates, hydraulic conditions, and disturbance regimes (Townsend 1989). The structure and diversity of biological communities in a rehabilitated stream depend on recovery of the mosaic as well as the individual patches. Streams are embedded in a watershed and linked to many other ecosystems including lakes, estuaries, forests, and grasslands. Hynes (1975) argued that these linkages are so strong that "the valley rules the stream." As such, rehabilitation of streams depends to some degree on the recovery of the surrounding watershed and the riparian and hyporheic ecotones.

Rehabilitation projects that do not take into account local ecological processes and thus fail to address the latent physical and biological controls on stream ecosystems will be short-lived, high maintenance, or simply ineffective (Kauffman et al. 1997). For example, stream rehabilitation projects frequently attempt to impose a specific channel form such as a meander bend or a pool rather than considering controlling mechanisms such as wood delivery to a channel, streamflow patterns, or sediment loads (Keller 1978; Brookes

1990). The recovery of missing elements does not necessarily produce the desired biological effects, particularly in urban streams where there are often multiple, interacting causes for stream degradation (Booth et al. 2000).

Urban-Stream Rehabilitation as Ecological Experiments

Due to the complexity of ecological controls and the multitude of anthropogenic constraints, urban-stream rehabilitation projects are largely ecological experiments that can be used to test whether the recovery of a missing or degraded element of a stream will provide broader ecological benefits. Evidence of the effectiveness of urban-stream rehabilitation projects is equivocal (Keller 1978; Brookes 1990; House 1996; Hilderbrand et al. 1997; Lasssonen et al. 1998; Center for Watershed Protection 2000). While some projects fail outright because of inadequate design or unforeseen events (floods), other projects are successfully implemented but do not provide broader benefits to the ecosystem (Chapman 1996; Larson 1999). In an experimental framework, the first step in stream rehabilitation is to diagnose the causes of degradation, which represents a hypothesis to be tested by implementing a project and monitoring its effects (see Chapter 9).

The most compelling diagnoses come from "natural" experiments along a stream where a longitudinal degradation or recovery in biological conditions is associated with a difference in a single, identifiable factor, such as a point discharge of wastewater (Baker and Sharp 1998). In urban areas, however, stream ecosystems are unlikely to be degraded solely as a result of a single stressor. Successful rehabilitation of urban streams requires diagnosis and a suite of techniques that address multiple stressors (Chapter 8).

The rehabilitation of an urban stream affected by multiple stressors is illustrated in Strawberry Creek, California (Charbonneau and Resh 1992). University of California at Berkeley staff diagnosed problems with water quality, habitat, and hydrologic regime in the creek. Direct discharges of sanitary sewers and runoff from areas likely to introduce high loads of nutrients and other urban pollutants were eliminated or controlled. Check dams and crib walls were constructed to control erosion, store sediment, and provide habitat for fish and aquatic insects. Stream banks were revegetated and native sticklebacks (*Gasterosteus aculeatus*) were re-introduced. Two years after implementing these projects, water quality improved, the macroinvertebrate assemblage was more diverse, and the native sticklebacks were successfully spawning.

Beyond Environmental Nostalgia in Urban-Stream Rehabilitation

Urban-stream rehabilitation is a compromise between social values aimed at making a landscape habitable for people and ecological processes (Keller and Hoffman 1977) that do not play favorites among species. Progress in rehabilitating urban stream ecosystems depends on the extent to which the attributes of stream ecosystems are compatible with the activities of people in urban areas and the extent to which human activities can accommodate the attributes of stream ecosystems. Schauman (1998) describes the pastoral aesthetic (imagine a garden and a red barn) as a form of nostalgia that does not acknowledge the "messy" aspects of agriculture and rural life. A parallel form of "environmental" nostalgia arises when the "messy" attributes of streams such as floods, channel migration, rotting salmon carcasses, windthrow trees, and large carnivores, such as bears, are neglected in stream rehabilitation plans. These messy attributes are elements of natural stream ecosystems, which may return to streams if rehabilitation efforts are successful or may foil attempts to rehabilitate streams through their absence.

The challenge for urban stream rehabilitation is to first identify which of the multitude of anthropogenic changes have significant ecological effects on the stream. Once these changes have been identified, then options for re-establishing elements of stream ecosystems can be developed. Where re-establishing an element is politically, financially, or otherwise socially acceptable in an urban landscape, rehabilitation projects must be developed to compensate for the missing or degraded element. These challenges require moving beyond a narrow, nostalgic view of streams and examining the physical and biological processes that regulate these ecosystems. Although ecological restoration may be infeasible for urban streams, their rehabilitation can be guided by an awareness of ecological processes. Because stream ecosystems are regulated by many factors, each of which can be modified in urban areas, urban stream rehabilitation will benefit from an experimental approach in which causes of degradation are hypothesized and tested by implementing projects and monitoring their effects (Chapter 9).

Acknowledgments

This chapter was supported by the Washington District, U.S. Geological Survey and the Water and Watersheds Program, U.S. Environmental Protection Agency under agreement R82-5484-010. The author appreciates Robert Black's review, which improved the chapter's clarity and content.

References

Allan, J.D. 1996. *Stream Ecology*. Chapman and Hall, London.

Allan, J.D., D.L. Erickson, and J. Fay. 1997. The influence of catchment land use on stream integrity across multiple spatial scales. *Freshwater Ecology* 37:149-162.

Athayde, D.N., P.E. Shelley, E.D. Driscoll, D. Gaboury, and G. Boyd. 1983. Results of the Nationwide Urban Runoff Program. NTIS PB84-185537, U.S. Environmental Protection Agency, Washington D.C.

Auble, G.T., M.L. Scott, J.M. Friedman, J. Back, and V.J. Lee. 1997. Constraints on the establishment of plains cottonwood in an urban riparian preserve. *Wetlands* 17:138–148.

Baker, S.C. and H.F. Sharp. 1998. Evaluation of the recovery of a polluted urban stream using the Ephemeroptera-Plecoptera-Trichoptera index. *Journal of Freshwater Ecology* 13:229-234.

Bisson, P.A. and J.R. Sedell. 1984. Salmonid populations in streams in clearcut vs. old-growth forests of western Washington. In W.R, Meehan, T.R. Merrell, Jr., and T.A. Hanley (eds.) *Proceedings, Fish and Wildlife Relationships in Old-growth Forests Symposium*. American Institute of Fishery Research Biologists, Asheville, North Carolina. pp. 121-130.

Black, R.W. and M. Silkey. 1998. Water-quality assessment of the Puget Sound Basin, Washington, summary of stream biological data through 1995. Water-Resources Investigation Report 97-4164. U.S. Geological Survey, Tacoma, Washington.

Bledsoe, L.J., D.A. Somerton, and C.M. Lynde. 1989 The Puget Sound runs of salmon—an examination of changes in run size since 1896. In C.D. Levings, L.B. Holby, and M.A. Henderson (eds.) *Proceedings of the National Workshop on Effects of Habitat Alteration on Salmonid Stocks* Canadian Special Publication of Fisheries and Aquatic Sciences 105:50-61.

Booth, D.B. 1990. Stream-channel incision following drainage-basin urbanization. *Water Resources Bulletin* 26:407-417.

Booth, D.B., C.P. Konrad, S.A. Morely, M.G. Larson, J.R. Karr, S. Schauman, and S.J. Burges. 2000. Evaluation and rehabilitation of urban streams in the Pacific Northwest. *EOS, Transactions of the American Geophysical Union* 81, Spring Meeting Supplement.

Borchardt, D. 1993. Effects of flow and refugia on drift loss of benthic macro-invertebrates: Implications for habitat restoration in lowland streams. *Freshwater Biology* 29:221-227.

Bortelson, G.C. and J.C. Ebbert. 2000. Occurrence of pesticides in streams and ground water in the Puget Sound Basin, Washington, and British Columbia,

1996-98. Water Resources Investigations Report 00-4118. U.S. Geological Survey, Tacoma, Washington.

Brookes, A. 1990. Restoration and enhancement of engineered river channels: Some European experiences. *Regulated Rivers: Research & Management* 5:45-56.

Buffington, J.M. and D.R. Montgomery. 1999. Effects of hydraulic roughness on surface textures of gravel-bed rivers. *Water Resources Research* 35:3507-3521.

Burges S.J., M.S. Wigmosta, and J.M. Meena. 1998. Hydrologic effects of land-use change in a zero-order catchment. *Journal of Hydrologic Engineering* 3:86-97.

Flosi, G., S. Downie, J. Hopelain, M. Bird, R. Coey, and B. Collins. 1998. *California Salmonid Stream Habitat Restoration Manual 3rd Edition.* California Department of Fish and Game, Sacramento, California.

Center for Watershed Protection. 2000. Urban Stream Restoration Practices: An Initial Assessment. Unpublished report prepared for U.S. Environmental Protection Agency, Office of Wetlands, Oceans, and Watersheds and Region V.

Chapman, D.W. 1996. Efficacy of structural manipulations of instream habitat in the Columbia River basin. *Rivers* 5:279–293.

Charbonneau, R. and V.H. Resh. 1992. Strawberry Creek on the University of California, Berkeley campus: A case history of urban stream restoration. *Aquatic Conservation: Marine and Freshwater Ecosystems* 2:293-307.

Dietrich, W.E., J.E. Kirchner, H. Ikeda, and F. Iseya. 1989. Sediment supply and the development of the coarse surface layer in gravel-bedded rivers. *Nature* 340:215-217.

Dinicola, R.S. 1990. Characterization and simulation of rainfall-runoff relations for headwater basins in western King and Snohomish Counties, Washington. Water Resources Investigations Report 89-4052. U.S. Geological Survey, Tacoma, Washington.

Ebbert, J.C., S.S. Embrey, R.W. Black, A.J. Tesoriero, and A.L. Haggland. 2000. Water Quality in the Puget Sound Basin, Washington and British Columbia, 1996-98. Circular 1216. U.S. Geological Survey, Denver, Colorado.

Ebisemiju, F.S. 1989. The response of headwater stream channels to urbanization in the humid tropics. *Hydrological Processes* 3:237-253.

Embrey, S.S. and E.L. Inkpen. 1998. Water-Quality Assessment of the Puget Sound Basin, Washington, Nutrient Transport in Rivers, 1980-93. Water Resources Investigations Report 97-4270, U.S. Geological Survey, Tacoma, Washington.

Finkenbine, J.K., J.W. Atwater, and D.S. Mavinic. 2000. Stream health after urbanization. *Journal of the American Water Resources Association* 36:1149-1160.

Fuchs, U. and B. Statzner. 1990. Time scales for the recovery potential of river communities after restoration: Lessons to be learned from smaller streams. *Regulated Rivers: Research & Management* 5:77-87.

Gore, J.A. 1985. Mechanisms of colonization and habitat enhancement for benthic macroinvertebrates in restored river channels. In J.A. Gore (ed.) *The Restoration of Rivers and Streams*. Butterworth Publishers, Stoneham. pp. 81–102.

Gorman, O.T. and J.R. Karr. 1978. Habitat structure and stream fish communities. *Ecology* 59:507-515.

Guy, H. and G.E. Ferguson. 1962. Sediment in small reservoirs due to urbanization, *Transactions American Society of Civil Engineers* 128:982-992.

Hartley, D.M. 2001. Detecting hydrologic change in high resource streams. *Proceedings of the Puget Sound Research Conference 2001*. Puget Sound Water Quality Action Team, Olympia, Washington.

Hession, W.C., T.E. Johnson, D.F. Charles, D.D. Hart, R.J. Horwitz, D.A. Kreeger, J.E. Pizzuto, D.J. Velinsky, J.D. Newbold, C. Cianfrani, T. Clason, A.M. Compton, N. Coutler, L. Fuselier, B.D. Marshall, and J. Reed. 2000. Ecological benefits of riparian reforestation in urban watersheds: study design and preliminary results. *Environmental Monitoring and Assessment* 63:211-222.

Hilderbrand, R.H., A.D. Lemly, C.A. Dolloff, and K.L. Harpster. 1997. Effects of large woody debris placement on stream channels and benthic macroinvertebrates. *Canadian Journal of Fisheries and Aquatic Sciences* 54:931-939.

House, R. 1996. An evaluation of stream restoration structures in a coastal Oregon stream, 1983-1993. *North American Journal of Fisheries Management* 16:272-281.

Hynes, H.B.N. 1975. The stream and its valley. *Verhandlungen der Internationalen Vereinigung fur Theoretische und Angewandte Limnologie* 19:1-15.

Jancarkova, I., T.A. Larsen, and W. Gujer. 1997. Distribution of nitrifying bacteria in a shallow stream. *Water Science and Technology* 36:161-166.

Karr, J.R., K.D. Fausch, P.L. Angermeier, P.R. Yant, and I.J. Schlosser. 1986. Assessing the biological integrity in running waters: A method and its rationale. Special Publication 5. Illinois Natural History Survey, Champaign, Illinois.

Kauffman, J.B., R.L. Beschta, N. Otting, and D. Lytjen. 1997. An ecological perspective of riparian and stream restoration in the western United States. *Fisheries* 22(5):12-24.

Keller, E.A. 1978. Pools, riffles, and channelization. *Environmental Geology* 2:119-127.

Keller, E.A. and E.K. Hoffman. 1977. Urban streams: Sensual blight or amentiy? *Journal of Soil and Water Conservation* 32:237-240.

Keller, E.A. and F.J. Swanson. 1979. Effects of large organic material on channel form and fluvial processes. *Earth Surface Processes* 4:361-380.

Kondolf, G.M and R.R. Curry. 1986. Channel erosion along the Carmel River, Monterey County, California. *Earth Surface Processes and Landforms* 11:307-319.

Konrad, C.P. 2000. The frequency and extent of hydrologic disturbances in streams in the Puget Lowland, Washington. Ph.D. dissertation. University of Washington. Seattle, Washington.

Konrad, C.P. and S.J. Burges. 2001. Hydrologic mitigation using on-site residential detention. *Journal of Water Resources Planning and Management* 127:99-107.

Konrad, C.P., S.J. Burges, and B.W. Jensen. 1995. An examination of stormwater detention and infiltration at the scale of an individual residence in the Sammamish Plateau of King County, Washington. Water Resources Series Technical Report No. 148. University of Washington, Department of Civil Engineering, Seattle, Washington.

Larson, M.G. 1999. Effectiveness of large woody debris in stream rehabilitation projects in urban basins. Master's thesis. University of Washington. Seattle, Washington.

Lasssonen, P., T. Muotka, and I. Kivijarervi. 1998. Recovery of macroinvertebrate communities from stream habitat restoration. *Aquatic Conservation: Marine and Freshwater Ecosystems* 8:101-113.

Lisle, T.E. and S. Hilton. 1992. The volume of fine sediment in pools: An index of sediment supply in gravel-bed streams. *Water Resources Bulletin* 28:371-383.

Lisle, T.E., F. Iseya, and H. Ikeda. 1993. Response of a channel with alternate bars to a decrease in supply of mixed-size bed load: A flume experiment. *Water Resources Research* 29:3623-3629.

MacArthur, R.H. and E.O. Wilson. 1967. *The Theory of Island Biogeography*. Monographs in Populations Biology 1. Princeton University Press. Princeton, New Jersey.

Matzke, G. 1993. Population. In P.L. Jackson and A.J. Kimerlings (eds.) *Atlas of the Pacific Northwest*. Oregon University Press, Corvallis, Oregon. pp. 18-24.

Montgomery, D.R, J.M. Buffington, R.D. Smith, K.M. Schmidt, and G. Pess. 1995. Pool spacing in forest channels. *Water Resources Research* 31:1097-1105.

Morely, S.A. and J.R. Karr. 2000. Urban streams in the Pacific Northwest: restoration with a biological focus. *EOS, Transactions of the American Geophysical Union* 81, Spring Meeting Supplement.

Moscrip, A.L. and D.R. Montgomery. 1997. Urbanization, flood frequency, and salmon abundance in Puget lowland streams. *Journal of the American Water Resources Association* 33:1289-1297.

National Research Council. 1992. *Restoration of Aquatic Ecosystems*. National Academy Press, Washington D.C.

Nehlsen, W., J.E. Williams, and J.A. Lichatowich. 1991. Pacific salmon at the crossroads—stocks at risk form California, Oregon, Idaho, and Washington. *Fisheries* 16(2):4-21.

Osborne, L.L and K.A. Kovacic. 1993. Riparian vegetated buffer strips in water-quality restoration and stream management. *Freshwater Biology* 29:243-258.

Odum, H.T. 1956. Primary production in flowing waters. *Limnology and Oceanography* 1:102-117.

Osher, P.N. and D.J. Shure. 1972. Effects of urbanization on the salamander *Desmognathus fuscus fuscus*. *Ecology* 53:1148-1154.

Poff, N.L. and J.D. Allan. 1995. Functional organization of stream fish assemblages in relation to hydrologic variability. *Ecology* 76:606-627.

Poff, N.L., J.D. Allan, M.B. Bain, J.R. Karr, K.L. Prestegaard, B.D. Richter, R.E. Sparks, and J.C. Stromberg. 1997. The natural flow regime: A paradigm for river conservation and restoration. *Bioscience* 47:769-784.

Poff, N.A., N.J. Voelz, J.V. Ward, and R.E. Lee. 1990. Algal colonization under four experimentally controlled current regimes in a high mountain stream. *Journal of the North American Benthological Society* 9:303-308.

Poff, N.L. and J.V. Ward. 1989. Implications of streamflow variability and predictability for lotic community structure: a regional analysis of stream flow patterns. *Canadian Journal of Fisheries and Aquatic Sciences* 46:1805-1818.

Ralph, S.C., G..C. Poole, L.L. Conquest, and R.J. Naiman. 1994. Stream channel morphology and woody debris in logged and unlogged basin of western Washington. *Canadian Journal of Fisheries and Aquatic Sciences* 51:37-51.

Rees, W.H. 1957. Effects of stream dredging on young silver salmon (*Oncorhynchus kisutch*) and bottom fauna. *Puget Sound Studies, Washington State Department of Fisheries* 1:52-65.

Reeves, G.H., J.D. Hall, T.D. Roelofs, T.L. Hickman, and C.O. Baker. 1991. Rehabilitating and modifying stream habitats. In W.R. Meehan (ed.) *Influences of Forest and Rangeland Management on Salmonid Fishes and Their Habitats*. Special Publication 19, American Fisheries Society, Bethesda, Maryland. pp. 591-558.

Reid, I. and J.B. Laronne. 1995. Bed load sediment transport in an intermittent stream and a comparison with seasonal and perennial counterparts. *Water Resources Research* 31:773-781.

Schauman, S. 1998. The garden and the red barn: The pervasive pastoral and its environmental consequences. *The Journal of Aesthetic and Art Criticism* 56:181-190.

Scott, J.B., C.R. Steward, and Q.J. Stober. 1986. Effects of urban development on fish population dynamics in Kelsey Creek, Washington. *Transactions of the American Fisheries Society* 115:555-567.

Sear, D.A. 1994. River restoration and geomorphology. *Aquatic conservation: Marine and freshwater ecosystems* 4:169-177.

Shelford, V.S. and S. Eddy. 1929. Methods for the study of stream communities. *Ecology* 10:382-391.

Sidle, R.C., A.J. Pearce, and C.L. O'Loughlin. 1985. *Hillslope stability and land use*. Water Resources Monograph 11. American Geophysical Union, Washington D.C.

Smock, L.E., J.E. Gladden, J.L. Riekenburg, L.C. Smith and C.R. Black. 1992. Lotic macroinvertebrate production in three dimensions: channel surface, hyporheic and floodplain environments. *Ecology* 73:876-886.

Thoms, M.C. 1987. Channel sedimentation within the urbanized River Tame, U.K. *Regulated Rivers: Research & Management* 1:229-246.

Townsend, C.R. 1989. The patch dynamics concept of stream community ecology. *Journal of the North American Benthological Society* 8:36-50.

Trimble, S.W. 2000. Trees versus grass on stream banks: Some geomorphological considerations. *EOS, Transactions of the American Geophysical Union* 81, Spring Meeting Supplement.

Vervier, P., M. Dobson, and G. Pinay. 1993. Role of interaction zones between surface and ground waters in DOC transport and processing: considerations for river restoration. *Freshwater Biology* 29:275-284.

Wesche, T.A. 1985. Stream channel modifications and reclamation structures to enhance fish habitat. In J.A. Gore (ed.) *The Restoration of Rivers and Streams*. Butterworth Publishers, Stoneham. pp. 103-163.

Wigmosta, M.S., S.J. Burges, and J.M. Meena. 1994. Modeling and monitoring to predict spatial and temporal hydrologic characteristics in small catchments. Water Resources Series Technical Report No. 137. University of Washington, Department of Civil and Environmental Engineering, Seattle, Washington.

Wolman M.G. and A.P. Schick. 1967. Effects of construction on fluvial sediment, urban and suburban areas of Maryland, *Water Resources Research* 3:451-464.

Wooton, J.T., M.S. Parker, and M.E. Power. 1996. Effects of disturbance on river food webs. *Science* 273:1558-1561.

12. Monitoring and Evaluating Fish Response to Instream Restoration

Philip Roni, Martin Liermann, and Ashley Steel

Abstract

Restoration efforts to recover Pacific salmon (*Oncorhynchus* spp.) populations will require monitoring programs to evaluate the effectiveness of restoration activities. Unfortunately, variability in fish abundance among years and streams makes quantitative assessments difficult and has led to inconclusive results in many previous studies of restoration effectiveness. Prior knowledge of spatial and temporal population variability, the expected magnitude of the response, and the time and resources available can be used to design studies that are most likely to detect a change in fish abundance. We review common study and monitoring designs, and discuss their usefulness at multiple scales and life stages. Using data from two recent restoration studies, we estimate sample sizes required to detect salmonid responses to restoration using a before-after (BA) or before-after-control-impact (BACI) design and an extensive post-treatment (EPT) design. Our analysis suggests that the BA and BACI designs may require 10 to more than 50 years of monitoring to detect a doubling of juvenile salmonid abundance. The EPT design, replicated in space instead of time, is particularly useful for evaluating reach-scale restoration activities (e.g., wood placement, riparian restoration). Our analysis also suggests that 20–30 sites are necessary to detect a doubling in juvenile coho salmon (*O. kisutch*) and steelhead (*O. mykiss*) numbers. The optimum study design for evaluating fish response to restoration activities will depend in part on the sample sizes needed to detect fish responses, but also on the project scale, species and life stage(s) of interest, and the level and duration of funding.

INTRODUCTION

> "...additional evaluation studies on stream improvement, especially with reference to the effect on the abundance of fish, are still urgently needed." —Clarence M. Tarzwell, Biologist, U.S. Forest Service (1937), p187

Hundreds of millions of dollars have been spent by federal, state, county, and municipal governments in recent years to restore and improve degraded salmon (*Oncorhynchus* spp.) habitat in Pacific Northwest rivers. Common restoration and habitat enhancement techniques include large woody debris (LWD) placement, removal of artificial barriers, estuarine restoration, road improvement, sediment reduction, bank stabilization, and riparian vegetation restoration. Evaluation of these and other restoration efforts is important to determine the physical and biological effectiveness, guide future restoration efforts, and assure that limited funds are spent wisely. Despite the importance of such evaluations and the large financial investment in stream restoration, research and monitoring to evaluate project effectiveness occurs infrequently and is often inadequate to quantify biological response (Reeves and Roelefs 1982; Reeves et al. 1991a; Beschta et al. 1994; Chapman 1996; Roni et al. 2002). The need for thorough monitoring and evaluation of instream (e.g., LWD and boulder placement) and other restoration techniques has been noted for decades (Tarzwell 1934; Reeves and Roelefs 1982; Reeves et al. 1991a). Given the current status of many salmonid stocks as threatened or endangered under the Endangered Species Act, and the large sums of money being allocated for instream restoration to improve salmonid population status, the need for evaluation of instream restoration efforts is particularly pressing.

Physical changes following instream restoration, such as increases in pool frequency, pool depth, woody debris retention, and sediment storage have been well documented (e.g., Crispin et al. 1993; Cederholm et al. 1997; Reeves et al. 1997). Biological responses to instream restoration are less predictable. Limited quantitative evaluations of project effectiveness exist (Roni et al. 2002), and little information exists on appropriate monitoring designs. Published evaluations provide conflicting conclusions about the biological effectiveness of restoration activities. Beschta et al. (1994) summarized several case studies on stream restoration in the intermountain western United States and concluded that few had produced the desired results. Chapman (1996) summarized data from instream restoration projects in the Columbia River basin and found little evidence that the projects were successful at increasing fish production. Conversely, others have reported large and significant increases in juvenile coho abundance following boulder and wood placement (House et al.1989; House 1996; Cederholm et al. 1997; Roni and Quinn 2001a).

The goals of this chapter are to review published evaluations of biological response to instream restoration projects in the Pacific Northwest and to provide guidance on designing research to evaluate fish responses to instream and other restoration techniques. The chapter is divided into four sections. In the first section, we review published evaluations of biological response to instream restoration projects. In the second section, we review common study (monitoring) designs used to determine fish response to restoration projects. In the third section, we present the results of sample size calculations for detecting juvenile salmonid response using various study designs. Finally, in the fourth section we discuss appropriate monitoring designs for different scales and salmonid life stages. We focus on monitoring of fish abundance, and on anadromous salmonids in particular, as most stream restoration projects in the Pacific Northwest are designed for salmonids, and little information exists for other species.

Published Evaluations of Instream Restoration

The vast majority of biological evaluations of stream restoration have focused on LWD, boulder placement, and other forms of instream restoration (Roni et al. 2002). We examined published evaluations of juvenile anadromous salmonid response to instream restoration projects in 30 Pacific Northwest streams (Table 1). Of these evaluations, statistically significant differences in juvenile salmonid abundance for at least one species or life stage were reported in 13 streams. The lack of significant findings may reflect a true lack of project success or may result from insufficient monitoring of highly variable phenomena. In only six of the streams was monitoring of salmonid populations carried out for more than five years (Figure 1).

Synthesis of published evaluations of juvenile anadromous salmonid response to instream restoration projects in western North America indicates that responses vary among species and life-stages. Across the studies and streams examined, coho salmon (*Oncorhynchus kisutch*) demonstrated the most consistent response. Juvenile coho salmon densities were consistently higher following instream restoration (18 of 19 streams); however, statistically significant results were reported in only 7 streams (Figure 2). Results for juvenile steelhead (*O. mykiss*), cutthroat (*O. clarki*), and trout fry (age 0+ cutthroat and steelhead) were less consistent. Differences in response among species may indicate true differences due to restoration actions or from

A

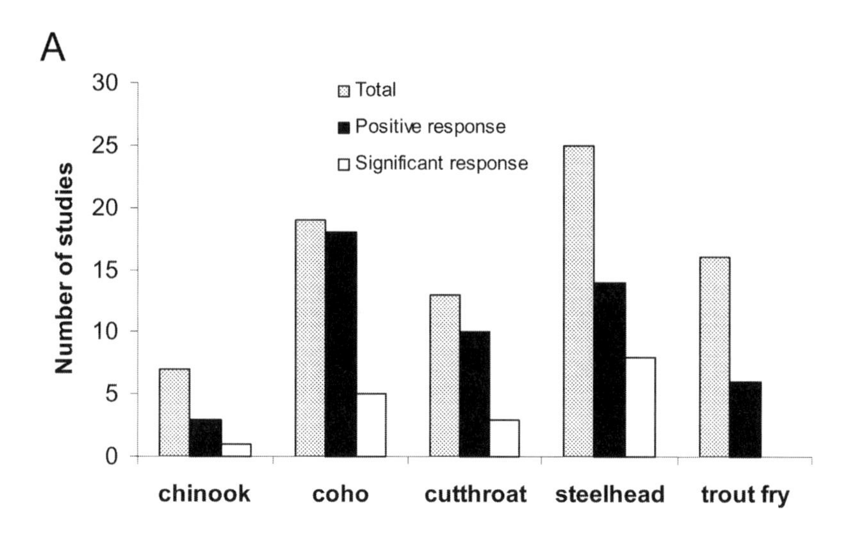

B

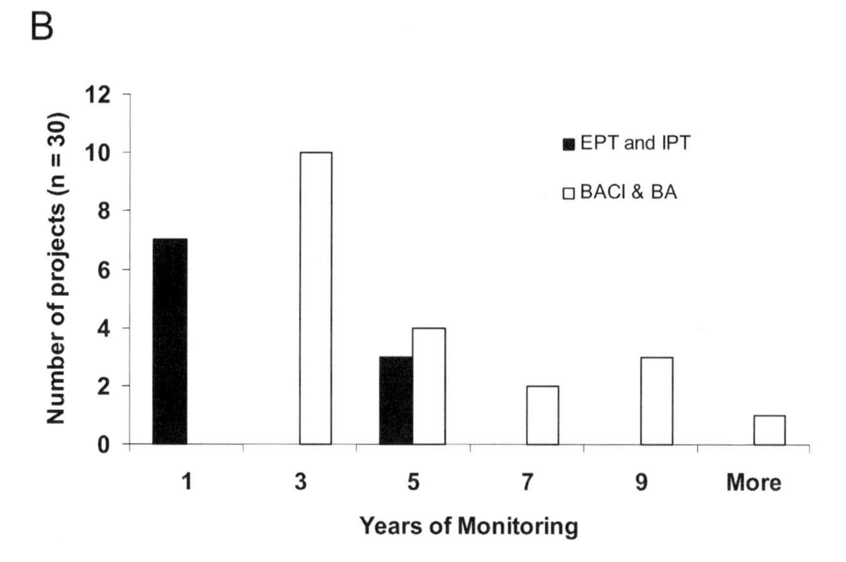

Figure 1. (A) Total number of published stream restoration projects that examined responses to instream restoration of juvenile chinook (*O. tshawytscha*), coho, cutthroat, steelhead, and trout fry (age 0 steelhead and cutthroat) and those that reported positive and significant response. (B) Number of years of monitoring for extensive post-treatment (EPT) and intensive post-treatment (IPT) designs and for before and after (BA) and before-after control-impact (BACI) designs. Data taken from Table 1.

Table 1. Summary of findings of Pacific Northwest research and monitoring of instream restoration projects for various species. 0 indicates no response, + indicates positive response, - indicates a negative response, and * indicates results were significant at the 0.05 level (modified from Roni et al. 2002). Structure types were categorized as LS (LWD structure), LN (naturally placed LWD), G (gabion), B (boulder clusters or structure), and AL (alcove or ponds). Sources: Ward and Slaney (1981); Moreau (1984); House and Boehne (1985, 1986); Fontaine (1987); House et al. (1989, 1996); Poulin et al. (1991); Slaney et al. (1994); Chapman (1996); Cederholm et al. (1997): Reeves et al. (1997); Solazzi et al. (2000); Roni and Quinn (2001a).

	State or		Structure		Juvenile Salmonid Response				
Study Design/Stream	*Province*	*Years*	*Type(s)*	*Season*	*Coho*	*Trout Fry*	*Cutthroat*	*Steelhead*	*Chinook*
Before and After (BA) design									
"J" Line Cr.	OR	5	LS, B	Su	+		-	-	
Fish Cr.	OR	13	LS, B, G	Su, Sp	-/0	-*[a]		+/+	
Little Lobster Cr.	OR	5	B	Su	+	0	-	-	
Lobster Cr.	OR	3	B, LS	Su	+	+	+	-	
Lower Elk Cr.	OR	4	LS, B	Su	+		0	-	
S.F. Lobster Cr.	OR	2	LS	Su		+	+		
Upper Lobster Cr.	OR	3	G	Su	+	+	+	+	
Upper Lobster Cr. B	OR	5	B, LS	Su	+	-	+	-	
Before-After Control-Impact Design									
Alsea River Tribs.	OR	8	LS, AL	Su, Sp	+*/+*	0[a]	0/+	0/+*	
Bonanza Cr.	BC	2	LN	Su	+*				
E. Beaver Cr.	OR	6	G	Su	+	+	+	0	
E.F. Lobster Cr. C	OR	9	B, G	Su	+*	0	+*	0	
Hurdygurdy Cr.	CA	2	B	Su				+	

Table 1 (continued).

Study Design/Stream	State or Province	Years	Structure Type(s)	Season	Juvenile Salmonid Response				
					Coho	Trout Fry	Cutthroat	Steelhead	Chinook
Before-After Control-Impact Design (continued)									
Keogh River	BC	3	LS, B, G	Su	+*			+*	
MacMillan Cr.	BC	3	LN	Su	+				
Nestucca River Tribs.	OR	8	LS, AL	Su, Sp	+*/+*	0[a]	0/+*	0/+*	
Porter Cr.	WA	6	LS, LN	Su, W, Sp	0,+*,+*	0,0,0		0,0,0	
Sachs Cr.	BC	2	LN	Su	+	-		+	
Southbay Dump Cr.	BC	3	LN	Su	+				
Tobe Cr.	OR	3	G	Su	+	+	+	+	
Extensive Post-treatment Design									
30 different streams	OR/WA	1	LS, LN	Su, W	+*/+*	+/+	+/+*	+/+*	
Crooked F. Lochsa R.	ID	1	LS	Su				0	0
Crooked River	ID	1	LS, LN, B	Su				+*	0
E.F. Papoose Cr.	ID	1	LS	Su		0		+*	
Lolo Cr.	ID	4	LS, B	Su				0	+
Nechako River	BC	1	LS, LN	Sp					+
Red River	ID	1	LS, B	Su				0	0
Steamboat Cr.	OR	1	LS, B	Su, W				+/+*	
Intensive Post-treatment Design									
Papoose Cr.	ID	4	LS	Su				+*	0
Squaw Cr.	ID	4	LS	Su				+*	+*

[a]summer data only

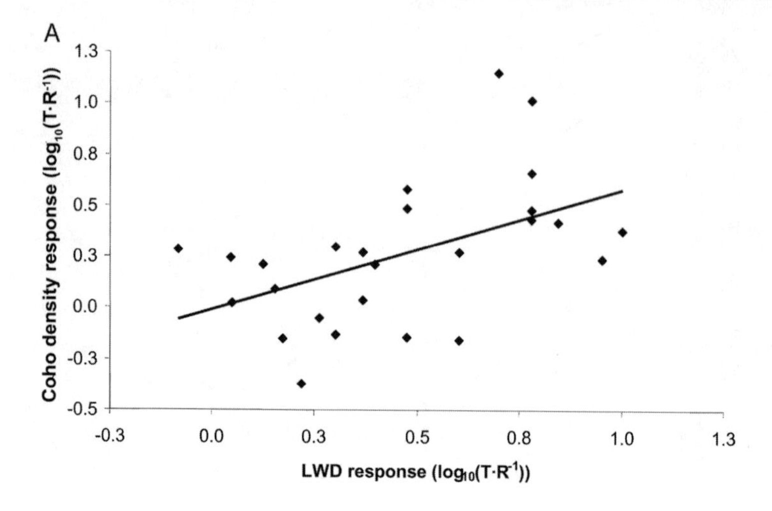

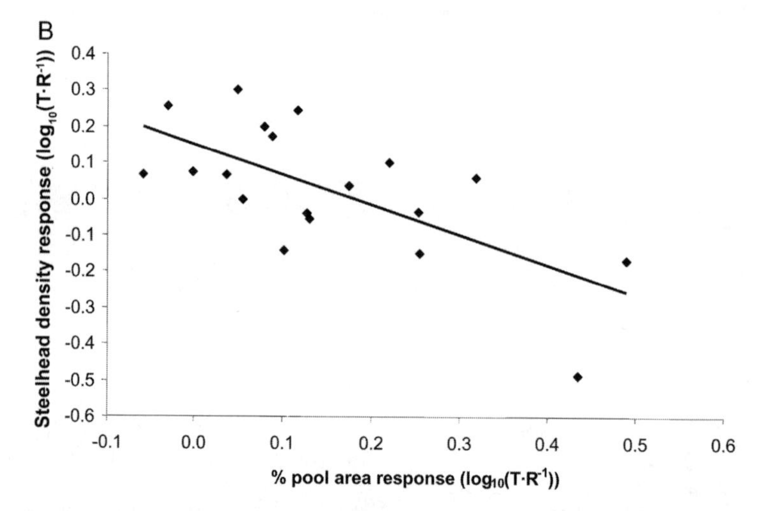

Figure 2. Examples of regression analysis that can be used with data collected from an extensive post-treatment study design. Relationship between (A) coho salmon response to restoration ($\log_{10}$ [treatment density/reference density]) and change in large woody debris (LWD) levels ($\log_{10}$[treatment/reference]) for 27 sites inhabited by coho during summer ($y = 0.59x - 0.01$; $p < 0.01$, $r^2 = 0.25$); and relationship between (B) age 1+ steelhead response to restoration ($\log_{10}$[treatment density/reference]) density and change in percent pool area ($\log_{10}$[treatment/reference]) for 20 sites containing 1+ steelhead during summer ($y = -0.83x + 0.15$; $p < .01$, $r^2 = 0.45$). Treatment (T) consisted of artificial placement of logs and log structures and reference (R) represent unaltered stream reaches. Data from Roni and Quinn (2001a).

interannual variation in abundance, making detection of responses easier for some species, life stages, or populations than others. In one of the few studies to quantify adult salmonid response to instream restoration techniques, Crispin et al. (1993) reported that coho spawner abundance in Elk Creek, Oregon, increased four-fold following placement of instream structures, while spawner abundance decreased elsewhere. Many studies may have failed to detect significant biological responses because monitoring designs were inadequate for detecting less dramatic changes in abundance. Korman and Higgins (1997) provide simulation results to suggest that even 10 years of monitoring may be insufficient to detect less than a two-fold increase in numbers of spawning adult salmon. While detecting changes in salmonid abundance will always be hampered by large interannual variation in fish populations, more consistent and comprehensive monitoring programs should be better able to tease out specific effects of stream restoration activities.

Comprehensive biological monitoring has been lacking for several reasons, including limited funds, an inability to follow individual projects for many years, and lack of guidance on monitoring (Reeves et al. 1991a). Funds are often allocated to restore or improve stream habitat, but relatively few dollars are allocated for research and monitoring of project effectiveness. The high costs of long-term monitoring necessary to detect fish response to restoration in an individual watershed or stream reach has limited thorough evaluations of many habitat restoration and enhancement techniques and projects. Moreover, little guidance exists on the level of effort needed to adequately demonstrate species-specific responses to projects at different scales.

Biological Monitoring and Study Designs

Hicks et al. (1991) outlined four basic experimental designs to evaluate the effects of land-use practices on streams and watersheds and described the pros and cons of each approach (Table 2). These designs can be divided into two categories: "before and after" treatment and "post-treatment" study designs. Many variations of these basic study designs have been used or proposed in monitoring of land use and habitat alterations (e.g., Johnson and Heifetz 1985; Walters et al. 1989; Bryant 1995). For example, Bryant (1995) proposed a pulsed monitoring system for evaluating stream restoration in which pulses of short-term intensive monitoring are separated by long periods of low intensity data collection. Most of these modifications can still be classified as either before-after or post-treatment study designs. Below, we discuss common modifications, and appropriate applications of these two general types of study design.

Table 2. Summary of advantages and disadvantages of the five major approaches to evaluating stream or watershed restoration or habitat alteration (modified from Hicks et al. 1991). Intensive study design includes sampling at a few study sites or streams, extensive is at many study sites or streams, and BACI equals before-after control-impact study design.

	Before and After			*Post Treatment*	
Attribute (pros and cons)	*Intensive*	*Extensive*	*BACI*	*Intensive*	*Extensive*
Ability to assess interannual variation	yes	yes	yes	yes	no
Ability to detect short-term response	yes	yes	yes	no	yes
Ability to detect long-term response	yes	no	yes	yes	yes
Appropriate scale (WA=watershed, R=Reach)	R/WA	R/WA	R/WA	R	R/WA
Ability to assess interaction of physical setting and treatment effects	low	high	low	low	high
Applicability of results	limited	broad	limited	limited	broad
Potential bias due to small number of sites	yes	no	yes	yes	no
Results influenced by climate, etc.	yes	yes	yes	yes	no
Length of time need to detect response (years)	10+	1-3	10+	5+	1-3

Before and After Studies

Several authors recommend long-term monitoring using before and after (BA) studies as a method for determining biological response to habitat alteration (e.g., Stewart-Owten et al. 1986; Reeves et al. 1991a; Smith et al. 1993). A BA study simply refers to a design where data are collected before and after treatment at the same location(s). They can be classified into three types depending upon observation intensity (number of study sites) and existence of a control site (Table 2). The simplest before and after study design includes the collection of data before and after treatment within a single stream site, reach, or watershed. This approach is often used for monitoring individual restoration projects. The most common BA approach, however, is the before-after control-impact (BACI) design in which a control site is evaluated over the same time period as the treatment (impact) site (Stewart-Oaten 1986). (For the purposes of this paper we define a site as a project area generally less than 200 m long, a reach as a geomorphically similar length of stream greater than 100 m in length, and a watershed as the contributing area upstream .) The addition of paired control sites, reaches or watersheds to the BA design is meant to account for environmental variability and temporal trends found in both the control and treatment areas, and thus increase the ability to detect treatment effects (Smith et al. 1993). Additional statistical power to detect treatment effects may be achieved with multiple control sites (Underwood 1994).

Recent evaluations of restoration projects using a BACI design include Cederholm et al. (1997) and Solazzi et al. (2000). While the BACI design is thought by many to represent an improvement over the BA study design, it suffers from a number of potential statistical problems and can be less powerful than the uncontrolled BA design (Hurlbert 1984; Smith et al. 1993; Conquest 2000; Murtaugh 2000). If the measurements are autocorrelated (correlated over time), the variance will be either under- or over-estimated, leading to incorrect conclusions about statistical significance and, potentially, treatment effects. False conclusions can also occur when the pretreatment trends in fish abundance are not similar between treatment and control reaches or watersheds. For these reasons, selection of appropriate controls is critical. Murtaugh (2000) suggested a two-stage statistical analysis that offers some improvement over the standard paired intervention analysis, but this method lacks statistical power. He and others have indicated that BACI studies should focus on graphical rather than statistical analysis (Reeves et al. 1997; Conquest 2000).

Replicating BA and BACI studies would address many of the problems inherent in these designs. However, it is seldom feasible to spatially replicate the restoration treatment due to limited resources, logistics, or project scale. For example, it would be difficult to replicate dam removals over several comparable rivers. Even where multiple reach-scale restoration projects do exist within a watershed, there may still be only one true replicate if the response of interest is at the watershed scale. Lack of true replication limits the statistical inference and application of results to the study site. The question of interest becomes, for example, "Did this particular habitat modification have an effect on local fish abundance?"; generalizations to effects at other sites are not statistically supported. In the absence of true replication, the results of both BA and BACI study designs should be interpreted with caution.

Post-Treatment Designs

In many situations, collecting data before treatment is not possible. This occurs, for example, when examining the physical and biological responses to past restoration activities or unplanned events such as debris torrents. In these situations, the treated reach or watershed is compared to areas thought to be similar in the absence of restoration activities. In essence, post-treatment designs are retrospective studies, replacing space for time. Many authors have emphasized the usefulness of these type of studies (Hilborn and Walters 1981; Hicks et al. 1991). Hall et al. (1978) and Hicks et al. (1991) defined two types of post-treatment design: intensive post-treatment (IPT), in which multiple years of data are collected at one or a few paired control and treatment sites; and extensive post-treatment (EPT), in which data are collected at many paired treatment and control sites over a shorter time period, usually 1-3 years. Grant et al. (1986) tested key assumptions of the EPT design and demonstrated that it is effective for examining the influence of habitat modifications on juvenile salmonids. The EPT design has most frequently been applied to evaluations of different timber harvest and riparian buffer widths (e.g., Murphy and Hall 1981; Hawkins et al. 1983; Grant et al. 1986). The EPT design is best suited for reach-scale projects because it is often difficult to locate suitable control sites at larger scales. It has, however, been applied at a watershed level to examine the effects of land use or other large-scale habitat modifications on fish communities (e.g., Reeves et al. 1993).

Data from an EPT design are particularly flexible. Regression analysis or other correlative techniques can be applied to such data to identify other factors determining fish response. For example, Roni and Quinn (2001a) found that juvenile coho salmon response in summer was positively correlated with pool-forming LWD. In contrast, juvenile steelhead numbers, which showed no overall response to restoration during summer when analyzed using a paired t-test, were negatively correlated with percent pool area (Figure 3). This flexibility is a strength of the EPT design not found in study designs that lack replication. The weaknesses of most post-treatment designs are the lack of pre-project data and the assumption that control reaches or watersheds are similar to pretreatment conditions (Table 2). As with the BACI design, the selection of appropriate control or reference sites is critical.

Sample Size Requirements for BACI and EPT Designs

It has long been recognized that long-term monitoring is necessary to detect treatment effects for populations with high levels of interannual variability in population size. Monitoring for more than a decade may be needed to determine statistically significant differences in juvenile or adult salmonid abundance before and after treatment. Chapman (1996) recommended 10–15 years of monitoring to detect treatment effects in salmonid populations; Bisson et al. (1992) suggested that more than 25 years of monitoring were needed to detect a 50% increase in salmonid abundance due to a specific treatment. Reeves et al. (1991a) suggested monitoring for a minimum of 2 generations of the species in question. More recently, Reeves et al. (1997) collected 14 years of data on a watershed restoration project and were not able to detect signifi-

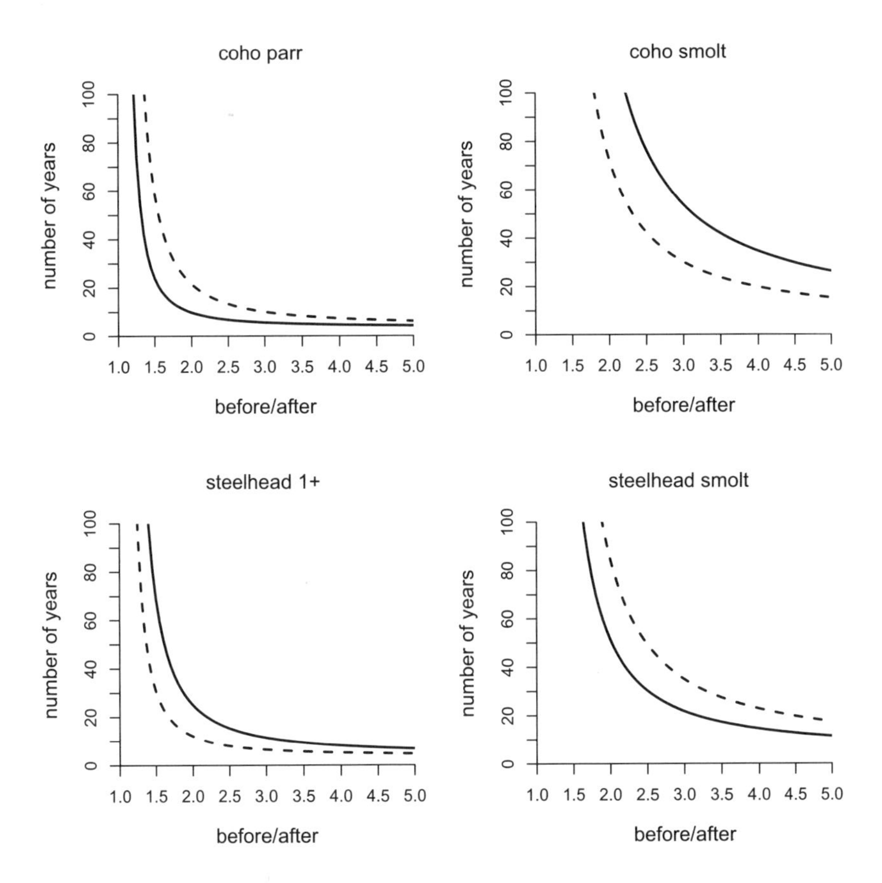

Figure 3. Sample (years of monitoring) needed to detect a change in coho and steelhead parr and smolt abundance using a two sided t-test, 0.05 level of significance and a power of 0.80 for both a before and after (BA, dashed line) and before-after-control impact (BACI, solid line) design. Based on data taken from Solazzi et al. (2000).

cant responses for coho or steelhead. They did detect small differences in salmonid abundance but indicated that longer monitoring was needed to determine if statistically significant differences existed. Few restoration projects have been monitored for more than 5 years (Figure 1B). Longer time periods are needed where smaller effects are expected (Bisson et al. 1992), where there are higher levels of natural variability in population size, or where there are larger errors in annual population estimates (Korman and Higgins 1997).

We examined data from two published studies to explore relationships between monitoring duration, number of study sites, salmonid species and life-stage, study design, and statistical power to detect treatment effects. The first data set we examined was from Solazzi et al. (2000). They measured the effect of increased over-winter habitat and LWD placement on steelhead and coho salmon populations in paired treatment and control streams: two streams in the Nestucca and two in the Alsea River basins in coastal Oregon. In each of the four study streams, they monitored abundance of coho salmon and steelhead parr and smolts for four years before and four years after habitat alterations. We estimated interannual variability for the two basins using their fish data and averaged the variance from the basins to estimate the duration of monitoring needed to detect an increase in parr and smolt abundance using either a BACI or a simple BA study design (Figure 4). For comparison with other study designs, we calculated the sample sizes necessary to detect a doubling of salmonid abundance.

If a simple BA design were used, approximately 20 years would be needed to detect a two-fold increase in coho parr using a two-sided t-test with an alpha level of 0.05 and statistical power equal to 0.80. Approximately 10 years would be required with the BACI design to detect the same effect. The estimated number of years necessary to detect a two-fold increase in coho smolt populations is greater than 70 years for both the BA and BACI designs (Figure 4). The power of the BACI design is reduced if trends in abundance between paired treatment and control streams differ. If, in addition, the interannual variability in fish abundance is low, the BA design can be more powerful than the BACI.

Similarly, analyses of steelhead data indicate that very different lengths of monitoring may be needed for different life-stages and study designs. To detect a two-fold increase in steelhead parr abundance would require approximately 12 and 25 years for the BA and BACI designs, respectively. Detecting the same level of effect for steelhead smolts would require approximately 85 and 50 years for the BA and BACI designs, respectively. Korman and Higgins (1997) found similar results in their modeling of adult salmonid data and indicated that the BACI design had a 10-15% lower probability of detecting a population change than a BA design unless the degree of covariation in survival rates between control and treatment stocks was strong. Although the results of these analyses represent data from one study and may not be readily applicable elsewhere, they demonstrate the importance of selecting appropriate control streams, the need for long-term monitoring, and species-specific evaluations of monitoring designs.

The second data set we examined is from Roni and Quinn (2001a). They used an EPT design to evaluate the response of stream fishes to instream restoration (LWD placement) in 30 streams with paired treatment and control

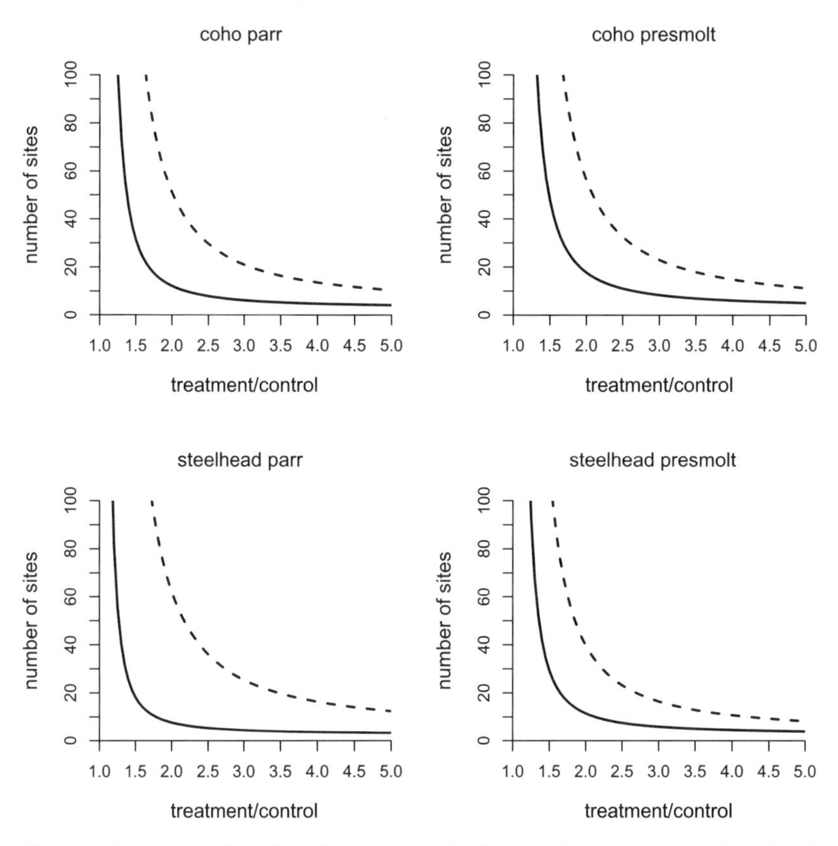

Figure 4. Sample size (number of sites) needed to detect a change in coho and steelhead parr and presmolts abundance using a two sided t-test, 0.05 level of significance and a power of 0.80 for an extensive post-treatment design. Solid line represents treatment and reference (control) reaches that are paired within a stream and dashed line represents no pairing. Data from Roni and Quinn (2001a).

(reference) reaches. They sampled each pair of sites within a stream once. Using their data, we estimated the number of replicates required to detect a range of increases in coho and steelhead parr (summer) and presmolts (winter) given an alpha level of 0.05 and a statistical power equal to 0.80. Here we discuss sample sizes needed to detect a two-or five-fold change in abundance. Approximately 12 paired sites would need to be sampled to detect a two-fold increase in coho parr, while a five-fold increase could be detected with as little as 5 replicates (Figure 4). If the control and treatment reaches are not paired within a stream (i.e., reaches are in different streams) the number of necessary replicates more than triples. We found similar results for winter coho salmon presmolt data (Figure 4). Juvenile steelhead numbers were less variable among streams and reaches. Our analysis indicates that a sample size of approximately 10 sites would be required to detect a two-fold increase for parr or smolts. As with coho salmon, the pairing of stream reaches greatly reduced the number of replications necessary to detect a significant steelhead response (Figure 4). Similar to the BA and BACI designs, much larger samples sizes are needed to detect smaller changes in abundance.

Considerations for Monitoring Design

The best monitoring design to determine fish responses to single or multiple restoration actions depends on the scale of the project and the species and life stage of interest. The scale of restoration effort is a key consideration in selecting the most appropriate design, as the BA or BACI and the EPT designs have strengths and weaknesses at different scales. Reeves et al. (1991b) and Everest et al. (1991) suggested that salmonid response to restoration should be monitored at a watershed rather than reach scale. However, Reeves et al. (1997), Solazzi et al. (2000), and other recent studies at a watershed scale have produced inconsistent results and suggest that some of the same concerns that exist at a reach scale (e.g., fish movement between reaches, variation in spawner escapement) exist for watershed-scale monitoring. If the objectives are to determine the effects of all activities throughout a watershed or basin then some type of BA design is likely most appropriate. The BA or BACI would also be most effective for determining the effectiveness of a single project regardless of scale. The EPT would be most useful for evaluating multiple site- or reach-scale projects, making inference for a group of similar projects rather than about an individual project.

Frequently, stream restoration projects are implemented with relatively vague objectives such as improved habitat complexity or increased salmonid abundance. Similarly, monitoring and evaluation efforts are often implemented with poorly defined objectives and hypotheses, or objectives that do not match those of the project design. In addition to statistical considerations, it is important that clear objectives and testable hypotheses are defined before designing and implementing biological monitoring (Chapters 9 and 13).

Reeves et al. (1991b) indicated that for salmonids there are three primary life stages that should be monitored to determine salmonid responses to restoration: summer rearing (parr), smolts, and adults. Specific project objectives will in part dictate the appropriate life stages for monitoring project success; often it will be necessary to measure multiple life stages. Egg-to-fry survival is an additional life stage that may require assessment, but it is more difficult to measure and specific only to those projects designed to increase or improve the quality of spawning habitat. Reeves et al. (1991b) and Everest et al. (1991) suggest that smolt production is the most efficient life stage to monitor for detecting the response of anadromous salmonids to habitat modifications. Monitoring watershed smolt production before and after restoration would allow evaluation of a suite of restoration activities. Smolt production can also be monitored at a reach scale (Cederholm et al. 1997), as can both summer and winter presmolt or parr counts, but monitoring smolt production at a reach scale requires multiple traps within a stream.

Smolt counts for coho and steelhead also may require longer term monitoring for BA and BACI designs than parr because data on smolts tend to be more variable (Figure 4). If the objectives of the restoration activity are to increase salmonid abundance by improving summer and winter habitat, then we recommend monitoring both parr and smolts regardless of the study design. However, if the ultimate objective of the restoration project is to increase overwinter habitat or smolt production or to evaluate a suite of restoration activities throughout a watershed, then smolt numbers alone would be most useful in determining project success.

Although the objective of many stream restoration projects is to increase the number of returning adult Pacific salmon, monitoring adult returns is rarely the best strategy for detecting population effects of changes in freshwater habitat. Adult anadromous salmonids are affected by numerous factors other than the freshwater environment, including ocean conditions and predators, fishing pressure, and barriers to migration. Marine survival of anadromous salmonids can vary by an order of magnitude from year to year (Pearcy 1992; Beamish et al. 2000). Marine and freshwater harvest rates vary among years and spawner counts are often imprecise, adding additional noise to adult abundance estimates (Korman and Higgins 1997). While juvenile numbers depend in part on adult abundance, they are inherently less variable and more precise than adult counts. Longer time frames or additional sites are needed to detect adult response to restoration or habitat alteration compared to parr or smolts (Bisson et al. 1992). Thus, the usefulness of adult salmon numbers to determine project success is limited in most cases.

Fish size and survival are two other parameters that are important in evaluating fish response to stream restoration. Roni and Quinn (2001a) found that while restoration involving LWD placement increased summer coho salmon densities, it led to smaller fish. Conversely, Reeves et al. (1997) found that coho and steelhead smolts in Fish Creek, Oregon, were larger after restoration than before. Parr and smolt size (length or weight) are positively correlated with overwinter survival and smolt-to-adult survival (Quinn and Peterson 1996). Thus, juvenile salmonid size and survival are important parameters to consider when evaluating fish response to restoration.

Few studies have examined the effects of restoring or creating habitat on fish movement. Monitoring fish abundance once or twice a year does not provide a complete picture of the extent fish use the project area, and monitoring the effects of restoration on fish movements provides additional information on project effectiveness. Large-scale movements of juvenile salmonids, particularly in fall and winter months, can affect both reach-scale and watershed-scale parr and smolt counts. Winter rearing habitat, in particular, may be used by fish throughout a watershed as juvenile fish are known to move several kilometers or more to find suitable overwintering habitat (Northcote 1992). Roni and Quinn (2001b) found little movement of presmolts between restored and unrestored reaches in a small Washington stream, but they suggested that additional research was needed to determine the influence of movement on fish response to restoration.

It is also important to consider the entire fish community when designing a monitoring and evaluation program. Restoration actions are often designed to benefit one or two species of salmonids and thus monitoring has focused on only a limited number of fish species. Comprehensive project evaluation will require that all affected species and life-stages are monitored, and we caution against using one or two species of fish as indicators of project success (Bisson et al. 1992; Roni 2000). For example, increasing pool habitat for one species, such as coho salmon, may negatively impact another, such as steelhead (Roni and Quinn 2001a). In addition to monitoring the response of the fish community, comprehensive evaluation of stream restoration should also include sampling of macroinvertebrates, nutrients, and other indicators of ecosystem health. Evaluating effects at the community level does not mean that the most efficient monitoring design will be the same for all species. Our analysis for coho and steelhead emphasizes differences among species as variability in steelhead parr and smolts was generally lower than that for coho. The sample-size calculations presented in this chapter suggest that differences in interannual population variability between species will require more years of data collection or more study sites to detect population changes in the entire fish community versus individual species. Regardless, it is important that monitoring efforts focus on multiple fish species to determine effects of restoration on biotic communities.

The ability of any study design to detect a change in abundance is influenced by the effect size, population variability, and the number of sites and years of monitoring. Many monitoring and evaluation programs suffer from either small samples sizes or short term monitoring. Most evaluations of stream restoration projects have only detected significant changes in fish abundance when relatively large changes (>2.0 fold) are observed. Detecting smaller changes (< 2.0 fold) will require increased replication, decades of monitoring, or improvements in detecting small treatment effects. The most appropriate design for a particular project will depend on the project scale and the target fish population. An ideal study design would include before-and-after components as well as spatial replication of treatments and well-selected controls (i.e., a combination of the BACI and EPT designs). Such a design is often difficult due to lack of large-scale replicates, inability to measure pre-treatment conditions, or costs. In conclusion, our analysis and review indicates that effective evaluation of stream restoration efforts needs to be replicated extensively in space (large number of sites) or time, or both.

References

Beamish, R.D. Noakes, G. Mcfarlane, W. Pinnix, R. Sweeting, and J. King. 2000. Trends in coho marine survival in relation to the regime concept. *Fisheries Oceanography* 9:114-119.

Beschta, R.L., W.S. Platts, J.B. Kauffman, and M.T. Hill. 1994. Artificial stream restoration—money well spent or expensive failure? In *Proceedings of Environmental Restoration*. UCOWR 1994 Annual Meeting, Big Sky, Montana. pp. 76-104.

Bisson, P.A., T.P. Quinn, G.H. Reeves, and S.V. Gregory. 1992. Best management practices, cumulative effects, and long-term trends in fish abundance in Pacific Northwest river systems. In R.J. Naiman (ed.) *Watershed Management*. Springer-Verlag, New York. pp. 189-232.

Bryant, M.D. 1995. Pulsed monitoring for watershed and stream restoration. *Fisheries* 20(11):6-13.

Cederholm, C.J., R.E., Bilby, P.A. Bisson, T.W. Bumstead, B.R. Fransen, W.J. Scarlett, and J.W. Ward. 1997. Response of juvenile coho salmon and steelhead to placement of large woody debris in a coastal Washington stream. *North American Journal of Fisheries Management* 17:947-963.

Chapman, D.W. 1996. Efficacy of structural manipulations of instream habitat in the Columbia River Basin. *Northwest Science* 5(4):279-293.

Conquest, L.L. 2000. Analysis and interpretation of ecological field data using BACI designs: Discussion. *Journal of Agricultural, Biological, and Environmental Statistics* 5:293-296.

Crispin, V., R. House, and D. Roberts. 1993. Changes in instream habitat, large woody debris, and salmon habitat after the restructuring of a coastal Oregon stream. *North American Journal of Fisheries Management* 43:96-102.

Everest, F.H., J.R. Sedell, and G.H. Reeves. 1991. Planning and evaluating habitat projects for anadromous salmonids. In J. Colt and R.J. White (eds.) *Fisheries Bioengineering Symposium*, American Fisheries Society Symposium 10, Bethesda, Maryland. pp. 68-77.

Fontaine, B.L. 1987. An evaluation of the effectiveness of instream structures for steelhead rearing habitat in the Steamboat Creek basin. Master's thesis. Oregon State University. Corvallis, Oregon.

Grant, J.W.A., J. Englert and B.F. Bietz. 1986. Application of a method for assessing the impact of watershed practices: Effects of logging on salmond standing crops. *North American Journal of Fisheries Management* 6:24-31.

Hall, J.D., M.L. Murphy, and R.S. Aho. 1978. An improved design for assessing impacts of watershed practices on small streams. *Internationale Vereinigung fur Theoretische und Angewandte Limnologie* 20:1359-1365.

Hawkins, C.P., M.L. Murphy, N.H. Anderson, and M.A. Wilzbach. 1983. Density of fish and salamanders in relation to riparian canopy and physical habitat in streams of the northwestern United States. *Canadian Journal of Fisheries and Aquatic Science* 40:1173-1185.

Hicks, B.J., J.D. Hall, P.A. Bisson, and J.R. Sedell. 1991. Responses of salmonids to habitat changes. In W.R. Meehan (ed.) *Influences of Forest and Rangeland Management on Salmonid Fishes and Their Habitat*, American Fisheries Society Special Publication 19, Bethesda, Maryland. pp. 483-518.

Hilborn, R. and C.J. Walters. 1981. Pitfalls of environmental baseline and process studies. *Environmental Impact Assessment Review* 2:265-278.

House, R. 1996. An evaluation of stream restoration structures in a coastal Oregon stream 1981-1993. *North American Journal of Fisheries Management* 16:272-281.

House, R. and P.L. Boehne. 1986. Effects of instream structures on salmonid habitat and populations in Tobe Creek, Oregon. *North American Journal of Fisheries Management* 6:38-46.

House, R. and P.L. Boehne. 1985. Evaluation of instream enhancement structures for salmonid spawning and rearing in a coastal Oregon stream. *North American Journal of Fisheries Management* 5:283-295.

House, R., V. Crispin, and R. Monthey. 1989. Evaluation of stream rehabilitation projects - Salem District (1981-1988). U.S. Department of Interior Bureau of Land Management. Technical Note, T/N OR-6. Portland, Oregon.

Hurlbert, S.H. 1984. Pseudoreplication and the design of ecological field experiments. *Ecological Management* 54:187-211.

Johnson, S.W. and H.J. Heifetz. 1985. Methods for assessing effects of timber harvest on small streams. NOAA Technical Memorandum NMFS F/NWC-73.

Korman, J. and P.S. Higgins. 1997. Utility of escapement time series data for monitoring the response of salmon populations to habitat alteration. *Canadian Journal of Fisheries and Aquatic Sciences* 54:2058-2067.

Moreau, J.K. 1984. Anadromous salmonid habitat enhancement by boulder placement in Hurdygurdy Creek, California. In T.J. Hassler (ed.) *Proceedings: Pacific Northwest Stream Habitat Management Workshop*. American Fisheries Society, Humboldt Chapter, Humboldt State University, Arcata, California. pp. 97-116.

Murphy, M.L. and J.D. Hall. 1981. Varied effects of clear-cut logging on predators and their habitat in small streams of the Cascade Mountains, Oregon. *Canadian Journal of Fisheries and Aquatic Sciences* 38:137-145.

Murtaugh, P.A. 2000. Paired intervention analysis. *Journal of Agricultural, Biological, and Environmental Statistics* 5:280-292.

Northcote, T.G. 1992. Migration and residency in stream salmonids—some ecological considerations and evolutionary consequences. *Nordic Journal of Freshwater Research* 67:5-17.

Quinn, T.P. and N.P. Peterson. 1996. The influence of habitat complexity and fish size on over-winter survival and growth of individually marked juvenile coho salmon (*Oncorhynchus kistuch*) in Big Beef Creek, Washington. *Canadian Journal of Fisheries and Aquatic Sciences* 53:1555-1564.

Pearcy, W.G. 1992. *Ocean Ecology of North Pacific Salmonids*. University of Washington Press, Seattle, Washington.

Reeves, G.H., F.H. Everest, and J.R. Sedell. 1993. Diversity of juvenile anadromous salmonid assemblages in coastal Oregon Basins with different levels of timber harvest. *Transactions of the American Fisheries Society* 122:309-317.

Reeves, G.H. and T.D. Roelofs. 1982. Rehabilitating and enhancing stream habitat: 2. Field applications. U.S. Forest Service, Pacific Northwest Forest and Range Experimental Station, General Technical Report PNW-140. Portland, Oregon.

Reeves, G.H., F.H. Everest, and J.R. Sedell. 1991b. Responses of anadromous salmonids to habitat modification: How do we measure them. In J. Colt and

R.J. White (eds.) *Fisheries Bioengineering Symposium*, American Fisheries Society Symposium 10, American Fisheries Society, Bethesda, Maryland. pp. 62-67.

Reeves, G.H., J.D. Hall, T.D. Roelofs, T.L. Hickman, and C.O. Baker. 1991a. Rehabilitating and modifying stream habitats. In W.R. Meehan (ed.) *Influences of Forest and Rangeland Management on Salmonid Fishes and Their Habitats*. American Fisheries Society Special Publication 19, American Fisheries Society, Bethesda, Maryland. pp. 519-557.

Reeves, G.H., D.B. Hohler, B.E. Hansen, F.H. Everest, J.R. Sedell, T.L. Hickman, and D. Shively. 1997. Fish habitat restoration in the Pacific Northwest: Fish Creek of Oregon. In J.E. Williams, C.A. Wood, and M.P. Dombeck (eds.) *Watershed Restoration: Principles and Practices*. American Fisheries Society, Bethesda, Maryland. pp. 335-359.

Roni, P. 2000. Responses of fishes and salamanders to instream restoration in western Oregon and Washington streams. Ph.D. Dissertation. University of Washington. Seattle, Washington.

Roni, P. and T.P. Quinn. 2001a. Density and size of juvenile salmonids in response to placement of large woody debris in western Washington and Oregon streams. *Canadian Journal of Fisheries and Aquatic Sciences* 58:282-292.

Roni, P. and T.P. Quinn. 2001b. Effects of wood placement on movements of trout and juvenile coho salmon in natural and artificial stream channels. *Transactions of the American Fisheries Society* 130:675-685.

Roni, P., T.J. Beechie, R.E., Bilby, F.E. Leonetti, M.M. Pollock, and G. Pess. 2002. A review of stream restoration techniques and a hierarchical strategy for prioritizing restoration in Pacific Northwest watersheds. *North American Journal of Fisheries Management* 22:1-20.

Slaney, P.A., B.O. Rublee, C.J. Perrin, and H. Goldberg. 1994. Debris structure placements and whole-river fertilization for salmonids in a large regulated stream in British Columbia. *Bulletin of Marine Science* 55:1160-1180.

Smith, P.E., D.R. Orvos, J. Cairns. 1993. Impact assessment using the before-after-control-impact (BACI) model: Concerns and comments. *Canadian Journal of Fisheries and Aquatic Sciences* 50:627-637.

Solazzi, M.F., T.E. Nickelson, S.L. Johnson, and J.D. Rodgers. 2000. Effects of increasing winter rearing habitat on abundance of salmonids in two coastal Oregon streams. *Canadian Journal of Fisheries and Aquatic Sciences* 57:906-914.

Stewart-Oaten, A., W.W. Murdoch, and K.R. Parker. 1986. Environmental impact assessment: "Pseudoreplication" in time? *Ecology* 67(4):929-940.

Tarzwell, C.M. 1934. Stream improvement methods. U.S. Bureau of Fisheries Division of Scientific Inquiry. Ogden, Utah.

Tarzwell, C.M. 1937. Experimental evidence on the value of trout stream improvement in Michigan. *Transactions of the American Fisheries Society* 66:177-187.

Underwood, A.J. 1994. On beyond BACI: Sampling designs that might reliably detect environmental disturbances. *Ecological Applications* 4:3-15.

V.A. Poulin and Associates Ltd. 1991. Stream rehabilitation using LOD placements and off-channel pool development. British Columbia Ministry of Forests, Land Management Report 61, Vancouver, British Columbia.

Walters, C.J., J.S. Collie, and T. Webb. 1989. Experimental designs for estimating transient response to habitat alteration: Is it practical to control for environmental interaction? In C.D. Levings, L.B. Holtby, and M.A. Henderson (eds.) *Proceedings of the National Workshop of Effects of Habitat Alteration on Salmonid Stocks*. Canadian Special Publication in Fisheries and Aquatic Sciences 105, Ottawa, Ontario. pp. 13-20.

Ward, B.R. and P.A. Slaney. 1981. Further evaluations of structures for the improvement of salmonid rearing habitat in coastal streams of British Columbia. In T.J. Hassler (ed.) *Proceedings, Symposium on Propagation, Enhancement, and Rehabilitation of Anadromous Salmonid Populations and Habitat in the Pacific Northwest*. California Cooperative Fishery Research Unit, Humboldt State University, Arcata, California. pp. 99-108.

13. Establishing a Standard of Practice for Natural Channel Design Using Design Criteria

Dale E. Miller and Peter B. Skidmore

Abstract

To date, no widely accepted standards of practice have emerged by which to conduct and evaluate natural channel design projects. Consistent application of design criteria to the channel design practice is an important first step toward the development of standards of practice. Design criteria serve the multiple purposes of clarifying objectives, allowing the design team to proceed with a high probability of meeting shared expectations, and defining the risk associated with design components. Natural channel design criteria should resolve the degree of channel deformability through time, the basic channel form desired, floodplain function, and objectives for providing aquatic habitat. A categorization of criteria is presented and includes: infrastructure protection, channel geometry, vertical stability, lateral stability, floodplain function, revegetation, and construction. Design criteria applied to the remediation of Silver Bow Creek, a Superfund site in Montana, illustrate the application of design criteria to clarify objectives from vague goals and numerous stakeholders. Additional applications include facilitation and review of project proposals, design review, and evaluation of project success. When multiple design criteria conflict, they must be resolved by determining the relative importance of various criteria.

INTRODUCTION

Formalized means and measures for appropriate conduct, practice, and procedures exist within most design professions. These standards translate into accepted methods for design and implementation in a wide variety of fields including manufacturing, medicine, electronics, and engineering. While standards provide a vehicle for establishing a regulatory framework, most standards are produced through a voluntary system comprised of government and industry, producers and consumers, institutions and individuals (Gross 1991). Indeed, standards of practice are generally developed through consensus, with open collaboration of affected parties in a balanced, transparent framework and due process to assure consideration of all views (ANSI 2001).

Regardless of the type of discipline, standards of practice serve a number of important functions. First, use of standards imparts consistent results from application to application. Kunich (1995, p. 3) states that "standards are the language of the engineer and scientist and capture their wisdom for society... without standards, the benefits are lost—no measurement, no data, no equation, no fit, no solution, no explanation." When consistent results can be expected from use of a particular standard or set of standards, then post-implementation monitoring is generally unnecessary. For example, bridges constructed to American Association of State Highway and Traffic Officials standards do not need to be monitored to determine whether they have been successful (although they are typically monitored as part of maintenance plans).

Second, standards are typically developed to promote public safety. Standards and codes provide the engineering and architectural communities with guidelines for design methodology, in part because of the consistency of results that can be expected. As such, standards of practice serve to manage the liability to designers for project failures. If commonly accepted standards of care are followed, then the direct liability to designers tends to be reduced (ECS 1999). Dam stability regulations are an example of standards developed to ensure structural integrity so as to minimize risk to property and human safety.

Third, standards provide the principle vehicle for the transfer of advancements from research into practice (Gross 1991), and as such they establish a framework to promote the development of a growing discipline. Fields that have a long history tend to demonstrate a universal acceptance of standards of practice. The American Society for Testing and Materials (ASTM) grew out of the railroad boom at the turn of the century, when train derailments began to take a toll on public safety and the economy. Material consistency was so poor that railroad companies began to import rails from Great Britain. It was only after development of testing methods and material standards for iron and steel that American railroads were again considered safe (Thomas 2001). ASTM now develops six types of standards: test methods, specifications, practices, terminology, guides, and classifications (ASTM 2001). In contrast, the National Committee for Information Technology Standards has existed (formerly under another name) for only a few decades, and their current efforts focus on developing standards for multimedia, storage media, security, and programming languages required by the burgeoning field of Information Technology.

Gross (1991) describes how standards have typically evolved. *Industry standards* are promulgated by a particular industry or industry group, reflect an agreement by the group, and are controlled by the group. Such standards are often the result of a limited participation process. Consequently, their longevity and use may be limited. In contrast, *consensus standards* are most often developed by organizations devoted to standards development, such as professional societies and industry trade associations. Consensus standards typically reflect the views of all interested and affected parties. Due to rigorous procedural requirements (for example, those promoted by ANSI) and the broad participation process, consensus standards generally receive the highest level of recognition, acceptance, and use.

The Lack of Standard Design Methodology for Channel Restoration

Channel restoration projects are implemented with increasing regularity, and while there have been successful restoration/rehabilitation projects documented in recent decades, the community lacks a standard approach to restoration design. We offer three explanations for the lack of commonly accepted standardization.

First, the wide array of physical variables that control river channels varies spatially and temporally, complicating any design process. Design in other engineering disciplines involves dealing with known controlling variables, such as material qualities (e.g., tensile strength, mass per unit volume, and conductivity) and readily quantified site conditions (e.g., temperature and humidity). In contrast, channel design must account for non-homogenous boundary materials (e.g., stratified soil layers with differing particle cohesion), hydrologic character (which is highly variable in magnitude and duration), sediment supply (which varies in size and volume, both spatially and temporally), and vegetative components (e.g., differing growing conditions, species variability, and plant succession). As a result, designers of natural channels must cope with natural variation from stream to stream, reach to reach, and site to site. For example, most natural channel design projects rely on the establishment and maturation of riparian vegetation to provide long-term stability. The rate of plant establishment, plant density, and plant health relates to soil conditions, the availability of water, and the length of seasons. While designers may have influence over soil quantity and quality, the unpredictable natural variability in hydrologic events can have a profound influence on project success.

Second, river restoration is still in its developmental and experimental stage (NRC 1992; Brooks and Shields 1996; Jonas 2001). Because of the relative youth of the practice, many practitioners are still experimenting, and this situation inherently limits standardization. Related to this is the fact that stream channel designers come from various disciplines, including biology, hydrology, geomorphology, and engineering. Practitioners within these varying disciplines are generally familiar only with the components of design that fall within the realm of their discipline and training. Consequently, their "standard approach" may vary significantly from approaches adopted by practitioners from other disciplines (Jonas 2001).

Third, widely differing levels of acceptable risk are typically associated with channel design projects. Risk may result from technically challenging scenarios, such as may be found with an incising channel in disequilibrium with extremely high shear stress exerted on steep, eroding banks. Risk may also result from the use of an experimental bank stabilization technique or a technique that may be successful only if a high magnitude flow does not occur before installed plants are expected to grow. An example of the latter includes the use of willow posts for bank protection as described by Shields et al. (1998). Therefore, it may appear that each project is unique (and thus standards are not applicable) because all projects address different levels of risk.

Lacking standards of practice, designers of natural channels encounter a number of difficulties throughout the design process, including:

- Identifying an appropriate design procedure(s) for differing restoration scenarios;
- Making choices of which techniques are most suitable for given conditions; and
- Ascertaining the level of documentation necessary to convey design analysis into design plans to ensure successful project implementation.

One of the first steps in developing standards of practice for river channel design is to encourage the application of design criteria to articulate specific project objectives and to monitor project performance relative to objectives. The use of design criteria may ultimately lead to standardization of channel design methods. As subsequent projects adopt and refine criteria that have been previously developed for other projects, approaches to address these criteria will likely become more standardized.

Role of Design Criteria in the Design Process

There are identified needs for "comprehensive and quantitative guidelines for selecting design criteria, methodologies, and specific practices" for stream restoration (TRB 1999, p. 1). One way to effectively navigate the numerous and often disparate forces molding a project is to identify and adopt clear, project-specific design methodology and criteria. Design criteria can facilitate the mutual understanding of expectations of property owners, project sponsors, designers, and regulatory agencies. Once specific design criteria have been established and agreed upon, the design can proceed with a high probability of meeting shared goals and expectations and a clear understanding of risk associated with selected criteria.

Design criteria ensure that project objectives are fully considered during the planning and design process of a project. As decisions are made, design criteria provide a reference for the applicability of relocation measures, stabilization techniques, or the dimensions of particular components. Design criteria depend on the scale and extent of a particular project. Projects with little ecological impact or social risk may require only a few design criteria, whereas projects presenting greater risk of ecological impact or potential social cost may require a complex suite of criteria.

Simple or complex, criteria should address design components associated with an entire system. Project managers must consider geomorphology, hydraulic engineering, fisheries biology, riparian ecology, as well as aesthetics, public safety, and cultural resources, among other interests. It is critical that designers consider the project reach within the context of the entire physical and biotic system, that is, its place within the watershed. In order to maximize and sustain natural and functional habitat, the appropriate geomorphic processes must be established (Gore 1985; Poff and Ward 1989).

In this chapter, we discuss the role of design criteria in the design process, describe the types of design criteria, and recommend minimum criteria that should be identified for natural channel design (with specific emphasis on describing desired functionality). We also provide recommendations for use of design criteria through proper documentation, multiple criteria decision making, and risk assessment.

Application of Design Criteria to Stream Projects

At the broadest level, clear measurable objectives are necessary to establish what features are needed for a project to function as desired within the physical, biological, and political setting. Specific criteria are then established to ensure that designs will provide for these objectives. Design criteria may also ultimately serve as a basis for measuring success of various project components. Without criteria on which designs are based, there is no way to legitimately evaluate the success of a project. The use of design criteria will ultimately allow for some degree of standardization in channel design processes and conversely allow for diversity in objectives among varying channel design projects.

What Are Design Criteria?

Design criteria are specific, *measurable* attributes of project components developed to meet objectives. Design criteria provide benchmarks for individual components of a design, specifying quantifiable limits of performance and tolerance for bank protection components and mitigation features. These performance limits may prescribe the mere presence of components (e.g., a requirement for the installation of a certain number of woody debris emplacements or use of native plants) or the requirements for preventing or minimizing the mechanisms of failure (e.g., a requirement for use of stone toe installed to the depth of calculated scour). There are a wide variety of mechanisms of failure that should be considered as part of channel design, many of which relate to the different causes of erosion (particle entrainment, bed scour, and mass failure of stream banks are three examples). Design criteria will specify the tolerances of the design to these mechanisms of failure.

Minimum Recommended Design Criteria

Criteria for natural channel design include details that describe the desired level of fluvial function and define the spatial and temporal nature of project objectives. Additionally, design criteria address constraints to full restoration, such as stream corridor limits, infrastructure protection, and flood control. At a minimum, the following process-related attributes should be addressed:

- Channel/planform deformability: Whether stream bank location is to be permanent, allowed to deform over time (at some identified rate or timeframe), or subject to some quasi-deformable framework (such as deformable channel features within limited rigid floodplain margins).
- Basic channel form: Whether a channel is to be a single thread, a multiple thread channel, or braided.
- Floodplain function: Whether a floodplain is to be regularly inundated, or site conditions dictate that a floodplain be constrained to limit the extent or magnitude of inundation.

- Aquatic habitat: Whether habitat is to be created to target specific species or life stages, allowed to evolve with fluvial disturbance processes, or both.

Design criteria for various project components, or objectives, can be related to hydrologic events, such as the design flood, bankfull flow, or low flow conditions. Most restoration projects will require two or even three different flows to define all criteria: 100-year discharge, bankfull (dominant) discharge, and a low-flow discharge. While most engineers are familiar with the 100-year requirement, and most geomorphologists are familiar with the bankfull or dominant flow (Williams 1978; Andrews 1980; Andrews and Nankervis 1995; Johnson and Heil 1996; Doyle et al. 1999), designers need to communicate effectively with fisheries biologists to accurately establish acceptable low flow levels, which have implications for habitat value.

Prescriptive Criteria and Performance Criteria

Design criteria essentially answer one or both of the key design questions: (1) *how* will a project be undertaken; and (2) *what* is a project to achieve? The former involves prescriptive criteria; the latter involves performance criteria.

Describing *how* a project will be undertaken through design criteria is an approach known as "means and methods." In the pre-design phase, these criteria call out specific design components and are often termed *prescriptive* criteria. In the design phase, plans and specifications prepared by engineers for implementation by contractors are almost always prescriptive criteria in that they provide explicit detail of project components. A few examples help define prescriptive criteria. In the first example, a criterion stipulates that 3-foot diameter rock will be used to resist bank erosion up to some flow event. This can be considered prescriptive in that it identifies that rock of a certain size will be used. In another example, where the objective is to provide juvenile salmonid habitat, a prescriptive criterion might state that 50 pieces of large woody material of a particular dimension be placed per 1,000 feet of channel, according to some stability criteria. As yet another example, a criterion for temporary stockpiling of excavated materials might indicate a specific 1,000 square foot area.

Describing *what* a project is to achieve is typically accomplished with *performance* criteria. A performance criterion "describes the required performance or service characteristics of the finished product or system without specifying in detail the methods to be used in obtaining the desired end result" (Clough 1986, p. 73). Restating the previous first example, performance criteria for bank stability would not identify the methods but only the level of erosion protection (in this case, to some specified flow event). Reframing the prior second example as performance-based, criteria intended to provide juvenile salmonid habitat may stipulate the creation of suitable habitat or channel conditions sufficient to support 600 fish of a certain size or age class per 1,000 feet of channel, at a given flow or range of flows, measured using a specified, standard habitat typing methodology. Similarly, using the third example, performance criteria for temporarily stockpiled materials might identify that any 1,000 square foot area could be used as long as it does not disturb delineated wetlands.

Performance criteria allow for creative application of methods or techniques, since it is the end result that is specified and not the means to achieve the end result. Indeed, performance criteria allow a designer to exercise "integrity, skill and experience to the fullest extent in achieving the desired result" (Clough 1986, p. 73). Designers guided by performance criteria have a greater opportunity to develop project components or alternatives that are more efficient or cost-effective than those that might result from prescriptive criteria, largely because they have the opportunity to more fully leverage their expertise. Furthermore, the use of performance criteria may reduce the likelihood of contradictory criteria, which can become problematic when attempting to address numerous prescriptive criteria simultaneously.

Design Criteria for Specific Elements Typical of Stream Projects

Both prescriptive and performance channel design criteria can be further categorized according their principle functional focus. Each of the categories listed below address components of channel design that are common to most comprehensive projects requiring channel reconstruction or significant modification. The minimum recommended design criteria addressed previously should be incorporated throughout these categories.

1. Infrastructure protection criteria address human-made features that in some way influence project design. These features may include buried utilities (water, sewer, gas, electricity), transmission towers, roadways, levees, embankments, property boundaries, rights-of-way, fence lines, river crossings (road, rail, and pedestrian bridges), buildings, and structures. These features generally restrict the location of a channel, require high levels of protection to maintain stability, and require containment of flood waters. Infrastructure protection criteria are often defined relative to the erosive forces or water surface elevations associated with the 100-year flood (1% probability of occurrence in any given year). Criteria within this category often conflict with criteria that define channel deformability, floodplain functionality, or riparian establishment.

2. Channel geometry criteria address the shape and dimensions of both channel cross-section and planform. Since channel geometries are related to discharge, these criteria are typically defined by dominant discharge. In addition, channel geometry criteria also address the degree of variability desired in channel geometry (both in section and in plan view).

3. Vertical stability criteria focus on design components that provide for maintenance of channel grade and may include criteria for determination of gradation, installation depth, and mobility of bed material. Should grade controls be employed, vertical stability criteria also describe the requirements for resistance to erosive energy resulting from a particular flow event (e.g., the shear stress expected to occur at a 100-year event).

4. Lateral stability criteria typically address the level of permanence or the rate and degree of deformability of channel banks (Miller and Skidmore 1998; Miller 1999). Such criteria may address the bank material stability and depth of installation; the determination of elevations of the bottom, top, and intermediate zones of bank protection; and the various types of treatment along the course of a stream (through straight reaches and bends). The U.S. Army Corps of Engineers (1999) recommends establishing criteria for effectiveness, environmental considerations, and economic factors for bank stabilization within the context of channel design.

5. Floodplain criteria address the geometry, configuration, and relative elevations of floodplain margins and terraces. These features are typically defined by the water surface elevations and/or erosive energies associated with one or more flow events. Criteria may also define the location and composition of channel migration zones (for example, floodplain boundaries may be estimated from topography or historic meander limits). Where appropriate, criteria may address the characteristics of multi-thread and high flow channels.

6. Revegetation criteria define all aspects of protecting existing vegetation, re-establishing vegetation following disturbance associated with channel restoration, and use of vegetation as a stabilization measure. These criteria may address plant growth forms or materials and species selection according to different elevations or zone stratification within the riparian corridor.

7. Construction criteria address design components related to project implementation, including sediment and erosion control, dewatering, access limitations, project sequencing, and other site controls.

Uses of Design Criteria in Project Development and Design

Design criteria should guide project design to achieve project objectives. Project designers typically develop design criteria with input from project stakeholders, including owners, regulators, funding agencies, and others affected by a project. Development of criteria often involves negotiation among stakeholders to achieve consensus in project goals and objectives. In addition to driving the design process, design criteria have the potential to serve a number of other important functions that contribute to project success, including requests for design proposals, instilling quality control, justifying design methodology, and monitoring projects after implementation. These additional functions tend to be largely underutilized.

Many channel restoration projects involve a design phase that is contracted to design consultants. Sponsoring organizations frequently request proposals from consultants who are asked to describe their proposed project approach. Proposals are typically ranked according to the thoroughness and applicability of their approach. Sponsors who utilize performance criteria to frame solicitations are more likely to receive proposals that meet project objectives. To use performance criteria, sponsors must clearly understand what they want to achieve, as well as the constraints associated with their projects. Some level of feasibility assessment, alternatives identification, or preliminary design is often undertaken by sponsors in order to develop realistic performance criteria. Proposals based on sponsor-identified performance criteria can be more readily compared, as they are all geared to achieving the same common objectives. Requests for proposals (RFPs) commonly do not include explicit performance criteria, and proposal reviewers are left to choose between widely divergent approaches with differing end products. Improvement in the communication of objectives, development of quality scopes of work, and equitable evaluation of proposals can be achieved when project sponsors identify design criteria in RFPs.

Design criteria also serve as a basis for review of draft designs. Design review may occur as an internal quality control procedure (either in the private or governmental sector), as value engineering to ensure that the designs are the best possible and most cost-effective, or merely to see that all objectives are satisfied. Design review also occurs at a regulatory level to see that regulatory objectives are met. Without design criteria, it is difficult for independent or objective reviewers to ascertain that objectives are achieved in a set of design plans.

Designers of stream channels are responsible for the success of their efforts. In many situations, they hold liability either as individuals (such as registered engineers who stamp design plans) or as corporations (who hold liability for errors and omissions). Indeed, we have observed that designers with natural science backgrounds tend to have less awareness of liability than engineers (the latter are trained to understand liability and exposure). As stated previously, standards of practice can reduce a designer's exposure to liability (ECS 1999), largely due to an accepted standard of care. Lacking a standard of practice, designers of natural channels can use design criteria to demonstrate a well thought out, step-by-step design process, which reflects an acceptable standard of care. Thus in projects that have the potential for litigation (for example, projects promulgated by natural resource damage provisions of Superfund, a permit violation, or those that run the risk of affecting adjacent property owners), use of design criteria may reduce liability to designers.

Lastly, and perhaps most importantly, good design criteria provide a basis for monitoring a project after it has been implemented. Simply put, design criteria provide the basis from which to measure whether designs have been successful in meeting those objectives. Rigorous post-implementation monitoring plans are based on measuring project attributes. For example, one might measure resistance to bank failure after a particular flood (surveying bank location and observing bank erosion), habitat created (area of salmonid juvenile rearing habitat), or vegetation success (number of woody stems of a certain caliber in a given area after some period of time). It is only by comparing these post-project measurements to pre-project design criteria that one can determine the success of a project. Indeed, to be most effective, the measurable attributes of monitoring programs should be directly related to design criteria. The importance of design criteria in measuring success is discussed further below.

Use of Design Criteria to Evaluate Project Success

Project success or failure should be evaluated in terms of both durability and project life. Durability refers to the ability of project components to withstand given disturbance-generating flows in any given year (this is commonly done by considering stability relative to some recurrence interval flow). Project life refers to the period of time over which success is to be measured. Note that engineers typically consider the intended lifespan of infrastructure; as designers of natural channels, we should also ask the question of anticipated project life span.

Success should be defined by the allowable variation from target conditions for the project life. Since change is not uniform, it is necessary to describe the magnitude of changes that constitute failure when criteria are not met. For example, consider a restoration project involving a 1-mile long relocation of a 35-foot wide stream. The stream is subjected to a 10-year flood the year after it was constructed. The design criterion for bank stability described an allowable rate of change of 10 feet of erosion in 5 years after the first 5 years of no movement (to allow plant establishment). If 5% of the bank margins eroded 10 feet in short sections no greater that 20 feet in length, would this constitute failure? Would it necessitate some level of maintenance? If the 5% of change occurred with a single meander adjustment and the reconstructed channel continues to provide functional resources, would that be considered failure? Similar questions can be asked regarding the incorporation of large woody material. Consider placement of wood within the channel to provide fish habitat in a situation where disturbance-generated habitat was expected. Would dislodged wood, transported and accumulated downstream causing bank erosion, constitute failure? Well-formulated performance and lifespan design criteria facilitate resolution of these questions.

Design Criteria in Practice

Effective communication of design criteria is paramount to the success of their application. A number of steps can be taken to ensure that criteria are adequately described, accepted by stakeholders, and utilized throughout the design process. Design criteria are usually developed after there is some level of understanding of the technical conditions and constraints of a project, typically during the preliminary design phase. Since design criteria are often described relative to hydrology, designers should have an estimate of the probability of flows that are expected (for geomorphic design, stability, and habitat). The first step then, is to characterize hydrology and determine all potential design flows. This may include dominant discharge (variously equated to effective discharge, a 1.5-year or 2-year flow, or bankfull discharge [Doyle et al. 1999]), base flow, mean annual flow, and various flood level events including the 5-year, 10-year, 50-year, and 100-year. In addition to hydrology, site constraints must be identified, which may include utilities, infrastructure, property boundaries, and other social or political factors that limit opportunities within the corridor.

With hydrologic statistics and site limitations in hand, stakeholders can develop criteria for each project objective. Every project objective should be paired with a criterion or suite of criteria to define the objective and compiled in a preliminary design report or technical memorandum. Such a document serves as a basis for discussion among stakeholders, particularly regulators and constituents. The design criteria technical memorandum can undergo a series of revisions to ensure stakeholder consensus. It is most effective if this discussion occurs prior to the design phase; in fact, it should occur prior to consideration of specific (or favored, controversial, or "standard") alternatives. In this way, the focus will be on desired results, rather than particular alternatives that are the responsibility of the designers.

Describing and Evaluating Risk and Uncertainty

All natural channel design projects have an element of risk and uncertainty. Risk can be defined as "a situation where the decision maker knows all the alternatives available, but each alternative has a number of (possible) outcomes" (USACE 1992, p. 7). Probabilities of occurrence are assigned to each outcome in order to evaluate risk (Miller and Skidmore 1998). In this manner, success can be related to the ability of a project component to continue to function under certain discharges, to which a probability of occurrence can be assigned. The risk of failure, then, is directly related to the probability of occurrence of a given discharge, which can alternatively be expressed as a return interval (the inverse of probability).

Uncertainty, on the other hand, can be defined as a situation where "probabilities cannot be assigned to outcomes" of alternatives (USACE 1992, p. 7). Design criteria can serve to define risk in terms of the probability, but cannot reduce uncertainty. There is a degree of uncertainty associated with design of natural channels, regardless of the criteria that are used to define objectives and risk. Designers may acknowledge uncertainty on some level, but they are typically reluctant to acknowledge the degree of uncertainty, particularly to the non-technical community (USACE 1992). An example of uncertainty is found in hydrologic and hydraulic modeling, where input data may be of poor quality (accuracy) or quantity (suitably representative), or where a model is used in a situation at the boundary of applicability (i.e., where equations may not be valid). Experts may base decisions on their best educated guess (which requires training, experience, knowledge, and judgment not available to the layperson). In order for stakeholders to be able to assess the appropriateness of design criteria for a particular project, they should also be aware of areas of uncertainty. To address this, channel design practitioners should openly discuss and record the uncertainty inherent in their work. Despite the fact that the public may not want to hear from experts that there is uncertainty, experts should be extremely careful of speaking in terms that denote certainty when such certainty does not exist.

Another aspect of project risk involves trade-off of risk, cost, and benefit. Two variations (USACE 1992) of this tradeoff are:

- If there are alternatives to achieve desired benefits, and each alternative has an acceptable cost, then choose the alternative with the least risk.
- Resources can be applied to a project to reduce risk to the point where the marginal cost of risk reduction equals the value of the reduction. In this case, the residual risk may be defined as the acceptable level of risk.

The development of design criteria, and the associated communication that is required to gain understanding and consensus, sets the stage for discussion of the variety of risks associated with natural channel design and how to manage or mitigate for these identified risks and uncertainties.

Making Decisions with Multiple Criteria

Complexity in the design process can arise from either a large number of criteria, conflicting criteria, or a combination of both. The probability of criteria being mutually exclusive or contradictory is closely related to the number of criteria applied. When there are less than 10 criteria, a decision maker is generally able to evaluate them simultaneously (Concilo et al. 1998), and conflicting criteria can be minimized.

Limitations Imposed by Application of Design Criteria

Working within the context of design criteria may restrict creative or experimental approaches to project design and implementation. While criteria serve an important function in formalizing a design framework, they should not necessarily limit opportunities or approaches. If alternative approaches are to be employed, they do not necessarily preclude use of design criteria. Rather, project designers and/or stakeholders can conduct an appropriate discussion of risk in order to frame the context of the approach relative to more commonly accepted programs. Furthermore, without tests to document reliability of otherwise untested methods or techniques, it is not possible to justify their use to satisfy criteria. For example, bioengineering methods that utilize plants as an integral stabilizing component certainly function well for bank reconstruction (Gray and Sotir 1996). However, there is little quantitative data on the resistance of either naturally vegetated banks or bioengineered banks to erosion (Hoitsma and Payson 1998), and therefore in most circumstances vegetation alone cannot satisfy numeric stability criteria.

A Case Study of Design Criteria in Practice

The Silver Bow Creek project involved the remediation of 8.4 km (5.2 miles) of the stream to meet the goals and guidelines established in the project's Record of Decision (MDEQ/EPA 1995). The project called for removal of tailings and impacted soils from within the boundaries of the 100-year floodplain. Tailings removal included excavation of the streambed and channel banks; the channel bed and stream bank were subsequently reconstructed.

The process of establishing design criteria greatly facilitated design for the remediation of Silver Bow Creek, a Federally listed Superfund site in southwestern Montana. Due to the technical challenges and complex, litigious nature of this project, establishing criteria to meet the objectives and mandates of numerous parties was paramount to progress and completion. Project goals cited in the Record of Decision (MDEQ/EPA 1995) were generally vague and subject to interpretation by numerous involved parties. Mandated project objectives were defined as (MDEQ/EPA 1995):

- Channel reconstruction to "...an appropriate size, shape and composition" and that the channel is "allowed to meander";
- "...suitable bedform morphology for aquatic habitat";
- "...a healthy riparian system to protect the remedy from high flows"; and
- "geomorphic stability."

An iterative design process was used on the project, which allowed for evaluation of an extensive number of criteria and resolution of conflicting criteria. The design process for Silver Bow Creek was streamlined by: (1) quantifying the independent variables of hydrology and sediment supply; (2) identifying the fundamental constraints (which included existing controls and other physical limitations); and (3) selecting preliminary values for the dependent variables that were consistent with the known constraints and were within the typical range of values for other similar streams. From this starting point, the design details were then refined by adjusting values for the dependent variables until the desired result was obtained.

The design criteria developed for the Silver Bow Creek remediation project illustrate thorough application of criteria-based design of a natural channel. Through considerable negotiation among involved parties, objectives were defined to meet project goals stated in the Record of Decision, and specific

Table 1. Design criteria developed for the reconstruction of Silver Bow Creek.

Infrastructure Protection Criteria

- Final hydraulic conditions will not reduce the stability of existing infrastructure.
- Grade control will be incorporated to protect bridges from scour and fill.

Channel Geometry Criteria

Cross-section Geometry

- A preliminary bankfull discharge of 200 cfs will be used in developing alternatives but may be refined based on further design analyses.
- Cross-sectional dimensions will be variable and will include pools and riffles.
- Average, but not necessarily, W/D ratio for the bankfull channel will fall within the range of 10 to 15 for riffles and straight reaches and between 7 and 12 for pools.
- Channel depth will meet geometry criteria without requiring additional floodplain fill.

Planform

- Design of planform geometry will not compromise channel grade criteria.
- A minimum ratio of bend radius (Rc) to channel width (W) > 4 will be applied.
- The channel bank will not be located within 50 feet of repositories, railroad grades, or other sensitive areas.
- Channel alignment through bridges will be modified to minimize backwater effects or channel instability.
- Planform modifications will be designed to minimize the number of crossings between existing channel and designed channel to minimize construction and dewatering time and costs.

Vertical Stability Criteria

Bed Material Gradations

- A maximum of two gradations will be selected for channel bed materials for the entire project reach.
- Material gradations for pool and non-riffle sections will approximate the average of bulk samples reported.
- Material gradations for riffle sections will approximate the average of gradation curves generated from Wolman pebble count analysis in existing riffles.
- Selected material gradations will consider the pit-run gradations available at local gravel sources.

Channel Grade

- Channel will be designed to ensure sediment continuity for flows encompassed by the mean flow duration curve.
- Existing channel bed elevations will not be lowered at existing bridges, unless bridge piers can be modified to accommodate changes in grade.

- Artificial grade control will be considered where sediment continuity, and thus channel bed elevation equilibrium, cannot be achieved within the bounds of these criteria and due to the physical limitations of the project area or to excessive costs.

Additional Grade Controls
- Grade controls will be designed to be stable and permanent up to the 100-year flow and will account for shear forces and modeled scour depths at these flows under existing bridge and capacity conditions.
- Grade controls will extend between the limits of the 100-year floodplain.

Lateral Stability Criteria

Non-Deformable Bank Treatments
- Non-deformable banks will be installed a minimum of 25 feet upstream of bridges and through bridge structures, wherever utilities and infrastructure warrant protection from channel migration, and wherever the channel location will be within 50 feet of railroad grades.
- Wherever non-deformable banks are required, the bank toe structure will be designed to withstand the shear forces anticipated to the depth of scour calculated for the 100-year flow event.

Deformable Bank Treatments
- The lower depth of the bank toe treatment will be set at the maximum depth of scour for that location.
- Bank toe structure will be designed to withstand the shear forces anticipated up to the 50-year flow for the life of the coir fabric (2–5 years) for the design channel.

Upper Bank, Deformable, and Non-Deformable Banks
- The lower boundary of the upper bank will correstpond to a flow duration that is not limiting for plant growth.
- Upper bank structure will be designed to withstand the shear forces anticipated up to the 50-yr flow for the life of the coir fabric (2-5 years) for the design channel geometry and grade.

Floodplain Critera
- Floodplain surfaces will have a minimum upward slope of 1.0% away from the channel banks.
- Terraces will be designed to contain a minimum of the 5-year flow at their crest surface elevation.
- Terrace design will not have an impact on in-channel hydraulics such that it requires modifications to channel or bank design.

Table 2. Prioritized design criteria for Silver Bow Creek arranged from highest to lowest priority. Higher priority criteria take precedence over lower priority criteria.

Infrastructure Protection Criteria:
- Protection of infrastructure from potential threats that are related to remedial actions.

Channel Geometry Criteria:
- Adopted bankfull discharge.
- Cross-sectional geometry.
- Lateral stability/channel planform.

Vertical Stability Criteria:
- Bed material transport capacity that is in approximate equilibrium with the sediment supply.
- Grade control where sediment continuity cannot be achieved.

Lateral Stability Criteria:
- Lateral deformation through time allowed within the context of dynamic channel equilibrium.

Floodplain Criteria:
- Lateral floodplain slope.
- Terrace design.
- Cost and constructability considerations.

Revegetation Criteria:
- Plant species selection.

Construction Criteria:
- Consideration of all design component costs.

criteria were established to address objectives in project design. The design criteria for this project are detailed in Table 1.

The problem of contradictory criteria was resolved by prioritizing criteria. Higher priority criteria simply took precedence over lower priority criteria, yet allowed for revision with approval of stakeholders. Specific design components were grouped, the order of which approximated the order of consideration in the design process (Table 2). Finally, interdependent criteria were satisfied through design iterations.

Summary: Successful Natural Channel Design

The practice of natural channel design, as a relatively new discipline, lacks a commonly accepted standard of practice upon which designers can rely. As a first step in developing *standards of practice* as means to improve the quality of channel design, we encourage designers to use design criteria to thoroughly articulate project objectives. As criteria are adopted and refined for subsequent projects, approaches to addressing criteria will become standardized.

We suggest that designers use performance criteria over prescriptive criteria where appropriate, so that creative solutions are not precluded during the criteria identification process. Categories of design criteria include: (1) infrastructure protection criteria, (2) channel geometry criteria, (3) vertical stability criteria, (4) lateral stability criteria, (5) floodplain criteria, (6) revegetation criteria, and (7) construction criteria. Project designers should identify specific criteria for each component of a project and "test" these criteria throughout project scoping and design. At a minimum, design criteria should describe the desired level of fluvial function and define the spatial and temporal nature of project objectives.

Complex projects may be faced with multiple, and at times conflicting, design criteria. Possible means to systematically evaluate multiple-criteria scenarios are suggested. We encourage the channel design community (practitioners, regulators, and funding agencies) to report on their application of criteria to projects, including the use of qualitative and quantitative decision methodology, development of criteria standards, and uses of criteria beyond design (such as proposal requisition, design review, and monitoring).

References

Andrews, E.D. 1980. Effective and bankfull discharges of streams in the Yampa River basin, Colorado and Wyoming. *Journal of Hydrology* 46:31-330.

Andrews, E.D. and J.M. Nankervis. 1995. Effective discharge and the design of channel maintenance flows for gravel-bed rivers. In J.E. Costa, A.J. Miller, K.W. Potter, and P.R. Wilcock (eds.) *Natural and Anthropogenic Influences in Fluvial Geomorphology*, Geophysical Monograph 89, American Geophysical Union. pp. 151-164.

ANSI (American National Standards Institute). 2001. National standards strategy for the United States. (www.ansi.org/public/nss.html).

ASTM (American Society for Testing and Materials). 2001. *ASTM Standards in Building Codes, 38th Edition*, West Conshohocke, Pennsylvania.

Brooks, A. and F.D. Shields, Jr. 1996. Perspectives on river channel restoration. In A. Brooks and F.D. Shields, Jr. (eds.) *River Channel Restoration*. John Wiley & Sons, New York.

Clough, R.H. 1986. *Construction Contracting, Fifth Edition*. John Wiley & Sons, New York.

Concilio, G., P. Korhonen, and M. Soismaa. 1998. Rank order for a rehabilitation program using multiple criteria. Interim Report IR-98-064. International Institute for Applied Systems Analysis, Laxenburg, Austria.

Doyle, M.W., K.F. Boyd, and P.B. Skidmore. 1999. River Restoration Channel Design: Back to the Basics of Dominant Discharge. *Second International Conference on Natural Channel Systems*. Niagara Falls, Canada.

ECS. 1999. *Managing Professional Exposures in the Environmental Services Industry*. ECS Risk Control, Inc. Exton, Pennsylvania.

Gore, J. A. 1985. Mechanisms of colonization and habitat enhancement for benthic macroinvertebrates in restored river channels. In J.A. Gore (ed.) *The Restoration of Rivers and Streams* Butterworths, Boston, Massachussets.

Gray, D.H. and R. B. Sotir. 1996. *Biotechnical and Soil Bioengineering Slope Stabilization*. John Wiley & Sons, New York.

Gross, J.G. 1991. Codes, standards, and institutions—pressures for change. *Journal of Professional Issues in Engineering Education and Practice*. 117(2):75-87.

Hoitsma, T.R. and E.M. Payson. 1998. The use of vegetation in bioengineered streambanks: Shear stress resistance of vegetal treatments. In *Proceedings of the ASCE Wetlands Engineering and River Restoration Conference* Denver, Colorado.

Johnson, P.A. and T. M. Heil. 1996. Uncertainty in estimating bankfull conditions. *Water Resources Bulletin* 32:1283-1290.

Jonas, M. 2001. The hydraulic engineer's role in stream restoration projects. In *Proceedings of the Seventh Federal Interagency Sedimentation Conference* Reno, Nevada.

Kunich, M.P. 1995. Technical Standards Manager Spotlight. *The Standards Forum*, U.S. Department of Energy 3(3):3.

Miller, D.E. 1999. Deformable stream banks: Can we call it a natural channel design without them? *1999 AWRA Specialty Conference*, June-July 1999, Bozeman, Montana.

Miller, D.E. and P.B. Skidmore. 1998. The concept of deformable banks for stream bank stabilization and reconstruction. In *1998 ASCE Bank Stabilization Mini-Symposium of the International Water Resources Engineering Conference* Memphis, Tennessee.

MDEQ/EPA (Montana Department of Environmental Quality and U.S. Environmental Protection Agency). 1995. Record of Decision, Streamside Tailings Operable Unit of Silver Bow Creek/Butte Area NPL Site, Silver Bow and Deer Lodge Counties, Montana.

NRC (National Research Council). 1992. *Restoration of Aquatic Ecosystems*. National Academy Press, Washington, D.C.

Poff, N.L. and J.V. Ward. 1989. Implications of streamflow variability and predictions of lotic community structure: A regional analysis of streamflow patterns. *Canadian Journal of Fisheries and Aquatic Sciences* 1805-1818.

Shields Jr., F.D., S.R. Pezeshki, and P.H. Anderson. 1998. Probable causes for willow post mortality. In *Proceedings of the ASCE Wetlands Engineering and River Restoration Conference* Denver, Colorado.

Thomas, J.A. 2001. NIST and ASTM share a century of innovation. *ASTM Standardization News*.

TRB (Transportation Research Board).1999. Problem 1: Designing and Restoring Stable Stream Channels. Group 2 Committee A2A03, Hydrology, Hydraulics, and Water Quality, Part 2 Hydraulics Research Problem Statements. (www.nas.edu/trb/publications/problems/a2a03ps2.pdf)

USACE (U.S. Army Corps of Engineers). 1992. Guidelines for risk and uncertainty analysis in water resources planning. Volume 1: Principles. IWR Report 92-R-1. Water Resources Support Center, Institute for Water Resources, Fort Belvoir, Virginia.

USACE (U.S. Army Corps of Engineers). 1999. Channel rehabilitation: processes design and implementation (draft). USACE, Washington, D.C.

Williams, G.P. 1978. Bankfull discharge of rivers. *Water Resources Research* 14:1141-1154.

14. Reference Conditions for Instream Wood in Western Washington

Martin Fox, Susan Bolton, and Loveday Conquest

Abstract

Stream channel assessments and enhancement efforts often associate salmonid habitat quality with the quantity and volume of woody debris. Existing wood targets used to assist resource managers do not adequately account for variations in quantity or volume due to differences in geomorphology, ecoregions, or disturbance regimes. To address this issue, field data on instream wood quantities and volumes from 78 stream segments draining unmanaged basins within western Washington State were used to develop reference conditions for restoration and management. Based on the sample distribution, these reference conditions are applicable to streams with bankfull widths between 1-100 m, gradients between 0.1–47%, elevations between 7-1,906 m, drainage areas between 0.7-325 km^2, glacial and rain/snow-dominated origins, and several other distinguishing physical and regional classifications. Assuming that streams draining unmanaged forest basins incorporate the range of conditions to which salmonids and other species have adapted, wood loads in these systems provide a reasonable reference for management. Due to both favorable and adverse conditions comprising wood loading ranges, we suggest that the 75th percentiles in each bankfull width class and ecoregion be used as an index of habitat quality to represent the lower limit for optimum wood quantities and volumes.

Introduction

The Importance of Instream Wood

The role of large woody debris (LWD) in Pacific Northwest streams is linked to channel processes that benefit salmonids. Woody debris plays an important role in controlling channel morphology, the storage and routing of sediment and organic matter, and the creation of fish habitat (Bisson et al. 1987). The geomorphic potential of the channel to process wood into features that benefit salmonids is often limited by the quantity and size of wood (Abbe and Montgomery 1996).

Large wood creates habitat heterogeneity by forming pools, back eddies, and side channels and by increasing channel sinuosity and hydraulic complexity (Spence et al. 1996). Pools are perhaps one of the most important salmon habitat features formed by large woody debris (Keller and Swanson 1979). Pools provide slow-water rearing habitat for juvenile anadromous and resident salmonids, especially coho and chinook, allowing access to drifting food organisms with less swimming effort (Fausch 1984). Large pools also serve as resting stations for fish as they migrate upstream to spawn (Bjornn and Reiser 1991). Abbe and Montgomery (1996) found that pools formed by log jams are deeper on average than free-formed pools; numerous woody debris accumulations increase pool frequency (Lisle and Kelsey 1982; Montgomery et al. 1996; Beechie and Sibley 1997).

Channel responses to wood vary with geomorphic character of the stream, such as gradient and confinement (Murphy and Koski 1989; Robison and Beschta 1990). In high energy channels, large woody debris functions to retain spawning gravel and can also provide thermal and physical cover for salmonids (Schuett-Hames et al. 1999). Log jams can create sections of low gradients with alluvial substrates in bedrock channels by storing sediment upstream of the jam (Montgomery et al. 1996; Massong and Montgomery 2000), which can provide localized low-gradient habitats in steep valley segments where none would have existed.

Wood indirectly serves as an important food source for salmonids by providing nutrients and insects to the stream (Naiman and Sedell 1979; Spence et al. 1996) or by the retention of salmon carcasses (Cederholm et al. 1989; Bilby et al. 1996). Wood serves as cover for juvenile salmonids, which are particularly vulnerable to predators when migrating (Larsson 1985). Certainly, wood is an important component of channel morphology and salmonid habitat, and its presence contributes to the biological and geomorphological processes of a stream.

Restoration activities in the Pacific Northwest today typically involve wood, often through placement of wood into a channel to provide habitat for salmonid use. To maximize the success of improving habitat, the amount of wood placed in a channel should be representative of wood quantities and volumes to which salmonids have adapted. Therefore, knowledge of the natural variation of instream wood loads among different stream types and regions should improve restoration activities as well as the scientific defensibility of regulatory thresholds.

Potential Sources of Variation for Instream Wood

The quantities and volumes of instream wood are highly variable. This variability is related to a number of factors, including channel morphology, disturbance, and climate. The following section elaborates on these potential factors.

Geomorphological Influences

Channel size influences the quantity of instream wood. Bilby and Ward (1989) found that mean length and diameter of wood pieces increased as channel width increased, and that the frequency of occurrence of pieces declined with increasing channel width in streams draining unmanaged basins. They also found that the frequency of instream wood ranged almost an order of magnitude, between 0.8 pieces/m in the smallest channels to 0.1 pieces/m in the largest systems.

Channel size, often synonymous with bankfull width (BFW), is a function of basin size. Basin size and bankfull width are highly correlated as a function of precipitation (Sternes 1969), which in turn influences discharge in Washington streams (U.S. Water Resources Council 1981; Sumioka et al. 1998; Pleus 1999). Consequently, bankfull width is often used as an indicator of basin size. However, caution should be heeded in channels where the hydrological or erosion processes have been altered by humans or where recent disturbances have occurred. Disproportionate widening of channels can result from land-use practices such as timber harvest (MacDonald et al. 1991; Pleus and Schuett-Hames 1998), urbanization (Booth 1991), recent debris flows or dam-break floods (Coho and Burges 1993), dredging or bank manipulation, and other factors not representative or typical of a channel in its natural state. The lateral boundaries used to measure bankfull width are identified by channel scour, and the presence, age, and species of adjacent vegetation (Pleus and Schuett-Hames 1998). Despite the difficulty of consistent BFW estimates, bankfull width is widely used as an indicator of stream size.

Bankfull width is a significantly better predictor of wood quantity and volumes per 100 m of channel length than basin size. However, due to the strong correlation between basin size and bankfull width, basin size may also be used to predict wood quantity and volume per 100 m if BFW measurements are unavailable (Fox 2001) or if BFW has been affected by land use.

Channel reach morphology (Montgomery and Buffington 1997) also influences instream wood loads in part because the presence or absence of wood helps determine channel type. Rot et al. (2000) found significantly more LWD pieces in forced pool-riffle channels than in bedrock or plane-bed channels; wood volume followed a similar trend. However, confinement was significantly related to LWD volume only in forced pool/riffle channels, where less wood was found in confined channels. Confinement had no effect on LWD volume in plane-bed channels.

Anthropogenic Influence

Differences in the distribution and characteristics of wood between managed and unmanaged basins have been clearly established. Wood can be limited due to riparian vegetation modifications (Ralph et al. 1991), whether due to forest practices, urban development, or agricultural practices. Unmanaged channels, often defined by streams draining un-roaded and unlogged basins, typically have more channel roughness due to instream wood than managed channels (Bilby and Ward 1991; Ralph et al. 1991), especially if the stream has been channelized. Where sediment deposits behind wood obstructions (Swanson and Lienkaemper 1978; Montgomery et al. 1996; Massong and Montgomery 2000), low-gradient steps are formed that reduce stream energy. Lower stream energy has less potential to mobilize wood than high stream energy (Braudrick and Grant 2000). These factors, especially if peak flows are exacerbated due to land uses, may lead to less retention of recruited wood in managed basins than in streams draining unmanaged basins.

Natural Disturbance

The quantity and volumes of instream wood vary over space and through time due to an array of natural processes. Instream wood may remain in the channel until it is lost to decay, achieving equilibrium over time with newly recruited wood. However, this balance is not often met due to the multiple types and rates of disturbances that influence this process. Riparian forest structure, composition, and spatial distribution through the network are driven by major disturbance processes (Fetherston et al. 1995). All channels have been affected by disturbance of some kind, whether historic or recent. Therefore, the characterization of wood from a single survey provides a temporal "snapshot," documenting a single point in the patterns of fluctuation. The accretion of wood may continue over time until capacities exceed an ecological, biological, or morphological threshold, some of which result in a catastrophic removal by disturbance. The amount of instream wood, therefore, represents how recent the last disturbance was and conditions during the recovery period. Four types of major disturbances are common in forested streams of western Washington: (1) fire, (2) floods, (3) debris flows, and (4) snow avalanches.

The return intervals for fires, which varies by ecoregion (Agee 1993), affect timber age (Henderson et al. 1992). Timber age influences mean tree diameter, which in turn influences the diameter of instream wood (Rot et al. 2000). Timber age also influences tree height (Henderson et al. 1992; Agee 1993), and wood recruitment distance is a function of height (McDade et al. 1990). Thus fire affects instream wood diameter and recruitment.

Floods entrain wood from areas adjacent to stream reaches. High flows associated with floods increase the shear stress upon instream wood and carry wood downstream or perhaps even completely out of a system. Braudrick and Grant (2000) found that wood entrainment is a primarily a function of piece angle relative to flow direction, the density of the log, and its length and diameter. Rootwads can inhibit LWD movement by anchoring logs to the streambed, increasing drag and thus decreasing mobility (Abbe and Montgomery 1996).

Debris flows and landslides are natural disturbances that affect stream channels and influence the quantity, quality, and distribution of instream wood. The often-violent mobilization of material in channels where this occurs may either transport wood out of a reach or bring in new wood from upstream sources. Debris flows tend to deposit wood on slopes of 3–6 degrees (approximately 5–10% gradient) (Ikeya 1981; Costa 1984). Due to bank and bed scour, riparian vegetation removal, wood deposition, and breakage of trees and wood pieces, debris flows often add variability to the size and distribution of instream wood.

Snow avalanches are also natural channel process that recruit wood into streams (Keller and Swanson 1979) and influence the riparian vegetation (Fetherston et al. 1995). Snow avalanche paths are typically less confined than debris flows, and they often form a broad fan where the channel gradient flattens, such as at the channel bottom intersecting with the floodplain of a larger system. Snow avalanches are most common in small, steep headwater channels (Keller and Swanson 1979). Due to the snow pack buffering the channel bed, substrates are often undisturbed following a snow avalanche; however, most trees larger than 10–15 cm in the path are sheared off at the level of snow depth. The loss of riparian vegetation is likely to influence instream wood quantities due to the disturbance of the recruitment source.

Riparian Influence

The characteristics of riparian trees also influence instream wood. Rot et al. (2000) found the diameter of instream LWD increased with riparian stand age, and that stand age and mean stem diameter were correlated. Tree age varies considerably within older Western Hemlock/ Douglas fir forests. Tappeiner et al. (1997) found age in old-growth stands ranged between 50 and 414 years at one site, and median age differences of 187 years from 10 sites of the same region. Timber on the Olympic Peninsula, often older than 700 years (Henderson, unpublished data), can produce large diameter instream wood. Indeed, in streams draining old-growth forests, McHenry et al. (1998) found a mean LWD diameter of 0.3 m and diameters up to at least 2.5 m. These differences in riparian characteristics are a combination of many influences including fire, climate, and species.

Regional Influence

Regional climatic variations that control the characteristics of forest vegetation can be grouped by a forest zone or forest series (Franklin and Dyrness 1973; Agee 1993), and they are hereafter referred to as "ecoregions." Ecoregions are characterized by climax species, tree size, and density of forest stands as influenced by climate and fire succession (Agee 1993). The distribution of tree species, tree heights, diameters, and stem densities in distinct ecoregions often differ due to variation in elevation, aspect, precipitation /soil moisture, and temperature (Henderson et al. 1992; Agee 1993). The four major forest types comprising the major ecoregions of western Washington State are the Sitka Spruce (*Picea sitchensis*), Western Hemlock (*Tsuga heterophylla*), Silver Fir (*Abies amabilis*), and Mountain Hemlock (*Tsuga mertensiana*) forests (Henderson et al. 1992; Agee 1993).

Applicability to Low Elevation Puget Sound Streams

Based on forest and climate characteristics, most streams of the Puget Sound area are within the Western Hemlock forest. Based on the assumption that the quantities and volumes of instream wood are influenced by the characteristics of the adjacent riparian forests, streams of similar morphologies within an ecoregion are likely to have similar instream wood characteristics. Therefore, streams within the Puget Sound region are likely to have evolved under the same riparian influences as streams sampled within unmanaged basins of Western Hemlock forests. Although most streams in the urban environment have been greatly modified by development, deforestation, flood control, or other factors since European settlement, the aquatic species in these systems adapted to habitats produced by natural processes in Western Hemlock forests.

Wood as an Indicator of Habitat Quality

Stream channel assessments often associate the size, distribution, and abundance of woody debris to salmon habitat quality based on data from old-growth forests as an "index of resource condition" for determining habitat condition (Peterson et al. 1992). Subsequently, several key assessment methods based on such target conditions were developed to evaluate the adequacy of wood quantities in the State of Washington. The National Marine Fisheries Service (NMFS) considers >80 pieces/mi. (>50 pieces/km) that are >50 ft (15.2 m) in length and >24 inches (0.6 m) in diameter necessary for western Washington streams to meet a "Properly Functioning Condition" for instream wood conditions in their Pacific Coast Salmon Plan (NMFS 1998), and to address ESA listed aquatic species such as salmon as part of their Matrix of Pathways and Indicators (NMFS 1996). The targets established in the diagnostics of the Washington Forest Practices Board (WFPB) Manual (1997) for conducting "Watershed Analysis" rate the condition of streams based on wood quantity in terms of "pieces per channel width," using >2 pieces as "Good," 1–2 pieces as "Fair," and <1 piece as "Poor" conditions for channels < 20 m BFW. A qualifying wood piece must be >10 cm diameter and 2 m in length. The WFPB manual criteria for "Key Pieces" are >0.3 pieces per channel width in streams <10 m BFW and >0.5 pieces per channel width in streams 10–20 m BFW. The WFPB (1997) describes key pieces as a necessary component of wood quantities for use in state Watershed Analysis, and it defines "key pieces" as a log and/or rootwad that is:

Table 1. WFPB definitions for volume of "key pieces," as defined in Table F-4 of the 1997 Methods Manual.

Bankfull Width (m)	*Minimum Volume (m^3)*
0–5	1
>5–10	2.5
>10–15	6
>15–20	9

1. independently stable in the stream bankfull width (not functionally held by another factor, e.g., pinned by another log, buried, trapped against a rock or bed form), and
2. retaining (or has the potential to retain) other pieces of organic debris.

Based on this definition of function, data from WFPB (1997) were used to develop a minimum piece volume criterion for a range of channel widths in western Washington (Table 1). Noteworthy is the fact that there are no minimum volume definitions for streams >20 m BFW.

Perspectives on Units Used to Quantify Wood Loading

Various units are used to describe existing instream quantities or volumes of wood to express wood targets. Some researchers express quantities and volumes of wood per unit channel length (Bilby and Ward 1989; Montgomery et al. 1995; Rot et al. 2000). Others (Montgomery et al. 1995; Beechie and Sibley 1997) have explored different means to express wood quantities, such as pieces per unit area (e.g., pieces per square meters of stream channel). The WFPB (1997) bases target references for wood quantities upon a sliding scale according to channel size, such as "pieces per channel width." Most methods scale these various units of wood with bankfull width as a means to control for channel size.

Fox (2001) compared three common methods of scaling wood abundance with bankfull width: (1) wood per 100 m of channel length (a unit of length that does not change with increases in BFW), (2) wood per square meter (a unit of area that increases exponentially with BFW), and (3) wood per channel width of length (a unit of length that increases linearly with BFW). He found that each of these relationships yields identical results for wood volume or quantity per a common unit of channel, although each incurs colinearity by having a factor common in both the x and y-axis (i.e., channel area). This supports the findings of Montgomery et al. (1996), who also explored these unit relationships. Fox (2001) also found that presenting the relationship in terms of wood per *channel width* affords the perception of the best "fit" (i.e., R^2 value) in relationship to bankfull width as the predictor variable, but using the scale of pieces per 100 m offers statistical advantages when grouping classes of different bankfull widths.

Potential Problems with Existing Reference Conditions

The LWD piece quantity targets now frequently used as management and restoration standards were developed with the most complete data available for relating wood frequency to channel width in Pacific Northwest streams (Peterson et al. 1992). However, Spence et al. (1996) notes that those targets do not fully consider potential sources of variation found throughout their application range, and they should only be applied to the types of streams for which they were derived. Because the currently existing targets do not fully account for this variation and are applied generically, they may be inappropriate for some channel types and regions outside the area where the targets were developed. For example, a stream enhancement project may place wood in a stream channel based on the quantities recommended by target references, but these efforts may not provide the quantities or volumes of wood representative of local conditions to which salmonids have adapted.

Development of Improved Reference Conditions

To better characterize natural quantities and volumes of instream wood within Washington State, survey sites were chosen within stream basins relatively unaffected by anthropogenic disturbance (Fox 2001). Selected basins are characterized by forests that are loosely termed as "natural" that also meet the following criteria: (1) no part of the basin upstream of the survey site was ever logged using forest practices common after European settlement; and (2) the basin upstream of the survey site contains no roads or human modifications to the landscape that potentially could affect the hydrology, slope stability, or other natural processes of wood recruitment and transport in streams. These basins will hereafter be referred to simply as "unmanaged basins," although it is acknowledged that some basins are "managed" to remain pristine, and that management may include fire suppression. Sites were chosen to represent a broad array of channel morphologies and hydrological origins (Table 2). This served to characterize the channel in relation to the dominant mechanism that

Table 2. Drainage area, confinement, and origin classes used to classify surveyed stream reaches

Gradient (%)	*Drainage Area (km²)*	*Confinement Classes*[1]	*Flow Origin*
<1	0–2	Confined (≤2 CW[2])	Snow melt/Rain
>1–2	>2–4	Moderately Confined (2–4 CW)	Glacial melt
>2–4	>4–8	Unconfined (>4 CW)	
>4–8	>8–20		
>8–20	>20–100		
>20	>100		

[1] (Pleus and Schuett-Hames (1998)

[2] CW=channel widths (as a function of distance) across the valley bottom

Table 3. The sample distribution within select gradient, elevation, and channel-width classes.

Channels < 6 m Bankfull Width

	Elevation (amsl)	*152–500 m*	*500–1,000 m*	*1,000–1,264 m*
Gradient	0.1–4%	n = 6	n = 3	n = 0
	4–36%	n = 1	n = 6	n = 3

Channels >6–30 m Bankfull Width

	Elevation (amsl)	*133–500 m*	*500–1,000 m*	*1,000–1,182 m*
Gradient	0.1–4%	n = 2	n = 12	n = 2
	4–21%	n = 8	n = 15	n = 4

Channels >30–100 m Bankfull Width

	Elevation (amsl)	*91–500 m*	*500–1000 m*	*1,000–1,174 m*
Gradient	0.1–4%	n = 8	n = 5	n = 2
	4–6%	n = 0	n = 0	n = 1

drives fluvial geomorphology and subsequent process on instream wood. Table 3 depicts the sample sizes by several gradient, elevation and channel width classes to help illustrate the regional and geomorphic distribution of our data.

Comparison of Existing and New Reference Conditions

To determine the appropriateness of current management targets towards meeting instream wood conditions found in unmanaged systems, current targets were compared to the data collected by Fox (2001). The two existing management targets were used to make these comparisons were the NMFS targets and the Washington Forest Practices Board targets for conducting Watershed Analysis.

RESULTS

Regional and Geomorphological Processes Affecting Instream Wood

Watershed and valley morphology play complex roles in the quantities and volumes of instream wood, but their influences are not overwhelming. Figure 1 shows that the volumes of instream wood per 100 m generally increases as

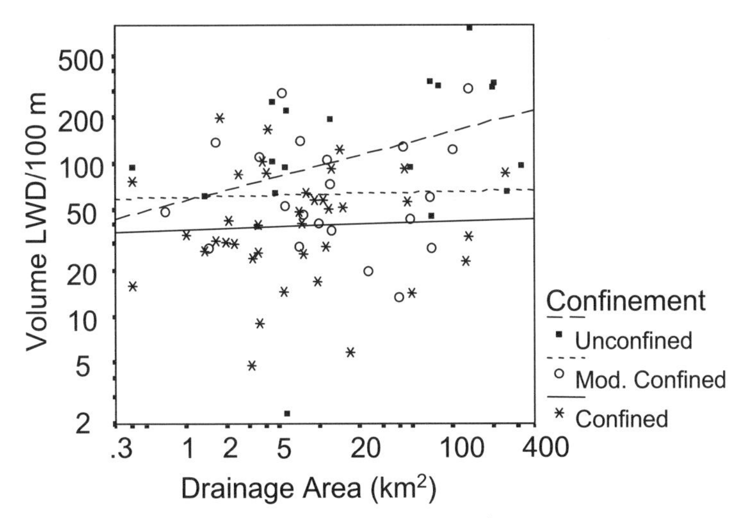

Figure 1. The effects of confinement upon the volume (m^3) of LWD per 100 m of channel length with respect to basin size (x-axis) for western Washington.

drainage area increases but only in unconfined channels. Figure 1 also suggests that more wood volume per 100 m is found as streams become less confined, particularly in watersheds greater than about 10 km^2 in drainage area. A relationship between wood volume and basin size is most evident at low gradients with no confinement (Figure 2). In all basin sizes, more wood volume is observed in alluvial channels as compared to bedrock channels (Figure 3A), a result that is generally applicable only to confined channels (Figure 3B) because over 90% of the bedrock channels surveyed are in confined valleys.

In basin drainages of 70 km^2 or more, streams originating from glacial sources have more wood volume per 100 m than streams fed predominantly with snowmelt and rain (Figure 4). This may be related to the larger number of side channels in streams originating from glacial sources, which averaged 3 per stream reach (n=7) as compared to only 1.8 in snow/rain-dominated channels (n=17). Although there is no significant relationship between bedform and volume of wood, pool/riffle channels, where lateral migration is typical,

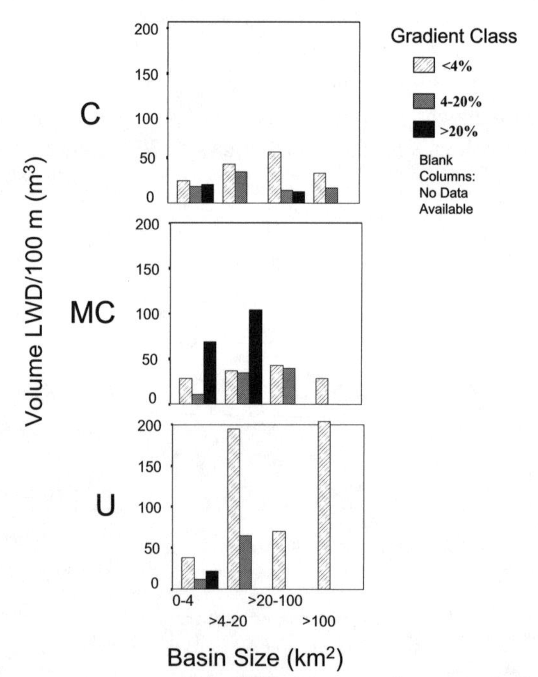

Figure 2. The combined effect of gradient and confinement upon the median volume (m^3) of instream wood per 100 m of channel length by basin size. C=Confined; MC= Moderately Confined; and U=Unconfined channels.

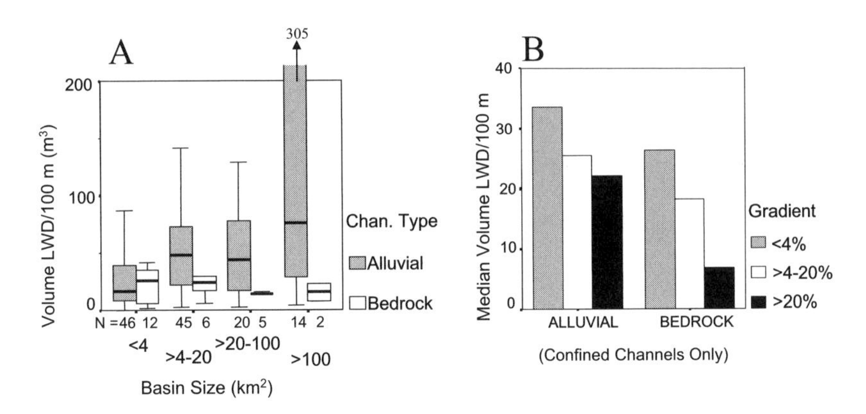

Figure 3. Comparison of instream wood volume between alluvial and bedrock channels, as grouped by basin size classes (A); and among gradient classes, as grouped by channel type for confined channels only (B).

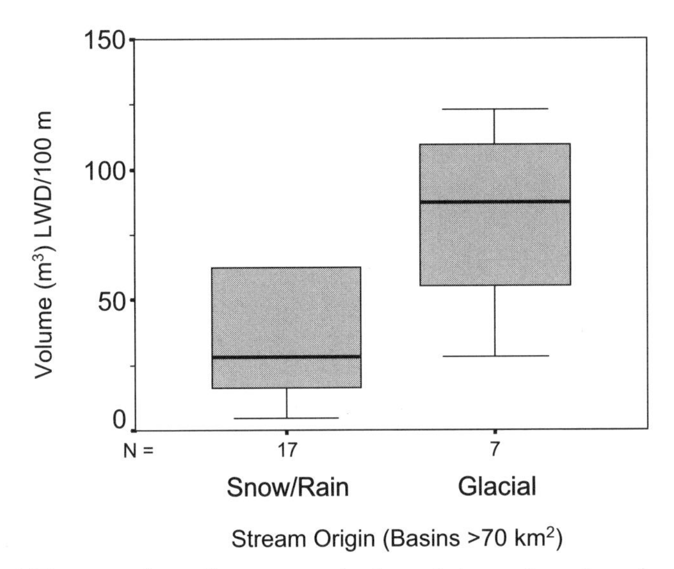

Figure 4. The comparison of instream wood volumes between channels predominantly originating from snow/rain-dominated systems and those predominantly originating from glacial sources.

commonly exhibit greater volume per 100 m than plane-bed, step-pool, or cascade morphologies (Figure 5).

Influences on Instream Wood by Channel Disturbance

Fire, as it affects riparian trees, was found to influence instream wood quantities and volumes. Regression analysis suggests that instream wood volumes increase with adjacent riparian timber age, as dictated by the last stand-replacement fire ($P=0.013$). Riparian characteristics such as mean tree diameter (dbh) and basal area (m^2/ha) are influenced by timber age, increasing as stands grow older (both with $P<0.001$).

Debris flows and snow avalanches also have an effect on instream wood. Trend analysis suggest that debris flows and snow avalanches reduce the quantity and volume of LWD per 100 m of channel length in channels >10% gradient as compared to channels without recent disturbance, but have nearly the same quantity of wood per 100 m in channels <6% gradient. In channels <6% gradient, wood volumes per 100 m are greater in channels with recent debris flows but less in channels with recent snow avalanches. However, these statistically significant differences could not be detected due to the small sample size for these types of disturbances (power <20% in most cases).

Recent floods did not appear to have a significant effect on instream wood in the streams surveyed. The comparison of regressions between channels

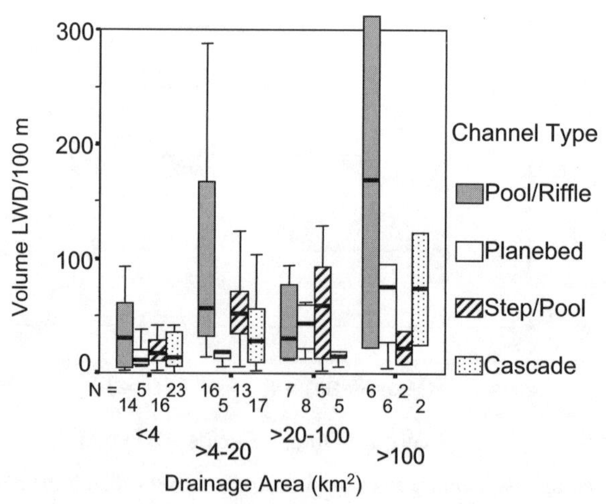

Figure 5. The percentile distribution of wood volumes (m^3) per 100 m of channel length for 4 types of channel morphologies (from Fox 2001).

with and without recent floods (≤ 10-years of survey and a ≥ 25-year recurrence) suggests that floods do not significantly decrease the quantity and volume of instream wood per 100 m ($P>0.6$ for both regression slopes and intercepts). Although this phenomenon is inferred by these data, the effects of floods depicted in these relationships are perhaps poorly defined due to the lack of equal replication of sites containing similar morphologies and regional characteristics. Without controlling for these variables, relationships are likely biased by one or multiple regional and geomorphic influences.

Reference Conditions for Instream Wood Quantity and Volume

Minimum Key Piece Volumes

As noted previously, there are currently no standards for minimum key piece volume for western Washington streams greater than 20 m bankfull width (WFPB 1997). Therefore, in order to develop targets for the quantity of key pieces, new minimums were established.

The length and diameter of key pieces are factors influencing buoyancy and mobility. Although some dimensional combinations may influence piece stability more than others as they interact with channel shape, we will assume that piece *volume* provides a reasonable representation of both length and diameter proportions necessary for stability.

Western Washington Channels >20 m BFW

The range of volumes for wood pieces meeting the definition for key piece function (WFPB 1997) in channels greater than 20 m BFW is presented in the form of percentile distribution plots (box plots) in Figure 6. From this distribution, the minimum volumes, as defined by the 25th percentiles, are approximately 9.7 m^3 for the 20–30 m BFW class, 10.5 m^3 for the 30–50 m^3 BFW class, and 10.7 m^3 for channels greater than 50 m BFW (see data summary table, Figure 6). A plot of these minimum volumes, including those currently defined by WFPB (1997), is presented in Figure 7.

The Influence of Rootwads

Of the pieces composing the volume percentile distributions presented in Figure 6 and the minimum volume curve in Figure 7, it would appear that the

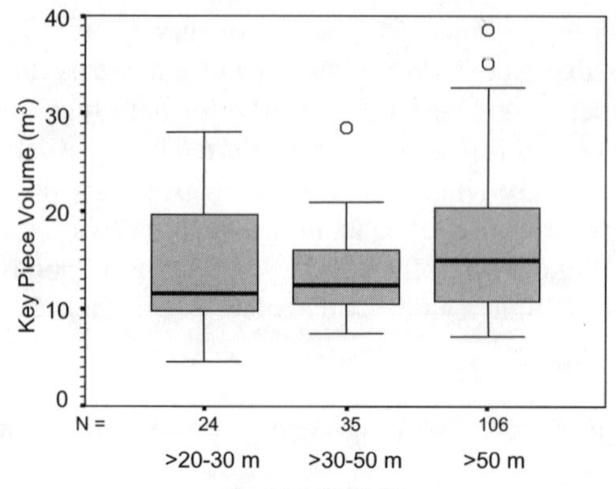

	Median	*Percentile 25*	*Percentile 75*
<20–30m	11.6	9.7	21
>30–50m	125	10.5	17.6
>50m	15.2	10.7	20.6

Figure 6. The percentile distribution of instream wood volumes for pieces meeting the definition of "independent stability" (WFPB 1997) for channels greater than 20 m BFW (current WFPB volume standards only apply to channels less than 20 m BFW). We define the minimum volume for key pieces in channels greater than 20 m using the 25th percentile.

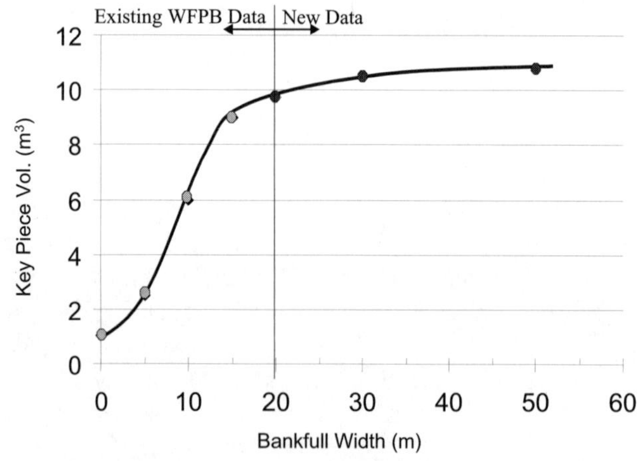

Figure 7. Plot of minimum wood volumes used to define key pieces. The points to the right of the vertical line represent the new minimum volumes defined in this analysis, and the points to the left represent existing minimum volume standards currently used in the State of Washington's Watershed Analysis (WFPB 1997).

minimum volumes defining key pieces are very similar in all channels with BFWs greater than 20 m. As channels become larger, one would expect that the mobility of wood also increases due to wood buoyancy and higher unit stream power. The reason that this is not reflected by an increase in the minimum key piece volumes in larger channels is likely due to the presence of rootwads, which compensate for stability in lieu of volume increases. Indeed, 96% of the wood pieces meeting the WFPB definition for key pieces in channels greater than 50 m BFW had rootwads attached to them. In channels with bankfull widths between 30–50 m, 91% of the pieces had rootwads, and in channels with bankfull widths between 20–30 m, 71% had rootwads attached.

Key Piece Quantities and LWD Quantities and Volumes

Overall, the quantity and volumes of LWD per 100 m of channel length increase with increasing BFW (Figure 8); however, the variance is only poorly

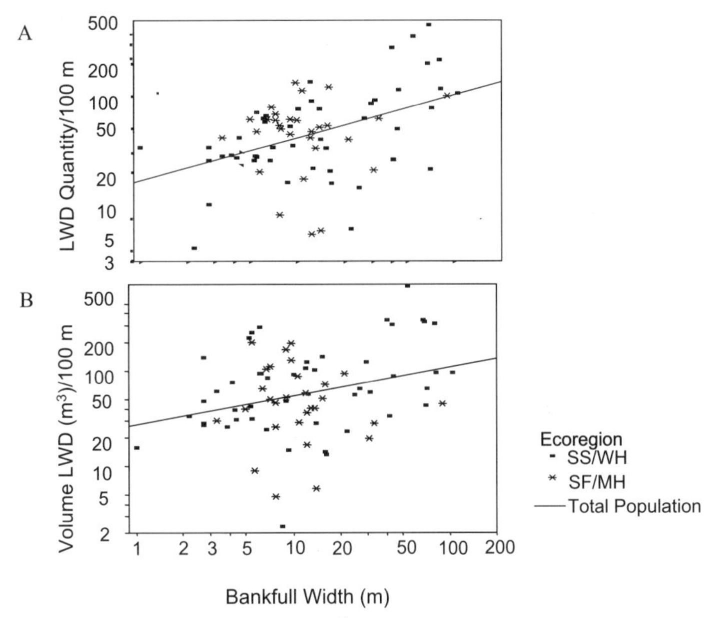

Figure 8. The quantity (A) and volume (m^3) (B) of LWD per 100 m of channel length depicted for all channel widths in both western Washington ecoregions. The total regression slope is significant ($P<0.001$) in each; however, the regression coefficient (R^2) is just 0.2 and 0.1, respectively. SS/WH=Sitka Spruce/Western Hemlock; SF/MH=Silver Fir/Mountain Hemlock

explained by the regressions (R^2=0.2 and 0.1, respectively). Therefore, a classification approach is more practical as a management tool than a regression, since a range of conditions is provided rather than merely a single point-estimate as predicted by an equation.

The percentile distribution of these data, distinguished by bankfull width classifications, provides reference conditions for wood quantity, key piece quantity, and wood volume in western Washington. Based on significant differences in log-normal means or variances, we identified three distinct bankfull width classes for LWD quantity per 100 m: 0–6 m, 6–30 m, and 30–100 m. The median quantity of LWD pieces is approximately 29, 52, and 106 for these groups, respectively (Figure 9). We also identified discrete bankfull width classes of <10 m and 10–100 m for key pieces per 100 m. The median quantity of key pieces per 100 m of channel length is approximately 6 for channels 0–10 m BFW, and 1 key piece per 100 m for channels 10–100 m BFW (Figure 10). Finally, we identified two discrete bankfull width classes of 0–30 m and 30–100 m for LWD volume per 100 m. The percentile distributions and other data summaries for LWD volume are presented in Figure 11.

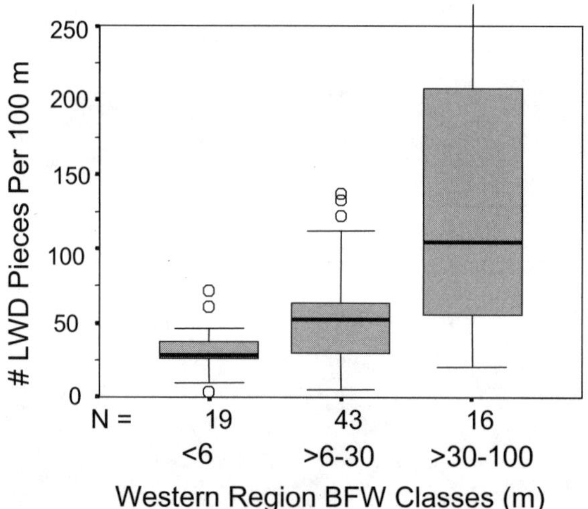

BFW	*Mean*	*Median*	*Percentile 25*	*Percentile 75*
<6 m	33	29	26	38
>6–30 m	52	52	29	63
>30–100 m	144	106	57	208

Figure 9. The percentile distribution of the quantity of LWD per 100 m for the western Washington region. BFW means bankfull width.

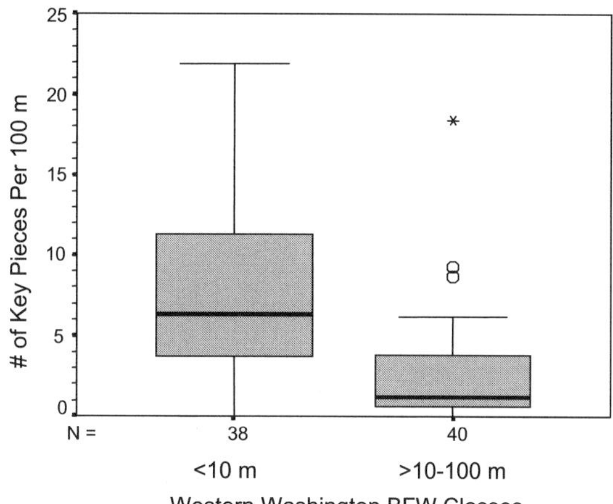

Key Pieces/100 m	*Mean*	*Median*	*Percentile 25*	*Percentile 75*
<10 m	8	6	4	11
>10–100 m	3	1	1	4

Figure 10. The percentile distribution of the quantity of key pieces per 100 m for the western Washington region. BFW classes are distinguished by significant differences between the variances. BFW means bankfull width.

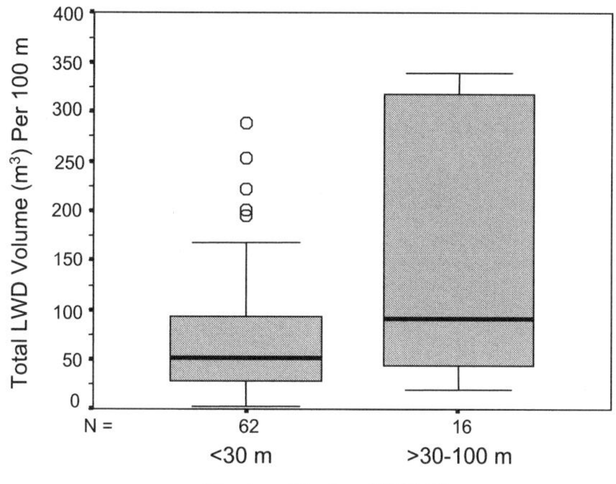

Volume LWD/100 m	*Mean*	*Median*	*Percentile 25*	*Percentile 75*
<30 m	71	51	28	99
>30–100 m	185	93	44	317

Figure 11. The percentile distribution of the volume (m^3) of LWD per 100 m for the western Washington region. BFW means bankfull width.

Comparison of Data to Existing Management Standards

As described earlier, various regulatory agencies use wood targets to define favorable habitat for salmonids. The following presents the results of comparing these target values with our data.

National Marine Fisheries Service: LWD Quantity

To achieve a "Properly Functioning Condition" (PFC) as defined by NMFS in western Washington there should be >80 pieces/m (>50 pieces/km). Qualifying wood must be >50 ft (15 m) in length and >24 in (0.6 m) in diameter. The distribution of the number of qualifying wood pieces per sampled site is presented in Figure 12 for western, or "Coastal Washington" (as stated by NMFS). Of the 78 streams sampled in western Washington streams, only 11 met the requirements of 80 pieces per mile put forth by NMFS for PFC. The mean of the data is significantly different from the PFC target ($P<0.001$). Percentile distributions and one-tailed t-tests suggest that the sample mean of qualifying wood pieces per mile is significantly lower than the NMFS target

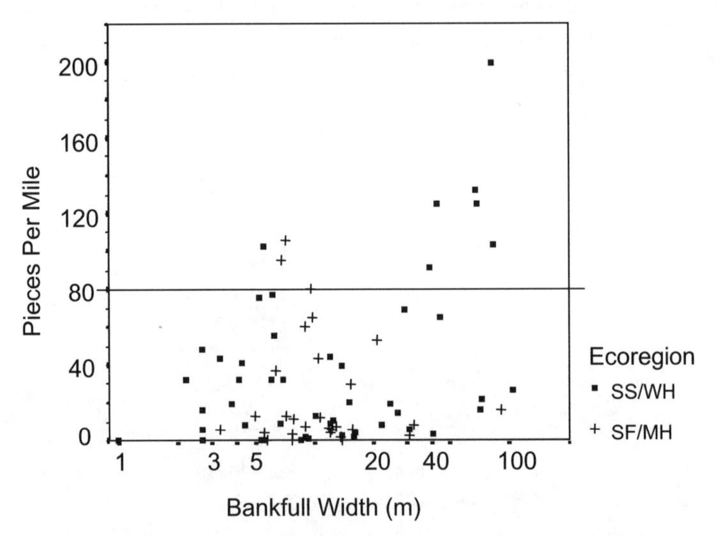

Figure 12. The distribution of sites (n=78) indicating the number of qualifying instream wood pieces that meet the NMFS criteria for coastal western Washington. The horizontal bar represents the lower threshold for streams meeting a "Properly Functioning Condition" (NMFS 1996). SS/WH=Sitka Spruce/Western Hemlock; SF/MH=Silver Fir/Mountain Hemlock

for western Washington (Figure 13). The data also suggest that the mean is similar to the PFC standard only in channels greater than 40 m BFW.

Washington Forest Practices Board: LWD Quantity

To achieve a "good" habitat quality rating in the WFPB manual (1997) there should be 2 pieces per channel width (<20 m BFW) >2 m length x 10 cm diameter. Comparing the mean of our data for instream LWD quantities in western Washington streams (channels <20 m BFW) to the WFPB target of 2 pieces per channel width, we found no significant difference (t-test) ($p=0.054$, $n=56$). The distribution of data (Figure 14A) suggests that this target is not

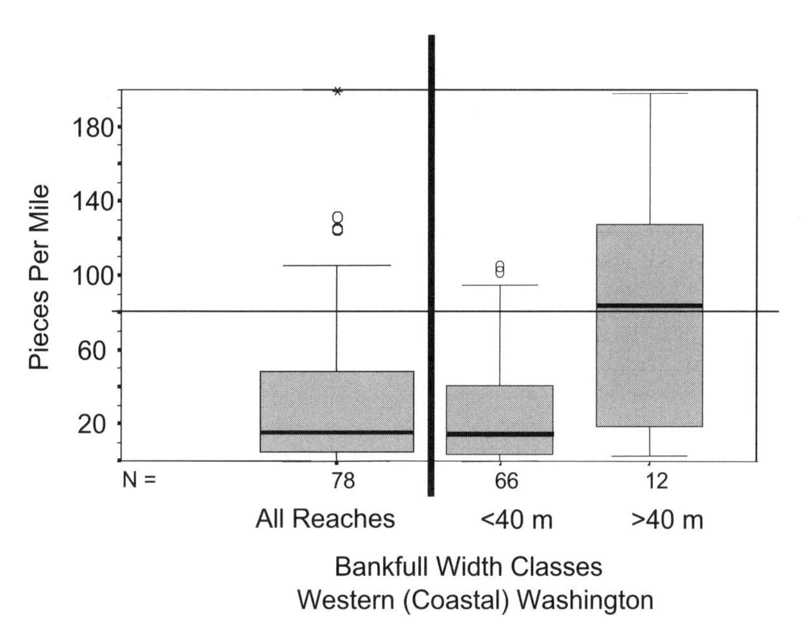

Figure 13. Box plots for the number of NMFS qualifying pieces per sampled site in western Washington. The mean of these samples is not equal to the standard of 80 pieces per mile as suggested by the results of a one-sample t-test ($P<0.001$) $n=70$ for log-normalized data. The analysis is conducted for all sites (left) and for two divisions of bankfull width classes (right) to illustrate disparities. The "extremes and outliers" depicted for "All Reaches" are statistically considered part of the range in the BFW class of ">40 m" due to the proportions related to the sample size of each class. The horizontal line represents the threshold for streams meeting a "Properly Functioning Condition" (NMFS 1996).

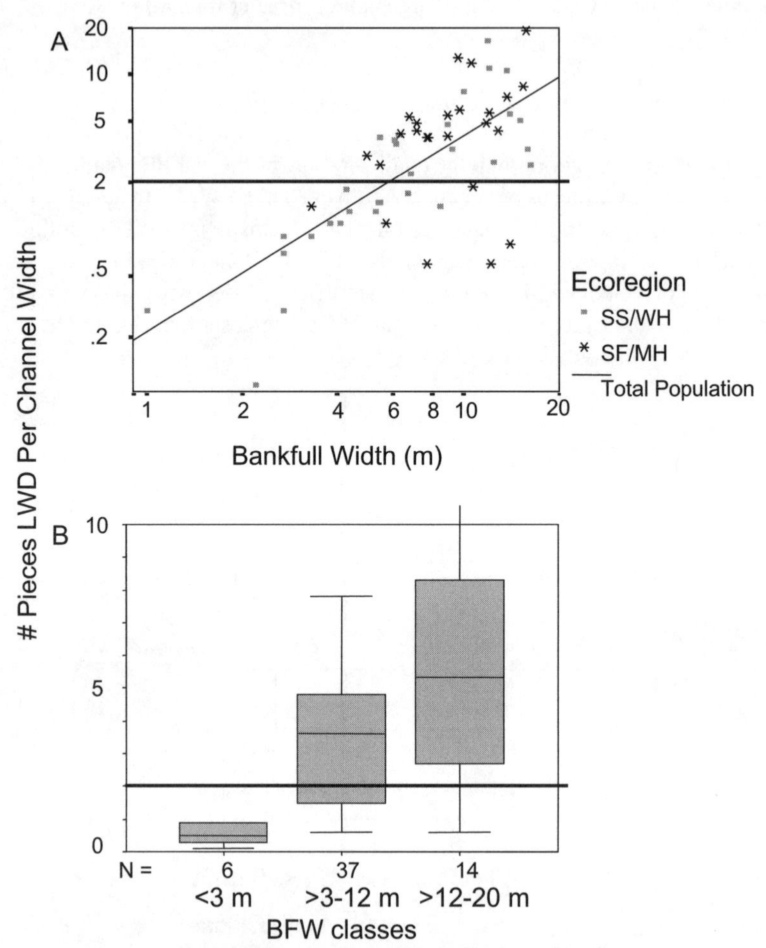

Figure 14. The number of LWD pieces per channel width by bankfull width for channels <20m in BFW. The target index of two pieces of LWD per channel width (WFPB 1997), as indicated by the horizontal line, is the quantity indicating "Good" habitat quality. T-tests with normalized data suggest that the mean is not significantly different from the WFPB standard (P=0.054, n=56). (A) Each data point represents the mean quantity per sample, labeled to identify discrete ecoregions. The slope of the regression through the points is significant (P<0.001). R^2 = 0.50, n=56. (B) The range of data illustrates non-uniform relationships to the target value among discrete bankfull width classes.

applicable for all channel widths <20 m due to the significantly positive regression slope ($P<0.001$), which is described by the equation:

$$Y=0.22x^{1.26} \tag{1}$$

where Y is the predicted number of LWD pieces per channel width and x is the bankfull width in meters. Using the percentile distribution of LWD quantity to define three bankfull width classes, we find that the WFPB target is higher than the data distributions for channels with a bankfull width <3 m, but the target is lower for channels >12 m BFW (Figure 14B).

Washington Forest Practices Board: Key Piece Quantity

To achieve a "good" habitat quality rating in the WFPB (1997) there should be 0.3 pieces per channel width (channels 0–10 m BFW) and 0.5 pieces per channel width (channels 10–20 m BFW) for pieces meeting the definition. T-tests suggest that the log-normal mean of our data is not significantly different from the WFPB target of 0.3 key pieces per channel width for channels 0–10 m BFW in western Washington ($p=0.897$); however, the mean for key pieces per channel width in channels 10–20 m bankfull width is significantly different from the WFPB target of 0.5 pieces per channel width ($p=0.001$). The percentile distribution (Figure 15) suggests the data mean in channels 10–20 m BFW is *less* than the WFPB target. The relationship of the number of key pieces per channel width to bankfull width is not significant ($p=0.625$).

Discussion

The Validity of Grouping the Sitka Spruce and Western Hemlock Forests

Although the mean diameters of trees in the Sitka Spruce forests appear to be significantly larger than in the Western Hemlock forests, the instream wood loads are similar (Fox 2001). Assuming instream wood quantity and volume is the "end result" of the wood recruitment process to these channels, the moderate difference in tree species may not be important for characterizing instream wood. Additionally, there is a paucity of information regarding the relative importance of different species to this process. The data suggest that both the quantities and volumes of wood in streams draining these forest types are similar.

Choice of Predictor Variables

Geomorphological Influence

Although channel type, bed-form, origin, gradient, and confinement all appear to influence instream wood quantities and volumes to some degree, the sample stratification in each geomorphic category is too small to isolate the effects of these factors for making statistical inferences. Greater certainty regarding these influences would require additional sampling of these morphologies. Based on (1) the results of the trend analysis with wood volumes with increasing basin size, (2) the correlation of bankfull width to basin size and cross-sectional area, (3) the demonstration that bankfull width has better predictive qualities than basin size for instream wood, and (4) the lack of significant influence by the previously discussed reach morphology influences, bankfull width is supported as the most significant geomorphic indicator for predicting instream wood volumes and quantities.

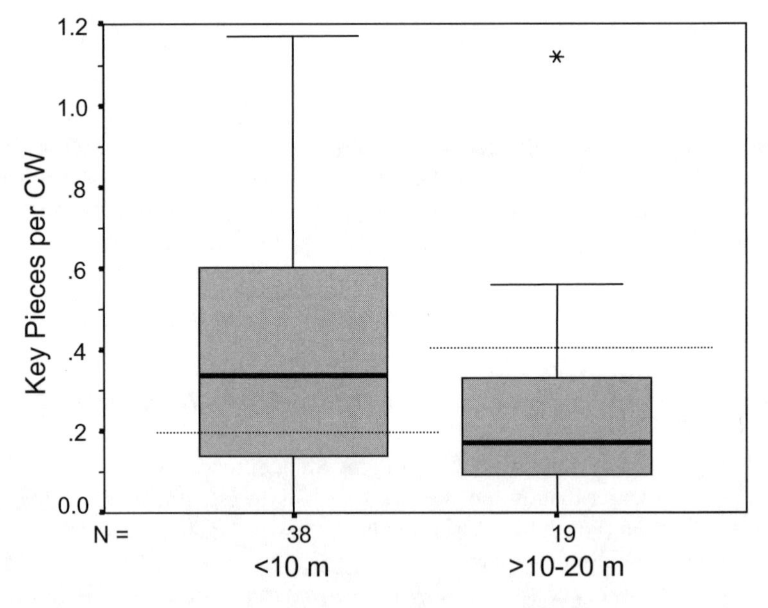

Figure 15. The data distribution as compared to the WFPB targets for key piece quantities per channel width in western Washington. The dashed horizontal line represents the WFPB target pertaining to each of these bankfull width classes. The means of the two populations are not significantly different (p=0.23).

Influence of Disturbance

Tree age, as influenced by natural fire history, increases with wetter climates. The rate of fire recurrence is a principle characteristic of ecoregions. Because the adjacent riparian trees influence instream wood, and the characteristics of riparian trees as influenced by fire recurrence vary by ecoregions, ecoregion, this grouping of forest zones is probably the best single indicator for predicting instream wood loads in relation to fire disturbance. However, due to the similarities in forest types of western Washington, distinction between the SS/WH and SF/MH ecoregions are insignificant (Fox 2001). From our data, floods do not appear to have a significant influence on wood and therefore are inconsequential to variable selection. Debris flows and snow avalanches perhaps have some influence on instream wood loads (Fox 2001); however, this influence could not be verified with statistical inferences due to the small number of disturbed sites compared to non-disturbed sites. We could not isolate any form of disturbance as a predictor variable of instream wood loads. However, the range of wood loads found within any one grouping likely reflects some level of natural disturbance that creates typical patchy stream habitat.

Setting Management Targets

The percentile (box-plot) distributions for LWD quantity, volume, and key piece quantity (Figures 9–11) represent the range of conditions found in streams draining unmanaged forests subject to a natural rate of disturbance (except for fire suppression). Assuming these data include both favorable as well as unfavorable habitat conditions as it relates to instream wood, the central 50% of these data (as defined by the 25th and 75th percentiles) may be used to represent a "fair" condition in which to base targets for habitat restoration, enhancement, regulation, and evaluation. The top of these distributions, the 75th percentile, is a logical point at which conditions begin to exceed this central range. Therefore, we recommend values at the 75th percentile and above represent the equivalent of a "good" condition as it relates to wood loads. Following this logic, wood quantities and volumes below the 25th percentile would represent the equivalent of a "poor" condition as it relates to instream wood loads.

The precise quantities and volumes of wood needed by salmonids for successful production are not well understood. Statistically sound studies to link instream wood loads to salmonid production would be expensive and have high levels of uncertainty due to the multiple variables influencing salmon production (Chapter 12). However, we do know that historic salmon populations were much higher than those found today, and as noted earlier, we assume that unmanaged forests offer the best source of information on wood loads as one component of habitat to which salmonids have adapted. In streams where management is needed to restore favorable conditions, wood loads are either no longer found in the upper distribution of these ranges, or the distribution is centered around a lower mean. Thus for management purposes intending to restore natural wood loading conditions, establishing instream wood targets based on the upper portion of the distribution (i.e., the 75th percentile) rather than the lower portion of the distribution is reasonable as well as prudent.

The Feasibility of the NMFS Targets

Western (Coastal) Washington

Figure 12 illustrates that few of the sampled streams in western Washington meet the minimum NMFS target to achieve "Properly Functioning Condition." The mean is significantly different from the target mean of 80 pieces per mile for the combined sampled streams. Indeed, the entire middle 50% of the sample is less than 52 pieces per mile (Figure 13), suggesting that the mean of the sampled data is significantly less than the NMFS target value. The highest proportions of streams that do meet the NMFS standard are greater than 40 m in bankfull width (42% greater than 40 m BFW). Realizing that not all streams in unmanaged forests are expected to be ideal habitat, an ideal "target" value should be located somewhere near the upper portion of the distribution (e.g., near the 75th percentile). This is the case only for streams greater than 40 m BFW, and therefore 80 pieces per mile seems to be an achievable target only for the larger streams.

The Feasibility of the WFPB Targets

Large Woody Debris Quantity

The mean of our sample data is not significantly different from the WFPB target of 2 pieces of LWD per channel width for channels <20 m in bankfull width. With the 75th percentile as a goal, this target is set too low. A more appropriate interpretation of the "Fair" condition as defined by WFPB would thus correlate with the range between the 25th and 75th percentiles.

Additional disparities exist among bankfull width classes when applying the WFPB target, because quantities vary significantly ($P<0.001$) with BFW (Figure 14A). The positive regression slope and the distribution of data suggest that the mean is < 2 pieces per channel width in the smaller channels and greater in the larger. Figure 14B illustrates these differences and suggests that application of the targets is not appropriate for all bankfull width classes. Because these data suggest that the application of the WFPB target should not be homogeneously applied to all stream bankfull width classes in western Washington, we conclude that this target will potentially lead to inappropriate wood quantity characterizations. Modifying the WFPB targets to adjust for variations in channel width and ecoregion would likely improve the outcome of management decisions based on these values.

Key Piece Quantity

Using the 75th percentile as a target value, the key piece target suggested by WFPB for channels <10 m BFW is set too low. This quantity, near the center of the percentile distribution, is within the expected range of conditions but perhaps not representative of "Good" habitat quality. Conversely, the standard set for channels 10–20 m BFW is too high. The relative location of each of these targets upon the data distribution almost implies that these values are reversed: the target of 0.3 pieces per channel width would be a more appropriate target for channels 10–20 m BFW, while 0.5 pieces per channel width better fits channels 0–10 m BFW. Due to the adjusting scale of minimum wood volumes defining key pieces, no clear relationships between key piece quantity and channel width could be established (Fox 2001).

Defining New Key Piece Minimum Volumes

The minimum volumes established in Figure 6 illustrate that the size of the pieces in channels greater than 20 m do not increase at the same rate as the minimum defined volumes in channels between 0–20 m. The change in rate is illustrated in Figure 7 as channels reach 15–20 m in bankfull width (i.e., 9 m^3) and suggests that the relationship between bankfull width (as representative of potential water depth and width) and wood volume (as a function of stability) is not linear. Certainly, one would expect that wood must be larger to counter the tendency to mobilize as channels become larger. This is not the case and is likely attributed to the presence of rootwads to help anchor logs. Clearly, this often compensates for the need of increased volume for stability.

Table 4. Summary of ranges for instream wood quantity and volumes according to BFW classes in western Washington based on the box plots in Figures 9-11. Wood quantities and volumes greater than the 75th percentile of the distribution are recommended as a target for "Good" conditions, the range between the 25th and 75th percentiles are recommended as a target for "Fair" conditions, and below the 25th percentile is recommended as a definition of "Poor" conditions. LWD is defined as a piece >10cm diameter and >2 m in length. Volumes are estimated by $\pi r^2 L$ where L is the piece length, and r is the piece radius at the mid-point.

LWD Piece Quantity: Number of Pieces Per 100 m of Channel Length

Region	*BFW Class*	*Good*	*Fair*	*Poor*
Western WA	0–6 m	>38	26–38	<26
	>6–30 m	>63	29–63	<29
	>30–100 m	>208	57–208	<57

LWD Volume: Cubic Meters Per 100 m of Channel Length

Region	*BFW Class*	*Good*	*Fair*	*Poor*
Western WA	0–30 m	>99	28–99	<28
	>30–100 m	>317	44-317	<44

Key Piece Quantity: Number of Pieces Per 100 m of Channel Length

Region	*BFW Class*	*Good*	*Fair*	*Poor*
Western WA	0–10 m	>11	4-11	<4
	>10–100 m	>4	1–4	<1

Minimum Piece Volume to Define Key Pieces

Bankfull Width Class	*Minimum Piece Volume (m^3)*
0–5 m	1*
>5–10 m	2.5*
>10–15 m	6*
>15–20 m	9*
>20–30 m	9.75
>30–50 m	10.5**
>50–100 m	10.75**

* Existing WFPB (1997) definitions

** Rootwads must be attached

This is illustrated by the increased prevalence of rootwads attached to key pieces as bankfull width increased, although the minimum volumes did not increase proportionately. The data suggest that without rootwads attached, the minimum volume required to meet the definitions for key pieces may indeed follow the near-linear relationship with bankfull width established by the WFPB in channels 0–20 m bankfull width (Figure 7). However, this relationship may not be fully realized due to the fact that samples for pieces this large without rootwads were rare (n=3).

Restoration and Management Recommendations

The percentile (box-plot) distributions for LWD quantity, volume, and key piece quantity (Figures 9 through 11) offer a range of reference conditions for habitat restoration, enhancement, regulation, and evaluation. Because these data represent a wide range of conditions found in streams draining unmanaged forests subject to a natural rate of disturbance (except for fire suppression), we base our recommendations for target values on instream wood loads at the 75th percentile, as previously discussed.

Minimum piece volumes used to define a key piece should also consider the role rootwads play in achieving stability. In channels greater than 30 m BFW, >91% of all key pieces had rootwads attached. Therefore, in order to meet the objective of defining a key piece, not only do the prescribed minimum volumes need to be met but also rootwads must be considered in this definition. Without rootwads to stabilize key pieces, the minimum volume needed for stability in large channels would be very large and are subsequently rare. Logs of this size are likely impossible to obtain for stream habitat enhancement projects, let alone transport and position into a channel. Therefore, we recommend that for channels greater than 30 m, a log must have a rootwad attached to be defined as a key piece, in addition to meeting the minimum volume requirements defined in Figure 6. Although having a rootwad attached to a log placed in a stream channel as part of restoration or enhancement efforts adds stability and longevity (Braudrick and Grant 2000), the data do not justify a requirement that all key pieces in the 20–30 m BFW class have an attached rootwad in addition to meeting the minimum volume requirement.

Table 4 summarizes our recommendations for instream wood loadings. These values offer targets and typical ranges of conditions for the quantities and volumes of wood found within the historical range of watershed conditions, given the natural disturbance regime for western Washington. We anticipate that wood loads within this range would form and maintain favorable habitat conditions; therefore, these quantities provide a reference to levels likely to promote natural stream processes that form salmonid habitat. These values can be used to: (1) assess current instream wood condition and ratings for the evaluation of stream habitat; (2) identify target wood loads levels for restoration, enhancement, and mitigation projects; and (3) develop land-use regulations, ordinances, and laws to protect and manage salmonid habitat.

References

Abbe, T.B. and D.R. Montgomery. 1996. Large woody debris jams, channel hydraulics and habitat formation in large rivers. *Regulated Rivers: Resources & Management* 12:201-221.

Agee, J.K. 1993. *Fire ecology of Pacific Northwest forests.* Island Press, Washington D.C.

Beechie, T.J. and T.H. Sibley.1997. Relationships between channel characteristics, woody debris, and fish habitat in northwestern Washington streams. *Transactions of the American Fisheries Society* 126:217-229.

Bilby. R.E., B.R. Fransen, and P.A. Bisson. 1996. Incorporation of nitrogen and carbon from spawning coho salmon into the trophic system of small streams: Evidence from stable isotopes. *Canadian Journal of Fisheries and Aquatic Sciences* 53:16 and 173.

Bilby, R.E. and J.W. Ward. 1989. Changes in characteristics and function of woody debris with increasing size of streams in western Washington. *Transactions of the American Fisheries Society* 118:368-378.

Bilby, R.E. and J.W. Ward. 1991. Characteristics and function of large woody debris in streams draining old-growth, clear-cut, and second-growth forests in southwestern Washington. *Canadian Journal of Fisheries and Aquatic Sciences* 48:24-2508.

Bjornn, T.C. and D.W. Reiser. 1991. Habitat requirements of salmonids in streams. In W.R. Meehan (ed.) *Influences of Forest and Rangeland Management on Salmonid Fishes and Their Habitats.* Special Publication 19. American Fisheries Society, Bethesda, Maryland. pp. 83-138.

Bisson, P.A., R.E. Bilby, M.D. Bryant, C.A. Dolloff, G.B. Grette, R.A. House, M.L. Murphy, K.V. Koski, and J.R. Sedell. 1987. Large woody debris in forested streams in the Pacific Northwest: Past, present, and future. In E.O. Salo and T.W. Cundy (eds.) *Streamside Management: Forestry and Fishery Interactions.* College of Forest Resources, University of Washington, Seattle, Washington. pp. 143-190.

Booth, D.B. 1991. Urbanization and the natural drainge system—Impacts, solution, and prognosis. *The Northwest Environmental Journal* 7:93-118.

Braudrick, C.A. and G.E. Grant. 2000. When do logs move in rivers? *Water Resource Research* 36:571-583.

Cederholm, C.J., D.B. Houston, D.L. Cole, and W.J. Scarlett. 1989. Fate of coho salmon (*Oncorhynchus kisutch*) carcasses in spawning streams. *Canadian Journal of Fisheries and Aquatic Sciences* 46:1347-1355.

Coho, C. and S.J. Burges. 1993. Dam-break floods in low order mountain channels of the PNW, Water Resources Series Tech Rep no. 138. Department Civil Engineering, University of Washington, Seattle, Washington.

Costa, J.E. 1984. Physical geometry of debris flows. In J.E. Costa and P.J. Fleisher (eds.) *Developments and Applications of Geomorphology.* Springer-Verlag, Berlin. pp. 268-317.

Delorme. 1999. 3-D TopoQuads™. Topographic mapping software. Yarmouth, Maine.

Fausch, K.D. 1984. Profitable stream positions for salmonids: Relating specific growth rate to net energy gain. *Canadian Journal of Zoology* 62:441-451.

Fetherston, K.L., R.J. Naiman, and R.E. Bilby. 1995. Large woody debris, physical process, and riparian forest development in montane river networks of the Pacific Northwest. *Geomorphology* 13:133-144.

Fox, M.J. 2001. A new look at the quantities and volumes of wood in forested basins of Washington State. Master's thesis. University of Washington. Seattle, Washington.

Franklin, J.F. and C.T. Dyrness. 1973. Natural vegetation of Oregon and Washington. USDA Forest Service. General Technical Report PNW-8.

Henderson, J.A., R.D. Lesher, D.H. Peter, and D.C. Shaw. 1992. Field guide to the forested plant associations of the Mt. Baker-Snoqualmie National Forest. USDA Forest Service, Pacific NW Region. Technical Paper R6 ECOL TP 028-91.

Ikeya, H. 1981. A method for designation forested areas in danger of debris flows. In T.R.H. Davies and A.J. Pearce (eds.) *Erosion and Sediment Transport in Pacific Rim Steeplands.* International Association of Hydrological Sciences, Publication 132. pp. 576-588.

Keller, E.A. and F.J. Swanson. 1979. Effects of large organic material on channel form and fluvial processes. *Earth Surface Processes* 4:361-380.

Larsson, P.O. 1985. Predation on migrating smolts as a regulating factor of Baltic Salmon (*Salmo salar*). *Journal of Fish Biology* 26:391-397

Lisle, T.E. and H.M. Kelsey. 1982. Effects of large roughness elements on the thalweg course and pool spacing. In LB. Leopold (ed.) *American Geomorphological Field Group Field Trip Guidebook.* Berkeley, California. pp. 134-135.

Massong, T.M. and D.R. Montgomery. 2000. Influence of sediment supply, lithology, and wood debris on the distribution of bedrock and alluvial channels. *Geological Society of America Bulletin* 112:591-599.

McDade, M.H., F.J. Swanson, W.A. McKee, J.F. Franklin, and J. Van Sickle. 1990. Source distances for coarse woody debris entering small streams in western Oregon and Washington. *Canadian Journal of Forest Resources* 20:326-330.

McDonald, L.H., A.W. Smart, and R.C. Wissmar. 1991. Monitoring guidelines to evaluate effects of forestry activities on streams in the Pacific Northwest and Alaska. EPA/910/9-91-001. Region 10. Seattle, Washington.

McHenry, M.L., E. Shott, R.H. Conrad, and G.B. Grette. 1998. Changes in the quantity and characteristics of LWD in streams of the Olympic Peninsula, Washington, USA (1982-1993). *Canadian Journal of Fish and Aquatic Sciences* 55:1395-1407.

Montgomery, D.R., T.B. Abbe, J.M. Buffington, N.P. Peterson, K.M. Schmidt, and J.D. Stock. 1996. Distribution of bedrock and alluvial channels in forested mountain drainages. *Nature* 381:587-589.

Montgomery, D.R. and J.M. Buffington. 1997. Channel-reach morphology in mountain drainage basins. *Geological Society of America Bulletin* 109:596-611.

Montgomery, D.R., J.M. Buffington, RD. Smith, KM. Schmidt, and G. Pess. 1995. Pool spacing in forest channels. *Water Resources Research* 31:1097-1105.

Murphy, M.L. and K.V. Koski. 1989. Input and depletion of woody debris in Alaska streams and implications for streamside management. *North American Journal of Fisheries Management* 9:427-436.

National Marine Fisheries Service. 1996. Making Endangered Species Act Determinations of Effect for Individual or Grouped Actions at the Watershed Scale. Environmental and Technical Services Division, Habitat Conservation Branch.

National Marine Fisheries Service. 1998. Draft proposed recommendations for Amendment 14 to the Pacific Coast Salmon Plan for Essential Fish Habitat. Northwest Regional Office, Seattle, Washington.

Naiman, R.J. and J.R. Sedell. 1979. Relationships between metabolic parameters and stream order in OR. *Canadian Journal of Fish and Aquatic Sciences* 37:834-847.

Peterson, N.P., A. Hendry, and T.P. Quinn. 1992. Assessment of cumulative effects on salmonid habitat: Some suggested parameters and target conditions. TFW-F3-92-001. Prepared for the Washington Department of Natural Resources and The Coordinated Monitoring, Evaluation and Research Committee. Timber Fish and Wildlife Agreement. University of Washington, Center for Streamside Studies, Seattle, Washington.

Pleus, A.E. and D. Schuett-Hames. 1998. TFW Monitoring Program Methods

Manual for the reference point survey. Prepared for the Washington State Dept. of Natural Resources under the Timber, Fish, and Wildlife Agreement. TFW-AM9-98-002.

Pleus, A.E. 1999. TFW Monitoring Program Method Manual for wadable stream discharge measurement. Prepared for the Washington State Dept. of Natural Resources under the Timber, Fish, and Wildlife Agreement. TFW-AM9-99-009.

Ralph, S.C., G.C. Poole, L.L. Conquest, and R.J. Naiman. 1991. Stream channel morphology and woody debris in logged and unlogged basins of western Washington. *Canadian Journal of Fisheries and Aquatic Sciences* 51:37-51

Robison, G.E. and R.L. Bestcha. 1990. Coarse woody debris and channel morphology interactions for undisturbed streams in southeast Alaska, U.S.A. *Earth Surface Process Landforms* 15:149-156.

Rot, B.W., R.J. Naiman, and R.E. Bilby. 2000. Stream channel configuration, landform, and riparian forest structure in the Cascade Mountains, Washington. *Canadian Journal of Fisheries and Aquatic Sciences* 57:699-707.

Schuett-Hames, D., A.E. Pleus, J. Ward, M. Fox, and J. Light. 1999. TFW Monitoring Program Methods Manual for the large woody debris survey. Prepared for the Washington State Dept. of Natural Resources under the Timber, Fish, and Wildlife Agreement. TFW-AM9-99-004.

Spence, B.C., G.A. Lomnicky, R.M. Hughes, and R.P. Novitzki. 1996. An Ecosystem Approach to Salmon Conservation. TR-4501-96-6057. ManTech Environmental Research Services Corp., Corvallis, Oregon.

Sternes, G.L. 1969. Climatological handbook: Columbia basin states precipitation. Vol 7. Pacific NW River Basins Commission, Meteorology Committee, Vancouver, Washington.

Sumioka, S.S., D.L. Kresch, and K.D. Kasnick. 1998. Magnitude and Frequency of Floods in Washington. U.S. Geological Survey Water-Resources Investigations Report 97-4277. Tacoma, Washington.

Swanson, F. J., and G. W. Lienkaemper. 1978. Physical consequences of large organic debris in Pacific NW streams. USFS General Technical Report PNW-69.

Tappeiner, J.C., D. Huffman, D. Marshall, T.A. Spies, and J.D. Bailey. 1997. Density, ages, and growth rates in old-growth and young-growth forests in coastal Oregon. Paper 3166 of the Forest Research Laboratory, Oregon State University, Corvallis, Oregon.

U.S. Water Resources Council. 1981. Guidelines for determining flood flow frequency: U.S. Water Resources Council Bulletin 17B.

Washington Forest Practices Board (WFPB). 1997. Board Manual: Standard Methodology for Conducting Watershed Analysis. Under Chapter 222-22 WAC. Version 4.0. Olympia, Washington.

15. Stream Enhancement Projects: A King County Perspective

John Bethel and Kathryn Neal

Abstract

Stream channels and riparian corridors form an extraordinarily complex component of Pacific Northwest ecosystems. The complexity of these systems has precluded a comprehensive understanding of either their dynamics in a pristine condition or their response when disturbed. Degradation of natural stream systems by human activities has become a matter of grave public concern, and efforts to restore these systems are widespread. The central challenge in attempting such enhancement work is to identify and carry out effective interventions despite an incomplete understanding of fluvial systems. Based on experience acquired by dealing with these issues, King County has identified a number of strategies that improve the likelihood that enhancement efforts will produce useful results. These include:

- Managing enhancement projects through self-directed, multi-disciplinary project teams that participate in the initial project definition and carry through design, construction, and assessment.
- Utilizing natural stream systems as templates for project design and using native materials for project construction.
- Critically assessing the results of completed enhancement projects to inform subsequent enhancement work.
- Conducting stream enhancement efforts in the context of a watershed level assessment of causes of habitat degradation and in concert with watershed-level land-use management.

Introduction

This chapter presents insights and perspectives on King County's past and continuing efforts to improve ecological function in degraded streams. The authors of this chapter are staff members directly involved in the design and construction of stream enhancement projects for King County. Case studies of five King County enhancement projects are presented and discussed. This discussion is intended both to share lessons learned during the course of these projects and to share some practical perspectives with non-practitioners on implementing enhancement projects.

Physical Setting

King County is located on the east side of the Puget Lowland in Washington State. The Puget Lowland is a north-south trending trough, with Puget Sound along its axis and the Cascade and Olympic Mountains bordering it to the east and west, respectively. The landscape of this region is largely a product of a series of continental glacial advances and, to a lesser extent, volcanic processes (Chapter 2).

King County has a maritime climate dominated by airflow from the north Pacific Ocean. Annual precipitation increases from west to east as a result of the orographic rainfall effect of the Cascade Mountains. Precipitation ranges from 900 mm near the shores of Puget Sound to 4,000 mm at the Cascade Crest. Typically, approximately 70% of precipitation falls between the first of November and the end of February. These climatic conditions support conifer-dominated forests that historically extended from the marine shoreline to the alpine tree line (Chapter 4).

The streams that drain this forested landscape support many aquatic organisms, including five species of salmon (pink [*Oncorhynchus gorbuscha*], coho [*O. kisutch]*, chinook [*O. tshawytscha]*, sockeye [*O. nerka*], and chum [*O. keta]*), two species of trout (rainbow [*O. mykiss*] and cutthroat [*O. clarki]*), and two species of char (Dolly Varden [*Salvelinus malma Walbaum*] and bull trout [*S. confluentus*]) as well as numerous other vertebrate and invertebrate species.

Stream Degradation

Pioneers of European and Asian descent began to settle in what is now King County in the mid-1800s. Agriculture and various forms of resource extraction, including logging, fishing, and coal mining, supported the economy of this new population. Cities and towns were established along with an associated infrastructure including roads, railroads, water supply diversions, flood control dams, and port facilities. Stream degradation was an unintended, but pervasive, consequence of this development.

As the intensity and extent of land development in King County increased, so did its effect on streams. Development of land for residential, commercial, and industrial use increased dramatically. This development occurred most rapidly and intensely in the western portion of the County and more slowly in its center, where limited agriculture still persists. The mountainous eastern portion of the County remains largely undeveloped and continues to be used for timber harvest and recreation.

According to the 2000 Census results, about 30% of the people who live in Washington State live in King County. Of the 1.7 million people in King County, about 564,000 live in Seattle and another 820,000 live in the surrounding cities, leaving about 353,000 residents in unincorporated King County. The suburban cities are growing faster than Seattle, at rates up to 40% over the past decade. This growth, and its associated extensive modification of the landscape, has had a variety of adverse effects on surface waters including:

- Natural channels have been dredged, diked, straightened, and/or cleared of large woody debris (LWD);
- Wetlands and marine estuaries have been filled and/or drained;
- Riparian zones have been cleared or overwhelmed by aggressive introduced plant species;
- A variety of fish-impassable structures, including culverts, weirs, and dams, have blocked anadromous fish access to hundreds of kilometers of stream channel;
- Water quality has been degraded by introduction of various pollutants; and
- Large areas of impervious surface have dramatically increased peak flows and probably decreased base flows.

King County's Role in Stream Enhancement

In recent decades, progressive degradation of King County's streams has become increasingly obvious (King County 1987). Degradation of these streams has economic, recreational, and aesthetic impacts that affect a broad spectrum of the populace. Concern over this degradation is an important issue for many King County citizens, who have looked to local government to participate in the protection and enhancement of threatened resources. As a result, King County has taken an increasingly active role in protecting and restoring the resource values of the County's streams and wetlands. Several groups within King County work on various aspects of stream enhancement. The projects discussed in this chapter were managed through the Surface Water Engineering and Environmental Services Section of King County's Department of Natural Resources (SWEES).

Within SWEES, a multi-disciplinary project design team is assembled after the project is identified and funded. Professionals on the design team typically include engineers and ecologists, often with support from a geologist and a landscape architect. Many habitat projects are generated through the comprehensive Basin Planning process that the County committed to in the 1980s. Projects are also identified through citizen input, King County staff observations, referrals from other public agencies, joint studies, and County Council requests. For a typical habitat enhancement project, one to three years are required from the time the project team begins work until design, permitting, and construction of earthwork and planting phases are complete. Monitoring and plant maintenance usually continue for an additional three to five years.

Working Definition of Stream Enhancement

The projects described in this chapter are intended to enhance the natural resource value of the subject streams. They all have two overall goals:

1. to establish the channel morphology appropriate to the topographic, geologic, and hydrologic setting, and
2. to establish the channel and riparian habitat that support a diverse native plant and animal community appropriate to the setting.

These goals, however, carry both implications and constraints for King County enhancement projects.

Streams are inherently dynamic. If a project is intended to restore even a fraction of the natural stream functions, the enhancement design must anticipate morphological and biological evolution of the system. Enhancement projects therefore should be designed to be appropriate for not only the existing but also the anticipated conditions at the project site. These are often markedly different from predevelopment conditions. Basin hydrology, in particular, appears to be permanently and significantly altered by even modest upstream development (Booth and Jackson 1997). Appropriate project designs are thus most likely to result in a "rehabilitated" stream that is very different from pre-European settlement conditions at the project site.

Case Studies

Five enhancement projects are described below. They illustrate typical situations in which King County undertakes stream enhancement work, some of the strategies used to address these situations, and some difficulties encountered in executing these strategies. These particular projects illustrate a progression of techniques employed to remedy undesirable channel erosion and sediment deposition. Table 1 provides a summary of project characteristics; Figure 1 shows project locations. Each project was initiated for the primary purpose of restoring aquatic and riparian habitat, not as an adjunct to development or infrastructure construction. In some cases, protection of downstream infrastructure was also an important design goal. Self-directed, multidisciplinary teams in SWEES designed each of the case study projects, and all were constructed by King County Roads Maintenance crews.

Each of these projects required at least a King County Clearing and Grading permit and Washington State Department of Fish and Wildlife (WDFW) Hydraulic Project Approval. Each was required to go through the Washington State Environmental Policy Act (SEPA) process, and each was constructed under one or more U.S. Army Corps of Engineers Nationwide Permits (USACE). In each case, the design engineer and the ecologist/scientist were on site during construction to direct details of habitat structure placement.

Madsen Creek LWD Placement (1992)

Madsen Creek is a tributary to the Cedar River east of the City of Renton in King County. The headwaters of Madsen Creek are located on a rolling till upland used primarily for single-family residential development. The stream descends to the floodplain of the Cedar River via a steep-sided ravine. Severe incision in the upstream section of the ravine generates abundant sediment that degrades downstream habitat by filling pools, burying riffles, and impairing water quality.

This project had two goals. The first was to increase channel stability, thereby improving salmon habitat and reducing sediment generation. The second goal was to evaluate the efficacy of unanchored LWD placement as an enhancement technique in urbanizing stream channel environments.

Logs and rootwads were placed along the mainstem of Madsen Creek. The length of the logs used was approximately 1.5 times the bankfull channel width in order to limit mobility (Lienkaemper and Swanson 1987). Woody debris was placed in or adjacent to the channel and not anchored mechanically. The intention was to simulate natural wood recruitment and for the LWD to be incorporated into the channel by natural fluvial processes.

The site had no existing vehicular access. To minimize construction impacts and preserve the dense existing riparian vegetation, debris was transported into the creek using a Bell UH 1A helicopter equipped for long-line, external-load operation. Following placement, LWD was adjusted to final positions by a Washington State Conservation Corps (WCC) crew using hand-carried equipment. Each piece was then marked with metal tags and the location was mapped.

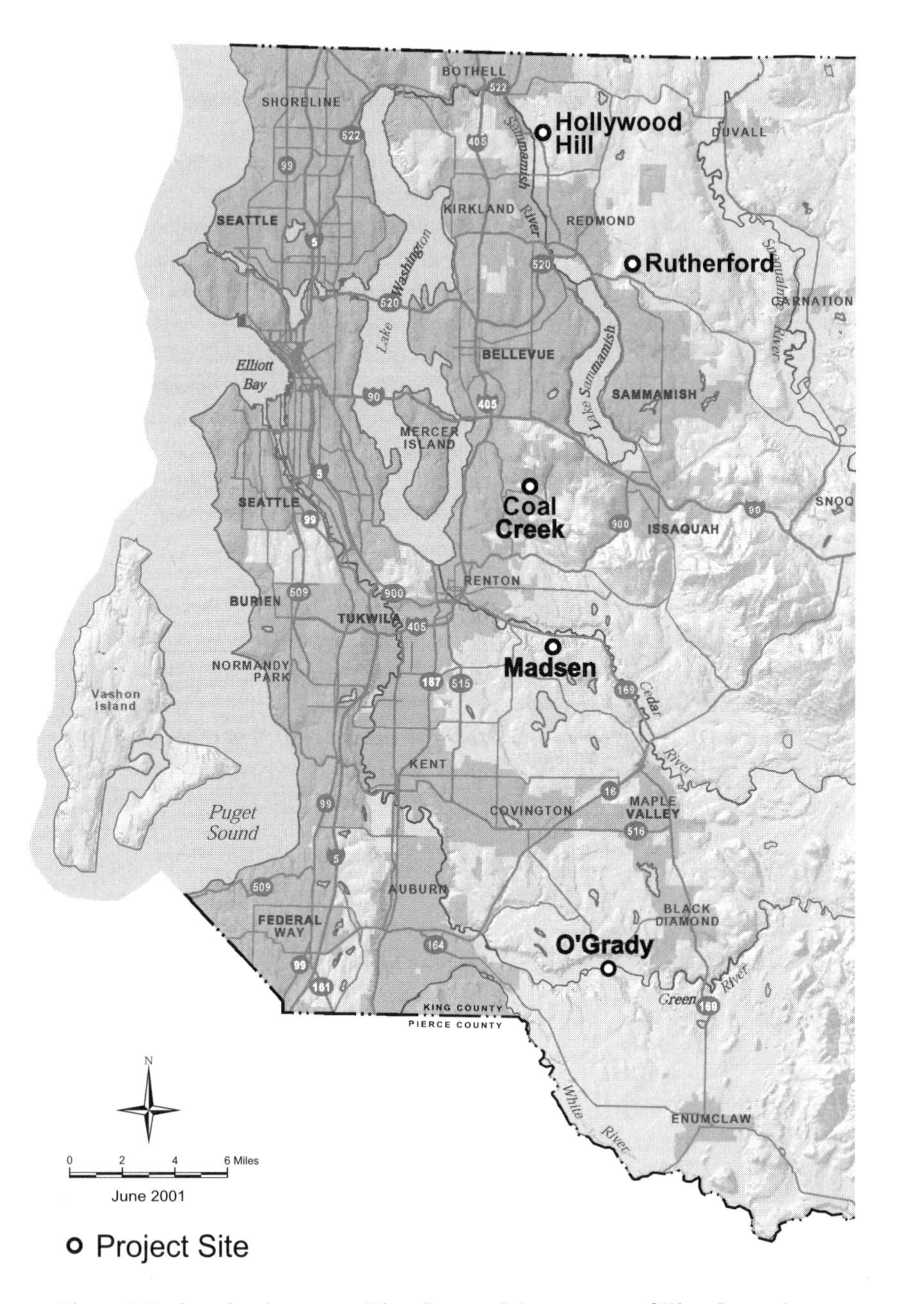

Figure 1. Project sites in western King County. (Map courtesy of King County.)

Table 1 Project information.

	Madsen Creek	*Hollywood Hill*
WRIA Tributary #	08.0305	08.0090
Basin Area (km)	5.4	2.4
Salmonid Use	Cutthroat	Cutthroat
Contributing Area (km^2)	4.8	2.2
Presenting Problem	Channel Instability	Channel erosion
Restoration Method	LWD placement	LWD and boulder placement
Project Length (m)	180	539
Stream Gradient through Project Reach	4.4%	3.3%
Project/mainstem channel length	4%	18%
Number of LWD pieces placed	38 (10 rootwads, 28 logs)	234 (220 rootwads, 14 logs)
Weight of Rock Placed (metric tons)	0	40
Construction Method	Helicopter	Mobile Crane
Year Built	1992	1995/96

Coal Creek	*Rutherford Creek*	*O'Grady Creek*
08.0268	08.0110	09.0107
19.0	5.5	0.4
Coho, Cutthroat	Coho, Cutthroat	Coho, Chum, Cutthroat
9.6	4.8	3.6
Bank erosion	Fish passage, channel incision	Fish Passage
Boulder and LWD placement	Channel reconstruction using gravel/cobble/boulder and LWD	Channel reconstruction using gravel/cobble/boulder and LWD. Excavation of floodplain swale and construction of channel.
134	185	80 – culvert phase 365 – alluvial fan phase
4.1%	3%	4% – culvert phase 4%–0.44% – alluvial fan phase
1%	5%	12%
89 (79 rootwads, 10 logs with rootwads)	41 (36 rootwads, 5 logs)	51 (logs w/rootwads) – culvert phase; >300 (snaller logs w/rootwads) –alluvial fan phase
136	1130	218 – culvert phase 20 – alluvial fan phase
Mobile Crane and Logging Skyline	Conventional construction equipment	Conventional construction equipment
1996/99	1998	1999/2000/2001

Table 2. LWD movement at Madsen Creek.

	Number of Pieces	*Max. Movement (m)*	*Min. Movement (m)*	*Mean Movement (m)*	*Median Movement (m)*
Logs w/o rootwads	25	94.3	0.0	16.1	1.8
Logs with rootwads	3	6.1	0.0	2.5	1.4
Rootwads	10	6.1	0.0	0.7	0.0

Post-construction monitoring revealed that 68% of the wood moved during high flows over the 5 years of monitoring. Table 2 summarizes the movement of several types of LWD. Wood was most mobile immediately after placement and appeared to stabilize over time.

Monitoring results did not definitely identify the effect of the project on channel stability or sediment supply. In some places, wood placement was associated with sediment deposition, but at other locations wood placement clearly resulted in local bank scour.

Evidence from tracking wood movement, along with observed patterns of channel erosion and sediment deposition, suggest that the channel responded dramatically to initial wood placement. Both the channel and the woody debris exhibited a high degree of instability immediately following placement, which appeared to decrease over time. It seems reasonable that the sudden introduction of a large quantity of debris into the channel was a disturbance that required a period of accommodation. This suggests that this type of intervention may initially cause a period of greater channel instability, followed by a period of gradually increasing stability as the introduced debris is incorporated into the channel structure.

Prior to placement of LWD, habitat in the project reach was homogeneous and dominated by fast water (mostly riffles). Following project construction, the number of habitat units increased from 3 to 8, and the volume of slow-water habitat increased by 13% (King County 2000). These results suggest that the Madsen Creek project produced a modest local improvement in instream habitat. This project, however, treated only 4% of the mainstem length of Madsen Creek. The project did not address the factors that led to degradation in this system, nor was it part of a coordinated basin-wide enhancement effort. Although interesting as a demonstration, the scope of the project was clearly too small to significantly affect the overall ecologic function of the Madsen Creek watershed.

Hollywood Hill Creek (1995-96)

Hollywood Hill Creek is a small tributary to the Sammamish River, which drains an upland area between the Cities of Redmond and Woodinville. The basin headwaters are on a rolling plateau with low-density residential development. The stream drops through a high-gradient, steep-sided ravine to the floodplain of the Sammamish River.

Flooding and out-of-bank sediment deposition by this stream damaged both public facilities and private property during severe storms over a number of years. The damage was a result of flooding, which in turn was a result of sediment accumulation and the consequent loss of conveyance capacity in the channel. Channel aggradation and flooding occurred, predictably, near the mouth of the ravine, in the area of the naturally-occurring alluvial fan. Although long-term deposition would be expected at this location, the rapid rate of sediment accumulation resulted from increased channel erosion in the ravine section. The channel expansion was a result of increased peak flows generated from impervious surfaces in the headwater area and from a deficiency of LWD in the channel.

The project was intended to reduce flooding and sediment deposition while improving habitat. Several options were considered including installation of a high-flow bypass pipe, upstream construction of a regional detention pond, and installation of a series of check dams in the channel. These solutions would have been costly and would involve construction and maintenance of additional artificial elements in this drainage system. The project team chose instead to modify the existing channel in the ravine to facilitate development of a natural morphology that would be self-maintaining and less vulnerable to erosion during high flows.

Installing woody debris and boulders in the channel was intended to provide structure that would allow development of a forced step-pool channel (Montgomery and Buffington 1997). The increased roughness in such a channel would dissipate kinetic energy more effectively than the existing plane bed channel, thus decreasing the energy available for erosion and sediment transport. The project involved only placement of wood and boulders; no direct reconstruction of the stream channel was proposed. It was anticipated that the channel would incorporate the added debris through natural fluvial processes, creating the desired channel morphology over a period of time. None of the debris was anchored or cabled in place, allowing installed debris to shift as the channel accommodated its presence. Woody debris was specified to be of a size such that it was unlikely to become fully buoyant during any anticipated flow event, and therefore unlikely to move a long distance downstream. Assuming woody debris has a specific gravity of 0.67, a log or stump will become buoyant when two thirds of its volume is submerged. Based on estimates of maximum flow depths and qualitative assessments of the geometry of the debris used for this project, it appeared unlikely that these pieces would ever become buoyant.

Most of the LWD pieces placed were rootwads, because experience had shown that rootwads were more likely to maintain close contact with the channel bed, especially in small channels in steep-sided ravines. Logs, in contrast, often end up in a position spanning the channel and not effectively interacting with flow. In addition, the root structure increased habitat complexity and tended to retain naturally supplied fine woody debris. The woody debris was generally placed in groups, in order to accelerate formation of functional debris jams. Boulders were included in a number of the groups to further stabilize adjacent LWD and to enhance bed stability in areas of locally steeper gradient.

To minimize construction impacts, especially to the dense existing riparian vegetation in the ravine, debris was placed in the channel using a crane operating from an adjacent roadway. The crane was able to reach over shrubs and between trees, placing wood and boulders with minimal disturbance to the channel and riparian zone.

There has been limited formal monitoring of the Hollywood Hills project. Prior to construction, a reconnaissance of the channel showed active erosion or erosion repairs along 48% of the channel in the project reach (Beck unpublished data). That reconnaissance showed no evidence of sediment deposition in the project reach.

Immediately following construction (1996), the entire project reach was photographed. Early the following year after a large flow event, photo-documentation was repeated (Figures 2 and 3). In 1999, a student from the University of Washington revisited the site and took a third set of photographs (Larson 1999). This sequence of photographs, along with observations of the site, clearly show that the extent of eroded channel banks has dramatically decreased, and that there are numerous areas of sediment deposition associated with installed debris. These observations suggest that the project has decreased the amount of sediment generated by erosion in the ravine. The subjective assessment of County biologists is that the increased bed stability, combined with the cover and complexity provided by the installed LWD, has likely improved instream habitat in the ravine.

Figure 2. Debris placement at Hollywood Hill, after construction. (Photo by John LaManna)

Figure 3. Same site as Figure 2, showing sediment accumulation following a large flow event. (Photo by John LaManna)

Coal Creek Channel Stabilization (1996-1999)

The Cinder Mine site is located on Coal Creek, a tributary to Lake Washington near the City of Newcastle. The project site was used for disposal of mine spoils from the Newcastle coal-mining district in the late 1800s and early 1900s. The spoils at the Cinder Mine site were unstable and subject to frequent failures, delivering sediment directly into Coal Creek. One factor contributing to instability of the spoils pile was erosion by Coal Creek into the base of the pile. The resulting sediment discharge was degrading instream habitat and causing flooding of residences downstream.

The goal of this project was to reduce erosion of the mine spoils in a way that enhanced local habitat conditions. The initial concept for the project was to construct a typical rock revetment along the toe of the spoils. On further review, however, the project team concluded that modifying the channel morphology could effectively dissipate stream energy and deflect flows away from the toe of the spoils while improving local instream habitat.

The design called for placing boulders and woody debris (mostly rootwads) to act as the key pieces to form a series of three boulder cascades in the channel. Boulders were sized to be stable in a 100-year recurrence flow event using a relationship developed by Costa (1983).

Boulders and rootwads were placed in bands perpendicular to the channel and extending across the floodplain. Woody debris was placed first, and then boulders were placed on, or downstream from, the LWD. It was intended that these debris bands would restrict flow and that sediment would be deposited upstream of them, leading to development of a stepped morphology. The use of wood and boulders together was modeled on the common occurrence of natural debris jams containing both boulder and woody debris in moderate- to high-gradient Puget Lowland streams.

The lowest debris band (Figure 4) was constructed in 1996 using a 50-ton mobile hydraulic crane that could reach the entire site via an existing access road. Because this placement method causes limited disturbance, construction was allowed in December, outside the normal instream construction window permitted by WDFW.

Funding constraints delayed completion of the project until 1999, when the two remaining upstream debris bands were installed. These sites were not accessible by crane, and a logging skyline was used to place the debris (Figure 5).

There has been no formal monitoring of the Cinder Mine channel stabilization project, and so assessment of the project is based on observations during post-construction site visits. Within a few days of completion of the lowest band, the Puget Sound area was subject to a major rainstorm, estimated to have been a 5-year 24-hour precipitation event in the Coal Creek Basin.

Figure 4. Lower debris band at Coal Creek during construction.

Figure 5. Logging skyline tower placing woody debris at Coal Creek.

Bedload mobilized during this event accumulated upstream of the complex, creating a well-defined longitudinal step in the channel profile and preventing erosion of the mine spoils in its vicinity.

There have been no flow events of sufficient magnitude to cause general bed mobilization subsequent to placement of the two upstream bands. As a result, the channel has not yet adjusted to their placement.

Observations of the site following the two construction phases show that the channel modifications appear to be successful in reducing erosion and slope failures in the adjacent mine spoils, keeping in mind that the two upper debris complexes have not experienced any large storms since construction. Fisheries biologists who have visited the site have expressed the opinion that the structures are not barriers to fish migration, and that they provide increased cover and habitat complexity compared with the unmodified channel.

Rutherford Creek (1998)

Rutherford Creek is a tributary to Evans Creek in the Bear/Evans basin, which drains to the Sammamish River in the vicinity of Redmond. Historically, Rutherford Creek had been an important spawning channel for coho salmon, with a median value of 335 spawners per mile in surveys conducted between 1976 and 1978 (King County 1989). Subsequently, however, the stream has been degraded by increased peak flows generated from residential development in the basin. In one area of previously high quality spawning habitat, the channel had incised as much as 1.5 m (Figure 6). Instream habitat in this area was severely degraded, and a 1-m-high erosional knickpoint at the upstream end of the project reach was a barrier to fish migration—a problem identified during development of the Bear Creek Basin Plan (King County 1990). In an initial attempt to halt the incision while minimizing construction impacts, check dams were constructed by hand in the project reach. The project failed shortly after construction because the hand-carried rock was too small to withstand flows. In much of the project reach, the bottom of the channel was becoming narrower as the incision deepened, resulting in increasing flow velocities. Cross-sections collected by King County as baseline monitoring of impacts of a large development proposed in the upper watershed confirmed the increasing erosion rate. The Basin Steward called the new data to the attention of the project team in February 1998. In this case, a project team had been assembled and assigned to the entire suite of projects identified in the Basin Plan, so it was possible to respond efficiently. The project was designed and constructed within the year. Due to Basin Plan recommendations, the project team and the Basin Steward were in place enabling King County to implement the project quickly.

The goals of the project were to restore fish passage, to control sediment mobilization, and to improve fish spawning and rearing habitat. In the project reach, the channel was severely degraded but the riparian zone was intact, with a second-growth conifer overstory and a diverse native shrub community. The project team considered use of a crane or helicopter to bring in material to rebuild the streambed while protecting the riparian vegetation, but lack of access and proximity to residences rendered these options infeasible. It had been noted that the stream goes dry for a period of time each summer in the project reach. Based on this information, the team decided to use the channel itself for bringing in equipment and materials. Streambed material was placed in the dry channel and used to fill it flush with the adjacent ground. Construction equipment was able to drive over the filled channel to bring in more streambed material and advance the fill (Figure 7). When the entire project reach had been filled, the construction equipment backed out, sculpting the channel as it withdrew. The roughened irregular channel, with its wider cross-section, was designed to slow velocities and improve salmon habitat. LWD was partially buried in the channel and banks to provide additional instream habitat complexity and channel roughness (Figure 8). To reduce riparian impacts, most of the overhanging vegetation

Figure 6. Rutherford Creek, showing incision prior to construction.

Figure 7. Rutherford Creek during construction. Note stump in this photo is the same one shown in Figure 6.

Figure 8. Rutherford Creek following construction. Stump on right is the same one seen in Figures 6 and 7.

was tied back and out of the way, rather than being cut. Following construction, the limited areas of riparian disturbance were planted with native vegetation.

Unlike the projects previously described, the Rutherford Creek project entailed completely reconstructing the channel through the project reach. The completed channel was intended to have a step-pool morphology. Boulder wedges, constructed of a well-graded mix that included rock up to 1 m in diameter, were incorporated into the fill at intervals of about 12 m (4 times the bankfull width) and were intended to act as catch-points in controlling the stream gradient. A cobble-gravel mix was used between the boulder wedges to shape the pools, which were expected to be reshaped by storm flows over time (Allan and Lowe 1997). Although the finished channel was entirely shaped by construction equipment, the channel was intended to be dynamic, and the design anticipated that large flow events would substantially rearrange the installed material. The size and distribution of the largest boulders and woody debris were intended to perpetuate the stepped morphology through subsequent channel evolution.

Monitoring has continued subsequent to project construction and includes several measured cross-sections (Figure 9). In the two years since project construction, there has been no evidence of notable channel expansion or the formation of knickpoints in the project reach.

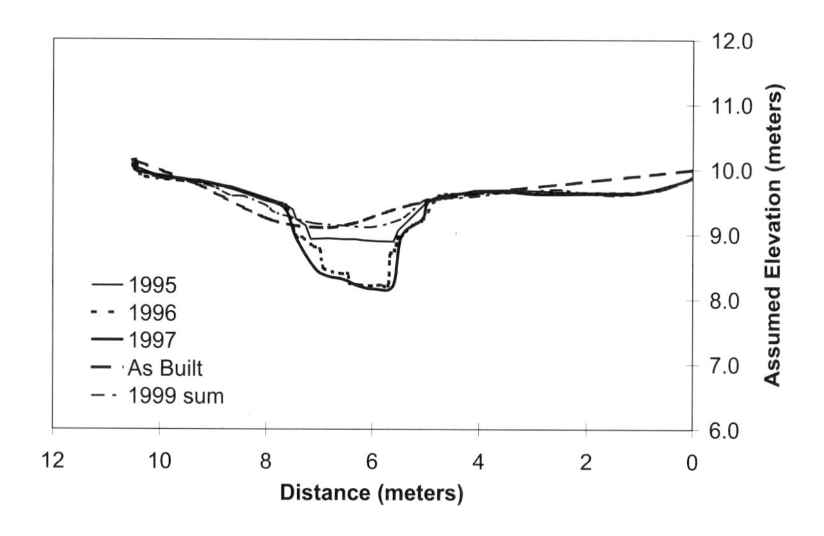

Figure 9. Rutherford Creek—Cross Section #173.

O'Grady Creek (1999-2001)

O'Grady Creek is a tributary to the Green River 63 km upstream of the river's mouth at the City of Seattle. The headwaters of O'Grady Creek drain a portion of the Enumclaw plateau, a rolling, agricultural landscape underlain by mudflow deposits from Mt. Rainier. The creek descends through a steep ravine system to the floodplain of the Green River where it has constructed a prominent alluvial fan. The fan and adjacent floodplain have a history of agricultural use. During this time a culverted road crossing was installed over O'Grady Creek, and the stream was diverted into an excavated channel along the margin of the fan. Over time, an incision developed at the downstream end of the culvert, leaving the culvert perched by 1 m and impassible to salmonids (Figure 10). King County Parks purchased the homestead over 30 years ago and manages it as open space. Sediment deposits in the excavated channel 100 m to 200 m downstream of the culvert have led to continuing out-of-channel discharge and frequent stranding of both juvenile and adult salmon on the pasture.

Stream enhancement on this property had two components: One component was to restore fish passage at the culverted road crossing by removing the existing 30 in culvert and replacing it with a 10-foot-wide bottomless concrete box culvert (Figure 11). Downstream of the new culvert, 60 m of

Figure 10. O'Grady road culvert before project.

channel was reconstructed to repair the incised reach, creating a step-pool morphology (Figure 12). Boulders, cobbles, and LWD were used to create irregular weirs in an adaptation of the design concept employed at Rutherford Creek the year before. The upstream sides of the weirs were sealed with fine-grained soil to create pools. It was anticipated that natural upstream sediment supplies would partially fill these pools over time, which limited the volume of gravel-sized sediment that needed to be imported to the project.

The second component addressed the problems resulting from episodic high sediment loads, which were leading to channel aggradation, out-of-channel flow, and fish stranding (Figure 13) on the pasture along the downstream reaches of the channel. Sediment deposition in the channel at the lower end of the alluvial fan was the inevitable result of the gentler gradient, and the project design anticipated that process. The goals of the project were to improve instream habitat and to improve riparian and wetland habitat by restoring a wooded floodplain within which the stream could move freely over time.

A broad swale was excavated and a new channel constructed along the axis of this swale. The swale and channel are intended to allow for long-term sediment accumulation and the resulting changes in the course of the stream channel while confining the channel sufficiently to maintain fish passage.

Figure 11. O'Grady road crossing immediately following construction.

Figure 12. Reconstructed step-pool channel downstream from new culvert on O'Grady Creek.

Figure 13. Adult chum salmon stranded due to channel aggradation on the O'Grady alluvial fan.

The swale varies in width from 30 m to 55 m and is typically about 1.5 m deep at the center. Approximately 10,000 m^3 of sediment was excavated to create the 370-m-long swale and inset channel. Woody debris was embedded in the banks of the new channel, and over 5,000 willow stakes were planted along the banks. Once the native vegetation is well established, this large project will sit unobtrusively in the landscape (Figure 14). Spoils from the excavation were placed in irregular mounds on the adjacent alluvial fan surface. These mounds provide topographic irregularity, which is intended to help define future channel alignments when O'Grady Creek eventually fills the excavated swale.

The earthwork was constructed in May 2000. Surrounding cottonwood trees set seed that month and blanketed the 9-acre site with seedlings. Community volunteers and the WCC planted about 3,300 additional native trees and 400 shrubs in November. The new channel was allowed to stabilize for about one year before stream flow was directed into it.

Originally the two components of this project, culvert removal and channel reconstruction, were to be constructed simultaneously. The design team began work in February 1999 with the goal of constructing both components that September. However, Puget Sound chinook salmon were listed as a threat-

Figure 14. Aerial view of the completed O'Grady stream relocation.

ened species under the Federal Endangered Species Act (ESA) in March. Both projects were therefore subject to Section 7 review under the ESA and required preparation of substantial supporting documentation. The ESA process required additional review by the USACE as well as concurrence by both National Marine Fisheries Service and the U.S. Fish and Wildlife Service. Because of increased liability under the ESA, all projects subject to the ESA, including O'Grady, were required to go through an additional detailed internal King County review process. The added process entailed a substantial level of effort that made it difficult to meet an already ambitious schedule. USACE concerns about wetland impacts caused the channel reconstruction work on the alluvial fan to be delayed until 2000.

The permitting complexities of this project highlighted the benefits of an interdisciplinary design team. Design input from staff with expertise in fisheries biology, wetland ecology, landscape architecture, and geomorphology resulted in a project that withstood detailed ESA review with no significant modifications. It must be noted that this interdisciplinary process carries costs as well as benefits to the project. Each hour of staff time, charged as a fully loaded rate, costs the project an average of $80. The stream/wetland system at O'Grady Park is complex, and King County works within an intricate permitting environment. The alluvial fan work, in particular, cost considerably more than was originally anticipated. Some of the additional costs are attributable to the accelerated schedule and to staff changes, and some must be allocated to responding to the expanded permitting process.

The culvert replacement and associated downstream channel reconstruction was completed as originally planned in September 1999. The following November, both coho and chum salmon adults were observed spawning in the project reach, and a pair of coho were observed on a redd 26 m upstream of the culvert. As of May 2001, the project remains passable, with no indication of renewed incision. Pools have partially filled in as expected, resulting in a step-pool morphology. O'Grady Creek was allowed to occupy the relocated channel in May 2001 and was utilized immediately by juvenile salmon. The vegetation is doing well, though deer predation is a problem. Plant maintenance (watering as required and invasive species control) and project monitoring will last five years. The focus of the monitoring is plant survival and documentation of fish use.

Discussion

Each of the projects described above share a central characteristic with other stream enhancement work conducted in King County and, we believe, with most such work conducted in the Pacific Northwest. These projects attempt to intervene in a stream system without a complete understanding of the complex physical and biological processes that make up that system, and therefore without being able to fully predict the results of the intervention. In a degraded watershed, it is generally possible to list numerous factors that may have had a role in degrading instream habitat, but it is seldom possible to clearly resolve the relative contribution of the various factors. Likewise, it is typically possible to identify numerous credible potential interventions, but there is little objective guidance in evaluating which are likely to be most effective at achieving habitat benefits. This is not a condemnation of the design process, but a recognition of our limited ability to understand stream function in detail. We must therefore rely on incomplete information in project selection, design, and assessment. This is the fundamental challenge in designing and implementing effective stream enhancement projects. Given that enhancement project design cannot follow deterministically from an assessment of stream conditions, experience in King County suggests that there are strategies that can improve the chances of success in stream enhancement.

Project Teams and Project Construction

Relatively independent teams comprised of individuals with complementary areas of expertise provide the best chance of developing and executing appropriate enhancement activities. In King County, such teams include members from some or all of the disciplines of engineering, wetland ecology, fisheries biology, geomorphology, and landscape architecture. Ideally, such a team is involved in initial project definition and retains responsibility for the project through design, construction, and assessment. The team should have the support necessary to fully investigate the potential project and the freedom and flexibility to consider a wide variety of possible design alternatives. The team should include members with construction experience and members should be actively involved in project construction.

For King County, there are typically two options for constructing an enhancement project. Work can be contracted out to a general contractor or it can be built by King County construction crews. When work is contracted out, a general contractor bids on the work based on a detailed set of project plans and specifications. The contractor is bound to construct the project exactly as shown on the plans at the bid price. Any changes during construction are subject to the contract terms for negotiating Change Orders and may significantly increase the construction cost. This process can be expensive and cumbersome for constructing habitat enhancement. Many habitat projects involve placement of natural materials in natural channels. Preparing plans that show the details of such placement is prohibitively expensive and allows little flexibility to respond to unexpected on-site conditions encountered during construction.

When County construction forces build enhancement projects, design plans can be much more conceptual (and less costly), as design team members are typically on-site to direct construction. Charges are based on time spent and materials used rather than on a fixed bid, so field changes are straightforward. Over time, County construction crews have become skilled in some of the unique construction techniques and equipment that are used for enhancement work, and they are able to act as partners in project implementation. The design team frequently consults construction supervisors during project development for advice on construction feasibility and equipment. This method of project management is similar to "design-build," in which a single organization is responsible for a project from concept development through construction. The advantages of a design-build model for habitat enhancement have been previously recognized (Moses et al. 1997).

Utilizing Natural Stream Systems as Templates

As our understanding of stream systems is incomplete, it is prudent to design projects using native materials and copying the morphology of natural channels. The character and arrangement of elements in a stable natural channel represents a dynamic equilibrium between the various processes active in that channel. If we identify appropriate natural templates, and then mimic these templates in our designs, we have some chance of producing a system with that desired equilibrium even if we lack full understanding all of the interactions that contribute to it. The inherent complexity of natural materials provides habitat niches for multiple species and life stages.

Assessment and Communication

Evaluating the success of a given enhancement project, and using that information to inform future designs, is a critical but still inconsistently-applied element in stream enhancement. Developing a monitoring plan that objectively assesses the ecological benefits of stream enhancement projects has proved an elusive goal for King County. This is true in spite of the fact that monitoring plans are developed and carried out for most enhancement projects.

Several factors have made substantive monitoring difficult. First, most urban stream systems are subject to multiple insults to healthy ecological function over much of their stream length. Projects are limited with respect to both the problems they are intended to address and the length of channel they affect. As a result, the beneficial effects of even a successful project may be small and difficult to detect.

Second, many King County projects include the addition of woody debris or boulders to the stream channel with the expectation that these elements will be incorporated into the system over time to create improved habitat. Such incorporation may take years and may depend on the occurrence of large, infrequent flow events. Most projects include planting native vegetation as an integral element. Growth of such plantings to the point that they substantially enhance habitat may take decades. Given large variability in many characteristics of natural systems, the likely delays for native biota to recolonize and fully utilize the improved habitat, and the sometimes subtle effects of a given project, it is often difficult to collect sufficient data to reach statistically defensible conclusions about project effects, even if those effects are real (Chapter 12).

Finally, monitoring is expensive, and it is hard to justify the expense when the cost of a multi-year monitoring program can approach the construction cost of a project.

The preceding are not intended to discourage monitoring but to point out some limitations on the extent to which post-construction monitoring can clearly determine a project's "success."

Watershed-Level Assessment and Land-Use Management

If physical stream enhancement projects are to be an effective element of a wider ecological enhancement effort, then the selection and design of such projects must be consistent with watershed-level enhancement planning. This planning in turn must be informed by watershed-level understanding of the relevant physical and biological conditions and of how future land-use changes will affect these conditions (Chapter 8).

King County has recognized the need for such informed and coordinated planning. This effort was initiated through the Basin Planning program and continues through the ongoing WRIA (Water Resource Inventory Area) planning program. Many of the enhancement projects that are carried out by King County originate through these watershed-level programs. Many other projects, however, are still designed and implemented without such a perspective. Only one of the five projects described above (Rutherford Creek) was developed through a watershed-level planning process.

Conclusions and Recommendations

King County has extensive experience and a broad spectrum of technical expertise in stream enhancement. Nonetheless, effectively identifying, designing, and constructing projects that demonstrably contribute to watershed-level resource enhancement remains problematic. This is true for a variety of reasons.

First, The resources available for physical enhancement within a watershed are often minute compared with the magnitude of the resource degradation in that watershed. Well-conceived, well-designed, and well-executed projects of limited scope may produce a positive but still ecologically insignificant benefit.

Second, the cost and time required to go through the permitting process for projects that include any substantive work in or immediately adjacent to a stream channel may render a valuable enhancement project prohibitively expensive compared to the potential benefits anticipated.

Moreover, our detailed understanding of the fluvial and riparian ecosystem is incomplete, as is our understanding of how the various manifestations of watershed degradation affect specific ecological functions. This incomplete understanding makes it difficult to use limited resources to identify and implement the most effective interventions.

In light of these constraints, we have concluded that the chances of effecting a successful enhancement project are increased if:

- Enhancement projects are conceived, designed, constructed and evaluated by a self-directed multi-disciplinary project team. The team members should be directly involved in working with construction crews in determining construction techniques and directing construction activities using a design/build project management approach.
- Projects are modeled on natural templates, are constructed using native materials, and use the least invasive construction method feasible. This design approach acknowledges the dynamic character of natural stream systems and accounts for that dynamic character in design and assessment.
- Each project is treated as a learning opportunity, and significant findings are documented and communicated to benefit subsequent design (Chapter 9).
- Stream enhancement efforts are carried out in the context of a watershed level assessment of causes of habitat degradation and in concert with watershed-level land-use management (Chapter 8).

References

Booth, D.B. and C.R. Jackson. 1997. Urbanization of aquatic systems: Degradation thresholds, stormwater detection, and the limits of mitigation. *Journal of the American Water Resources Association* 33:1077-1090.

Costa, J.E. 1983. Paleohydraulic reconstruction of flash-flood peaks from boulder deposits in the Colorado Front Range. *Geological Society of America Bulletin* 94:986-1004.

King County Department of Natural Resources. 2000. 1999 Annual Report, Wastewater Treatment Division, King County Department of Natural Resources, CIP Monitoring Program. Seattle, Washington.

King County Surface Water Management Division. 1990. Bear Creek Basin Plan. Seattle, Washington.

King County Surface Water Management Division. 1989. Bear Creek Basin, Current and Future Conditions Report. Seattle, Washington.

King County Department of Public Works. 1987. Basin Reconnaissance Program Summary. 3 Volumes. Seattle, Washington.

Larson, M.G. 1999. Effectiveness of large woody debris in stream rehabilitation projects in urban basins. Master's thesis. University of Washington. Seattle, Washington.

Lienkaemper, G.W. and F.J. Swanson. 1987. Dynamics of large woody debris in old-growth Douglas-fir forests. *Canadian Journal of Forest Research* 17:150-156.

Montgomery, D.R. and J.M. Buffington. 1997. Channel-reach morphology in mountain drainage basins. *Geological Society of America Bulletin* 109:596-611.

Moses, T., S. Morris, and D. Gorman. 1997. Institutional constraints to urban stream restoration, Parts 1 and 2. *Public Works* June pp. 36-39 and July pp. 40-43.

16. Use of Long-Line Cabled Logs for Stream Bank Rehabilitation

Roger A. Nichols and Sallie G. Sprague

Abstract

Declining salmonid populations in the Pacific Northwest motivates re-examination of conventional river engineering practices. We discuss experiences with the novel approach of using long-line cabled logs to decrease river bank erosion without the detrimental effects on fish habitat seen after extensive riprap treatments. The technique involves cabling and anchoring logs in slack water eddies along the eroding banks so that they remain floating during bankfull conditions. At three sites where the floating log structures have been in place for two to seven years, long-line cabled logs protected eroding stream and river banks. As flood waters subsided during high flow events, the long-line cabled logs trapped additional floating woody debris and promoted sediment and gravel deposition around the captured wood. The main channel flow moved away from the eroding bank as deposition occurred. In addition, channel complexity was increased and potential fish habitat was improved or created on all three treated sites. Essential to the success of this technique are matching the size of the logs, the length of the cables, and the weight of the ballast anchors with the velocity of the surrounding water. Moreover, the technique is economical because it utilizes river transportation of material instead of importation of large trees.

Introduction

Roads placed in close proximity to rivers can degrade fish habitat through bank failures and loss of woody debris input, illustrating conflicts between human access and fish habitat. Sometimes, relocating the road at risk can solve both problems. Where relocation is not possible, large rock or riprap is often used to line eroding stream banks. This approach not only has a dismal long-term success record for protecting riparian land but also limits or degrades fish habitat (Beamer and Henderson 1998; Lister 1998; Peters et al. 1998; DeHaven 2000). With greater emphasis on fish habitat in the new century, alternative methods of bank protection are needed (Slaney and Zaldokas 1997).

Large log jams were once a common and important feature of rivers in the Pacific Northwest before land-use changes in the last century removed much of the mature forests (Abbe and Montgomery 1996; Chapter 4). Removal of large woody debris may affect stream channel stability (Bilby 1984). The smaller trees in and along large rivers today are readily washed out during high flows. In contrast, large trees that fall into streams remain to capture additional material, form log jams, and add complexity to the channel. Today the cost associated with procurement, transport, and installation of large trees into log jams that mimic those found historically in local rivers is high and often prohibitive for landowners. A technique that accumulates the small wood remaining in the river systems would not only avoid the cost and difficulties associated with the transport of large trees but also allow the stream or river to (re-) develop its own channel structure over time.

Since 1993, we have used a technique for protecting eroding stream banks in Skagit County, Washington, in order to prevent damage to roads that are located on the margins of stream channels. Faced with the continuing failure of extensive riprap repairs of river banks to protect adjacent property, landowners and government agencies have placed long-line cabled logs along the eroding banks to help retard bank erosion of the streams and rivers in question. The key to the success of this technique is the capture and utilization of the small woody debris that remains in a watershed or river system after the removal of large trees during intensive timber harvest, farming, and land development. We describe the long-line cabled log technique as it developed over three sites and seven years as part of larger efforts at each site.

Design

The design challenges for these projects were: (1) to reduce the erosive energy of water on adjacent stream and river banks, and (2) re-create the channel complexity seen in rivers with natural log jams in a way that would be both self-sustaining and economically feasible for landowners. Because large trees are lacking in many river systems after timber harvest and are frequently expensive to acquire ($2000 to $2500 per tree), logs cabled to ballast anchors were used as a substitute for large fallen trees (Figure 1). Using cables long enough to maintain the anchored logs at the surface of the water at bankfull conditions would minimize the resistance the log presented to the flowing water and allow the logs to collect floating woody material as the water level falls after high flow events (Figure 2). Long-line cabled logs placed at specific locations along eroding banks should mimic the presence of a tree fallen into the channel following erosion of the bank, slow water flow in the immediate area of the logs, deflect water flow away from the eroded bank, and allow natural bank stabilization processes to work in the system.

We also hypothesized several advantages to this approach. Placement of long-line cabled logs along a river bank would allow the river system to recruit additional materials to the site after each high flow and use the river's energy to change the channel condition. Project costs would be lower than those of comparable methods by using logs smaller than full-size trees, and natural channel processes would be simulated. Subsequent habitat development would more closely reflect that seen in natural systems than follows riprap bank treatment but occur over a shorter time frame than if one waited for adjacent trees to mature and fall in the channel.

METHODS

Logs were attached at two points, using cables long enough to allow the logs to remain at the water's surface at bankfull conditions, and placed along the eroding banks. Cables from each log were attached to ballast material (rocks or concrete ecology blocks) on the bottom of the channel. Essential for the success of long-line cabled logs were:

1) understanding or creating appropriate hydraulic controls for the specific project reach,
2) allowing enough time for the river and long-line cabled logs to interact,
3) calculating the length of the cables for bankfull depths at the site, and
4) matching the size/weight of the log with the appropriate weight of ballast for the anticipated bankfull water velocity.

Plan and cross-sectional views of single long-line cabled logs show their placement relative to the stream bank and the bankfull water level (Figure 1). The desired log position is maintained by placing the ballast rocks upstream

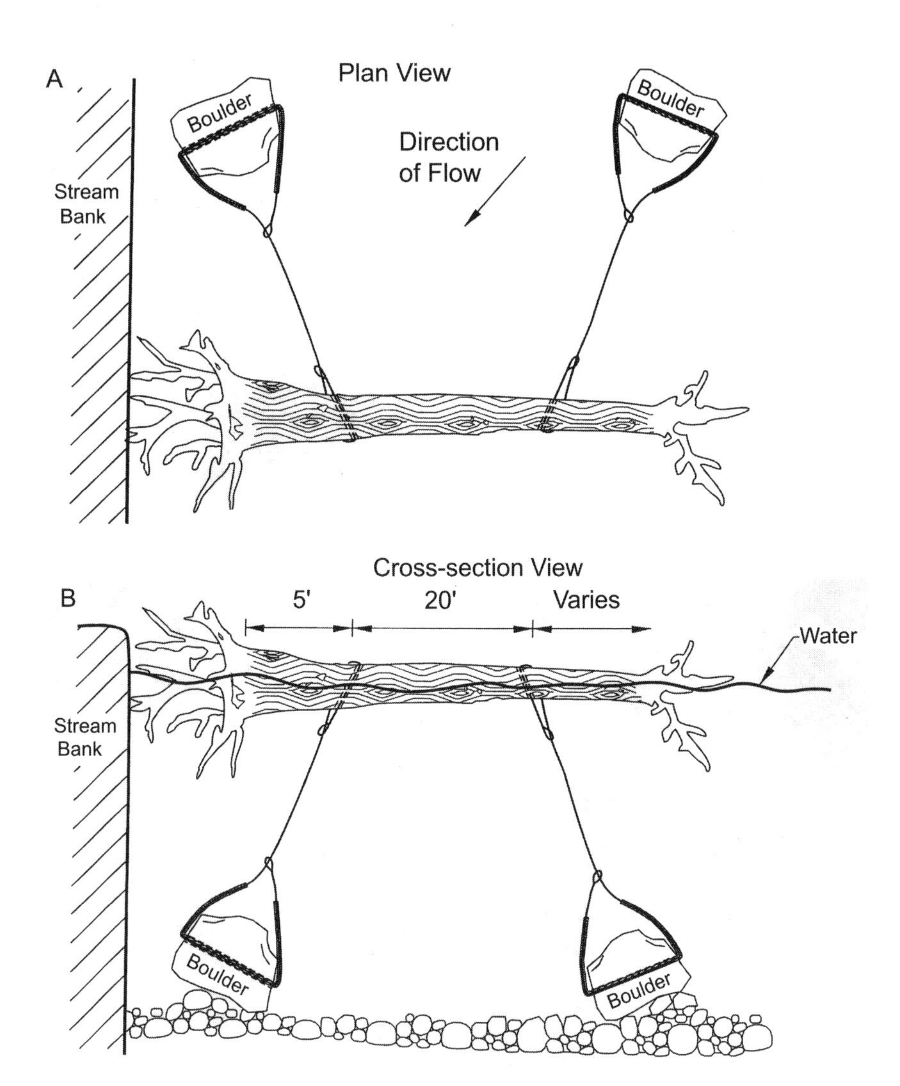

Figure 1. Long-line cabled log placement details. (A) Plan view showing placement of log and ballast in relationship to the current and the stream bank. Arrow indicates direction of stream flow into the eroding bank prior to installation of the long-line cabled log. (B) Cross section of long-line cabled log position at bankfull conditions. Cables are long enough for the log to remain at the water surface at bankfull. Logs are placed adjacent to but not embedded in the bank.

Figure 2. Long-line cabled logs in place. (A) Long-line cabled logs installed two and three years prior to this photo are visible along the formerly eroding bank. The logs in the foreground are at the downstream end of a 330 m section of eroding bank and have collected 1/2 to 1 m of gravel. Deposition of sediment and gravel extend at least 50 m from the former eroding bank (on the right side of the photo) to the new channel which is out of the photo to the left. Cable wraps and staples are visible around the logs and marked by arrows. (B) Three long-lined cabled logs are shown with captured wood and sediment two years after installation. Arrows indicate wrapped cable anchoring logs to ballast. One ballast rock is visible (#) in the lower right of the photo. Gravel deposition is visible downstream from the installed logs at the top of the photo.

from the log and more widely separated than the holes drilled for attaching cables to the log. Strength and flexibility of the cable need to be appropriate for the conditions at each site. For smaller streams (channel width <20 m, [66 ft] or discharge <142 m^3/s, [5000 cfs]), 13 mm (1/2 in) cable with a breaking strength of 9,072 kg (20,000 lb) was used. In larger systems, 16 mm (5/8 in) cable with 18,144 kg (40,000 lb) breaking strength provided additional strength while retaining the needed flexibility for proper installation and function.

Log lengths and structure location for each project were determined in a manner similar to that for spur deflectors. Log lengths vary on small streams from one-quarter to one-half the width of the stream (Klingeman et al. 1984), and on large streams, from 10 to 15% of the bankfull channel width on straight reaches, long radius turns and braided channels (Richardson and Simons 1974). Hydraulic conditions and stream erosion varied widely from one site to another due to the differences in local stream characteristics, such as flow conditions, bed and bank material, channel geometry, and perpendicular projection of woody material into the channel (Klingeman et al. 1984). Based on detailed local flow patterns and soil conditions for these projects, log lengths were between 6.2 and 8.2 m (20 to 30 ft) and diameters 0.5 to 1.0 m (18 to 36 in). Actual log sizes used were determined by the availability of woody material near the site or by importing wood of pulp market quality. Logs were specified to be >60% sound wood to meet the 25–50 year project design life estimated for re-establishment of natural processes. Cable lengths were determined by the scour depth of the channel at bankfull. Ballast weights for the logs and the stream conditions were determined according to D'Aoust and Millar (1999, 2000) using estimated stream flow velocity data for each site.

In practical terms, the weight of the ballast that could be placed in a channel more than a meter deep was limited by the reach and lift capacity of the readily available equipment. For a 50–ton gross weight track excavator to place its maximum load 7 m from its base, the ballast material was limited to boulders of 1.9 m^3 (2.5 yd^3) or approximately 1.4 m (4.5 ft) in diameter. This limit in ballast weight meant that long-line cabled logs of the size specified for the projects would remain in place only if the water velocity remained below 2.7 m/s (9 ft/s). To accommodate the maximum water velocity dictated by the maximum load and reach of the equipment, it was necessary to create slack-water eddies along the eroding banks. In these projects, rock deflectors (hard points) were installed along the eroding banks, and long-line cabled logs were then placed in the slack water below each deflector.

Deflectors are structures extending outward from the streambank into the channel. Generally constructed of large, 0.765–1.53 m^3(1–2 yd^3), riprap rockfill, deflectors are keyed into the bank a minimum of 3 m (10 ft) and 1.6 m (5 ft) below the channel bed. After installation, deflectors are partially exposed at most water levels and slope at 6:1. This slope reflects the typical contour of stable gravel bars in rivers without rapid erosion and deposition. The main effects of a deflector are on local sediment scour, deposition, current velocity and flow patterns in a river. Deflectors reduce and break up the concentration of current near the bank, redirecting the water from the portion of the deflector that is in the channel (Klingeman et al. 1984; Sutton Corp. 1985). Deflectors are most effective at very high water, such as during floods. Deflectors on the Lower Sauk River site varied in length from 10 m (35 ft) to 36 m (120 ft); those at Lower Finney Creek and Ovenell's Slough were approximately 10 m (35 ft). Spacing between deflectors was determined by experience and judgment of the project designer using information from Sutton Corp. (1985).

Site Descriptions

The three sites in Skagit County, Washington, treated with long-line cabled logs represent very different river and flow conditions (Figure 3).

Lower Sauk River

The Lower Sauk River is free-flowing and has a base discharge of 453 m^3/s (16,000 cfs, USGS Station 12189500 discharge data). Peak discharges have been as high as 2773 m^3/s (98,000 cfs, 1980, flood of record). At the project site, the channel was >100 m (330 ft) wide, and had a pool depth of approximately 1.5–3 m (5–10 ft), as measured by sacrificing a 1.9 m (6 ft 4in) fisheries biologist in waders.

After flooding removed most of Bryson's field and threatened the fill for the highway and the approach to the bridge, the Lower Sauk River site was treated in 1993. Additional rock was placed along the toe of the eroding left bank of the 1993 channel, and a dozen paired long-line cabled logs downstream from rock deflectors installed in the 1993 channel (Figures 3 and 4a). Paired logs (y-logs) were installed with the lower log perpendicular to the eroding bank to help move the channel away from the bank and the upper log positioned to direct the main water flow to the center of the highway bridge. Log pairs were cabled together and took two concrete ecology blocks per log with long-line cables to maintain their orientation to each other, the bank, and the bridge. Through their position and buoyancy, the y-logs were designed to trap floating wood mobilized during high flows as the water subsided and to deflect water flow away from the eroding bank and toward the

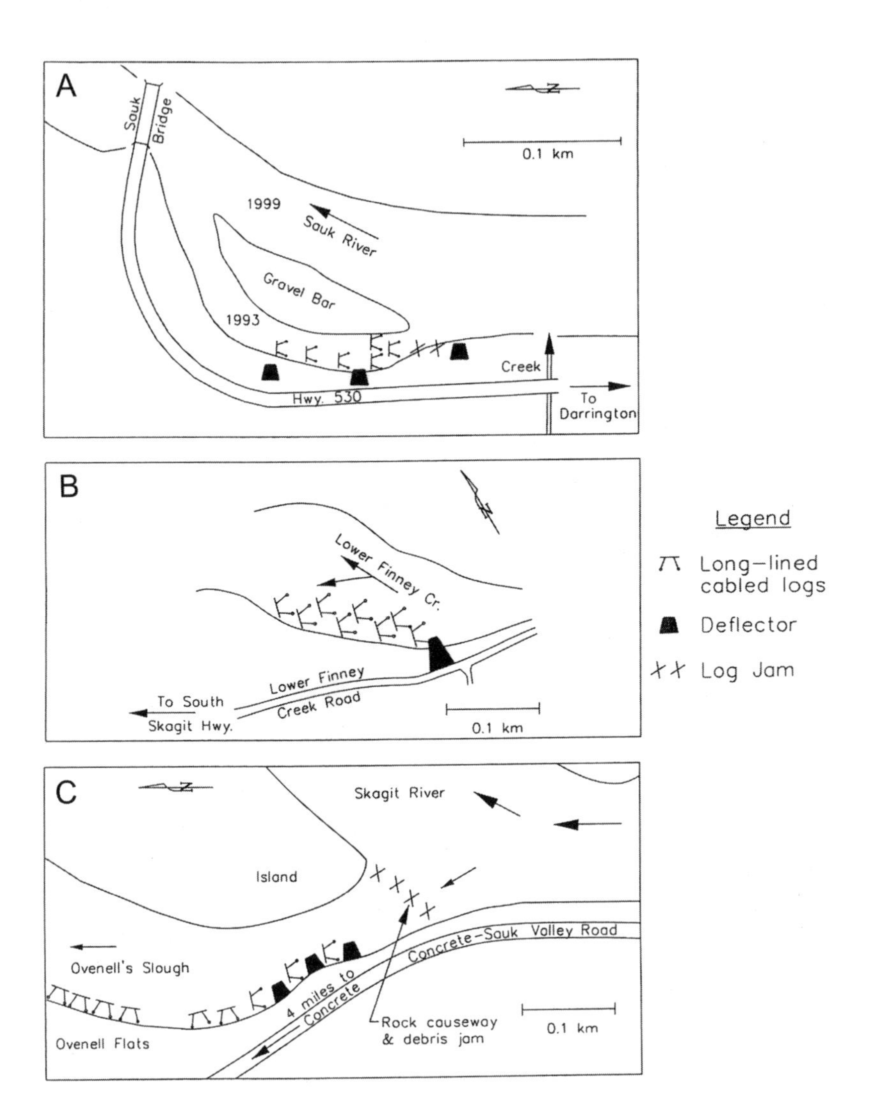

Figure 3. Site diagrams. (A) Lower Sauk River site upstream from the bridge on Highway 530. (B) Lower Finney Creek received a staggered array of long-line cabled logs in 1997 over 330 m of eroding bank to protect the road and a stand of conifers between the road and the creek. (C) Ovenell's Slough received emergency bank stabilization in May, 1998, after the rock causeway debris jam broke during the winter of 1997–1998.

next hydraulic opening. The y-logs would collect debris, thus adding to the structure and roughness strengthening the left bank.

Lower Finney Creek

Lower Finney Creek drains steep valleys intensively harvested for timber and had daily discharges between 0.45 m^3/s (15.9 cfs) and 82.9 m^3/s (2931 cfs) between 1941 and 1948. Lower Finney Creek is free-flowing and subject to flashy discharges after heavy rainfall (personal observation). At the project site, the channel was 12–75 m (40–250 ft) wide with an active flood plain and a pool depth of approximately 1.5 m (5 ft).

Lower Finney Creek received a staggered array of 40 long-line cabled logs along 308 m (1000 ft) of eroding bank (Figures 3 and 5a) in 1997. In 1998, a second array was installed over the first one, which was buried by gravel and sediment deposition during one winter season. The 1997 long-line cabled log structures were buried under accumulated wood, gravel, and sand—a marked contrast to the pre-installation incision observed along this reach. A rock deflector at the base of the road marked the upstream hydraulic control and upper end of the array.

Ovenell's Slough

Annual peak flows are high on the Skagit River and ranged from 990 m^3/s (35,000 cfs) to 4188 m^3/s (148,000 cfs) between 1980 and 1999. As a side channel, Ovenell's Slough receives the maximum stream flows of the main Skagit River channel and often sees a 6.7 m (20 ft) fluctuation in depth during an average year. Ovenell's Slough is also subjected to surge flows from debris jam failures at its upstream end, which can create a 1.5 m (5 ft) head wave. The channel width varies from approximately 20–30 m (66–100 ft) along the project reach with a scour depth of approximately 1.5–4.5 m (5–15 ft) as determined by site survey.

Ovenell's Slough received emergency bank stabilization in May, 1998, after the rock causeway debris jam broke during the winter of 1997–1998. The old rock causeway snags wood during normal river flows and creates a debris jam. When the jam reaches a certain size and can no longer resist the energy of the water during elevated river levels, a jam break occurs. The subsequent flood flow through the slough, caused serious erosion of the left bank and threatened the highway. Long-line cabled logs were placed below each of three rock deflectors at Ovenell's Slough and along the eroding bank downstream from the deflectors (Figures 3 and 6a). Farther downstream the long-line cabled logs were installed on their own to help stabilize the channel as it adjusted to the new deflectors. Essential to the decision to not use toe rock along the edge of the field was the cooperation of the landowner, the US Forest Service. One of the possible river adjustments that was envisioned was the development of a new channel through a field purchased to protect habitat as part of the Wild and Scenic River designation for the Skagit River.

Installation

Spacing of the long-line cabled log structures in a bank protection scheme is a function of length, angle, and permeability, as well as a channel bend's degree of curvature. Calculations for positioning log structures were in accordance with those described for spur or deflector placement (FHA 1983; Sutton Corp. 1985). Final spacing between deflectors was determined by experience and judgment of the project designer at <2.5 times the structure's extension outward from the streambank into the channel.

Landowners (private timber company) or governing agencies (State Department of Transportation, County Public Works Department) supervised crews during the installation of structures. None of the supervisors or their crew members were familiar with the installation of log structures before their project began. Logs were at least 60% sound wood, >45 cm (18 in) dbh, and either easily accessible nearby or financially feasible for transport on standard log trucks. Boulders (Ovenell's Slough and Lower Finney Creek) or concrete ecology blocks (Lower Sauk River) were used as ballast (anchors) for the logs. Cables of 19 mm (3/4 in), >27,200 kg (80,000 lb) breaking strength (Lower Finney Creek, 1997 array), 13 mm (1/2 in), 9,072 kg (20,000 lb) breaking strength, or 16 mm (5/8 in), 18,140 kg (40,000 lb) breaking strength (Ovenell's Slough) were used to attach logs to ballast material. Boulders were drilled to allow attachment of eyed cable. Logs were either drilled so that eyed cable ends could be passed through the logs and shackled back to themselves (Ovenell's Slough), or wrapped with three turns of cable, and were cinched tight and stapled to the log (Lower Sauk River and Lower Finney Creek).

The weight of the ballast material that could be placed in the channel was limited by the capacity of the largest commonly available track excavators (50.8 metric ton or 100,000 lbs gross weight). To place ballast 6.2–7.7 m (20–25 ft) from the base of an excavator perched on the tilted surface of a rock deflector, boulders could not exceed 1.9 m^3 (2.5 yd^3) or approximately 1.4 m (4.5 ft) in diameter. The effective size of ballast material can be increased by "necklacing" several rocks together, although this can be logistically difficult at some sites. At Ovenell's Slough, a model 400 Hitachi excavator was used to construct the deflectors and to place the long-line cabled logs in the channel. Three separate actions were needed to place first a ballast rock, then the log, and finally the second ballast rock in the proper location and orientation in the channel or along the bank. Ballast was buried at the scour depth of the channel (Lower Sauk, Lower Finney 1997 array) or set on the gravel bar (Lower Finney 1998 array). Due to the changing water depth and velocity in May along the Skagit River, the long-line cabled logs attached to their ballast rocks were dropped or flung into their approximate positions in Ovenell's Slough by the excavator.

Project sites were observed and monitored by repeat visits from the authors and the landowners. Photos were taken to document changes in the channel and eroding banks. Deposition of material was calculated from measurements of the area and estimates of the depth of new gravel, sediment, and wood. Interest in the long-line cabled log technique prompted fisheries biologists from Washington State Department of Fish and Wildlife and U.S. Fish and Wildlife Service to review all three sites.

Results

The long-line cabled logs performed as anticipated on all three sites. They floated at the water surface at bankfull conditions, remained in place, and collected wood carried downstream during high flows. Sediment deposition occurred downstream of recruited woody debris, eventually changing mobile wood to fixed structures on gravel bars. The main channels moved away from the eroding banks. Two years after installation of long-line cabled logs along

Figure 4 (facing page). Lower Sauk River along Bryson's Field. (A) 1993: Before the project began, the new channel on the left bank of the river was clearly visible through the area that was Bryson's field. The main river flow had moved to this new position from the former channel, which was immediately out of view along the right edge of the photo. The Washington DOT purchased Bryson's field to protect the road. A pair of conifers along the highway is indicated by ** in each photo as a reference for the viewer. The entrance to the 1993 channel is marked by >. (B) 1995: Two years after installation of the long-line cabled logs in the 1993 channel, the main water flow had moved back to its pre-1993 position. The 1993 channel had filled in significantly. A gravel bar with deposited woody debris was visible along the left bank of the main channel. (C) 1998: Five years after installation, vegetation was established along the left bank of the Lower Sauk River, almost obscuring the entrance to the side (1993) channel.

A
**
1993
B
**
1995
→ Sauk River →
C
**
1998
Sauk R →

the Lower Sauk River, the 1993 channel was reduced to an overflow channel (Figure 4B). Wood and gravel had collected on site, and the entrance to the 1993 channel was barely visible during the summer months. Recruitment of wood by the long-line cabled logs was more effective after bigger storms than after smaller ones along the Lower Sauk River. By 1998, the 1993 channel was almost invisible through the established vegetation (Figure 4C). Inspection of the site revealed pools among the vegetation, several of which were used by juvenile coho and spawning chum salmon in the fall of 1998. The orientation of the paired long-line cabled logs proved difficult to maintain despite the dramatic success of the structures to accumulate woody debris as anticipated. The increased bank roughness that developed from the trapped wood appeared sufficient to deflect the water flow so that long-line cabled logs were installed singly in subsequent projects.

The long-line cabled log arrays installed at Lower Finney Creek along the left bank of the 1997 channel (Figure 5A) were visible on top of and embedded in a gravel bar in June, 2000 (Figure 5B and 5C). The 1998 array of 40 long-line cabled logs was perched on top of a gravel bank in the former channel of 1997. The first array of 40 long-line cabled logs, installed in 1997, was buried in a terraced gravel bar in one year. The second 40-structure array was installed in 1998 and further developed the gravel bar. In February 2000, approximately 11,100 m^3 of material had been deposited along the eroding bank (300 m of bank, average width of depositional area 37 m, average depth of deposition 1 m). The deposited material had changed from gravel smaller than 2 inches in diameter to cobbles greater than 6 inches in diameter. From our 30 years of experience watching rivers in northwest Washington following flood damage and channel migration (Skagit and its forks, Sauk, Suiattle,

Figure 5 (facing page). Lower Finney Creek long-line cabled log array. (A) 1997: Upstream view of the creek before installation of the first long-line cabled log array. The main channel of the creek flows along the left bank. (B) 2000: Upstream view showing long-line cabled logs on top of an extensive gravel deposit along the former left bank of the creek. The lower ends of the 1997 and 1998 arrays are visible in the foreground. The main channel is now almost invisible in the photo, close to the right bank of the creek and considerably narrower than in 1997. A marker tree is indicated in this and the previous view, as well as the road as it emerges from the trees. Photographer's position is not identical because the foreground foliage in (A) would have obscured the view shown in (B) in 2000. (C) 2000: Downstream view from the road of the protected bank shows the extent to which the upstream ends of the long-line cabled logs were buried in recruited wood and trapped sediment. The material collected by the uppermost members of the 1998 array can be seen. The upper end of the 1997 array was buried in the gravel and is not visible in the photo.

A
Road
Marker tree
1997
B
Road
Marker tree
2000
C
Marker tree
2000

Whitchuk, Cascade, Nooksack, and Stilliguamish Rivers), this amount of deposition would have taken many years to develop without the addition of the long-line cabled logs. The movement of the channel away from the eroding bank in 1997 (Figure 5A) was clearly seen in June 2000 (Figure 5B). The bottom-most long-line cabled logs were partially embedded in gravel (Figure 5B). The uppermost long-line cabled logs from the 1998 array were completely covered in trapped wood (Figure 5C). In addition to the wood trapped by the floating logs, trees and stumps were visible along the gravel bar that marked the left bank of the channel (Figure 5C).

Photos at Ovenell's Slough during installation of the rock deflectors in May 1998 show the absence of wood along the eroding bank or associated with the deflectors (Figure 6A). By February 2000 wood accumulation was observed at Ovenell's Slough (Figure 6B). The annual water level fluctuation (6.2 m, 20 ft) on the Skagit River required 12.3 m (40 ft) cables between logs and their ballast rock. After one high flow event, the long-line cabled logs were re-positioned on top of the rock deflector immediately downstream. The ballast rocks remained in place and attached to the logs. Additional wood was trapped on the upstream side of the deflector in association with the long-line cabled logs. The height of the deflector was increased while the structure remained in contact with the channel (Figure 6B).

Positioning requirements for the ballast rock used to anchor the long-line cabled logs evolved during the installation and monitoring of the three sites. At the Lower Sauk River project, ecology blocks (0.76 m^3, 1723.7 kg or 1 yd^3, 3800 lb) were used to ballast the long-line cabled logs, which were installed as paired structures. The design called for installation of the ballast at the scour depth of the channel, approximately 1.5 m (5 ft) for water depth and bed scour. In practice, the ballast was installed only about 1 m (3 ft) below the channel bed. Observations of the site showed that the ballast blocks were uncovered due to the channel scour and re-buried during high river flows. Ballast rocks at Lower Finney Creek were placed at the scour depth of 1.5 m (5 ft) in 1997 and on the newly formed gravel bar in 1998. In Ovenell's Slough where the long-line cabled logs were flung or dropped into place, one structure landed in the channel and out of the eddy below a rock deflector. This one structure is submerged during high flows and not functioning as designed at bankfull conditions.

Costs

Installation of long-line cabled logs requires materials, transportation, equipment, and labor. If wood is available on site and can be scavenged, then

Figure 6. Ovenell's Slough. (A) 1998: Upstream view of the site during installation of rock deflectors shows the very soft, wet sand on the left bank of Ovenell's Slough. The position of the rock causeway and frequent debris jam is indicated by the white arrows. The main flow of the Skagit River is from right to left in the photo, on the far side of the debris jam and causeway. The flow through Ovenell's Slough is indicated by the arrows in the channel. The middle rock deflector is the one most prominent in the photo. (B) 2000: Downstream view in February of the middle rock deflector shows cabled log and other wood recruited half way through the second winter. The captured wood has added to the complexity of the habitat and the size of the deflector.

purchase costs for logs can be avoided or reduced. Based on a pulp price in 2000 of $750 per 28.3 m^3 (1000 board feet) and 10–15 logs per truck load of 99–142 m^3 (3500–5000 board feet), suitable logs can be obtained for $175–$375 each. At $26.16 per m^3 ($20 per yd^3) of rock delivered, ballast boulders cost approximately $90 each. Suitable cable with eyed ends (13 or 16 mm, 1/2 or 5/8 in) can be obtained for approximately $20 per cable. Each long-line cabled log requires two boulders for ballast and two eyed cables. Thus each assembled long-line cabled log structure requires between $200 and $500, if pulp quality logs are purchased. In contrast, a large diameter, 18.5 m (60 ft) full-size tree with attached rootwad typically costs $2400 to $3000, which includes both the purchase price and the transportation expenses associated with over-sized loads ($500/hr versus $75/hr for log loads). The price of full-size trees is likely to increase as availability decreases. In contrast, the more available and local pulp quality logs avoid the expenses related to scarcity and over-size transportation charges.

Rock deflectors require 191–382 m^3 (250–500 yd^3) of rock and can add $5,000 to $10,000 to the cost for materials. Diminishing the amount of rock needed by using wood in fixed and floating structures could decrease this expense in the future. Large, cabled log assemblages have been created in Boyd Creek, a side channel of the North Fork Nooksack River in Whatcom County, Washington, at less expense and with similar success in protecting the eroding bank.

Equipment costs are determined by the size of the track excavator needed to place the structures along the eroding bank and site conditions that dictate access by the excavator. As a deep water site, Ovenell's Slough represented a maximum in equipment expense for installation of long-line cabled logs (14 days @ $125–150/hr for a 50-ton excavator). The long-line cabled log array (40 structures) along 308 m (1000 ft) of Lower Finney Creek was placed in one day after the materials were assembled on a nearby gravel bar. The wet sand bank and deep water at Ovenell's Slough required construction of a platform for the equipment, where an existing dry gravel bar served on Lower Finney Creek.

Discussion

At all three sites, long-line cabled logs have performed as anticipated in protecting river banks from erosion and moving main water flow away from the eroding bank. The use of the long-lined cabled logs reduced the amount of rock armoring along the eroding bank and, by collecting and retaining small woody debris, improved the habitat potential (Peters et al. 1998) in the project area for aquatic organisms. The long-line cabled logs trapped additional wood in a manner similar to that observed for mature trees fallen into a stream following erosion of the bank. This technique has the potential to simulate wood recruitment and aid in the lateral stability of a channel in the absence of large woody debris recruitment.

Installation by diverse crews and landowners has demonstrated that the technique is repeatable at different sites. The floating logs have recruited local woody material mobilized during high flow and trapped it along the bank. Eddies created by the anchored or trapped wood slowed flood waters, which then dropped gravel and sediment. Gravel bars developed along the shore, like those seen in Pacific Northwest rivers before the wholesale removal of wood and the harvest of many of the mature trees, which would have served the same purpose in years past. In contrast to most fixed structures, the long-line cabled logs were able to comb additional material from the flow and increase their effect on the channel with each high flow event. For the Lower Sauk River, the amount of material recruited increased with the intensity of the high flow. For Ovenell's Slough, which is often protected from large wood (but not from large fluctuations in water depth) by the upstream causeway and debris jam, wood recruitment has been slower than at the other sites.

Observation of these sites since 1993 has led to modification of the installation design. Most significantly, the projects have demonstrated that properly-sized ballast rock can be placed on the channel bed rather than buried at the scour depth. This decreases the time and cost of the project as well as lessening construction effects on the aquatic environment. Secondly, the first array at Lower Finney Creek demonstrated that cable wraps are more effectively completed with smaller and more flexible cable. In each case, 13 and 16 mm (1/2 and 5/8 in) cables were flexible and strong enough to anchor logs to ballast boulders. Drilling logs for the use of eyed cables decreased the amount of cable used and provided a tighter hold on the logs than the wrap, cinch, and staple method. The risk of losing logs was decreased with shackled, eyed cables, which also eliminated the problem encountered at Lower Finney Creek, where additional staples were needed in 1997 and at the Lower Sauk River site, where only one structure was lost.

Most critically, the weight of the ballast material must match both the size of the log and the local anticipated water conditions (D'Aoust and Millar 1999, 2000). For example, a similar long-line cabled log project was built on the Hoh River in 1997 by the Department of Transportation using the Lower Sauk River design. Log structures were installed in both the main channel and a side channel of the Hoh River. By February 1998, the logs in the main channel were missing and presumed washed downstream. The logs installed in the side channel remained in place until 1999, and one was still on site at the end of that year (Thompson, personal communication, 1999). Review of the design for the Hoh River suggests that the ballast weight did not match the log size or the anticipated channel velocities. Much larger wood was installed on the Hoh River, while the ballast was the same as that specified for the slack water areas of the Lower Sauk River. Cable lengths used on the Hoh River may have been too short for the tethered logs to remain floating at the water surface during bankfull events. Logs submerged during high flow would offer greater resistance to the current and be more apt to drag ballast downstream.

There is potential for further decreasing the use of riprap by constructing deflectors from woody material around rock cores (Feduck 1997) rather than riprap, as was done in these projects. Combinations of fixed, ballasted wood and ballasted long-line cabled logs may be able to provide the slack water areas now created behind the rock deflectors. It will take time and willing landowners to test the potential of wood deflectors to further enhance fish habitat without losing property to endlessly unraveling stream banks.

Summary

In all three projects, long-line cabled log structures contributed to:

- a decrease in erosion of the stream or river bank,
- on-going capture and accumulation of wood and deposition of gravel along the eroding bank,
- development of terraces along the banks similar to those seen naturally when debris collects on vegetation,
- movement of the main channel flow away from an eroding bank (with a decrease in channel width at Lower Finney Creek),
- an increase in local channel roughness (coarse gravel to cobble and additional woody debris on banks) and bank complexity, and
- the protection or growth of riparian vegetation.

At all three sites, time has been—and continues to be—a necessary part of the solution. Structures were added to a stream or river channel, and then the channel was given time to work with the new structures. The river delivered more wood, gravel, and sediment, which were captured by the long-line cabled logs at no additional cost to the landowners or agencies.

In addition to the effectiveness in decreasing bank erosion, the long-line cabled log technique creates an alternative to the extensive riprap efforts that have plagued rivers in the past. The benefits to the natural characteristics include increased bank roughness and complexity that can be of benefit to all aquatic species. Long-line cabled logs now provide an economical bank stabilization technique that protects riverside property and enhances, rather than limits, river habitat for fish.

Acknowledgments

We would like to thank landowners, agencies and organizations for their roles in completing the work described in this report and in making the results of the work available to the larger community involved in stream protection: Washington Department of Transportation, John Hancock, Campbell Group, Olympic Resources, Skagit County Public Works Department, University of Washington Center for Streamside Studies, USDA Forest Service Pacific Northwest Research Station, Washington Department of Ecology, and Skagit Conservation District.

References

Abbe, T.B. and D.R. Montgomery. 1996. Large woody debris jams, channel hydraulics and habitat formation in large rivers. *Regulated Rivers: Research & Management* 12:201-221.

Beamer, E.R. and R.A. Henderson. 1998. Juvenile salmonid use of natural and hydromodified stream bank habitat in the main stem Skagit River, northwest Washington. Report to the US Army Corps of Engineers, Seattle District, Washington.

Bilby, R.E. 1984. Removal of woody debris may affect stream stability. *Journal of Forestry* 82:609-613.

D'Aoust, S.G. and R.G. Millar. 2000. Stability of ballasted woody debris habitat structures. *Journal of Hydraulic Engineering* (American Society of Civil Engineers) 126:810-817.

D'Aoust, S.G. and R.G. Millar. 1999. Large woody debris fish habitat structure performance and ballasting requirements. Watershed Restoration Management Report No. 8, Watershed Restoration Program, Ministry of Environment, Lands and Parks, and Ministry of Forests, British Columbia.

DeHaven, R.W. 2000. Impacts of riprapping to ecosystem functioning, Lower Sacramento River, CA. Report by the Sacramento office of the U.S. Fish and Wildlife Service for the Sacramento District, US Army Corps of Engineers, as part of the U.S. FWS Fish and Wildlife Coordination Act Report and Biological Opinion for proposed bank protection work.

Federal Highway Administration. 1983. Laboratory investigation of flow structures for highway stream crossings. U.S. Department of Transportation,

(Available through the National Technical Information Service, Springfield, Virginia.)

Feduck, M.D. 1997. Application of river spur design methods to watershed restoration projects. 2nd Annual U.S.–B.C. Watershed Restoration Technical Transfer Workshop, Watershed Restoration Program, Ministry of Environment, Lands and Parks, Nanaimo, British Columbia.

Klingeman, P.C., S.M. Kehe, and Y.A. Owusu. 1984. *Streambank Erosion Protection and Channel Scour Manipulation Using Rockfill Dikes and Gabions (WRRI 98)*. Water Resources Research Institute, Oregon State University, Corvallis, Oregon.

Lister, B. 1998. Shears and eddies: Habitat use by juvenile chinook salmon and steelhead trout in Thompson River, B.C. In *Abstracts and Program*, American Fisheries Society, North Pacific International Chapter, Annual General Meeting, Union, Washington.

Peters, R.J., B.R. Missildine, and D.L. Low. 1998. *Seasonal Fish Densities Near River Banks Stabilized with Various Stabilization Methods*. First Year of the Flood Technical Assistance Project, US Fish and Wildlife Service, North Pacific Coast Ecoregion, Western Washington Office, Aquatic Resources Division, Lacey, Washington.

Richardson, E.V. and D.B. Simons. 1974. Spurs and guide banks. Open File Report, Colorado State University Engineering Research Center, Fort Collins, Colorado.

Slaney, P.A. and D. Zaldokas (eds.). 1997. Fish habitat rehabilitation procedures. Watershed Restoration Technical Circular No. 9. British Columbia Watershed Restoration Program, Ministry of Environment, Lands and Parks, and Ministry of Forests.

Sutton Corp. 1985. Design of spur-type stream bank stabilization structures. Federal Highway Administration, McLean, Virginia. U.S. Department of Transportation Catalogue No. PB86-186830. U.S. Government Printing Office (1985-461-816:20141).

17. Integrating Engineered Log Jam Technology into River Rehabilitation

Tim Abbe, George Pess, David R. Montgomery, and Kevin L. Fetherston

Abstract

Reach-scale river rehabilitation projects using Engineered Log Jams (ELJs) were implemented successfully in four demonstration projects in western Washington from 1995 through 1999. ELJ technology is founded on the premise that river management can be improved by understanding, emulating, and accommodating natural processes using sound science and engineering practices. The ELJ demonstration projects were developed as part of river rehabilitation efforts in which reach analyses were crucial for providing information about historical channel dynamics and revealing opportunities and constraints that helped refine project objectives and improve designs. Each ELJ demonstration project constructed to date improved salmonid habitat and addressed traditional problems constraining habitat rehabilitation, such as bank and bridge protection. The projects described here offer examples of instream structures compatible with rehabilitating and maintaining aquatic and riparian habitat in fluvial corridors throughout the Puget Sound.

Restoration of Forest Rivers

Rivers of the Puget Sound region, as elsewhere across North America, have been severely impacted by land development. In particular, the role of large woody debris as a principal structural component of forest streams has been almost eliminated during the last century throughout the Pacific Northwest (Chapter 4). In the Puget Sound, as in many other regions, the removal of woody debris has reduced the physical and ecological complexity of streams and rivers (e.g., Marzolf 1978; Shields and Nunnally 1984; Harvey and Biedenharn 1988; Smith and Shields 1990; Hartopo 1991; Maser and Sedell 1994). This is of particular concern today as the physical habitat created by woody debris provides important habitat for salmon and other aquatic species (e.g., Tschaplinski and Hartman 1983; Swales 1988; Pearsons et al. 1992; Lonzarich and Quinn 1995).

Despite the widespread recognition of woody debris as a principal physical and biological component of forest streams, and despite extensive wood reintroduction programs aimed at channel restoration, little has been done to develop engineering guidelines for wood placement. Guidelines that have been developed assume that wood must be artificially anchored to remain stable (D'Aoust and Millar 1999). The engineering analysis of such studies is sound, but the underlying assumptions ignore the mechanics that underpins the stability of natural snags, which, of course, do not benefit from artificial anchoring (e.g., Abbe et al. 1997; Brauderick and Grant 2000). Random placement of woody debris without an understanding of the geomorphology (e.g., mechanics of wood stability, hydraulic conditions, sediment transport, natural woody debris supply, channel dynamics) and social context (e.g., local land use, infrastructure, recreational activity in rivers) can significantly increase the potential for unanticipated consequences, including habitat degradation, property loss, and injury.

Initially, stream channels were cleared of stable wood to improve navigation and later because it was assumed that instream woody debris reduced flow conveyance and increased flood risks. Recent studies, however, have shown that instream woody debris can block up to 10% of a channel's cross-sectional area without significantly reducing conveyance (Gippel 1995; Shields and Gippel 1995). Channel clearing was not the only practice in traditional river engineering that degraded fluvial environments. Traditional river engineering focuses on straightening, impounding, and generally simplifying channel conditions. Common bank protection measures do not emulate natural conditions and processes and dramatically reduce the habitat and hydraulic complexity found in natural forest rivers. Traditional measures such as rock revetments provide little beneficial habitat for most salmonids when compared to unprotected banks with vegetation or woody debris. Incorporation of vegetation into bank protection measures, such as bioengineering, has been widely used to reduce environmental impacts, but many of these measures amount to cosmetic treatments on traditional structures (e.g., Thorne 1990; Shields et al. 1995). Restoration efforts have long attempted to create "natural" structures in streams for habitat and to stabilize channels (Tarzwell 1934) but have not been based on accommodating processes and conditions typical of forested systems, and the resulting structures bear little, if any, resemblance to natural structures that accomplish the same effect.

Forested alluvial river valleys undisturbed by humans have high levels of morphological and biological complexity (e.g., Hawk and Zobel 1974; Sedell and Frogatt 1984). The upper Sauk River north of Darrington, Washington, offers an example of a relatively intact channel migration zone of a large forested alluvial river and exhibits a complex anastomosing channel with numerous log jams (Figure 1A). The White River southeast of Auburn, Washington, and east of Lake Tapps has a significantly simpler channel form with fewer secondary channels and lower sinuosity (Figure 1B). Both the Sauk and White Rivers are low-gradient (<0.01) unconfined gravel bedded rivers. Industrial forestry within the depicted portion of the White River valley has reduced the quantity of functional woody debris (i.e., large trees) capable of forming log jams vital to maintaining an anastomosing system and a complex forest structure. Agricultural development has had even greater impacts, as illustrated by the channelization of the Snoqualmie River north of Duvall, Washington (Figure 1C). This portion of the Snoqualmie is a very low gradient river that once had secondary channels, extensive wetlands, and a diverse riparian forest. All of these have been lost as the river has been channelized into a fraction of its original corridor. Ultimately human development can transform a river valley from a dynamic complex mosaic of forest, wetlands, and channels into a static channel with an impermeable floodplain, such as found in urban areas along the Green River in south Tukwila, Washington (Figure 1D). Here, cultural constraints leave little opportunity for restoration other than improving channel-boundary complexity to improve aquatic refugia for migrating fish.

Current river management often precludes a reach-based, scientific approach because much of the funding to maintain infrastructure along rivers comes from state and federal emergency response programs that require rapid response and often involve replacement of the original structure. Such emergency response actions almost always fail to incorporate environmentally sustainable solutions. The cumulative effect of river management actions arising from emergency response can significantly impact aquatic ecosystems through progressive confinement of a channel by successive rock revetments. Throughout much of the Puget Sound, human activity has transformed complex anas-

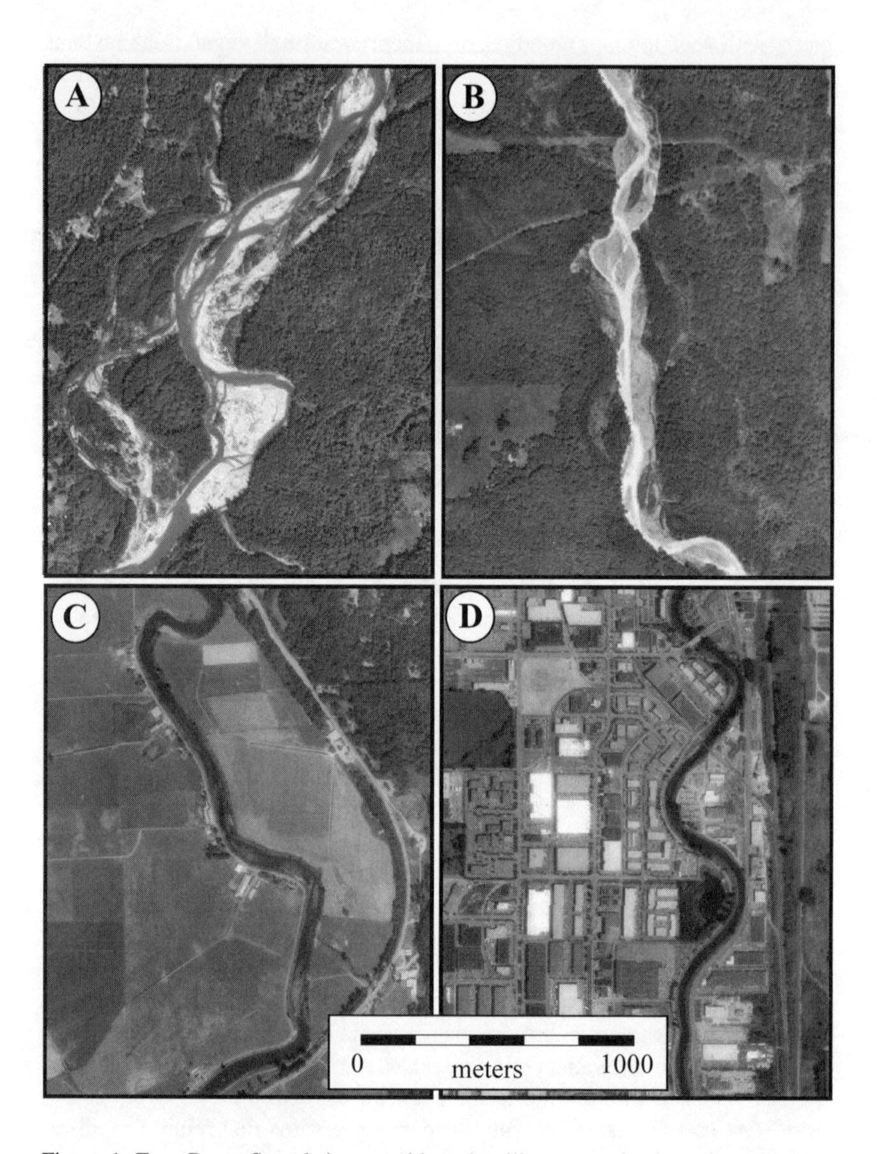

Figure 1. Four Puget Sound river corridors that illustrate reduction of geomorphic complexity and salmonid habitat with loss of woody debris and progressive encroachment on the fluvial corridor. All photos are from the U.S. Geological Survey and are identical in scale. (A) The upper Sauk River north of Darrington (09-07-89). (B) The White River southeast of Auburn and east of Lake Tapps (07-20-98). (C) The Snoqualmie River north of Duvall (08-04-90). (D) The Green River in south Tukwila (07-10-90).

tomosing forest channel systems with abundant woody debris and diverse habitat into simple single-thread channels with little complexity and cover (Figure 2).

Engineered Log Jam (ELJ) Technology

ELJ technology is based on the premise that the manipulation of fluvial environments, whether for traditional problems in river engineering (e.g., flood control, bank protection) or for habitat restoration, is more likely to be sustainable if it is done in a way that emulates natural landscape processes. The concept of ELJs began with the observation that natural log jams can form "hard points" that provide long-term forest refugia (Abbe and Montgomery 1996). Such natural hard points create stable foundations for forest growth within a dynamic alluvial environment subject to frequent disturbance. Log

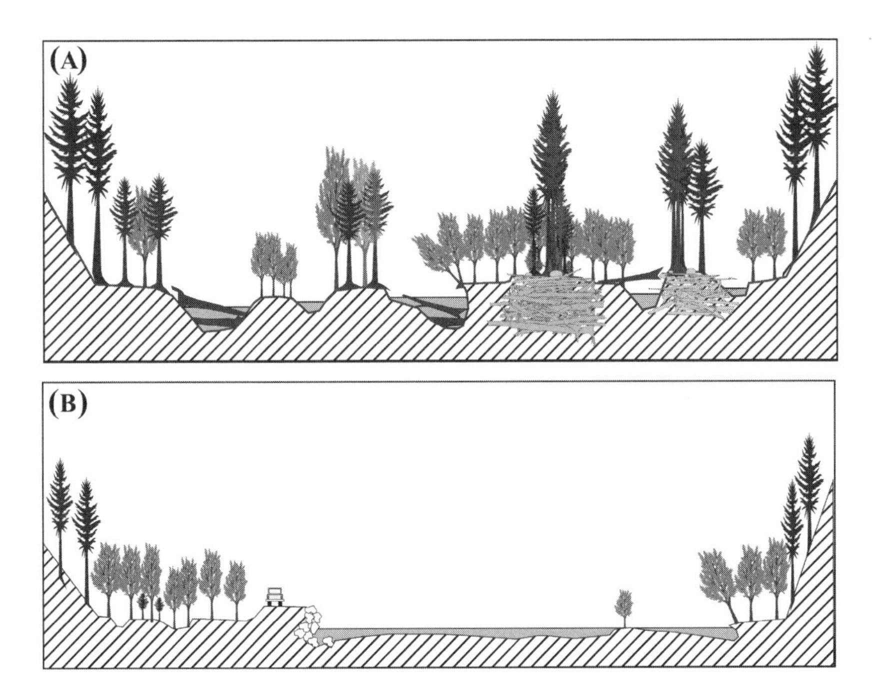

Figure 2. (A) Natural anastomosing forest river valley with abundant instream woody debris, complex mosaic of channels, and forest structure associated with regions such as found in the upper Sauk River (Figure 1A). (B) Degradation of forest rivers due to direct (e.g., channel clearing and confinement) and indirect (e.g., removal of riparian trees, increase in sediment supply or discharge associated with upland disturbance) human disturbance, such as the White River (Figure 1B).

jams thereby enable the development of trees large enough to continue forming stable log jams. Scientific and engineering studies of both woody debris and other types of flow obstructions contributed to the development of ELJ technology, such as the effect of boundary roughness on flow conditions, channel migration, and bed surface grain size (e.g., Raudkivi 1990; Pitlick 1992; Buffington and Montgomery 1999a), the effect of bluff body obstructions on flow deflection and scour (e.g., Garde et al. 1961; Raudkivi and Ettema 1977; Miller et al. 1984; Hoffmans and Verheji 1997), the impacts of debris accumulation at bridge piers (e.g., Melville and Sutherland 1988; Melville and Dongol 1992; Richardson and Lagasse 1999), and the hydraulic and geomorphic effects of natural snags (e.g., Shields and Gippel 1995; Abbe and Montgomery 1996; Gippel et al. 1996; Wallerstein et al. 1997).

Distinct types of log jams, or instream woody debris accumulations, are found in different parts of a channel network (Abbe et al. 1993; Wallerstein et al. 1997). Using observations from the Queets River basin on the Olympic Peninsula in Washington, distinct types of log jams have been classified based on the presence or absence of key members, source and recruitment mechanism of the key members, jam architecture (i.e., log arrangement), a jam's geomorphic effects, and patterns of vegetation on or adjacent to the jam (Abbe et al. 1993). Six jam types (Figure 3) provide naturally occurring templates for ELJs intended for grade control and flow manipulation (Figures 4-9). Jam types primarily applicable to grade control include log steps and valley jams; those types more applicable to flow manipulation include flow deflection, bankfull bench, bar apex, and meander jams.

Channel planform and flow obstructions can result in significant changes in water surface topography, locally raising water elevations enough to inundate secondary channels and portions of the floodplain during flows that otherwise would not engage the floodplain (Miller 1995). ELJs can create the same effect as they obstruct flow and control channel planform, thus serving as one of the principal mechanisms of connecting secondary channels and wetlands within floodplains to the mainstem channel.

The design process recommended for ELJs (Figure 10) begins with analysis of the watershed context within which the project is set, then follows with reach analysis and assessment. If opportunities are identified for potential ELJ applications, then appropriate types of natural log jams are selected based on the project objectives and constraints. After the general reach-scale strategy and ELJ layout are refined, individual structures are designed and specifications for logs and jams are prepared. Finally, the structures are constructed and evaluated over time.

Logs used to construct individual ELJs fall into three basic structural categories. *Key members* are individual logs with rootwads, which are unlikely to

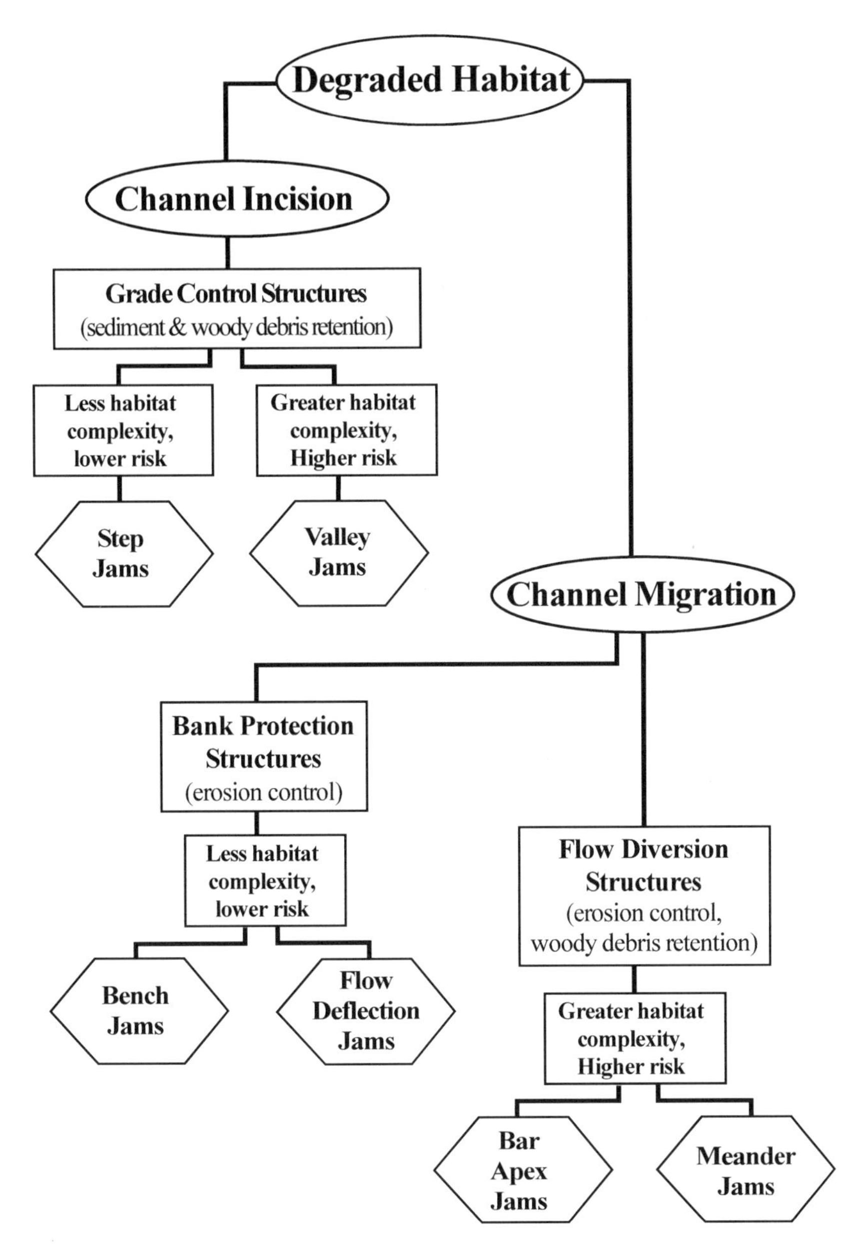

Figure 3. Classification of engineered log jam structures appropriate for treating different problems associated with habitat degradation. Two basic categories of habitat degradation involve vertical (incision) and lateral (migration) changes in channel position.

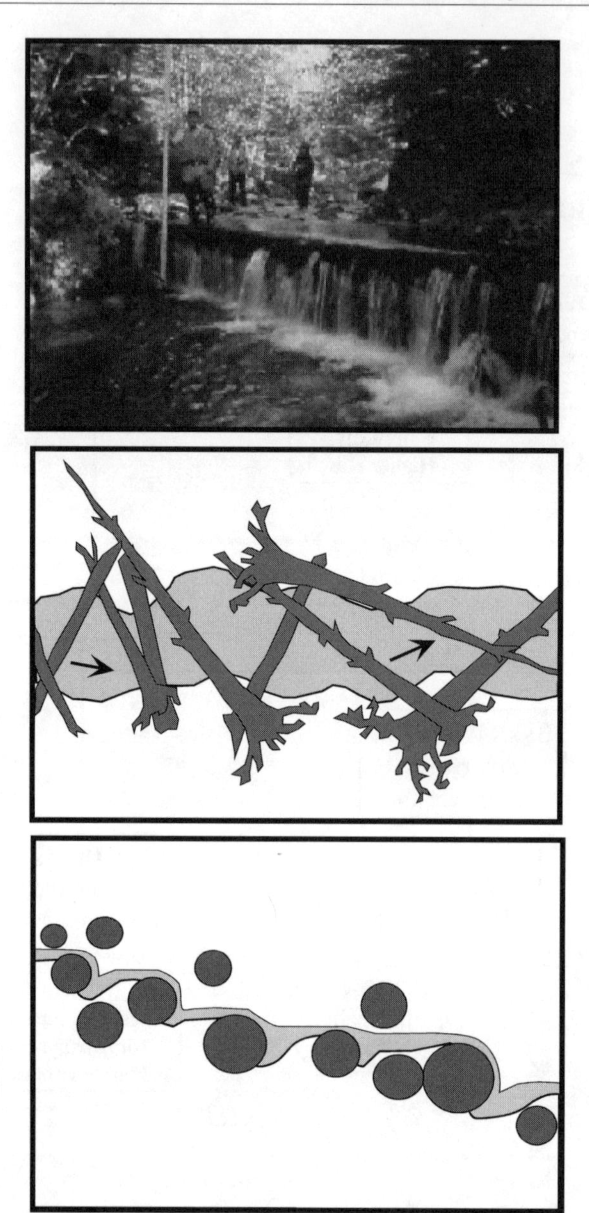

Figure 4. Step jams or multi-log log weirs are found in relatively small channels with a wide range of gradients. These structures can account for more than 80% of the head loss in a channel (Abbe 2000) and almost all of the hydraulic and habitat diversity within the channels where they occur.

Figure 5. Valley jams are large, complex grade control structures found in channels with gradients ranging from 2 to over 20%. These structures are typically composed of tens or hundreds of trees, can raise the channel bed over 5 m, and transform plane-bed and step-pool channels into pool-riffle channels (Abbe 2000). These structures are also responsible for creating a complex channel network across the valley bottoms in which they occur.

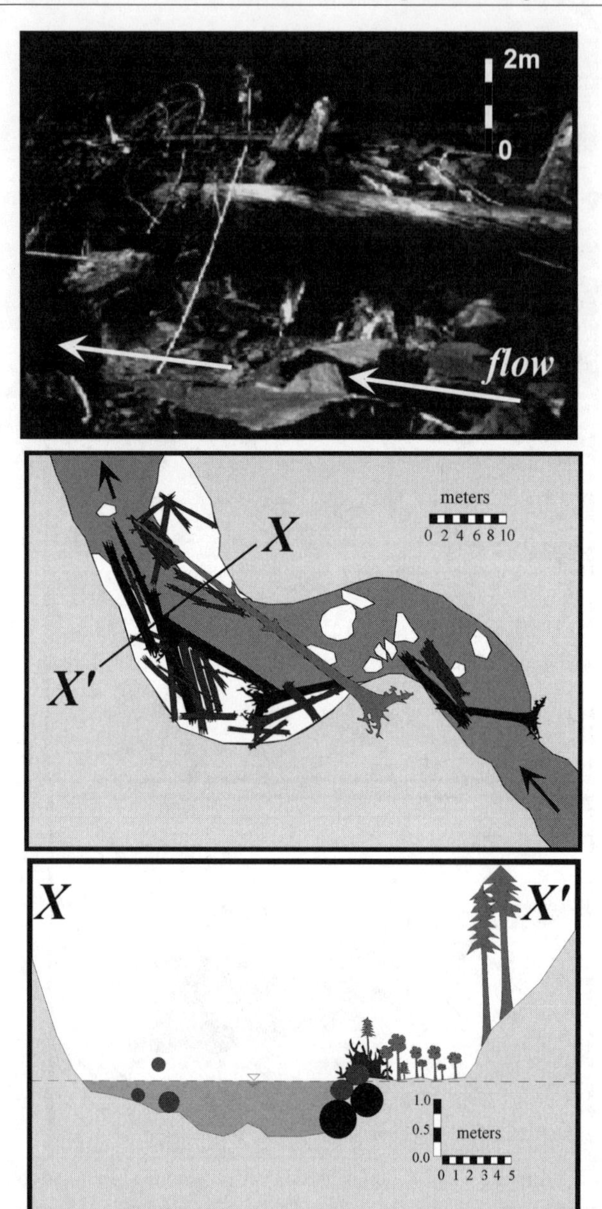

Figure 6. Bench jams are typically found in relatively small, steep channels (slopes >2%) where large logs become wedged into the margins of a channel and create local revetments protecting floodplain deposits and vegetation. Where these structures occur, wood forms the stream bank and prevents erosion of alluvium stored behind them.

Figure 7. Flow deflection jams are found in relatively large channels with moderate gradients. These structures form initially when large trees (key members) fall into the river and deflect flow. But with time these structures become integrated into a new river bank and are thus classified as bank protection or revetment type structures as opposed to flow diversion structures.

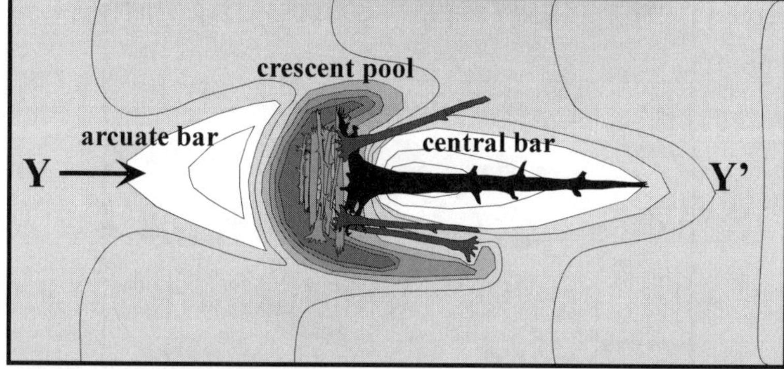

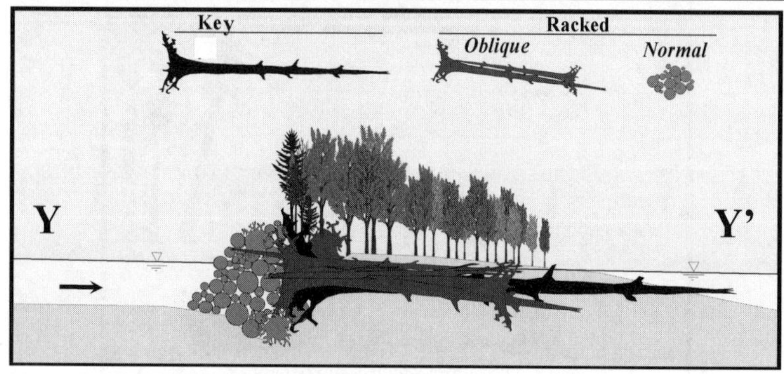

Figure 8. Bar apex jams are bi-directional flow diversion structures found in large channels with low to moderate gradients. These structures create forest refugia in dynamic channel migration zones and are responsible for much of the channel complexity and pool formation in these systems. Bar apex jams are a principal mechanism contributing to the formation of anastomosing channel systems in the Pacific Northwest.

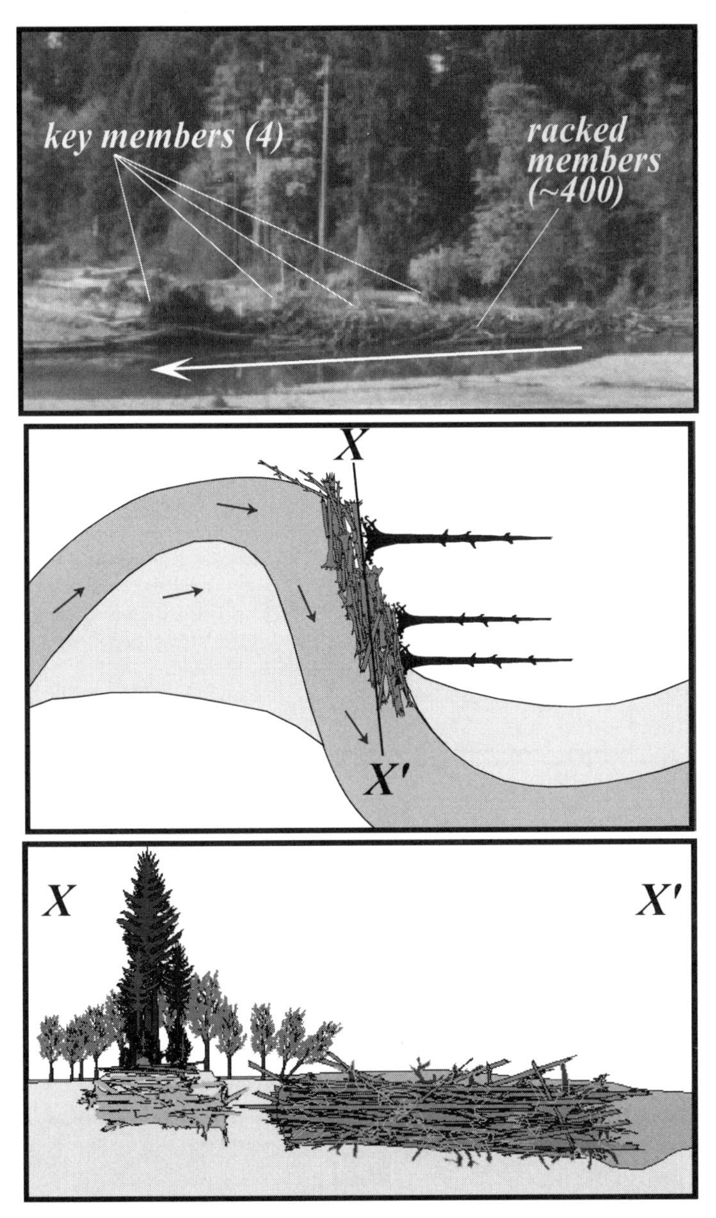

Figure 9. Meander jams are large flow diversion channels found in large alluvial rivers. These structures offer a model that has been successfully emulated to limit channel migration, protect banks, and restore aquatic habitat and riparian forests. Natural meander jams are a principal cause of channel avulsions in Pacific Northwest rivers.

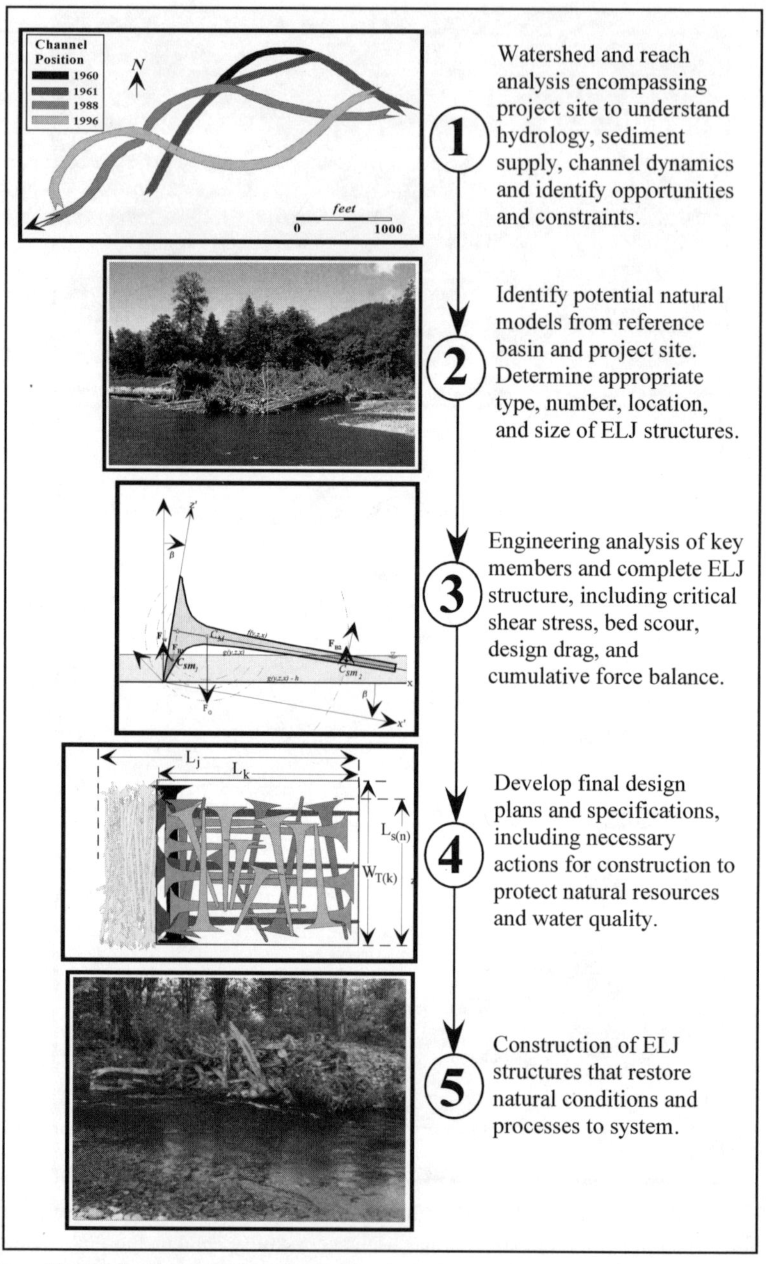

Figure 10. Five basic steps recommended for designing and implementing an ELJ project.

move during a bankfull flow, and are used as the foundation of all ELJs. In alluvial channels, key members are usually set deep into the channel substrate. In bedrock channels, key members are situated on the channel bed between pre-existing roughness elements or opposing banks. Properly situated, key members can transform a bedrock channel into an alluvial channel (Montgomery et al. 1996). *Stacked members* are slightly smaller than key members and are used in some ELJs to supplement key members. Stacked members are laid down in two or more layers that link individual members together and increase the integrity of the structure. Most stacked member logs should retain a substantial rootwad. *Racked members* include the smallest logs, with the largest range in sizes, and are often the only logs visible after construction is completed, depending on the type of ELJ. Racked members form a dense, chaotic pile of debris extending from well below the channel thalweg to above the bankfull elevation. Racked members act to decrease the permeability of and deflect flow around the structure.

No artificial materials are necessary to construct an ELJ. Native trees and alluvium at the site are all that is needed if the trees meet the design specifications for size and shape. Most projects will import trees to the site because it is usually preferable to preserve existing riparian trees, and an adequate local supply of large trees is rare. Trees large enough to act as key members may need to be cut for transport and then glued and bolted together at the site before placement. The stability of ELJs is founded on how snags interact with alluvium and instream flows. The shape and size of individual logs is critical, as are the architecture of the ELJ and its size and position within the river system. Long-term contributions to stability come from trees growing on top of ELJs, due to both root cohesion in alluvium under which the structure is buried and from the weight of the trees themselves.

Most ELJ projects involve a series or array of structures within the channel or extending across the channel migration zone (CMZ). The appropriate type, size, and position of ELJs will depend on a thorough geomorphic, hydrologic, and hydraulic analysis of the project site sufficient to characterize the river's dynamics and predict the likely range of future conditions. Such studies should include historical analysis of the changes the river has undergone and, if possible, what conditions were like prior to human development. These site assessments are referred to as reach studies and are recommended for any project that will manipulate the boundary conditions in and along a river.

Between 1995 and 1999, thirty ELJ structures were constructed in four demonstration projects in western Washington (Figure 11). The objectives of these projects ranged from bank protection to habitat restoration and illustrate a wide range of applications for this technology in Puget Sound rivers and streams.

Reach-Based Design

Before attempting to design ELJs, it is important to understand a river's physical boundary conditions and the relationship of those boundary conditions to fluvial processes and habitat. A reach analysis must be done at spatial and temporal scales adequate for describing these relation ships. With this understanding, ELJs can be designed and placed to achieve the desired goals, accommodate natural processes, and in some cases even diminish risks associated with human infrastructure and property. In a reach analysis, physical and human constraints are identified and demarcated. These areas are then incorporated into design alternatives; for example, differentiating areas within the channel migration zone (CMZ) where the mainstem channel can freely move, areas in the CMZ where only secondary channels are acceptable, areas which can tolerate inundation but no channels, and those areas where no inundation is acceptable.

A reach analysis is linked to changing conditions and disturbance patterns in the watershed. For example, industrial forestry can significantly increase

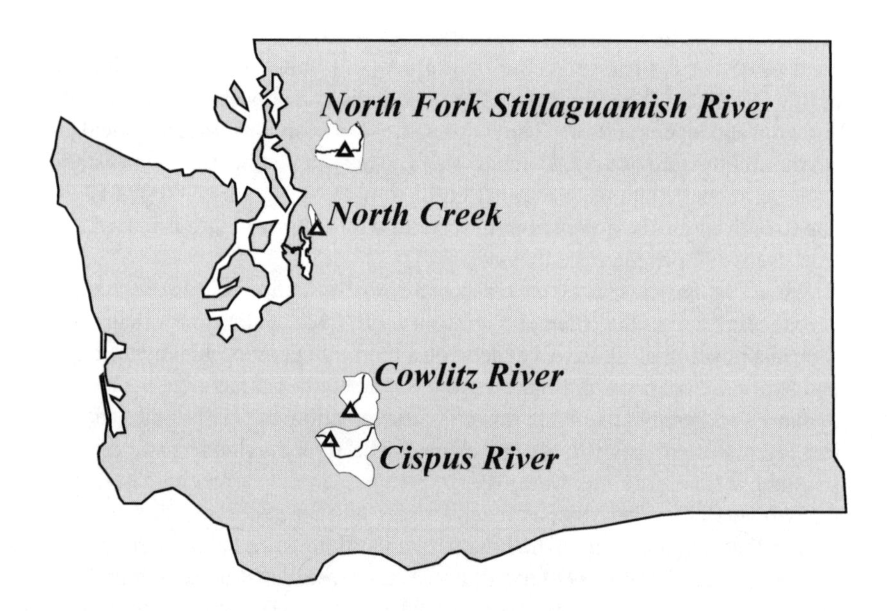

Figure 11. Locations of ELJ demonstration projects constructed in western Washington between 1995 and 1999: North Fork Stillaguamish River (1998), North Creek (1998), Upper Cowlitz River (1995), and Cispus River (1999).

sediment delivery to the river system (e.g., Kelsey 1980), which in turn can result in channel aggradation (Stover and Montgomery 2001) and textural fining (Buffington and Montgomery 1999b). The removal of instream woody debris and riparian forest may also increase the frequency and magnitude of peak flows and lead to significant geomorphic changes such as channel incision (e.g., Brooks and Brierly 1997). The most dramatic increases in the frequency and magnitude of peak flows are associated with rapidly urbanizing watersheds (e.g., Hammer 1972; Graf 1975; Booth and Jackson 1997; Moscrip and Montgomery 1997). Because these types of watershed disturbances will ultimately influence fundamental conditions within a project reach, they should be accounted for in design strategies.

The nature of these reach analyses and subsequent designs are illustrated by four ELJ demonstration projects constructed from 1995 to 1999. The overall goal of each project was to help restore fluvial environments in the contexts of natural processes and existing human constraints. Goals specific to each project are discussed in detail in the following sections.

Case Studies

Upper Cowlitz River

Three unanchored ELJ structures emulating meander jams were installed in December 1995 to halt erosion and reduce property loss from channel migration along 430 m of privately owned land along the upper Cowlitz River, Washington. Cost was a substantial constraint to the landowner, who nonetheless expressed a clear desire to maintain or improve aquatic and riparian habitat. The unvegetated width of the channel at the site is 195 m; the average bank erosion rate from 1990 to 1995 was 15 m/yr. Erosion along the landowner's shoreline from 1992 through 1995 resulted in as much as 50 m of bank retreat and the loss of about one hectare of forest land. After bank erosion associated with a 12-year recurrence interval flow in November 1995, the landowners became concerned they would lose the entire riparian corridor and inquired about erosion control alternatives that could retain as much of the habitat and aesthetic qualities of the site as possible. The high cost of a rock revetment or rock barbs (groins), together with the desire to salvage woody debris along the channel, led the landowners to pursue the experimental use of ELJs.

The floodplain adjacent to the site consists of timberlands that have been selectively harvested since the 1930s. Present forest cover is dominated by a 50–80 year old mixed conifer and deciduous forest with basal stem diameters up to 2.2 m and averaging about 0.4 m. Bank erosion along the Upper Cowlitz is common, and several large, conventional bank revetment projects have been constructed (and reconstructed) since the 1960s. Analysis of historical aerial photographs revealed northward channel migration and progressive widening of the Cowlitz River since 1935.

The three ELJ's built along the Upper Cowlitz River (summer 1996) were based on bar apex and meander jams (Abbe and Montgomery 1996) common in large alluvial channels and naturally occurring in the Cowlitz River. Both jam types consist of large key member logs with rootwads facing upstream and boles aligned with bankfull flow. Bar apex jams are usually relatively narrow structures with 1 or 2 key members that direct flow to either side of the jam. Meander jams usually are considerably wider with 3 to 6 key members, and they are situated such that they force a change in channel direction.

Five weeks after construction, the project experienced a 20-year recurrence interval flow of approximately 850 m^3/s (Abbe et al. 1997). Each ELJ remained intact and transformed an eroding shoreline into a local depositional environment. In addition, approximately 93 tons of woody debris that was in transport during the flood was trapped by the ELJs, which helped to increase the stability of the ELJs and alleviate downstream hazards. Enhancement of physical habitat included creation of deep pools at each ELJ. Because enough trees were found at the site (local landowner) and costs for design and permitting were extremely low, this project cost less than 1% of a traditional rock revetment project along an upstream meander. The cost of the ELJ project for a 430 m long reach was $10,000, or $23 per meter, whereas the cost of rip rap for a 683 m long project was $999,253, or $1464 per meter. This experimental project demonstrates that ELJs can meet local bank erosion control objectives while helping to rehabilitate riverine habitat in a large alluvial river.

Cispus River

In 1998-1999, the United States Forest Service (USFS) and Lower Columbia Fish Recovery Board (LCFRB) collaborated on an ELJ project on two side channels in the Cispus River near Randle, Washington. The project objectives were: (1) to protect a USFS road damaged in 1996, and (2) to create habitat complexity for adult and juvenile anadromous fish in a morphologically simplified stretch of the river. The Cispus River, a tributary to the Cowlitz River, had the potential to support salmonids after a program was begun in 1993 to reintroduce three species of anadromous fish to the upper Cowlitz River Basin and evaluate and improve habitats where possible.

Two sets of ELJ structures (revetments) were constructed along the Cispus River in 1999. Four ELJs were constructed directly adjacent to Forest Road (FR) 23 at Cispus River Mile (RM) 20 (Site B) and another set of three ELJs was constructed upstream at RM 21 (Site C) (Figure 12). All the structures were part of a strategy to protect FR 23 (Figure 13), because the February 9, 1996 flood, reportedly a >100-year recurrence interval event (Brenda Smith, USFS, personal communication), destroyed several hundred feet of the road at Site B and threatened the road at Site C. Pre-existing rock revetments failed at both sites. An emergency rock revetment was constructed along Site B as part of replacing the road washout. Reach analysis commenced in the summer of 1996 and the seven ELJs were constructed in September 1999.

The goal to improve fish habitat focused primarily on the placement of woody debris structures and debris jams into two side channels. Plans for the upstream site (Site C) included the placement of three large structures. The downstream site (Site B) called for the placement of four debris jams (Figure 14). The goal was to place these structures in a manner to protect the road during periods of high runoff while providing habitats for both juvenile and adult anadromous fish. The intent was to provide holding pools for upstream migrating adults and rearing habitats for juveniles during higher flows. It was anticipated that high flows would deposit the scoured materials downstream of the structures, sorting out gravels that may be used for adult spawning. The sites were completed and monitoring began in the fall of 1999.

By the fall of 1999 and the spring of 2000, it was apparent that winter high flows had scoured the base of the structures at Site B but had little effect on Site C. Numerous adult coho salmon were observed holding in the pools at the

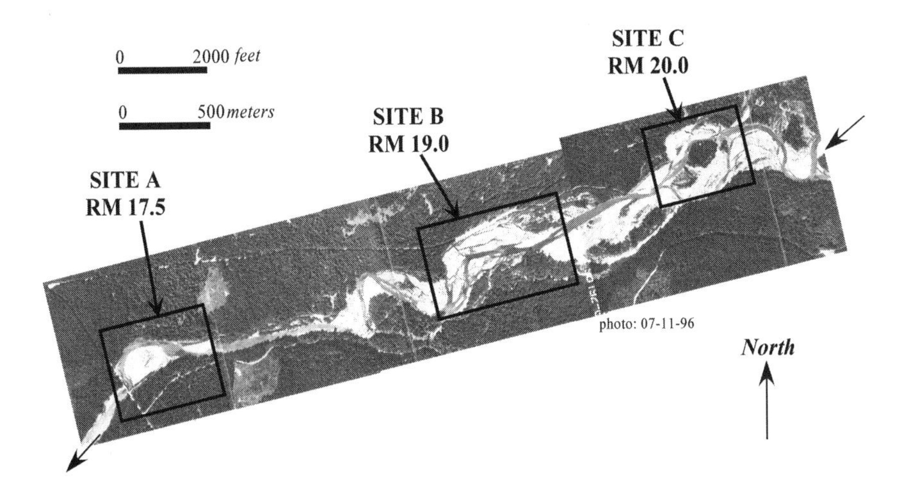

Figure 12. Cispus River sites A, B, and C. River flows from right to left. Forest Road 23 is on the north side of river.

Figure 13. Photographic illustration of the differences between traditional blanket rock revetment (left) and ELJ solution (right) to protecting Forest Road 23 along Cispus River at Site B. A series of 4 ELJs was constructed to protect the road, enhance aquatic habitat, and establish a riparian buffer between the road and river. Each of the structures is approximately 7 m in height with about 4 m exposed above the low water table.

Figure 14. Oblique aerial photograph of the Cispus River ELJ project site B at River Mile 19. Arrows indicate flow direction and B1 through B4 indicate ELJ locations.

structures at both sites. Over 100 redds were counted in the reaches between the jams at Site B. Twelve redds were observed at Site C, but these were located above and below the construction site. One steelhead redd was observed at Site B in 2000. In the spring of 2001 (a period of lower than normal flows), only twelve redds were observed at Site B and none were observed at Site C.

Snorkeling surveys were performed in cooperation with Washington Department of Fish and Wildlife (WDFW) staff in July and August of 2000 to evaluate site utilization by juveniles. Observations indicate extensive use of the structures by young of the year coho. The scouring effects near the structures at Site B provided cover and depth for the juvenile fish. Juvenile coho use was limited in the reach between sites B and C because of local sediment deposition that reduced flow depth. At Site B, 92% of the young of the year coho observed were associated with the structures, and only 8% were found in the area above or below the structures at Site B. Many of the observed fish between the structures were juvenile steelhead. At Site C, 61% of juveniles were located in pools associated with the ELJs, even though these pools account for only a small percentage of the surface area of the stream. The cost of constructing Cispus sites B and C was approximately $300,000.

North Fork Stillaguamish River

The North Fork Stillaguamish River project site is about 8 km east of Oso, north of Washington State Highway 530 and upstream of the C-Post bridge (Figure 15). The project was first conceived in 1996 for enhancement of salmon habitat.

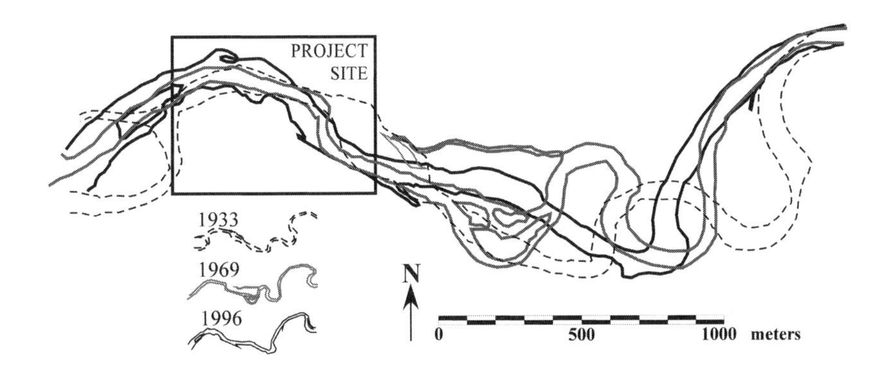

Figure 15. Selected historical planforms of North Fork Stillaguamish River ELJ Project Reach (River Miles 21-23): 1933, 1969, and 1996.

This goal was based on a comprehensive assessment of habitat conditions and historic change that identified a need to develop and maintain pool habitat as a key to recovery efforts for chinook salmon (*Oncorhynchus tshawytscha*) (Pess et al. 1998).

Chinook salmon are large-bodied fish that spend months in deep, cool pools during low flow prior to spawning. A key observation is that chinook spawning location strongly correlates to pool frequency and size; more than 80% of the chinook spawning nests (redds) surveyed in the North Fork Stillaguamish occurred within one channel width of a pool (Pess et al. 1998). Furthermore, twice as many redds were associated with pools formed by log jams versus pools with no wood, which also had three times as much instream cover (Pess et al. 1998). Historically, log jams were abundant and played a significant role in the morphology of the Stillaguamish River (Secretary of War 1931). The combination of these factors led to the proposal to construct engineered log jams in the North Fork Stillaguamish to create and enhance summer chinook holding pools.

The ELJ project reach has a drainage area of approximately 300 km^2 and is a low-gradient (<0.01) meandering gravel-bed channel that has repeatedly migrated across the floodplain during the past century (Figure 15). Natural log jams historically stabilized gravel bars in the North Fork Stillaguamish, allowing vegetation to take hold and create in-channel "islands" that resulted in an anastomosing channel network. Gravel bars and forest encompass most of the floodplain, but some homes and pastures are located along the lower portion of the surveyed reach. Estimates of the one- and five-year recurrence interval peak flows at the USGS gage at Arlington, Washington, are 258 cfs (7.3 cms) and 425 cfs (12 cms), respectively.

The upper North Fork Stillaguamish (above RM 15) has gone through large-scale channel changes over the last 70 years. A four- to five-fold increase in hillslope sediment input (primarily as landslides) between 1978 and 1983 from the upper portion of the North Fork Stillaguamish basin above RM 35 is likely to have contributed to an expansion of the unvegetated channel width and rapid changes in channel position. Many of the landslides were associated with logging and road-building in steep headwaters (Pess et al. 1998). A large increase in the sediment supply of a river can result in channel aggradation and extensive infilling and loss of pools (e.g., Kelsey 1980; Lisle 1995). Channel aggradation and widening, combined with the loss of pool-forming structures such as log jams, is thought to have reduced the quantity and quality of large pool habitat for adult and juvenile salmonids in the North Fork Stillaguamish. The lack of high quality pool habitat has altered migration and spawning timing for steelhead and possibly summer chinook (Curt Kraemer, WDFW, personal communication).

Project Objectives, Constraints, and Opportunities

The primary goal of the project was to increase quality and quantity of holding pool habitat for spawning summer chinook in the project reach. While evaluating the system, a number of additional objectives, constraints, and opportunities were also identified.

Objectives

- Maintain an active channel migration zone
- Increase the quality and diversity of aquatic and riparian habitat
- Increase linkages between channel system and riparian floodplain forest and wetlands by:
    - Maximizing the length of perennial channels
    - Maximizing linkages between channel system and floodplain

Constraints

- Accommodate existing infrastructure encroachment into channel migration zone
- Avoid increasing flood peak water elevations
- Protect property along southern margin of project reach
- Maintain or increase protection to downstream bridge by:
    - Minimizing woody debris accumulation at bridge
    - Minimizing threat of channel avulsion around bridge

Opportunities

- Introduce a multiple channel system for both perennial and ephemeral flow conditions
- Incorporate ELJ structures to:
    - Emulate instream structures representative of a low-gradient Puget Sound river
    - Limit channel migration at sensitive locations
    - Stabilize and help sustain secondary channel system
    - Increase physical and hydraulic complexity within the channel
- Increase bank protection in specific locations using an approach that emulates naturally occurring structures (e.g., log jams) and incorporates natural physical processes (e.g., channel migration, wood accumulation).

Implementation

In the summer of 1998, five ELJs were constructed upstream of the C-Post Bridge (Figure 16). Four of the ELJs were meander type jams designed to deflect flow on only one side. The remaining ELJ was a bar apex type designed to accommodate flow around either side. Each ELJ is completely inundated

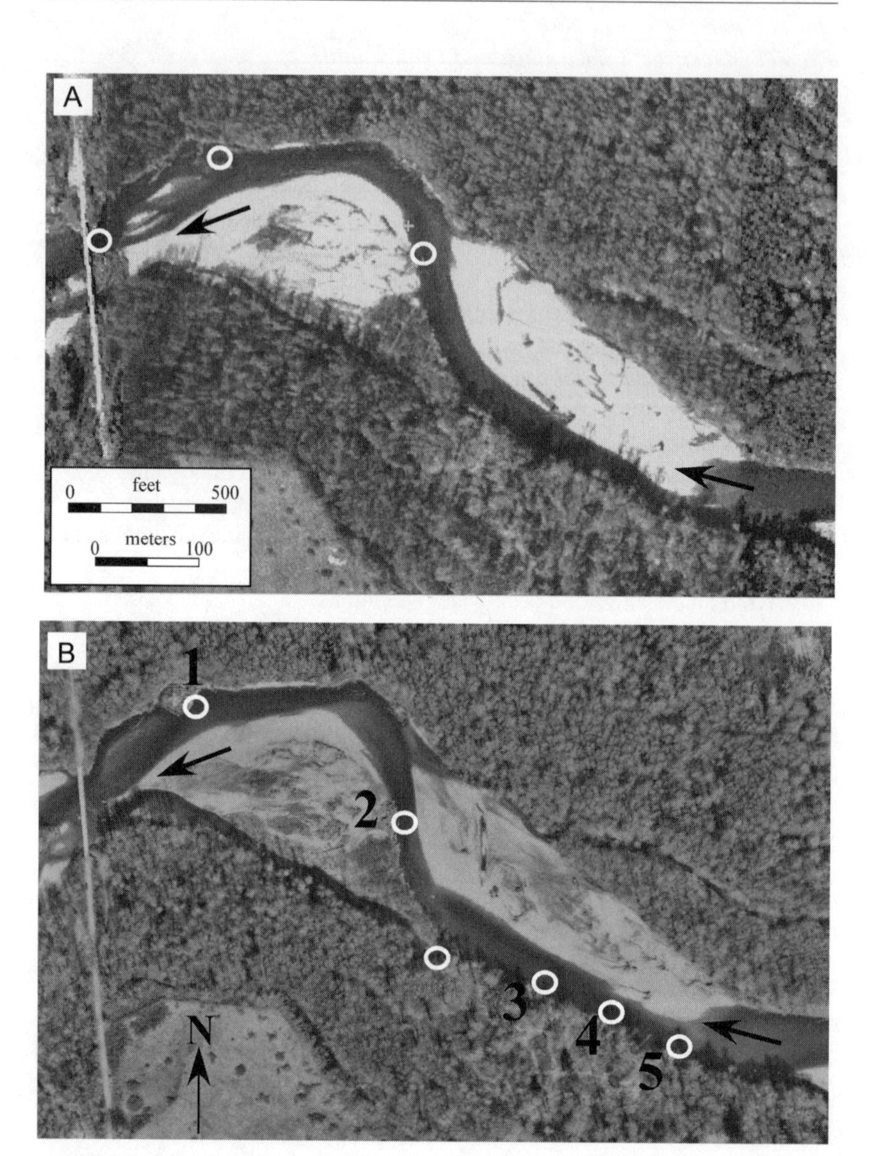

Figure 16. North Fork Stillaguamish River 1998 ELJ project site. March 1998 prior to construction (A) and two years after construction in March 2000 (B). Principal pool locations are noted by circles and ELJ location are numbered. Note the large increase in drift directly upstream of ELJ 2 between 1998 (A) and 2000 (B). ELJs 1, 3, 4, and 5 simulate "meander jams" and ELJ 2 simulates a "bar apex jam."

during bankfull flow. The North Fork Stillaguamish ELJ project also included the acquisition of 29 hectares of conservation easement within the channel migration zone. This area is set aside to permit natural migration of the channel and migration induced by the installation of the ELJs. The project also included installation of arch culverts at side channels passing beneath the C-Post Bridge road. The total project cost was approximately $400,000.

In 1997 and 1998, we collected information on characteristics of wood naturally occurring within the project reach and of wood for ELJ construction. Post-construction wood surveys conducted in 1999 included a field reconnaissance of approximately 10 km of river downstream of the project site. Natural and imported logs were given identification tags and cataloged with data that included species, location, rootwad dimensions (minimum and maximum diameters), basal trunk diameter (equivalent to diameter at breast height), crown diameter, length, and physical condition (state of decay). Imported logs also included measurements of cut geometry when applicable. These data were used to measure the stability, movement, and recruitment of individual logs, structural integrity of the ELJs, and evaluate ELJ performance relative to the project design and objectives.

Results to Date

Between September 1999 and February 2000, at least fourteen flows equaling or exceeding bankfull stage occurred (Figure 17). All five ELJs remained in place. During the first high flows in November and December of 1998, ELJ 1 was damaged when one of the structure's seven key members was lost. Significant scour occurred beneath the outer upstream corner of the ELJ and undercut the key member in question. With nothing to support the saturated log from beneath, it sank, broke in half, and was carried 10.5 km downstream to where it was found in the summer of 1999. The loss of ELJ 1's outer key member was only confirmed when the structure was inspected from below, since there was almost no change in the structure visible from above (Figure 18). Even with the loss of a key member, ELJ 1 remained in place and continued to perform as predicted. Each of the five ELJs have formed and maintained scour pools ranging from 2–4 m in depth. Sand deposition has occurred downstream of all five ELJs. Designed as a series of flow deflectors, the three upstream ELJs (3, 4, and 5) have prevented further bank erosion along the south bank.

All of the structures except for ELJ 3 experienced a net increase in woody debris or drift, particularly ELJ 2, which collected over 500 pieces of woody debris exceeding 2 m in length. Drift accumulation upstream of ELJ 2 effectively increased the structure's breadth by six-fold and contributed to the development of a perennial secondary channel south of the mainstem channel, thereby

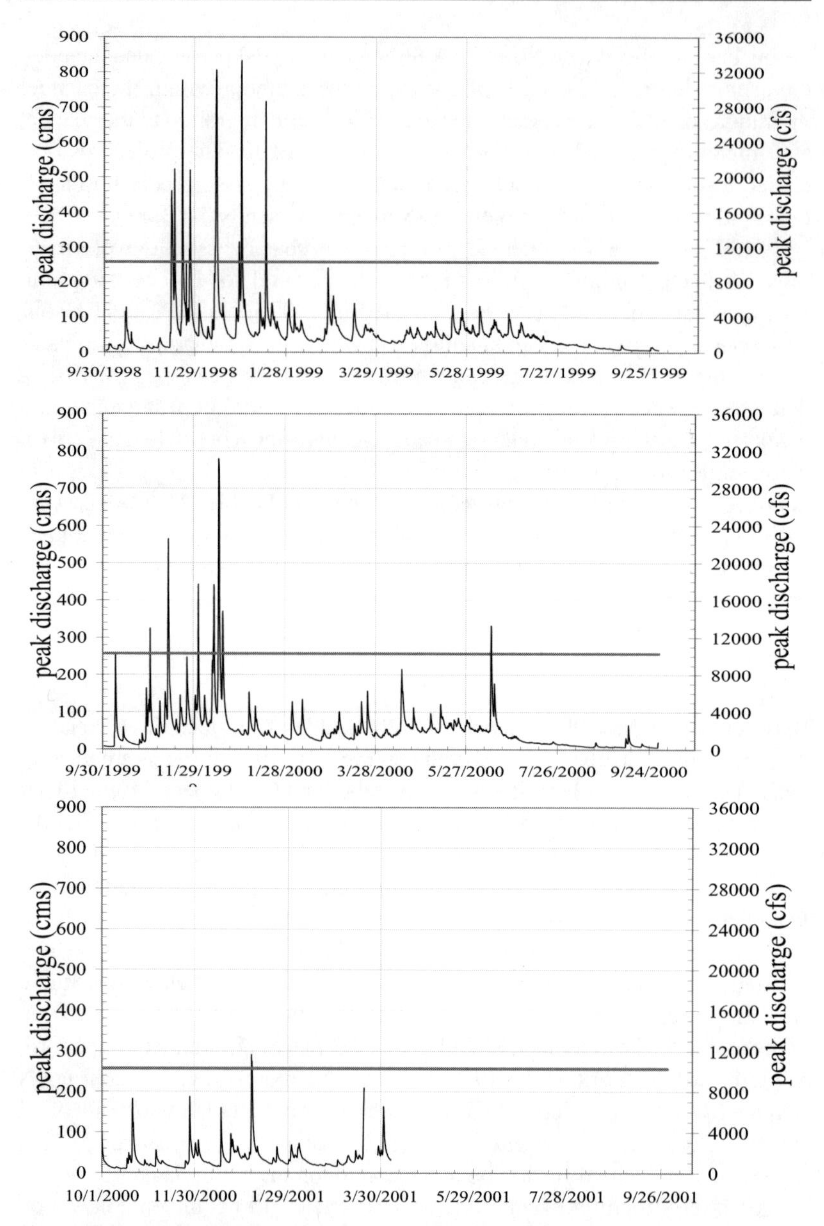

Figure 17. Annual hydrographs for Water Years 1999, 2000, and first half of 2001, North Fork Stillaguamish River, USGS Gage 12157000 near Arlington, Washington. Bankfull stage at the 1998 ELJ site (horizontal line) corresponds to approximately 10,000 cfs at Arlington gage.

Figure 18. Photographs looking downstream at ELJ 1 (C-Post Bridge is in background). As-built conditions in (A) September 1998; (B) November 1998 during a peak flow cresting bankfull stage and over topping the ELJ; (C) June 1999 after 8 peak flows equal to or exceeding bankfull stage; and (D) in August 2001 after 16 peak flows equal to or exceeding bankfull stage.

creating a forested island. The effectiveness of the North Fork Stillaguamish ELJs in collecting drift is revealed by data collected on log displacement distances during Water Year 1999. Of the logs that moved, those that had to pass at least one ELJ had average displacement distances an order of magnitude less than those logs that passed downstream of the C-Post bridge (Figure 19). The North Fork Stillaguamish downstream of the C-Post bridge is a relatively simple, clear channel lacking stable log jams. From field surveys in September 1999, we estimate that 98% of the approximately 350 logs used in the five ELJs remained in place through eight peak flows equal or exceeding bankfull stage.

Reduction of the drift accumulation at the C-Post bridge was to be accomplished by: (1) trapping drift that might otherwise accumulate at the bridge; and (2) deflecting flow to improve channel alignment nearly orthogonal to the bridge, thereby providing for more efficient conveyance past the bridge. The large drift accumulation formerly lodged on the bridge's center pier was removed during ELJ construction and as of spring of 2001 no drift has yet to lodge on the bridge (Figure 20).

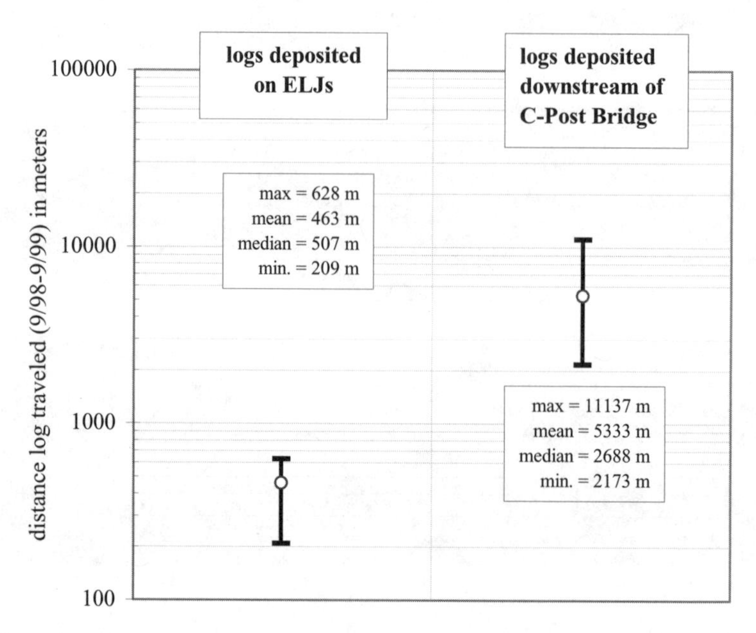

Figure 19. Displacement distances of tagged logs that moved in Water Year 1999: logs which had to travel past at least one ELJ had a significantly lower distance traveled than those logs that moved downstream of the C-Post Bridge, where few major flow obstructions were encountered all the way to Puget Sound.

The biological response to ELJ construction was evaluated by comparing baseline physical habitat and fish population information to post-construction surveys. Baseline data includes adult chinook and other salmonid population estimates through snorkel surveys, quantitative measures of habitat characteristics (e.g., number of pools, residual pool depth), and qualitative measures of habitat quality (e.g., amount of in-channel cover). Preliminary monitoring data suggest that changes in habitat condition have led to redistribution in adult chinook within the treatment reach. ELJs in the North Fork Stillaguamish have increased pool frequency, pool depth, and in-channel wood cover. Pool frequency increased immediately after ELJ construction from 1 pool/km to 5 pools/km and has remained at that level. Residual pool depth in the treatment reach also increased after ELJ construction, increasing from an average of 0.4 m to 1.5 m. The total number of pools in the area shown in Figure 16 increased from 3 to 6 after the project, but residual pool depths increased significantly. Most (80%) of chinook salmon utilization within the project reach was concentrated at the C-Post bridge in the largest, deepest pool within the reach; the remaining 20% was observed in a small pool adjacent to a natural log jam situated where ELJ 2 was constructed (Figure 21). Chinook response was immediate and consistent over the three years following construction. Instead of congregating in one pool (80% found in the C-Post Bridge pool) prior to ELJ construction, chinook redistributed throughout the treatment reach, utilizing the increase in pool availability and quality.

Lower North Creek

North Creek runs through the new University of Washington Bothell-Cascadia Community College (UWB-CCC) Campus in Bothell, Washington. The North Creek catchment is situated at the north end of Lake Washington northeast of Seattle. Restoration of North Creek is the result of a political, environmental, regulatory, and ecological design process that began in 1989 when the Washington State legislature authorized the design and construction of the branch campus. The restoration project was intended to mitigate for impacts to wetlands resulting from construction of the campus buildings and infrastructure. The State of Washington committed to a restoration design of the North Creek channel and floodplain that was significantly greater in scope, complexity, and cost than required by federal regulatory agencies.

The North Creek watershed is approximately 7,300 hectares and extends 20 km north of the Sammamish River. The watershed experienced intensive timber harvest at the turn of the century, which was followed by a long period of agricultural development. Present estimates of percent impervious area within

Figure 20. C-Post Bridge directly downstream of the 1998 ELJ project on the North Fork Stillaguamish River. Prior to constructing five ELJs upstream of the bridge, drift (mobile woody debris) accumulation was a chronic problem requiring frequent maintenance (A). Drift was removed in 1998 when the ELJs were built to test the hypothesis that ELJs could reduce drift accumulation by collecting drift upstream and improving channel alignment with the bridge to facilitate drift conveyance beneath the bridge. In the first year, there were eight flow events that equaled or exceeded bankfull stage without any drift accumulation on the bridge (B). The bridge remained clear after two years and 8 more flows equal to or exceeding bankfull stage (C). Only one peak flow equal to or exceeding bankfull stage occurred in the third year (Water Year 2001) and the bridge remains clear of drift.

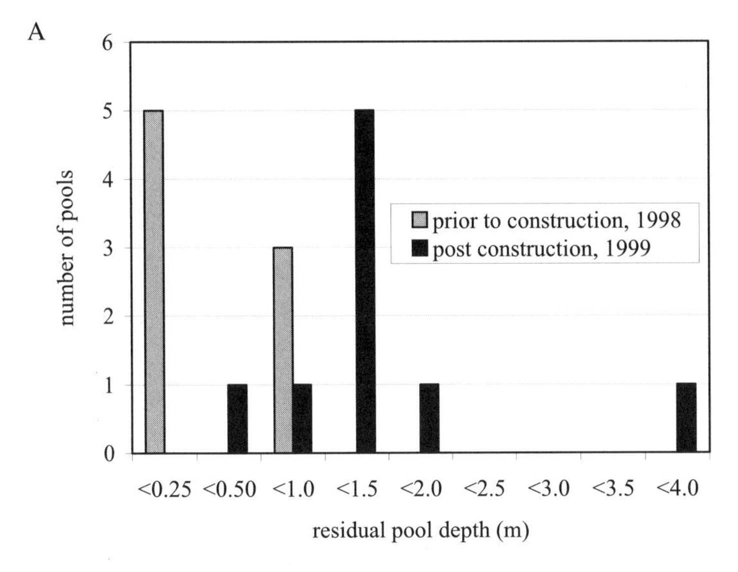

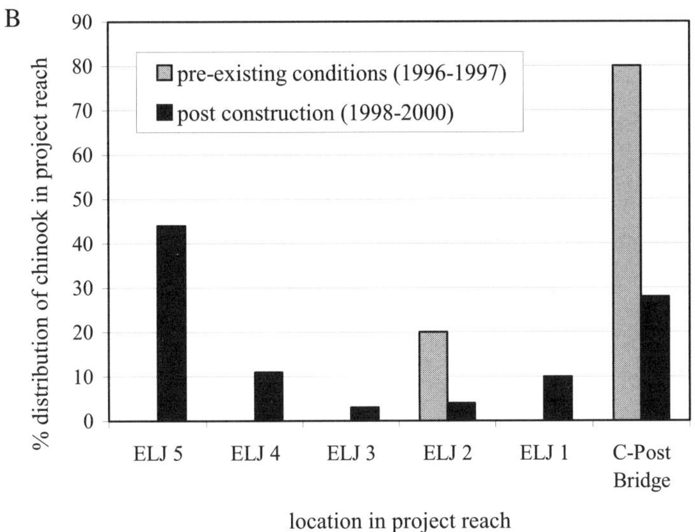

Figure 21. Results of the 1998 ELJ project in the North Fork Stillaguamish River. (A) The total number of pools only increased from 8 to 9 after the project, but residual pool depths increased significantly. (B) 80% of chinook salmon utilization within the project reach was concentrated at the C-Post Bridge in the largest, deepest pool within the reach; the remaining 20% was observed in a small pool adjacent to a natural logjam situated where ELJ 2 was constructed. Chinook distribution dispersed significantly after construction, correlating directly to the presence of ELJs.

the North Creek watershed vary from 14% to 27%. The estimated 100-year flood in lower North Creek is 41 m^3/s based on 16% effective impervious area.

The project site is situated just upstream of North Creek's confluence with the Sammamish River and covers approximately 24 hectares and 1,000 m of the lower creek channel (Figure 22). Historically, the landscape of the North Creek and Sammamish River confluence was a complex mosaic of very low gradient floodplain channels, depressional ponds, and marsh, scrub-shrub, and forested wetlands. The pre-settlement floodplain vegetation reflected the physical diversity of the landscape, with conifer dominated patches, scrub-shrub thickets of small trees and shrubs, and open water ponds fringed by emergent marsh vegetation, all set within a valley bottom deciduous forest matrix comprised of cottonwood and red alder. By the early twentieth century, the site was logged and the North Creek channel was straightened and leveed along the valley margin. An extensive network of ditches was excavated to dewater the forested wetland. These alterations effectively decoupled North Creek from its floodplain, drastically reduced the total channel length, and transformed the native emergent, shrub, and forested wetlands into a pasture. Prior to construction in 1998 the site was covered by Reed Canary Grass (*Phalaris arundinacea*). The net result of this historic land use was to significantly diminish salmonid habitat quality and abundance in North Creek.

Project Objectives, Constraints, and Opportunities

The UWB-CCC reach of North Creek is typical of many urbanized, low-gradient stream and floodplain environments in the Puget Sound region. The rehabilitation design was constrained by single points of channel entry and exit to the campus property and a floodplain limited in extent by the Highway 405 and 522 road corridors. Given the degraded status and inherent physical constraints of the campus site, the goal of the design was to restore as much as possible the site's hydrologic, biogeochemical, and habitat functions. The restoration design was based upon historic site information, hydrologic modeling, and an extensive sampling effort to characterize ecosystem structural characteristics of similar Puget Sound lowland riverine reference sites.

Objectives

- Hydrologically reconnect North Creek with its floodplain
- Reintroduce both in-channel and floodplain large wood
- Restore native floodplain forest plant community
- Increase the quantity, quality and diversity of aquatic and terrestrial habitat
- Provide visual access from both the campus and highway corridors

A

B

Figure 22. North Creek channel and floodplain restoration site: (A) pre-existing conditions in November 1997 with creek channelized at northern margin of floodplain and (B) after construction of new channel and floodplain system in January 2002. ELJs constructed at the North Creek site include flow deflection jams, a bar apex jam (at inlet to secondary channel) and log crib revetments. Photographs courtesy of Soundview Aerial Photography, Arlington, WA. Flow is from right to left in both images.

- Increase linkages between channel system and riparian floodplain forest and wetlands by:
    - Maximizing length of perennial channel system
    - Maximizing contact time between water and wetlands
    - Maximizing linkages between channel system and floodplain

Constraints

- Limit the area of flood inundation and channel migration on urbanized site
- Accommodate increased peak flows resulting from urbanization of the upstream watershed
- Allow no export of drift downstream of project area
- Protect critical infrastructure beneath and adjacent to the project area (storm sewer pipe and university campus buildings)

Opportunities

- Introduce a multiple channel system for both perennial and ephemeral flow conditions
- Maximize tolerance for channel change (i.e., lateral channel movement)
- Incorporate ELJ structures to:
    - emulate instream structures representative of a low gradient Puget Sound stream
    - limit channel migration at sensitive locations
    - stabilize and help sustain secondary channel system
    - increase physical and hydraulic complexity within the channel

It was decided that a more natural stream channel morphology would be returned to North Creek by constructing a new channel system that provided a greater diversity of habitat such as found in pristine, low-gradient sites in the Puget Lowland. In particular, the new stream channel system was constructed to allow overbank flow to occur on an approximately 1-year return interval. This approach seeks to restore the linkage between channel and floodplain components of the North Creek ecosystem. The new main channel was designed with bed and bank features and a variety of in-channel habitats, including pools, riffles, and large wood. Secondary channels were designed to engage at different flow stages.

Project Design

The North Creek project involved construction of a sinuous new mainstem and a perennial side channel, four types of ELJs incorporating approximately 1200 unanchored logs, and an aggressive revegetation plan. Infrastructure constraints mandated that channel migration be controlled. The overall project goal of improving aquatic habitat with respect to this constraint was achieved by using ELJ structures to limit bank erosion, contain channel migration, and create beneficial instream habitat. Engineered log jams emulating flow deflection jams were used along many of the channel meanders. At the inlet to the secondary channel a bar apex jam was constructed and inside the inlet a set of log steps were placed to prevent incision and dissipate energy. These jams were integrated with flow deflection jams to protect banks of the channel. Toward this end, tree bole revetments and crib structures were used to stabilize the critical banks and meanders; a bar apex type ELJ was built to locally raise water elevations at the secondary channel inlet; and a complex multiple log weir was set beneath the bed of the inlet channel to reduce the probability of the secondary channel becoming the mainstem channel.

The restoration design for the floodplain plant community was based upon quantitative characterization of similar floodplain forests at 58 Puget Sound reference sites. Based on these reference site data, 25 distinct plant communities were designed and planted at North Creek. The goal of the North Creek plant community restoration was to set the stage for the development of a compositionally and structurally representative Puget Lowland floodplain forest. The newly constructed channel reach was not engaged upon initial construction in order to allow riparian vegetation to become established along the channel banks. During the vegetation establishment period from August 1998 to August 2001 the project site was inundated several times due to backwater effects of the Sammamish River during winter high flows. The cost for the entire North Creek restoration project was approximately $6 million.

Results to Date

The new creek channel system was opened to the full discharge of Lower North Creek in August 2001. During the winter of 2001-2002 the creek experienced several peak flows that inundated the floodplain. Students from the Center for Streamside Studies surveyed twenty-five channel cross-sections in October 2001 and re-surveyed them again in January 2002. At the cross-sections, the channel has experienced some net scour and no significant change in width or location. All of the engineered log jams remain intact and are associated with deep pools. The North Creek project shows that a large-scale project involving rehabilitation of a complete channel and floodplain reach is feasible in urbanized areas if sufficient land is available. The project also suggests that unanchored logs can be incorporated into engineered log jams as an integral part of stream restoration, even in an urban stream, although the long-term consequences of increasing channel discharges with progressive watershed urbanization have yet to be evaluated.

Conclusion

River rehabilitation in large portions of fluvial landscapes, including areas within naturally defined channel migration zones, can be severely constrained or even precluded due to agriculture, industry, commercial forestry, residential development, and transportation infrastructure. Because human development affects so much of the fluvial landscape and is likely to continue to do so, meaningful rehabilitation of fluvial ecosystems will require strategies that integrate technology that not only re-establish and sustain natural processes but also maintain infrastructure and protect human life and property. Consequently, strategies are most likely to succeed if based on multi-disciplinary collaboration of physical and biological scientists, civil engineers, planners, and community representatives. Traditional engineering problems can be solved with non-traditional approaches, such as ELJs, that provide specific benefits together with habitat enhancement. In this context, ELJs are versatile in that they can be used for both habitat enhancement as well as general river engineering. However, in the implementation of ELJ projects, it is important to clearly delineate objectives and constraints, establish the spatial and temporal scale of the project, and document what ultimately happens on the ground. The potential risks of applying ELJ technology without adequate scientific assessment and engineering design can threaten not only the success of a single project but also human welfare and future policy decisions regarding the management of instream woody debris. The success to date of ELJ projects in western Washington highlights the potential benefits of this experimental technology for enhancing fluvial ecosystems while protecting infrastructure and property within fluvial corridors.

Acknowledgments

Many individuals from the public and private sectors and representing a wide range of disciplines contributed to implementing the projects described and the development of ELJ technology. We thank Selene Fisher for her contributions in reviewing and editing. Funding and in-kind support for the projects has come from an almost equally diverse group. We would like to acknowledge Ms. Donna Ortiz de Anaya, Mr. Greg Arkle, and Mr. Tim Lofgren, the three private landowners who implemented the first ELJ project on the Cowlitz River in 1995. We are especially grateful to Mary Lou White of Washington Trout for her work supporting the North Fork Stillaguamish River project and data on wood monitoring. Snohomish County, the Tulalip and Stillaguamish Tribes, the Washington Interagency Committee on Outdoor Recreation (IAC), Washington Trout, the U.S. Environmental Protection Agency, the Cascade Land Conservancy, the Center for Streamside Studies at the University of Washington, and the U.S. Army Corps of Engineers all contributed funding or services to the North Fork Stillaguamish project. The Cowlitz Valley Ranger District of Gifford Pinchot National Forest funded and supported the Cispus River projects. We thank Brenda Smith, U.S. Forest Service, and Mike Kohn, Cowlitz Valley Public Utility, who both contributed valuable information for the Cispus project. L.C. Lee & Associates implemented ELJ design recommendations into the North Creek project. We thank Peter Hrynyshyn of Mortenson Construction for the use of the North Creek aerial photography. Finally, we extend our sincere gratitude to the outstanding construction crews who built all of these complex structures.

References

Abbe, T.B. 2000. Patterns, mechanics, and geomorphic effects of wood debris accumulations in a forest river system. Ph.D. dissertation. University of Washington. Seattle, Washington.

Abbe, T.B. and D.R. Montgomery. 1996. Large woody debris jams, channel hydraulics, and habitat formation in large rivers. *Regulated Rivers: Research & Management* 12:201-221.

Abbe, T.B., D.R. Montgomery, K. Fetherston, and E.M. McClure. 1993. A process-based classification of woody debris in a fluvial network: preliminary analysis of the Queets River, WA. *EOS, Transactions of the American Geophysical Union* 73(43):296.

Abbe, T.B., D.R. Montgomery, and C. Petroff. 1997. Design of stable in-channel wood debris structures for bank protection and habitat restoration: an example from the Cowlitz River, WA. In S.S.Y. Wang, E.J. Langendoen, and F.D. Shields Jr. (eds.) *Proceedings of the Conference on Management of Landscapes Disturbed by Channel Incision*. University of Mississippi, Oxford, Mississippi. pp. 809-816.

Booth, D.B. and C.R. Jackson. 1997. Urbanization of aquatic systems: degradation thresholds, stormwater detention, and the limits of mitigation. *Journal of the American Water Resources Association* 33:1077-1090.

Braudrick, C. A. and G. E. Grant. 2000. When do logs move in rivers? *Water Resource Research* 36:571-583.

Brooks, A.P. and G.J. Brierly. 1997. Geomorphic responses of lower Bega River to catchment disturbance, 1851-1926. *Geomorphology* 18:291-304.

Buffington, J.M. and D.R. Montgomery. 1999a. Effects of hydraulic roughness on surface textures of gravel-bed rivers. *Water Resources Research* 35:3507-3521.

Buffington, J.M. and D.R. Montgomery. 1999b. Effects of sediment supply on surface textures of gravel-bed rivers. *Water Resources Research* 35:3523-3530.

D'Aoust, S.G. and R.G. Millar. 1999. Large woody debris fish habitat structure performance and ballasting requirements. Province of British Columbia, Ministry of Environment, Lands and Parks and Ministry of Forests. Watershed Restoration Management Report No. 8.

Garde, R.J., K. Subrananya, and K.D. Nambudripad. 1961. Study of scour around spur dikes. *Journal of Hydraulics Division (ASCE)*, 87(HY6), pp. 23-27.

Gippel, C.J. 1995. Environmental hydraulics of large woody debris in streams and rivers. *Journal of Environmental Engineering* 121:388-395.

Gippel, C., I.C. O'Neill, B.L. Finlayson, and I. Schnatz. 1996. Hydraulic guidelines for the re-introduction and management of large woody debris in lowland rivers. *Regulated Rivers: Research & Management* 12:223-236.

Graf, W.L. 1975. The impact of suburbanization on fluvial morphology. *Water Resources Research* 11:690-692.

Hammer, T.R. 1972. Stream channel enlargement due to urbanization. *Water Resources Research* 8:1530-1546.

Hartopo. 1991. The effect of raft removal and dam construction on the Lower Colorado River, Texas. Master's thesis. Texas A&M University. College Station, Texas.

Harvey, M.D. and D.S. Biedenharn. 1988. Adjustments of the Red River following removal of the Great Raft in 1873. *EOS, Transactions of the American Geophysical Union* 68:567.

Hawk, G.M. and D.B. Zobel. 1974. Forest succession on alluvial landforms of the McKenzie river valley, Oregon. *Northwest Science* 48:245-265.

Hoffmans, G.J.C.M. and H.J. Verheji. 1997. *Scour Manual.* A.A. Balkema, Rotterdam, Netherlands.

Kelsey, H.M. 1980. A sediment budget and an analysis of geomorphic process in the Van Duzen River basin, north coastal California, 1941-1975. *Geological Society of America Bulletin*, Part II 91:1119-1216.

Lisle, T.E. 1995. Effects of coarse woody debris and its removal on a channel affected by the 1980 eruption of Mount St. Helens, Washington. *Water Resources Research* 31:1797-1808.

Lonzarich, D.G. and T.P. Quinn. 1995. Experimental evidence for the effect of depth and structure on the distribution, growth, and survival of stream fishes. *Canadian Journal of Zoology* 73:2223-2230.

Marzolf, G.R.1978. The potential effects of clearing and snagging of stream ecosystems. U.S.D.I. Fish and Wildlife Service, OBS-78-14, Washington D.C.

Maser, C. and J.R. Sedell. 1994. *From the Forest to the Sea: The Ecology of Wood in Streams, Rivers, Estuaries, and Oceans.* St. Lucie Press, Delray, Florida.

Melville, B.W. and A.J. Sutherland. 1988. Design method for local scour at bridge piers. *Journal of Hydraulic Engineering* 114:1210-1226.

Melville, B.W. and D.M. Dongol. 1992. Bridge pier scour with debris accumulation. *Journal of Hydraulic Engineering* 118:1306-1310.

Miller, A.C., S.N. Kerr, H.E. Reams, and J.P. Sartor. 1984. Physical modeling of spurs for bank protection. In Elliott, C.M. (ed.) *River Meandering*. Proceedings of the ASCE Conference on Rivers. New Orleans, 24-26 October 1983. American Society of Civil Engineers, New York. pp. 996-1007.

Miller, A.J. 1995. Valley morphology and boundary conditions influencing spatial patterns of flood flow. In Costa, J.E. et al. (eds.) *Natural and Anthropogenic Influences in Fluvial Geomorphology*. American Geophysical Union, Geophysical Monograph 89, Washington D.C. pp. 57-81.

Montgomery, D.R., T.B. Abbe, N.P. Peterson, J.M. Buffington, K.M. Schmidt, and J.D. Stock. 1996. Distribution of bedrock and alluvial channels in forested mountain drainage basins. *Nature* 381:587-589.

Moscrip, A.L. and D.R. Montgomery. 1997. Urbanization, flood frequency, and salmon abundance in Puget lowland streams. *Journal of the American Water Resources Association* 33:1289-1297.

Pearsons, T.N., H.W. Li, and G.A. Lamberti. 1992. Influence of habitat complexity on resistance to flooding and resilience of stream fish assemblages. *Transactions of the American Fisheries Society* 121:427-436.

Pess, G.R., T.B. Abbe, T.A. Drury, and D.R. Montgomery. 1998. Biological evaluation of engineered log jams in the North Fork Stillaguamish River, Washington. *EOS, Transactions of the American Geophysical Union* 79 (45): F346.

Pitlick, J. 1992. Flow resistance under conditions of intense gravel transport. *Water Resources Research* 28:891-903.

Raudkivi, A.J. 1990. *Loose Boundary Hydraulics*. Pergamon Press, Oxford.

Raudkivi, A.J. and R. Ettema. 1977. Effect of sediment gradation on clear water scour. *Proceedings of the American Society of Civil Engineering* 103(HY10):1209-1213.

Richardson, E.V. and P.F. Lagasse (eds.). 1999. *Stream Stability and Scour at Highway Bridges*. American Society of Civil Engineers, Reston, Virginia.

Secretary of War. 1931. Report from the Chief of Engineers on the Stillaguamish River, Wash., covering navigation, flood control, power development, and irrigation. 71st Congress, 3rd Session. House of Representatives Document No. 657. (House Documents Vol. 31). Government Printing Office, Washington D.C.

Sedell, J. R. and J. L. Frogatt. 1984. Importance of streamside forests to large rivers: the isolation of the Willamette River, OR, U.S.A., from its floodplain by snagging and streamside forest removal. *Verhandlungen-Internationale Vereinigung für Theorelifche und Angewandte Limnologie* 22:1828-1834.

Shields, F.D., Jr., C.M. Copper, and S. Testa, III. 1995. Towards greener riprap: environmental considerations from microscale to macroscale. In C.R. Thorne, S.R. Abt, F.B.J. Barends, S.T. Maynord, and J.W. Pilarczyk (eds.) *River Coastal and Shoreline Protection: Erosion Control Using Riprap and Armourstone*. John Wiley & Sons, Chichester. pp. 557-576.

Shields, F.D., Jr. and C.J. Gippel. 1995. Prediction of effects of woody debris removal on flow resistance. *Journal of Hydraulic Engineering* 121:341-354.

Shields, F.D., Jr. and N.R. Nunnally. 1984. Environmental aspects of clearing and snagging. *Journal of Environmental Engineering* 110:152-165.

Smith, R.H. and F.D. Shields, Jr. 1990. Effects of clearing and snagging on physical conditions of rivers. *Proceedings of the Mississippi Water Resources Conference, Jackson, Mississippi*. Water Resources Institute, Mississippi State University, Starkville, Mississippi. pp. 41-51.

Stover, S.C. and D.R. Montgomery. 2001. Channel change and flooding, Skokomish River, Washington. *Journal of Hydrology* 243:272-286.

Swales, S. 1988. Utilization of off-channel habitats by juvenile coho salmon (*Oncorhynchus kisutch*) in interior and coastal streams in British Columbia. *Verhandlungen-Internationale Vereinigung für Theorelifche und Angewandte Limnologie* 23:1676.

Tarzwell, C.M. 1934. The purpose and value of stream improvement method. Presented at the Annual Meeting of the American Fisheries Society in Montreal, Quebec, September 12. Stream Improvement Bulletin R-4. Ogden, Utah.

Thorne, C.R. 1990. Effects of Vegetation on Riverbank Erosion and Stability. In Thornes, J.B., *Vegetation and Erosion*. John Wiley & Sons, Chichester.

Tschaplinski, P.J. and G.F. Hartmann. 1983. Winter distribution of juvenile coho salmon (*Oncorhynchus kisutch*) before and after logging in Carnation Creek, British Columbia, and some implications for overwinter survival. *Canadian Journal of Fisheries and Aquatic Sciences* 40:452-461.

Wallerstein, N., C.R. Thorne, and M.W. Doyle. 1997. Spatial distribution and impact of large woody debris in Northern Mississippi. In S.S.Y. Wang, E.J. Langendoen, and F.D. Shields, Jr. (eds.) *Proceedings of the Conference on Management of Landscapes Disturbed by Channel Incision*. University of Mississippi, Oxford, Mississippi. pp. 145-150.

18. Restoration of Puget Sound Rivers: Do We Know How to Do It?

Susan Bolton, Derek B. Booth, and David R. Montgomery

> A woodsman riding through the woods spots a herd of pigs rooting for acorns beneath an oak tree. The woodsman notices that the pigs have unearthed and damaged the roots of the oak tree. He warns the pigs "Be careful, you are killing the tree." The pigs reply, "We don't care about the tree, we only care about the acorns."

As is so common in scientific endeavors, the answer to the title question is "it depends." It depends on what we mean by restoration and whether that goal is even attainable for the river and the watershed under consideration. If the goal *is* attainable, then it further depends on the chosen approach; on who undertakes the project; on how certain we wish to be that we will be successful on the first attempt; and on how soon we expect results.

Fish are often regarded as the target for river restoration, but focusing on fish disregards the necessary role of the ecological processes that sustain the fishery and often results in neglecting the role of the processes that sustain the system (Tockner and Schiemer 1997). Single-interest "restoration," especially those that focus only on fish, may appear to provide local improvements to habitat structure but are likely to prove unsustainable (Boon 1998).

For at least a decade, calls have been made for addressing causes and not symptoms, assessing watersheds as a whole before beginning restoration projects, restoring ecosystem processes, and using monitoring to learn and improve on what we do (e.g., Stanford and Ward 1992; Ziemer 1996; Frissel 1997). Yet widespread, on-the-ground implementation of these ideas has not taken hold. The difficulties of implementation are many, but the costs of continuing with uncoordinated "restoration" projects are huge.

Humans depend on a wide range of natural goods and services provided by the structure, function, diversity, and dynamics of ecosystems (Wyant et al. 1995). Inordinate focus on single species can blind us to our dependence upon intact ecosystems and the goods and services they provide. Despite calls for ecosystem or watershed restoration, rather than single-species restoration, some in the region call for more explicit proof that restoration actions increase salmon productivity. If this means continuing to do research to understand ecosystem processes more completely, then it serves as a valuable reminder to continue learning more about our systems. If it used as an excuse to delay implementing the required changes in institutions and society that are needed to restore aquatic ecosystems, then it is a policy and value statement masquerading as science.

Implementing River Restoration

This volume brings together current thinking and methodologies for river and watershed restoration developed for the Puget Sound region. Several messages are repeated throughout various chapters in this volume: integrate ecological knowledge into design and monitoring of projects, restore ecosystem processes and not individual species, and have clear and measurable objectives. In 1992, Naiman et al. (1992) identified five factors that define the fundamental elements of ecologically healthy watersheds in the Pacific Northwest: geology, hydrology, water quality, riparian forests, and habitat features. These factors have been expanded on here to bring the current knowledge of watershed processes and river restoration together in one place for use by scientists, managers, and policy makers. The methodologies can be used in any watershed in any region, but in this volume we have used local information to provide a template for local restoration.

The answers to several questions should guide any approach to restoration, and they will determine the likelihood of success.

What Is the Physical Template Upon which Restoration Will Take Place?

Booth et al. (Chapter 2) reminds us of the complex effects of recent glaciation on the region and of the importance of this glacial legacy on evaluating the effects of human actions on contemporary channel processes and conditions. For example, the use of geologic maps can help locate areas of low versus high soil permeability (high infiltration) and areas where groundwater influences are more or less likely to occur. Buffington et al. (Chapter 3) describe the range of fluvial processes that are encountered in Puget Sound. They address how the type and setting of a channel set bounds on channel characteristics, and they show how these constraints can help in developing restoration objectives for specific channel reaches. The unique functions of woody debris in the forested and once-forested rivers and streams of the Pacific Northwest (Chapters 10 and 14) are particularly important in these assessments and subsequent actions.

Is the Watershed Urbanized, Agricultural, or Forested?

The mix of land use in a watershed will affect the extent to which restoration versus rehabilitation can be realized. Urban streams are limited to rehabilitation, because project approaches and goals will be constrained by concerns for human and infrastructure safety. Unless the changes in flow regime caused by urbanization can be corrected, the highest goal of urban stream rehabilitation is probably to provide sustainable habitat under the new flow regime (Chapter 11). Agricultural lands in Puget Sound are largely located on lowland areas adjacent to large rivers and have also been substantially modified (Chapters 4 and 10). Basin-by-basin identification of current stream conditions, critical fish habitat, and changes in habitat availability over time will help guide stream and habitat improvement in these areas (Chapter 8). Forestry has altered the structure of regional forests, which in turn has changed many aspects of watershed processes. Much of the past regulatory effort for protecting aquatic systems has focused on forestry (Chapter 1). Other than national parks and wilderness, however, forestry is also the land use that retains the landscape most similar to its original state. Compared to agricultural and urban settings, more habitat-forming processes are intact in forested areas, and so restoration is much more feasible here.

Is the River Being Restored Large or Small?

Due to the extensive channel migration that large rivers typically exhibit in their natural state, extensive restoration is much less common on large rivers, and mostly local rehabilitation projects are undertaken. Most restoration and rehabilitation projects have taken place on smaller streams and rivers, and therefore more information exists for these types of sites (Gore and Shields 1995).

This dichotomy is evident in this region's practice. In Puget Sound, many projects on small streams are completed or underway (Chapter 15), but the number of projects on large rivers is much smaller (Chapters 16 and 17). On small streams, direct intervention and establishment of riparian buffers may substantially address restoration concerns, and the level of engineering design is relatively low compared to that needed on large rivers.

Is There a Thorough Watershed Assessment that Identifies Historic and Current Habitat-Forming Processes and Fish Distribution?

Without knowing what condition our watersheds are in, we cannot formulate how to improve them. Information on historic and current conditions including vegetation and fish distribution can be used to identify priorities for restoration and rehabilitation. Methodologies have been developed that can be widely applied to acquire this information (Chapters 4, 5, and 8). The current state of our rivers and streams, and of their associated fauna, is the result of more than a century of human activity; it will take time to recover some of the lost ecosystem functions.

Has a Monitoring Plan Been Developed in Concert with the Planned Restoration Action?

Given the complex and dynamic nature of stream ecosystems and the added complexity of salmon anadromy, all stream enhancement projects are experiments. Acknowledging the experimental nature of restoration actions allows us to learn from our actions and improve on them in the future (Chapter 9). Without having measurable objectives (i.e., quantifiable outcomes, such as 100 m^2 of spawning area instead of 'improve habitat quality'), we cannot rigorously evaluate the effectiveness of our efforts. Miller and Skidmore (Chapter 13) describe ways to improve project designs and implementation through the development and standardization of design criteria. Using measurable performance criteria in project design, rather than prescriptive criteria, should allow innovations in restoration to continue. Roni et al. (Chapter 12) describes methods that may allow us to evaluate projects that were not initially implemented with such criteria in mind, or indeed with any monitoring component at all. "Adaptive management" is the rubric under which such monitoring falls, but many have questioned its implementation to date (e.g., Fischer 1990; Moir and Block 2001). Ralph and Poole (Chapter 9) make a case for reviving the original meaning of this term and for integrating monitoring into a project before its implementation, not afterwards.

Lessons from Past Experiences

"Resource problems are not really environmental problems," but rather they are *human* problems that have been created at many times in many places under a variety of political, social, and economic systems (Ludwig et al. 1993). Yet the history of fisheries management does not bode well for recovery efforts based on voluntary behavior and unenforced regulations (Chapter 1). However, government cannot do it all. There are many conflicting social needs and demands (Chapter 7), and individuals must take an active role for restoration to be successful (Chapter 6). Moreover, solutions to many of society's most pressing resource and restoration issues (e.g., population growth, overconsumption, endangered species, and pollution) are more social than technical in nature (Wood et al. 1997).

It is wise to remember that all decisions and actions are based on facts (science) *and* values (ethics). This may not appear to be the case in routine natural resource management decisions, because the values are widely shared and go unnoticed. When a new way of resource allocation or management is suggested, however, it is often labeled as an ethic (value judgement) and then dismissed as insubstantial or a matter of opinion (Callicott 1991). For example, the National Research Council (1996) identified a number of outmoded institutions that control water and water rights, including subsidized federal reclamation projects, whose special interests are not always the same as the larger public interests (Johnson 1989). Deciding whose interests take priority in policy changes, or choosing to implement restoration actions that often result in gains for some and losses for others, cannot be resolved by science alone (Chapter 6).

Popular and political support for cleaning up environmental catastrophes such as the Exxon Valdez oil spill is generally easier to rally than support for long-term chronic degradation of watersheds and local species (Wood et al. 1997). In the Pacific Northwest, however, we do have a rich opportunity to accomplish restoration. The combination of cultural, economic, and ecological value of salmon creates a setting for restoration that is broadly supported. Many tribal cultures and livelihoods rely on salmon; commercial and recreational fisheries depend on salmon; aquatic and terrestrial organisms use salmon carcasses and their decomposition products for food. There are few, if any, other places where an ESA-listed species can generate support across such a diverse group of interests. That does not mean conflicts do not exist—they do. But if river restoration in concert with recovery of salmon cannot happen here, where can it?

Time to Act

We have a good, basic understanding of the ecological processes in aquatic systems, and we know enough to be acting effectively now. Five years ago, the National Research Council stated that "Because habitat loss is widely acknowledged to have contributed to the decline of virtually every species of Pacific salmon in western North America (Nehlsen et al. 1991), the lack of precise knowledge of relationships between various types of habitat change and salmon populations need not be a barrier to improved environmental management" (NRC 1996, pp. 165-166).

The template for recovery is well described for Puget Sound. The preceding chapters bring together a suite of information that is directly relevant to Puget Sound river and stream restoration efforts. It is clear that we don't know everything, but it is equally clear that we know enough to make progress. Many enhancement options exist, ranging from preservation of existing systems that are still intact, to better methods of re-establishing floodplain forests, to "habitat-friendlier" bank stabilization. Key to all methods used to restore or to rehabilitate our streams and rivers, and the species that depend on them, is to identify current and former conditions in the watershed wherever possible; to focus on addressing the causes of stream degradation and not just the symptoms; to move the system in the direction of being more self-sustaining; and to integrate monitoring into the design and implementation of restoration actions.

Our hope is that the time is right to finally incorporate current scientific thinking into restoration actions. Much of what we call for has been called for before (e.g., NRC 1996, 1999; Williams et al. 1997), but only in a few select locations has the vision been fully implemented. To date, that has largely *not* happened here.

Yet restoration must be undertaken with humility; precise future trajectories of complex natural systems are impossible to predict (McQuillan 1998). The unbridled technological optimism of the nineteenth and twentieth centuries did not solve ecological problems. As Angermeier (1997) notes, technological fixes often lead to unanticipated ecological damage even as they sustain the myth that technology can solve complex ecological problems. Thus a better question about restoration than "Do we know how to do it?" may be "Will we try our best to do it?" And if not here, where? And if not now, when?

References

Angermeier, P.L. 1997. Conceptual roles of biological integrity and diversity. In J.E. Williams, C.A. Wood, and M.P. Dombeck (eds.) *Watershed Restoration: Principles and Practices*. American Fisheries Society, Bethesda, Maryland. pp. 49-65.

Boon, P. J. 1998. River restoration in five dimensions. *Aquatic Conservation: Marine and Freshwater Ecosystems* 8:257-264.

Callicott, J.B. 1991. Conservation ethics and fishery management. *Fisheries* 16(2):22-28.

Fischer, F. 1990. *Technocracy and the Politics of Expertise*. Sage Books, Newbury Park, California.

Frissell, C.A. 1997. Ecological principles. In J.E. Williams, C.A. Wood, and M.P. Dombeck (eds.) *Watershed Restoration: Principles and Practices*. American Fisheries Society, Bethesda, Maryland. pp. 96-115.

Gore, J.A. and F.D. Shields Jr. 1995. Can large rivers be restored? *BioScience* 45:142-152.

Johnson, R.W. 1989. Water pollution and the public trust doctrine. *Environmental Law* 19:485-513.

Ludwig, D., R. Hilborn, and C. Waters. 1993. Uncertainty, resource exploitation and conservation: Lessons from history. *Science* 260:17,36.

McQuillan, A.G. 1998. Defending the ethics of ecological restoration. *Journal of Forestry* 1:27-31.

Moir, W.H. and W.M. Block. 2001. Adaptive management on public lands in the United States: Commitment or rhetoric? *Environmental Management* 28:141-148.

Naiman, R.J., T.J. Beechie, L.E. Benda, D.R. Berg, P.A. Bisson, L.H. MacDonald, M.D. O'Connor, P.L. Olson, and E.A. Steel. 1992. Fundamental elements of ecologically healthy watersheds in the Pacific Northwest coastal ecoregion. In R J. Naiman (ed.) *Watershed Management: Balancing Sustainability and Environmental Change*. Springer-Verlag, New York. pp. 127-188.

Nehlsen, W., J.E. Williams, and J.A. Lichatowich. 1991. Pacific salmon at the crossroads: stocks at risk from California, Oregon, Idaho and Washington. *Fisheries* 16(2):4-21.

NRC (National Research Council). 1996. *Upstream: Salmon and Society in the Pacific Northwest.* National Academy Press, Washington, D.C.

NRC (National Research Council). 1999. *New Strategies for America's Watersheds.* National Academy Press, Washington, D.C.

Stanford, J.A. and J.V. Ward. 1992. Management of aquatic resources in large catchments: Recognizing interactions between ecosystem connectivity and environmental disturbance. In R.J. Naiman (ed.) *Watershed Management.* Springer-Verlag, New York. pp. 91-124.

Tockner, K. and F. Schiemer. 1997. Ecological aspects of the restoration strategy for a river-floodplain system on the Danube River in Austria. *Global Ecology and Biogeography Letters* 6:321-329.

Williams, J.E., C.A. Wood, and M.P. Dombeck (eds.). 1997. *Watershed Restoration: Principles and Practices.* American Fisheries Society, Bethesda, Maryland.

Wood, C.A., J.E. Williams, and M.P. Dombeck. 1997. Learning to live within the limits of the land: Lessons from the watershed restoration case studies. In J.E. Williams, C.A. Wood, and M.P. Dombeck (eds.) *Watershed Restoration: Principles and Practices.* American Fisheries Society, Bethesda, Maryland. pp. 445-458.

Wyant, J.G., R.A. Meganck, and S.H. Ham. 1995. A planning and decision-making framework for ecological restoration. *Environmental Management* 19:789-796.

Ziemer, R.R. 1996. Temporal and spatial scales. In J.E. Williams, C.A. Wood, and M.P. Dombeck (eds.) *Watershed Restoration: Principles and Practices.* American Fisheries Society, Bethesda, Maryland. pp. 80-95.

Contributors

Tim Abbe
Herrera Environmental Consulting
Seattle, WA

Eric Beamer
Skagit System Cooperative
La Conner, Washington

Timothy J. Beechie
Watershed Program
Northwest Fisheries Science Center
Seattle, WA

Dean Rae Berg
Silvicultural Engineering
Edmonds, WA

John Bethel
King County Dept. of Natural Resources
Seattle, WA

Robert E. Bilby
Weyerhaeuser Company
Tacoma, WA

Susan Bolton
Center for Streamside Studies
University of Washington
Seattle, WA

Derek B. Booth
Earth and Space Sciences, and
Center for Urban Water Resources
Management
University of Washington
Seattle, WA

John M. Buffington
Department of Civil Engineering
University of Idaho
Boise, ID

Brian D. Collins
Earth and Space Sciences
University of Washington
Seattle, WA

Loveday Conquest
Center for Streamside Studies
University of Washington
Seattle, WA

As of September 2002, the Center for Streamside Studies and the Center for Urban Water Resources Management merged to form the Center for Water and Watershed Studies.

Ralph A. Haugerud
U.S. Geological Survey
Seattle, WA

Lisa Holsinger
Watershed Program
Northwest Fisheries Science Center
Seattle, WA

Sarah M. Jensen
College of Forest Resources
University of Washington
Seattle, WA

Christopher P. Konrad
U.S. Geological Survey
Tacoma, WA

Martin Liermann
Watershed Program
Northwest Fisheries Science Center
Seattle, WA

John Lombard
King County Dept. of Natural Resources
Seattle, WA

Gino Lucchetti
King County Dept. of Natural Resources
Seattle, WA

Michael J. Maki
Agroforestry Associates
Hoquiam, WA

Arthur McKee
Department of Forest Science
Oregon State University
Corvallis, OR

Dale E. Miller
Mainstream Restoration, Inc.
Bozeman, MT

David R. Montgomery
Center for Streamside Studies
University of Washington
Seattle, WA

Kathryn Neal
King County Dept. of Natural Resources
Seattle, WA

Roger A. Nichols
U.S. Forest Service
Sedro Woolley, WA

George Pess
Watershed Program
Northwest Fisheries Science Center
Seattle, WA

Geoffrey C. Poole
Eco-metrics, Inc.
Tucker, GA
and
Institute of Ecology
University of Georgia
Athens, GA

Stephen C. Ralph
National Park Service
Seattle, WA

Philip Roni
Watershed Program
Northwest Fisheries Science Center
Seattle, WA

Clare M. Ryan
Center for Streamside Studies
University of Washington
Seattle, WA

Amir J. Sheikh
Earth and Space Sciences
University of Washington
Seattle, WA

Peter B. Skidmore
Inter-Fluve
Bozeman, MT

Sallie G. Sprague*
Department of Biology
Western Washington University
Bellingham, WA

Ashley Steel
Watershed Program
Northwest Fisheries Science Center
Seattle, WA

Kathy Goetz Troost
Earth and Space Sciences
University of Washington
Seattle, WA

Richard D. Woodsmith
PNW Research Station
USDA Forest Service
Juneau, AK

* Now at Department of Forest Sciences, Colorado State University, Ft. Collins, CO

Index

D

E

F

G

H

I

S

T